DRUG–DNA INTERACTIONS

DRUG–DNA INTERACTIONS
STRUCTURES AND SPECTRA

Kazuo Nakamoto
Wehr Professor Emeritus of Chemistry
Marquette University

Masamichi Tsuboi
Emeritus Professor of Pharmaceutical Science
University of Tokyo

Gary D. Strahan

A JOHN WILEY & SONS, INC., PUBLICATION

About the Cover: The cover was created by G. Strahan, and is composed of the circular dichroism (CD) spectrum and a molecular model of a complex formed when the drug Distamycin binds as a dimer to an AT-rich DNA sequence (see Chapter 3). The molecular model was generated by the program AMBER9, using the methods outlined in Chapter 1 and Appendix 2, and visualized by the program UCSF Chimera.

Copyright © 2008 by John Wiley & Sons, Inc. All rights reserved

Published by John Wiley & Sons, Inc., Hoboken, New Jersey
Published simultaneously in Canada

No part of this publication may be reproduced, stored in a retrieval system, or transmitted in any form or by any means, electronic, mechanical, photocopying, recording, scanning, or otherwise, except as permitted under Section 107 or 108 of the 1976 United States Copyright Act, without either the prior written permission of the Publisher, or authorization through payment of the appropriate per-copy fee to the Copyright Clearance Center, Inc., 222 Rosewood Drive, Danvers, MA 01923, (978) 750-8400, fax (978) 750-4470, or on the web at www.copyright.com. Requests to the Publisher for permission should be addressed to the Permissions Department, John Wiley & Sons, Inc., 111 River Street, Hoboken, NJ 07030, (201) 748-6011, fax (201) 748-6008, or online at http://www.wiley.com/go/permission.

Limit of Liability/Disclaimer of Warranty: While the publisher and author have used their best efforts in preparing this book, they make no representations or warranties with respect to the accuracy or completeness of the contents of this book and specifically disclaim any implied warranties of merchantability or fitness for a particular purpose. No warranty may be created or extended by sales representatives or written sales materials. The advice and strategies contained herein may not be suitable for your situation. You should consult with a professional where appropriate. Neither the publisher nor author shall be liable for any loss of profit or any other commercial damages, including but not limited to special, incidental, consequential, or other damages.

For general information on our other products and services or for technical support, please contact our Customer Care Department within the United States at (800) 762-2974, outside the United States at (317) 572-3993 or fax (317) 572-4002.

Wiley also publishes its books in a variety of electronic formats. Some content that appears in print may not be available in electronic formats. For more information about Wiley products, visit our web site at www.wiley.com.

Library of Congress Cataloging-in-Publication Data:
Nakamoto, Kazuo
  Drug-DNA interactions : structures and spectra / Kazuo Nakamoto, Masamichi Tsuboi, Gary D. Strahan.
      p. ; cm.
   Includes index.
   ISBN 978-0-471-78626-9 (cloth)
 1. DNA-drug interactions–Research–Methodology. 2.   DNA–Spectra. 3. Drugs–Spectra.   I. Tsuboi, Masamichi, 1925- II. Strahan, Gary D. III. Title.
   [DNLM: 1. DNA–drug effects. 2. Chemistry, Analytical–methods. 3. DNA–chemistry. 4.   Pharmaceutical Preparations–chemistry. 5. Structure-Activity Relationship. QU 58.5 N163d 2008]
   QP624.75.D77N35 2008
   572.8'6–dc22
                                                               2007050457

Printed in the United States of America
10 9 8 7 6 5 4 3 2 1

*The authors wish to dedicate this book to their wives,
Kimiko Nakamoto, Yo Tsuboi, and Kellie Hom.*

CONTENTS

Preface		xi
Introduction		xiii
1. DNA Structures and Spectra		1
1.1.	DNA Structures	2
	1.1.1. Primary (Chemical) Structure	2
	1.1.2. Bond Lengths and Bond Angles	5
	1.1.3. Canonical Structure of DNA: B-DNA	9
	1.1.4. Deoxyribose Moiety as a Flexible Joint	13
	1.1.5. Variations of B-DNA Structure	15
	1.1.6. Participation of Water Molecules in B-DNA Structure	16
	1.1.7. Partially Dehydrated Structure: A-DNA	17
	1.1.8. Internal Dihedral Angles	19
	1.1.9. Left-Handed Duplex: Z-DNA	21
	1.1.10. Superhelical Form of DNA	23
	1.1.11. Other Duplex Structures	25
	1.1.12. Higher-Order DNA Structures	29
1.2.	Electronic Spectra	34
	1.2.1. Ultraviolet Absorption Spectra	34
	1.2.2. Fluorescence Spectra	37
	1.2.3. Linear and Circular Dichroism	38
1.3.	Vibrational Spectra	42
	1.3.1. Infrared Spectra of Bases	42
	1.3.2. Raman Spectra of Bases	49
	1.3.3. Vibrational Spectra of Nucleosides	54
	1.3.4. Vibrational Spectra of Nucleotides	57
	1.3.5. Resonance Raman Spectra	59
	1.3.6. Vibrations of the Phosphodiester Linkage	59
	1.3.7. Infrared Spectra of DNA	61
	1.3.8. Raman Spectra of B-DNA	63
	1.3.9. Comparison of Raman Spectra of A- and B-DNA	67
	1.3.10. Raman Spectra of Z-DNA	70

1.4. NMR Spectra   72
    1.4.1. NMR Signals and Magnetization Relaxation   73
    1.4.2. Proton NMR Spectra of Bases, Nucleosides, and Nucleotides   76
    1.4.3. Correlation Analysis, or "Spin–Spin Coupling"   80
    1.4.4. Nuclear Overhauser Effect   85
    1.4.5. Residual Dipolar Coupling   94
1.5. Electron Spin Resonance Spectra   95
1.6. X-Ray Crystallography   96
    1.6.1. Bragg Diffraction   97
    1.6.2. Diffraction of Helical DNA   98
1.7. Molecular Modeling and Molecular Mechanics   99
    1.7.1. Molecular Modeling   99
    1.7.2. Modeling of Experimental Data   100
References   102

2. Intercalating Drugs   119

2.1. Acridine Dyes   122
    2.1.1. X-Ray Diffraction Studies of Acridine–Dinucleoside Monophosphate Crystals   122
    2.1.2. Complexes with Longer DNAs: X-Ray and Modeling Analyses   128
    2.1.3. Unwinding Angle Measurements   131
    2.1.4. NMR Studies   134
    2.1.5. Fluorescence Spectra   137
    2.1.6. Flow Linear Dichroism Spectra/Studies   140
    2.1.7. Circular Dichroism Studies   141
    2.1.8. Biochemical Implications—a Few Examples   142
2.2. Ethidium Bromide   143
    2.2.1. X-Ray and Modeling Analyses of Ethidium Bromide–Dinucleoside Monophosphate Crystals   143
    2.2.2. Fiber X-Ray Diffraction Studies   145
    2.2.3. Visible Absorption Spectra   148
    2.2.4. Circular Dichroism Spectra   150
    2.2.5. Fluorescence Spectra   152
    2.2.6. Raman Spectra   154
    2.2.7. Electrophoresis Studies of DNA Unwinding   159
    2.2.8. Atomic Force Microscopy   161
2.3. Aclacinomycin   165
    2.3.1. Absorption Spectroscopy   165
    2.3.2. ^{31}P NMR   167
    2.3.3. Exchangeable ^{1}H NMR Spectra   168

2.3.4. Nonexchangeable ^{1}H NMR Spectra	170
2.3.5. Raman Spectroscopy	175
2.3.6. Unwinding	176
2.4. Sequence Preference	178
2.5. Bis- and Tris- Intercalators	194
References	199

3. Groove-Binding Drugs — 209

3.1. Netropsin and Distamycin	210
3.2. Derivatives of Netropsin and Distamycin	217
3.3. Hoechst 33258, SN6999, and Their Derivatives	224
3.4. Chromomycin, Mithramycin, and Other GC Binders	230
3.5. Groove Binding and Intercalation	234
References	240

4. Covalent Bonding Drugs — 245

4.1. (+)CC1065 and Related Drugs	245
4.2. Anthramycin and Tomaymycin	254
4.3. Ecteinascidins	256
4.4. Mitomycins	259
4.5. Intercalating Alkylators	264
References	271

5. Strand-Breaking Drugs — 275

5.1. Bleomycins	275
5.1.1. Metal-Free Bleomycins	275
5.1.2. Iron Complexes of Bleomycin	279
5.1.3. Cobalt and Zinc Complexes of Bleomycin	283
5.1.4. Model Compounds of Bleomycins	288
5.2. Enediyne Antibiotics	290
5.2.1. Calicheamicins	291
5.2.2. Esperamicins	294
5.2.3. Neocarzinostatin	296
References	299

6. Metal-Containing Drugs — 303

6.1. Cisplatin	303
6.1.1. Intrastrand Crosslinking Complexes	304
6.1.2. Interstrand Crosslinking Complexes	316
6.2. Cisplatin Derivatives	316

6.2.1. Oxaliplatin and Carboplatin	316
6.2.2. Platinum Complexes of Monofunctional Ligands	318
6.2.3. Platinum Complexes of Chelating Ligands	321
6.3. Transplatin and Derivatives	327
6.4. Monofunctional Platinum Complexes	332
6.5. Polynuclear and High-valent Platinum Complexes	336
6.6. Complexes Containing Other Metals	345
6.6.1. Palladium(II) and Gold(III)	346
6.6.2. Other Metals	347
References	347

Appendixes   357

A.1. Raman Tensor-Depolarization Ratio	357
A.2. NMR Spectroscopy	359
A.2.1. Origins of the Phenomenon	359
A.2.2. Chemical Shift	361
A.2.3. Magnetization Relaxation Processes	362
A.3. Molecular Mechanics	364
References	366

Index   367

PREFACE

Drug discoveries are made serendipitously or systematically. In the latter case, a large number of derivatives of a candidate drug molecule are prepared and screened first by cytotoxicity tests *in vitro* since no definitive structure–activity relationships are available. This process is inefficient and time-consuming, in spite of the recent advent of combinatorial chemistry and laboratory automation. Hence, the ultimate goal of this book is to contribute to our knowledge of the structure–activity relationship found in drug–DNA interaction by coalescing much of the basic structural information, thereby improving the efficiency of the drug discovery process.

First, however, we must discuss the structures of drug–DNA complexes that were determined by a variety of physicochemical and biochemical techniques. Although some of these structural studies have already been reviewed by other authors, they tend to focus on structural information on the atomic level. In this book, we have emphasized global information obtained by a variety of spectroscopic techniques. It is hoped that this book will serve as a guide for students in chemistry, biochemistry, and pharmacology who are interested in learning the nature of drug–DNA interactions and as a reference book for academic and industrial researchers who are doing research in this field.

The authors would like to express their sincere thanks to the staffs of John Raynor Library of Marquette University for their help in collecting references,to Professor Jung-Ja Kim of Medical College of Wisconsin for her critical review of our manuscript, and to many authors and publishers who kindly gave us permission to reproduce their figures.

Kazuo Nakamoto
Masamichi Tsuboi
Gary D. Strahan

INTRODUCTION

Cellular DNA is the major target of most antitumor drugs, as well as many antiviral and antibacterial agents. Depending on the mode of interaction, these drugs can be classified as intercalating, groove-binding, covalent bonding, and strand-breaking drugs. They can distort the double-helix structure of DNA, alkylate or cleave a DNA strand, thereby inhibiting DNA replication and transcription, which are the necessary preconditions for cell division. Previously, the nature of the drug–DNA interaction was investigated extensively using a variety of physicochemical and biochemical techniques. The results obtained from these studies have begun to shed light on the underlying principles of the structure–activity relationship, which is indispensable for understanding the mechanism of the drug–DNA interaction and for designing new and more effective drugs with fewer side effects.

Although many review articles on drug–DNA interactions (e.g., see Refs. G1–G8 listed at the end of this Introduction) have been published, comprehensive reviews combining structural and spectroscopic studies have not yet been available. The aim of this book is to provide such information using examples chosen from a large number of references. Cytotoxicity data are quoted whenever available and appropriate. It is hoped that any unbalanced presentation may be compensated by the review articles cited throughout the book. Drug–DNA interactions on the cellular and clinical levels and drug–protein interactions are not included because detailed structural and bonding information on these systems is limited by their complexity.

In this book, the terms *DNA* and *drug* are used loosely to cover a wide range of chemicals. Thus, *DNA* includes its components such as purine and pyrimidine bases, nucleosides as well as synthetic oligonucleotides, and natural DNA. Likewise, a *drug* (or a ligand) refers to any (natural and synthetic) chemical with potential pharmacological activity. Most of these are anticancer drugs, although antiviral, antibacterial drugs and a wide range of other chemicals such as biological stains and metal complexes are included.

This book focuses largely on structural and bonding information obtained by physicochemical methods, namely, X-ray crystallography, NMR spectroscopy, ESR spectroscopy, and optical spectroscopy such as UV–visible absorption, fluorescence, circular dichroism (CD), flow linear dichroism (FLD), and infrared (IR) and Raman (R) spectroscopy. Among these methods, X-ray crystallography and NMR spectroscopy are the most important when atom-level structural information is desired. However, X-ray crystallography is limited to drugs bound to relatively short oligonucleotides *in the crystalline state*, because it becomes more difficult to obtain single crystals of diffraction quality as the nucleotide sequence becomes longer. Although multidimensional NMR spectroscopy coupled with computer-aided molecular modelling techniques can provide structural information on the atomic level *in solution*, this method is also limited to drug molecules bound to relatively short nucleotides since NMR signals tend to overlap in longer oligonucleotides. It should be noted that the molecular structure determined in the crystalline state suffers from the lattice effect, which could influence the molecular conformation. Although the solution structure obtained by NMR spectroscopy may be less precise than X-ray structure, it is closer to the structure in a biological environment and can provide insight into molecular flexibility. Furthermore, it is possible to study the effects of changing the pH, temperature, and drug–DNA mixing ratio on the drug–DNA intraction in solution and even dynamic equilibria between different conformers.

In contrast, the other spectroscopic techniques mentioned above are applicable to drug molecules complexed to long oligonucleotides such as poly(dG-dC)$_2$, poly(dA-dT)$_2$, and natural DNA, and provide structural and bonding information on the molecular level. It should be noted, however, that detailed structural information obtained by X-ray and NMR studies may not always be directly applicable to those with long oligonucleotides, because the structure of the latter is affected by the flexibility and base-pair sequence of their strands.

Since the main objective of this book is to review structural and spectroscopic information obtained by the aforementioned techniques, we provide only a very limited discussion of their basic theories and experimental methods. This information is found mainly in Chapter 1, with some expanded details presented in the appendixes. More thorough information on these topics may be found in many monographs and review articles cited (e.g., see Refs. (G9–G14).

As stated earlier, and discussed at greater length in the following chapters, the modes of the drug–DNA interaction can be classified into the categories of *intercalation, groove binding, covalent-bonding and strand breaking* as well as a variety of their combinations. Furthermore, their bindings are often reinforced by hydrogen bonding, Coulombic forces, and van der Waals interactions. These are discussed throughout this book, which consists of six chapters.

Chapter 1 provides basic knowledge about the structures and spectra of bases, nucleosides, and nucleotides, and is vital to the understanding of the nature of drug–DNA interactions discussed in later chapters. It explains all of the conventional nomenclatures regarding DNA structures for the reader who is not familiar with DNA structures. For more information, the reader should consult those books and review articles on DNA structures that are abundantly available. Chapter 1 also provides an overview of how this structural information can be derived from the experimental methods.

Chapter 2 deals with drugs bound to DNA *via intercalation*, which was discovered in 1961 and is the best known mode of drug–DNA binding. Acridines, aclacinomycins, actinomycins, and ethydium bromide are typical examples. More than 20 other intercalating drugs are listed together with their sequence preferences. Recently, a variety of bis and tris intercalating drugs have been prepared to attain stronger binding and better selectivity of binding sites, and their modes of binding are discussed.

Chapter 3 focuses on *groove-binding* drugs such as netropsin and distamycin, which interact with AT-rich region of DNA in the minor groove via hydrogen bonding and Coulombic and van der Waals forces. Sequence-specific drugs have been prepared by arranging pyrrole and imidazole rings in a specific order since pyrrole binds to A/T base whereas imidazole prefers to interacts with the G base. Some derivatives of netropsin and distamycin exhibit antitumor and antiviral activities much stronger than those of distamycin. Chromomycin, mithramycin, and their derivatives represent another type of *groove-binding* drugs that interact with the GC-rich region of DNA. A variety of hybrid drugs in which groove binders are connected to intercalators have been synthesized and their modes of interactions with DNA investigated.

Chapter 4 describes *covalent bonding* drugs (e.g., alkylating agents) such as (+) CC-1065, anthramycin, ecteinascidins, and mitomycins. Active-carbon atoms of these drugs form *covalent bonds* with the N3 atom of adenine or guanine, and with the N7 atom and exocyclic NH_2 (N2 amine) group of guanine. Hedamycin and atromycin B are known as *intercalating alkylators*. Although metal-containing drugs such as cisplatin and its derivatives also form *covalent bonds* with DNA, they are discussed separately in Chapter 6, because their interactions with DNA have been studied much more extensively than other covalent bonding drugs.

Chapter 5 deals with *strand-breaking* drugs. Bleomycins and enediyne antibiotics such as calicheamicins, esperamicins, and neocarzinostatins are able to cleave either single- or double-stranded DNA. Bleomycins consist of a DNA-binding domain and a metal-binding domain that are connected by a linker. The former brings the latter close to the DNA strand, while the latter, in the presence of the Fe(II) ion and O_2, produces "activated BLM," which is responsible for DNA *strand breaking*. Green BLM-Co (III)-OOH serves as an ideal model of the "activated BLM" because it is a stable, diamagnetic compound and the structure of its complex with a short oligonucleotide can be determined by NMR spectroscopy. Enediyne antibiotics contain highly strained enediyne rigns that produce phenylene biradicals in the presence of a nucleophile such as thiol. Simultaneous cleavage of both strands of DNA by these biradicals is responsible for their strong antitumor activities.

Chapter 6 covers a large number of metal-containing drugs. Among them, cisplatin is best known, and its binding to DNA has been studied most extensively. The major products of cisplatin–DNA reaction are *1,2-intrastrand crosslinking* adducts formed by *covalent bonding* of the Pt atom with the N7 atom of guanine or adenine The minor products are *1,3-instrastrand and 1,2-interstrand crosslinking* adducts between two guanine bases. Thus, the main interest of research is on *1,2-intrastrand crosslinking* adducts, particularly those involving two guanine bases. To search for more potent drugs with fewer side effects, a large number of cisplatin analogs have been synthesized, and their cytotoxicities have been tested. These include cisplatin derivatives in which its NH_3 ligands are replaced by a variety of monodentate and bidendate (chelating) ligands, and transplatin and monofunctional complexes such as $[Pt(NH_3)_3Cl]^+$. Dinuclear platinum complexes can form *1,4-interstrand crosslinks* with DNA, and some of them are highly cytotoxic and more effective against cisplatin-resistant cell lines. Other metal-containing drugs included are those of Pt(IV), Pd(II), Au(III), Ru(II,III), Rh(III), Cu(II), and Zn(II), the cytotoxicities of which are reported.

Appendixes include discussion of several topics on Raman and NMR spectroscopy and molecular modeling that are not described in the main text of Chapter 1. This appendix material is intended to help the interested reader understand some of the more technical aspects of these methodologies.

GENERAL REFERENCES

Drug–DNA Interaction

G1. Wilman, D. E. V., ed, *The Chemistry of Antitumour Agents*, Blackie & Sons, London, 1990.

G2. Kallenbach, N. R. ed, *Chemistry and Physics of DNA-Ligand Interactions*, Adenine Press, Schenectady, NY, 1990.

G3. Prospt, C. L. and Perun, T. J. eds., *Nucleic Acid Targeted Drug Design*, Marcel Dekker, New York, 1992.

G4. Krush, T. R. "Drug-DNA interactions," *Curr. Opin. Struct. Biol.* **4**: 351–364 (1994).

G5. Hurley, L. H. and Chaires, J. B. eds., *Advances in DNA Sequence Specific Agents*, Vol. 2, JAI Press, London, 1996.

G6. Yang, X.-L. and Wang, A. H.-J., "Structural studies of atom-specific anticancer drugs acting on DNA," *Pharm. Ther.* **83**: 181–215 (1999).

G7. Chaires, J. B. and Waring, M. J., eds., "Drug-nucleic acid interactions," in *Methods in Enzymology*, Vol. 340, Academic Press, San Diego, 2001.

G8. Demeunynck, M., Bailly, C., and. Wilson, W. D., eds., *DNA and RNA Binders: From Small Molecules to Drugs,* Vols. 1 and 2, Wiley-VCH, Weinheim, Germany, 2003.

Experimental Methods

G9. Cantor, C. R. and Schimmel, P. R., *Biophysical Chemistry, Part II. Techniques for the study of Biological Structure and Function, W. H. Freeman*, San Francisco, 1980.

G10. Campbell, J. D. and Dwek, R. D., *Biological Spectroscopy*, Benjamin/Cumming, Menlo Park, CA, 1984.

G11. Sauer, K., ed., Biochemical Spectroscopy, in *Methods in Enzymology*, Vol. 246, Academic Press, San Diego, 1995.

G12. James, T. L., ed., "Nuclear magnetic resonance and nucleic acids," in *Methods in Enzymology*, Vol. 261, Academic Press, San Diego, 1995.

G13. Perun, T. J. and Propst, C. L. eds., *Computer-Aided Drug Design: Methods and Applications*, Marcel Dekker, New York, 1989.

G14. Sheehan, D., *Physical Biochemistry: Principles and Applications*, Wiley, New York, 2000.

CHAPTER 1

DNA Structures and Spectra

The study of the physico-chemical bases of the interaction of drugs with DNA, as well as the expression of genes and the genetic inheritance of biological characteristics, began in earnest with the quest for the structure of DNA itself, which was ultimately "solved" in 1953. The earliest research on DNA structure used optical spectroscopic methods, which are still used for studying drug–DNA interactions. By 1950, Maurice Wilkins and Ray Gosling began to study DNA structure using X-ray diffraction and obtained the first diffraction images of DNA fibers. With help from Alec Stokes, their colleague at Kings College, they proposed that the patterns resulted from a helical structure.

In 1951, Rosalind Franklin also joined the faculty at Kings College. Franklin worked closely with Wilkins and Gosling, but by using her own techniques, she obtained better X-ray images of DNA than anyone had done previously. These images clearly showed the helical structure of DNA, and enabled the identification of the locations of the phosphate sugars. These data also revealed that the DNA had two forms, depending on environmental humidity, which Franklin and Wilkins called the A and B forms of DNA. Meanwhile, Francis Crick and James Watson of Cambridge University were also working on DNA structure. They obtained some of Franklin's data and, by combining her data with their own analyses, built the first correct DNA model. In 1953, their paper was published in *Nature* [1], along with papers by Wilkins, Stokes, and Wilson [2], and Franklin and Gosling [3]. Sadly, Franklin succumbed to cancer in 1958, but Wilkins, Watson, and Crick shared the Nobel Prize in Physiology and Medicine in 1962.

It is likely that modern molecular genetics would not exist without this basic structural knowledge, and since that time, spectroscopic studies have continued to expand our physical understanding of DNA structure. There remains growing need to understand the structure of DNA and its interactions with drugs, mutagens, RNA, and protein ligands. DNA is no longer regarded as a rigid cylinder, but as a highly plastic and adaptable molecule, capable of many structural variations, each having unique and important biological consequences.

Drug–DNA Interactions: Structures and Spectra by Kazuo Nakamoto, Masamichi Tsuboi, and Gary D. Strahan
Copyright © 2008 John Wiley & Sons, Inc.

2 DNA STRUCTURES AND SPECTRA

The aim of this chapter is to explain the most fundamental concepts of DNA structure, especially the terminology and factors that describe its primary, secondary, tertiary, and quaternary structures. It will describe the structures and spectroscopic properties of DNA, as well as relate these to its biophysical characteristics.

1.1. DNA STRUCTURES

1.1.1. Primary (Chemical) Structure

The monomeric units of nucleic acids are called *nucleotides*. These consist of a phosphate, a sugar, and a special aromatic moiety called a purine (*pur*) or pyrimidine (*pyr*) base. Figure 1.1 shows the chemical structures and names of the common bases and sugars and also the numbering of the ring atoms (C and N) of bases. This IUPAC [4] numbering system is used throughout this book. All of the carbon atoms in the DNA bases and sugars are given unique position numbers. This is also true of the nitrogen atoms in the aromatic rings of the bases. All other atoms or groups (H, O, NH_2, and P) are numbered according to the carbon or nitrogen atoms to which they are directly bound. The thymine methyl group is given the unique number of C7 by IUPAC. (An alternate number of C5M is used in the PDB database, but will not be used in this book.) The sugars have the β-D-stereoisomeric configuration shown in Figure 1.1, which is a reference to the relative orientations of the OH groups. The atoms belonging to the

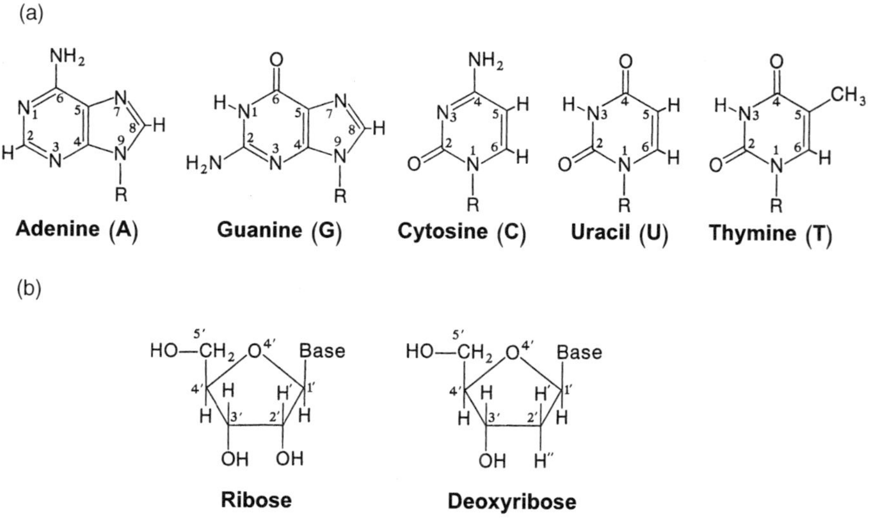

FIGURE 1.1. Chemical structure of nucleic acid constituents, the normal bases (a) and sugars (b). In DNA and RNA, the R group is the attached sugar ring. Isolated bases have R=H. The IUPAC [4] nomenclature for naming atoms are given here, and is used exclusively in this book. All C and aromatic N have unique atom position numbers, whereas the other atoms and groups (H, O, NH_2, and P) are numbered according to the carbon or nitrogen atoms to which they are directly bound. The thymine methyl group is given the unique number of C7 by IUPAC.

sugar moiety are distinguished from those of the base by one or two superscript prime marks (i.e., ' or ") on the atom number. The chemical unit that contains only the sugar and base is called a *nucleoside*; the phosphate ester of a nucleoside is called a *nucleotide*.

Nucleosides are abbreviated by combining an uppercase letter representing the base and a lowercase letter representing the sugar. For example, adenosine, which has a ribose sugar and an adenine base, is called "rA" (see Fig. 1.2). Deoxycytidine, with a 2'-deoxyribose sugar and a cytosine base, is "dC" (Fig. 1.2). Nucleotides are denoted by adding a "p" to represent each phosphate, as shown in Fig. 1.2; thus, for 5'-nucleotide, the abbreviation is "pN," and for 3'-nucleotide, "Np." If the base in Fig. 1.2 (bottom left) is guanine, it is referred to as guanosine-5'-monophosphate, often abbreviated as 5'-GMP.

Natural nucleic acids, RNA and DNA, are polynucleotides, the backbone of which consists of alternating sugars and phosphodiesters, as shown in Fig. 1.3. In RNA (ribonucleic acid) the sugars are ribose, and four nucleosides, rA, rU, rC, and rG, predominate. In DNA (deoxyribonucleic acid), the sugars are all 2'-deoxyribose, and four nucleosides, dA, dT, dC, and dG, make up essentially the total structure of all known DNAs. For both of these polynucleotides, each sugar has two ester linkages at the 5'O and 3'O atoms. The DNA chain is described as having a "direction," which is defined as the contiguous set of atoms starting at the 5'O and going to the 3'O (5' → 3').

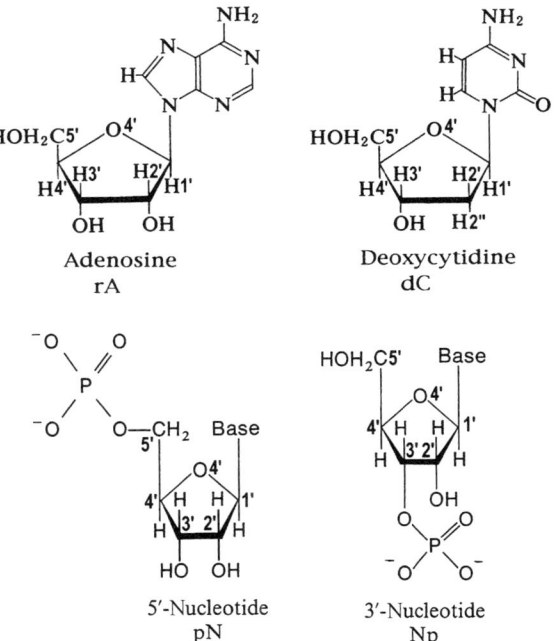

FIGURE 1.2. Chemical structure of nucleoside and nucleotide. A common nucleoside analog is inosine, which is similar to similar to guanosine, except that the $-NH_2$ group is replaced by an H.

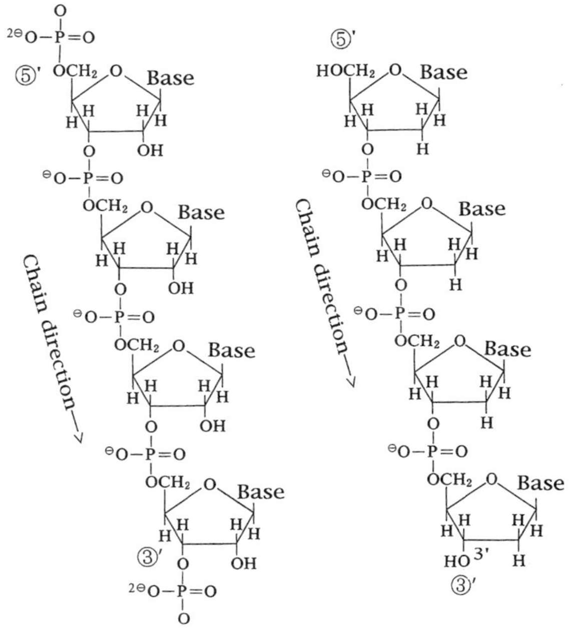

4 DNA STRUCTURES AND SPECTRA

FIGURE 1.3. Chemical structure of polynucleotide. The terminal oxygen atoms, O5' and O3', are named according to the carbon atoms to which they are directly bound.

(Note that these oxygen atoms are designated as 5' and 3', according to the carbon atom of the sugar to which they are bound.) This set of atoms is called the "backbone" of the DNA. Thus, a polynucleotide is abbreviated as ...pNpNpNpNpNpNp..., where the left end is called the "5' end" and the right end, the "3' end." These are also known as "the five-prime end" and "the three-prime end."

All the common DNA bases are uncharged at neutral pH. At lower pH, however, adenine (A) and cytosine (C) have positively charged protonated NH^+ forms, as shown in Fig. 1.4. Uracil (U) and thymine (T) have negatively charged deprotonated forms (N^-) at higher pH, while guanine (G) has both a positively charged protonated form (at lower pH) and a negatively charged deprotonated form (at higher pH; see Fig. 1.4).

In contrast, all the phosphate residues in nucleotides or polynucleotides are negatively charged at the pH at which one normally works with these compounds. This means that there must always be sufficient counter-cations present in the environment to neutralize the electronic charges on the DNA. The monoester and diester forms of phosphoric acid exist in an ionization equilibrium, as shown in Fig. 1.5. Therefore, in every nucleotide, at either end of a polynucleotide (5' or 3'), the phosphate has a doubly negatively charged form ($DNA-O-PO_3^{2-}$) at pH higher than 7 (i.e., at pH >7). On the other hand, all of the phosphodiester linkages in the polynucleotide chain have a single negative charge ($-O-PO_2^--O-$) at neutral pH.

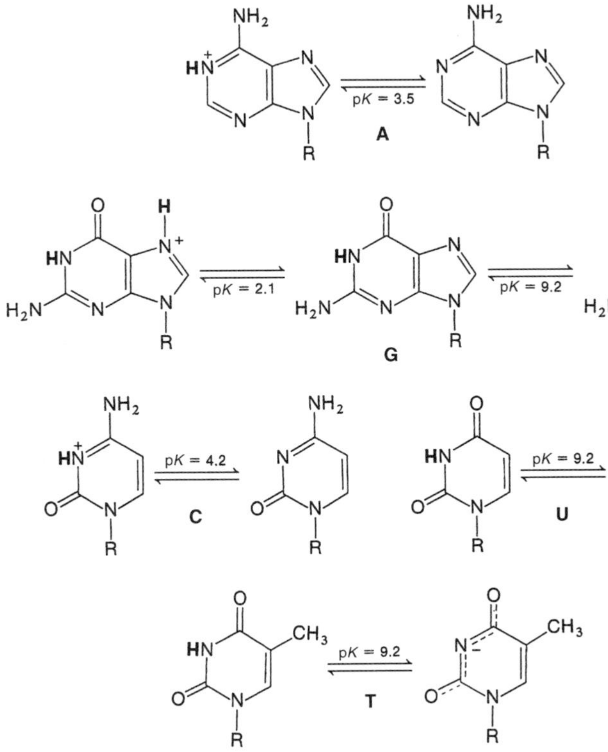

FIGURE 1.4. Ionization of aromatic bases (nucleosides) in DNA and RNA.

It should be noted that the two nonester P—O bonds are nearly equivalent to each other, so that each of these bonds has semi-double-bond character, and the single negative charge is shared between the two oxygen atoms.

1.1.2. Bond Lengths and Bond Angles

The secondary structure of DNA (or any polynucleotide) consists of three-dimensional regions that have ordered, locally symmetric, structures. The local

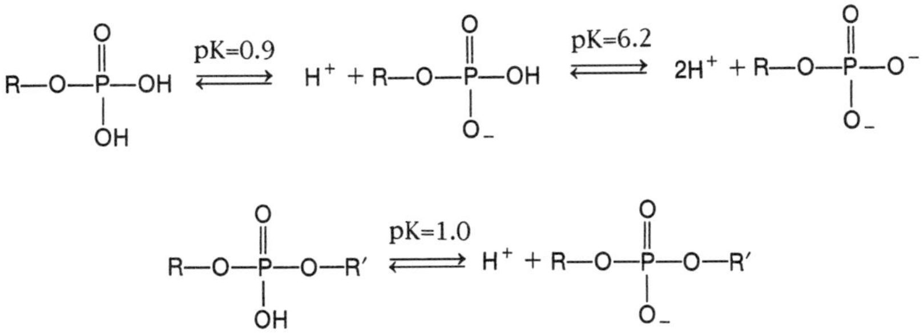

FIGURE 1.5. Ionization of phosphate.

three-dimensional structure is given by a list of atomic coordinates (see discussion below and Table 1.2, at the end of Section 1.1.3).

The three-dimensional structure of a DNA (or polynucleotide) molecule can be fully defined by listing either the (1) atomic Cartesian coordinates, or (2) the bond lengths, the bond angles, and the internal dihedral torsion angles. While the former type of list (1) is simpler to write out, the latter (2) better clarifies the physico-chemical properties, and has the advantage of listing characteristics that are not limited to a specific location in space, but that can be freely translated or rotated. Dihedral torsion angles are explained more fully in Section 1.1.8. The term *dihedral torsion angles* is often shortened in the literature, to either *torsion angles* or *dihedral angles*. In this book, we will use the term *dihedral angles* throughout.

All nucleic acid base residues are nearly planar, having deviations from planarity of less than 0.2 Å. Typical bond lengths and angles for the nucleoside base residues are given in Fig. 1.6 [5]. These aromatic moieties are central to the genetic significance of DNA, and contribute significantly to its energetic stability through the inter-residue van der Waals attractive forces between the π electrons in the aromatic rings. "Stacking interactions" of this type are revisited throughout this book.

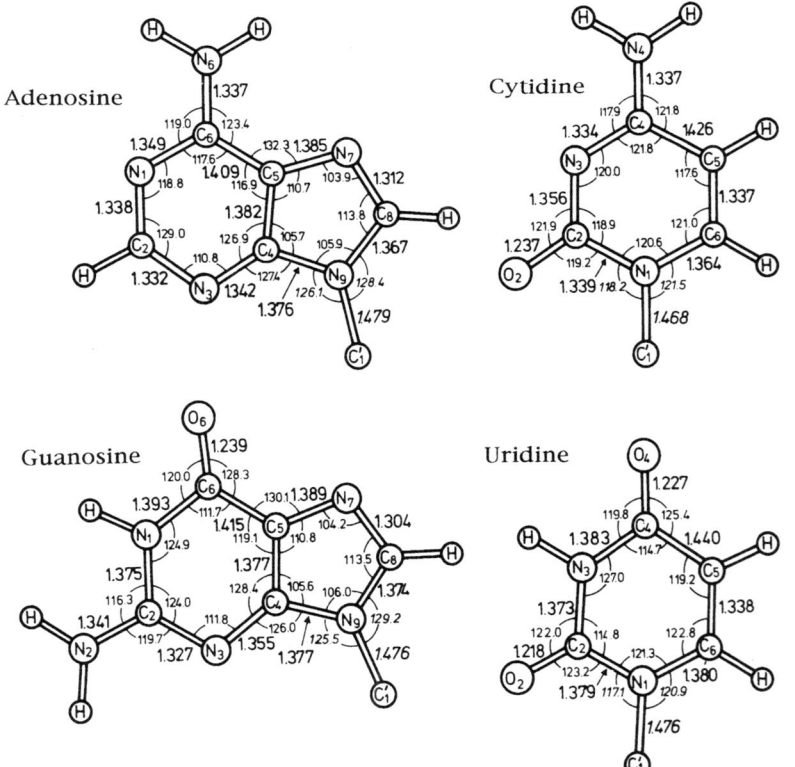

FIGURE 1.6. Average bond lengths (Å) and bond angles (°) in 1-substituted uracil and cytosine and in 9-substituted adenine and guanine [10].

DNA STRUCTURES 7

Unlike the nucleotide bases, the furanose sugar rings are not planar, but have out-of-plane puckering in either an "envelope" or a "chair" conformation (Section 1.1.4). To better understand this puckering, it is instructive to examine the conformation of tetrahydrofuran (Fig. 1.7a), which is the unsubstituted prototype of the five-membered sugar rings in nucleosides. According to electron diffraction [6,7], microwave [8], and theoretical [9] studies, this simple five-membered ring is also not planar, but is puckered. The bond distances agree well with standard, alkane-like C—C and C—O values (see Fig. 1.7), but it is the bond angles that cause the nonplanarity. These angles are

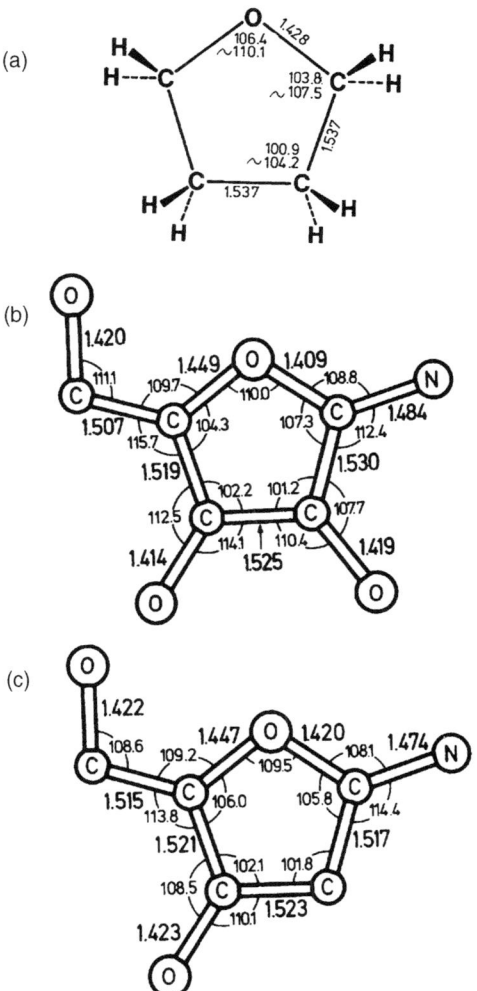

FIGURE 1.7. (a) Geometry of the unsubstituted furanose five-membered ring, tetrahydrofurane, as obtained from electron diffraction data [6–9]; (b) geometric data for ribose in nucleosides with C3'-endo puckering; (c) geometric data for deoxyribose in nucleosides with C2'-endo puckering [10].

significantly smaller than both the standard sp^3-tetrahedral geometry (109.5°) and the expected internal angles for a planar five-membered ring (540°/5 = 108°). In addition, the bond angles are not fixed, but fluctuate in the ranges shown in Fig. 1.7a. Much of the flexibility of the DNA backbone is attributable, at least in part, to these sugars.

The C—C and C—O bond distances in nucleoside furanose sugars are nearly the same as those in tetrahydrofuran, whereas more pronounced differences are seen in the bond angles (Fig. 1.7). The five-membered sugar ring can acquire a variety of puckering structures, which will be described in detail in Section 1.1.4, following discussion of the canonical three-dimensional structure of DNA in Section 1.1.3.

The phosphate groups bridging between nucleosides are phospho*diesters*, in which the ester (P—O) "single" bonds have some double-bond character due to π-electron delocalization. This is reflected in the fact that the P—O bonds in these mono- and diesters range within 1.59–1.63 Å (Fig. 1.8) [5,10], which is an interatomic distance

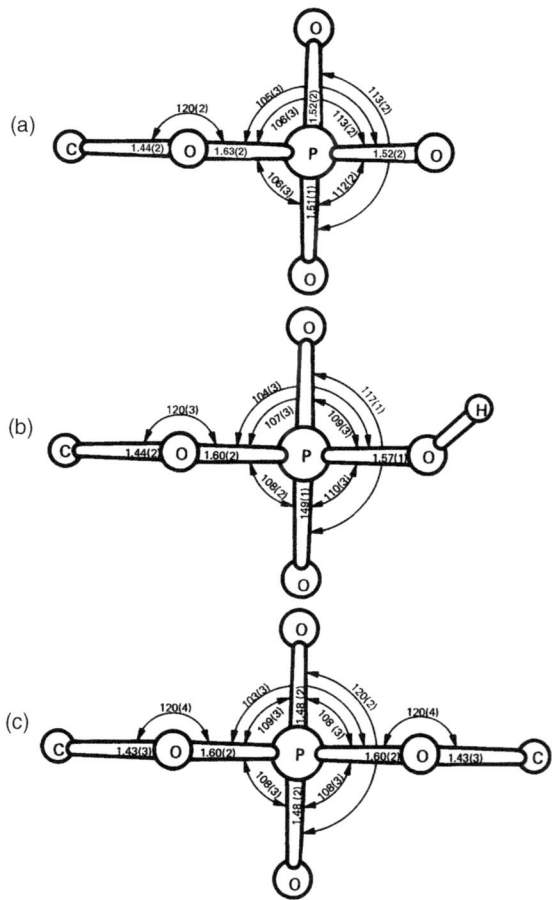

FIGURE 1.8. Geometric data for phosphoric acid esters: (a) monoester in higher pH; (b) monoester in lower pH (protonated); (c) diester [10].

that is slightly shorter than that of standard single bonds. Likewise, the valence P–O–C bond angle is ~120°, which is wider than expected for an oxygen atom with sp^3 character. Since these are single bonds with some double-bond characteristics, they have the effect of allowing a somewhat reduced amount of torsional flexibility in this portion of the DNA backbone.

In the single protonated phosphomonoesters (Fig. 1.8b), which occur at the very ends of the DNA chain, the P–OH bond distance is close to 1.57 Å. This is shorter than the P–O ester bond but longer than the two P–O bonds that carry the negative charge. The two unesterified P–O bond distances (1.46–1.56 Å) vary as a result of environmental influences in the crystalline state and are distinctively longer than the P=O double bond of 1.40 Å. If the phosphomonoester is not protonated but exists as a dianion (Fig. 1.8a), then the three unesterified P–O bonds are electronically equivalent. Likewise, the two unesterified P–O bonds in the diester anion (Fig. 1.8c), which links the DNA residues, are also electronically equivalent.

1.1.3. Canonical Structure of DNA: B-DNA

DNA has significant structural plasticity and can adopt a wide range of three-dimensional conformations. DNA rarely exists as a single strand, but prefers to associate into double-stranded, or "duplex" structures, the most common of which are known as the "A," "B," and "Z" conformations. More than 90% of known DNA sequences have three-dimensional conformations that fall into a family of structures called "B-DNA." For example, the cellular DNA found in *Escherichia coli*, virus P22, and the chromatin of all higher organisms are all B-form structures. As described in Section 1.1, the polynucleotide chain is a heavily charged polyanion that is neutralized by cations such as K^+ and Na^+. These cations can have a significant effect on DNA structure, but as a first approximation, one usually considers the generalized B-DNA structure to be independent of the kind of counter-ion.

The polynucleotide chain of B-form DNA is characterized by a "right-handed" helix. This term derives from the fact that when one points along the central, helical axis with one's right thumb, the fingers curl in the direction of the twist of the strand. For a right-handed helix, the strands curl counterclockwise when the helical axis is pointing upward. In B-form DNA, two such helices run in antiparallel directions with respect to each other, and are interwound coaxially (Fig. 1.9b). The 5' end of one strand is paired with the 3' end of the other, in a head-to-tail orientation. The two helices are held together through hydrogen bonds between two base residues, one from each strand, forming specific base pairs. The B-form helix makes one complete turn in about 10 residues, has a diameter of about 24 Å, and has a "pitch" of about 35.5 Å (Table 1.1). The pitch of a helix is the amount of rise necessary for one complete turn of 360°.

The standard base pairs consist of guanine to cytosine (G-C) and adenine to thymine (A-T). These are called *Watson–Crick base parings* (Fig. 1.10) (for other possible pairings, see Section 1.1.11). These bases are aromatic moieties with internal π-electron delocalization. This delocalization both flattens the individual bases and stabilizes the vertical stacking of these bases. Such stacking interactions are an

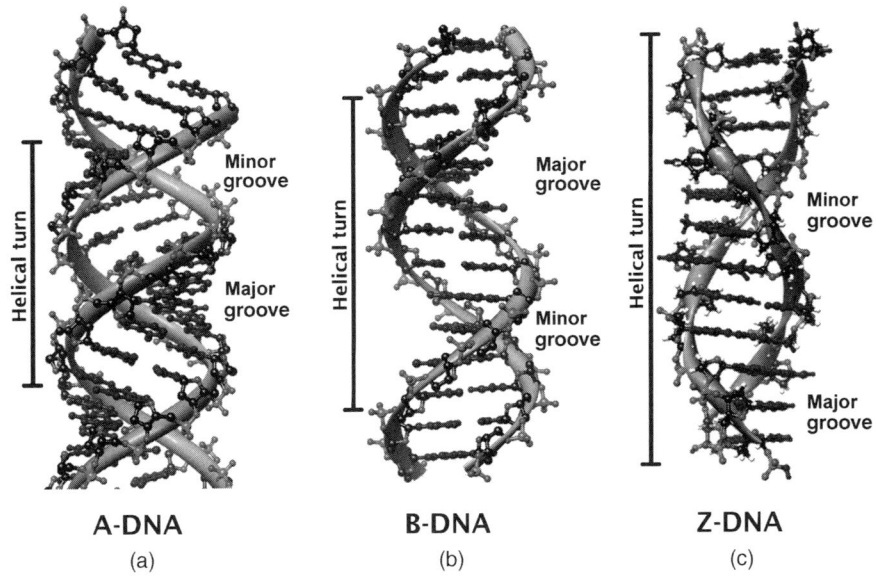

FIGURE 1.9. The side views of idealized fragments of (a) A-DNA, (b) B-DNA, and (c) Z-DNA, along with their approximate helical turns [276]. The A form has the tightest helical turn and hence is the shortest. Its base pairs also have the greatest amount of tilt with respect to the helical axis. The helical turn for the Z form is the least compact and stretches the longest distance. Note that the alternation of *syn* and *anti* glycosidic dihedral angles within the Z form is emphasized here as a twisting of the backbone ribbon. The B form has a moderate helical turn and length, with regular glycosidic dihedral torsion angles. (See color insert.)

TABLE 1.1. Geometric Characteristics of Different DNA Helical Forms

Structural Characteristic	Form		
	A	B	Z
Helical sense	Right-handed	Right-handed	Left-handed
Diameter	25.5 Å	23.7 Å	18.4 Å
Repeating unit	1 BP	1 BP	2 BP
Rotation per repeating unit (t_g)	32.7°	35.9°	−60°
Base pairs (BP) per turn (mean)	10.7	10.5	12
Tilt of BP relative to helical axis	+19°	−1°	−9°
Rise/BP along axis (p)	2.56 Å	3.38 Å	3.71 Å
Pitch/turn of helix (P)	28.2 Å	35.5 Å	44.6 Å
Mean propeller twist	+18°	+16°	0°
Sugar pucker	C3′-*endo*	C2′-*endo*	G: C3′-*endo* C, T, A: C2′-*endo*
Glycosidic dihedral torsion angle (χ)	*anti*	*anti*	G: *syn* C, T, A: *anti*
Minor groove	Wide and shallow	Narrow and deep	Narrow and deep
Major groove	Deep and narrow	Deep and wide	Shallow

Source: Data compiled from Refs. 10, 67, 198 and 277.

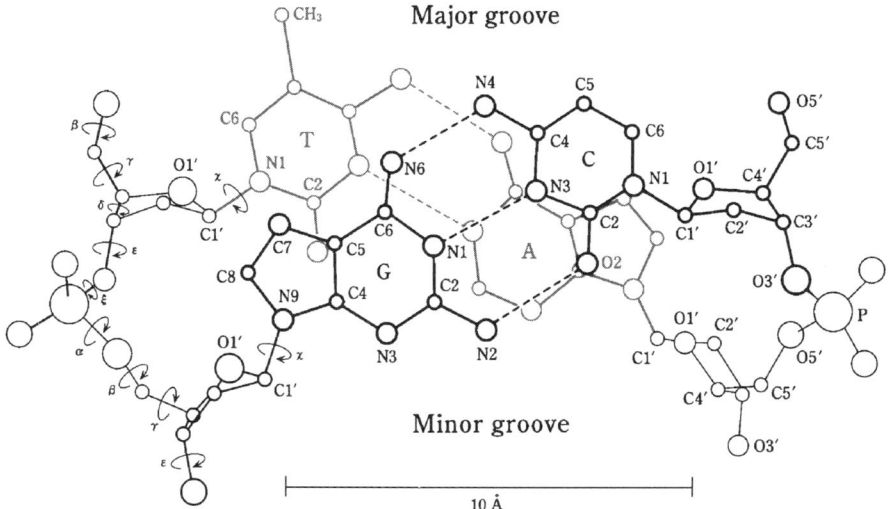

FIGURE 1.10. B-DNA structure, viewed along the helical axis. A G-C base pair is "above" an A-T base pair. (See color insert.)

important source of the energetic stability of DNA. The base pairs form planes that are stacked on top of one another along the axis of the double helix, and are oriented roughly perpendicular to the helical axis. As will be discussed later, the base pairs rarely form perfect planes; rather, the two bases are usually slightly twisted with respect to each other in a propeller-like manner. Not surprisingly, the angle between the planes of the two bases is a measure of a phenomenon called the "propeller twist."

As noted above, the phosphate groups of DNA are negatively charged, and as there are many phosphate groups in the DNA strands, the overall molecule has a very large net charge. The electrostatic repulsion between two DNA strands is sufficient to prevent their association into a duplex. To counteract this tendency, a sufficient concentration of cations must be present to maintain electroneutrality. The primary cations in the cellular environment are the monovalent ions, K^+ and Na^+, which are hydrated in solution. They stabilize DNA by electrostatically clustering around the phosphate groups, forming a loose spine along the strand. This clustering is referred to *salt condensation*. Divalent cations, such as Mg^{2+}, can also stabilize DNA, but often do so by bridging between the two phosphate groups on opposite strands. For this reason, divalent cations can provide greater stability to DNA, sometimes stabilizing nonstandard DNA structures, such as distorted or triple helices.

Many of the most important structural characteristics of B-form DNA, as well as those of A- and Z-form DNA (see later sections), are summarized in Table 1.1. The atomic coordinates in B-DNA were given first by Wilkins and his collaborators [11–13], and later revised by Arnott and others [14–16]. Table 1.2 lists the atomic coordinates of B-DNA given by Arnott and Hukins [15]. These are the geometrically preferred structural parameters (bond lengths, bond angles, dihedral angles, etc.) describing a DNA structure that is linear and not distorted by drug binding.

TABLE 1.2. Cylindrical Polar Coordinates of a Nucleotide Residue in B-DNA[a]

Atom		r (Å)	θ (°)	z (Å)
Phosphate	O1	8.75	97.4	3.63
	O2	10.20	91.1	1.86
	O3	8.82	103.3	1.29
	P1	8.91	95.2	2.08
	O4	7.73	88.0	1.83
Deoxyribose	C5	7.70	79.8	2.77
	O5	6.22	66.0	1.83
	C4	7.59	69.9	2.04
	C3	8.20	69.9	0.64
	C2	7.04	73.2	−0.24
	C1	5.86	67.4	0.47
Adenine	N9	4.63	76.6	0.42
	C8	4.84	93.0	0.50
	N7	3.95	105.4	0.43
	C5	2.74	94.0	0.28
	N6	1.83	154.0	0.14
	C6	1.41	107.2	0.15
	N1	0.86	40.1	0.03
	C2	2.17	30.6	0.04
	N3	3.24	47.0	0.16
	C4	3.33	70.5	0.28
Guanine	N9	4.63	76.6	0.42
	C8	4.82	93.2	0.50
	N7	3.92	105.7	0.42
	C5	2.70	94.0	0.28
	O6	1.71	154.6	0.13
	C6	1.39	109.3	0.15
	N1	0.92	37.9	0.03
	N2	3.01	4.2	−0.10
	C2	2.28	28.7	0.03
	N3	3.29	46.7	0.16
	C4	3.33	70.3	0.28
Cytosine	N1	4.63	76.6	0.42
	C6	4.99	92.2	0.52
	C5	4.35	107.0	0.47
	N4	2.76	136.6	0.27
	C4	2.94	110.0	0.32
	N3	2.31	83.9	0.22
	O2	3.69	47.9	0.18
	C2	3.40	67.4	0.27
Thymine	N1	4.63	76.6	0.42
	C6	5.01	92.3	0.52
	Me	5.40	119.8	0.58
	C5	4.38	106.9	0.47
	O4	2.82	136.3	0.27
	C4	2.98	111.9	0.32
	N3	2.36	85.2	0.23

TABLE 1.2. (*Continued*)

Atom		r (Å)	θ (°)	z (Å)
	O2	3.64	47.8	0.18
	C2	3.42	67.3	0.27

[a]Coordinates for successive nucleotide residues may be obtained by adding 36.0° to θ and 3.38 Å to z. All base coordinates refer to one residue. To generate coordinates of the dyadically related base pair, change the signs of θ and z.
Source: Ref. 15.

1.1.4. Deoxyribose Moiety as a Flexible Joint

To be biologically relevant, a DNA duplex must maintain a sufficiently regular structure to be recognized by proteins as a line of letters spelling genetic information. At the same time, however, it must also have enough flexiblity to allow its strands to dissociate and associate during transcription and replication. One source of this flexibility is a peculiarity of the deoxyribose moiety. The sugars act as a junction between the phosphate groups and the sidechain (the base residue) and can function as a partial "universal joint," which provides just the right amount of flexibility to the whole DNA structure.

As mentioned in Section 1.1.2, a saturated, five-membered ring (like deoxyribose ring) is not planar, but has an envelope (E) or a twist (T) geometric conformation (Fig. 1.11). In an E conformation, one of the member atoms (C or O) protrudes from the plane formed by the other four atoms; in a T conformation, two adjacent members are out of plane, with one above the plane (up) and the other below it (down). Ten possible E forms of the deoxyribose ring are shown in Fig. 1.12: C1′-*endo*, C2′-*exo*, and so on. Here, "C1′-*endo*" refers to a geometry in which the C1′ atom is projected outward and upward (toward the same side on which the glycosidic bond is located) with respect to the C2′C3′C4′O plane. It also is denoted as ^{1}E, where the "E" refers to the envelope form and "1" refers to C1′. The fact that the "1" is raised (superscripted), reminds us that the C1′ is oriented *upward*, or in the *endo* conformation. The term "C2′-*exo*" refers to a geometry in which the C2′ atom is projected downward (toward the side opposite to the location of the glycosidic bond) from the C1′C3′C4′O plane. It has

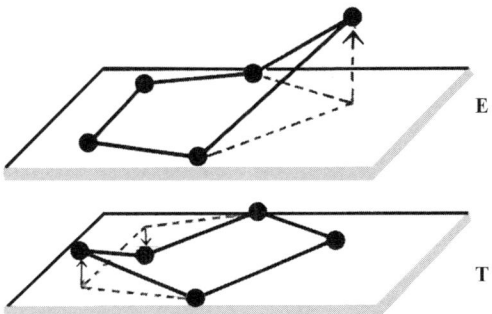

FIGURE 1.11. Two possible conformations of a saturated five-membered ring (or a deoxyribose ring) (E—envelope form, T—twist form).

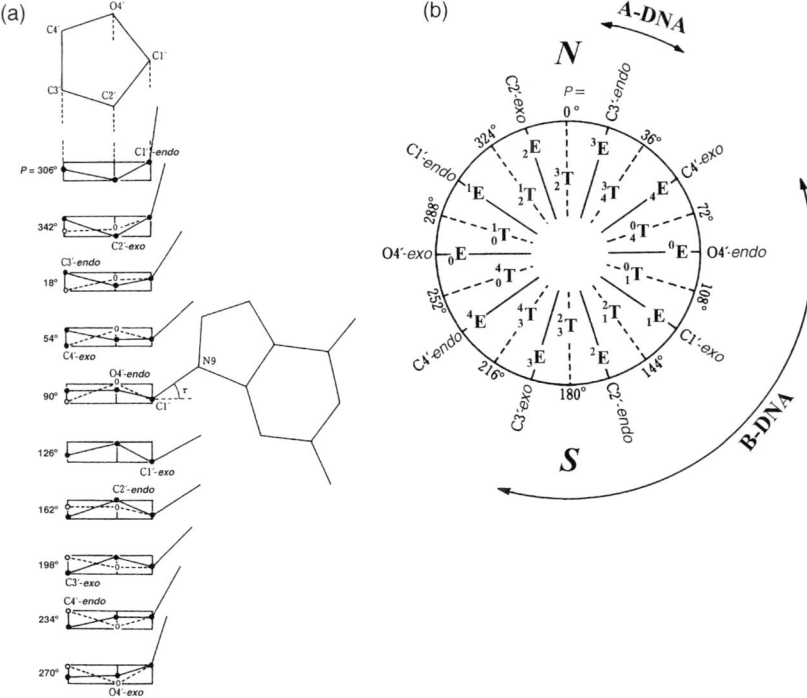

FIGURE 1.12. Conformations of the deoxyribose ring. (a) Side views of each of the 10 defined envelope forms, with the associated P (= angle of pseudorotation) value. Carbon atoms and bonds are indicated by solid lines and dots (●) if they are on the reader's side of the ring, while dashed lines and hollow dots (○) indicate bonds and atoms located on the far side of the ring. An "O" indicates an oxygen atom located in the far side of the ring. (b) The relation between the pseudorotation angle P and the conformation of deoxyribose ring. E is an envelope conformation. N indicates "north," or $P \sim 0$, and S indicates "south," or $P \sim 180°$.

the label of $_2$E, indicating that it is an envelope form with the C2′ oriented *downward*, or in the *exo* conformation.

Since the internal dihedral angles within the deoxyribose ring are all interdependent, a change in one of the angles must affect all the others. Only two parameters are needed to define the conformations of all the dihedral angles (z_i) within the ring. The first parameter ($z_{\max}$), defines the maximum out-of-plane pucker, or maximum dihedral angle within the ring. The other parameter is the pseudorotation phase angle, or P. Every E or T form is regarded as a "stop" in the *pseudorotation* cycle of the five-membered ring. For any given value of P, the sum of all of the positive dihedral angles in the ring must be equal to the sum of all the negative dihedral angles: $z_1 + z_2 + z_3 + z_4 + z_5 = 0$.

One can also write an equation for the out-of-plane coordinate z_j for the atom j, given any value of P and $z_{\max}$:

$$z_j = z_{\max} \sin(144°j + P) \tag{1.1}$$

By definition, $P = 0°$ when the sugar ring of a nucleotide is halfway between the C2′-*exo* ($_2$E) and the C3′-*endo* (^{3}E) forms. This is a symmetric twist form, written as $^3_2$T, where the 3 and 2 refer to the neighboring *endo* and *exo* conformations (notice that the vertical positions of the numbers is important). Since the pseudorotation cycle resembles a compass, the sugar pucker angles that are close to the $^3_2$T form are often called "north" or "*N*." Similarly, when the sugar ring has a pseudoration angle of $P = 180°$, it is halfway between the C3′-*exo* ($_3$E) and the C2′-*endo* (^{2}E) forms. This is the twist form $^2_3$T, which is called "south" or "*S*."

The *P* values corresponding to different deoxyribose ring conformations can be seen in Fig. 1.12. In this description, the E and T conformations alternate every 18° as one moves around the pseudorotation cycle. Real sugar rings need not fall within these discrete categories, but may exist at any energetically allowed *P* value. Thus, the conformation of every deoxyribose moiety in a DNA chain can be represented by an angle *P*. It is noteworthy that these different conformations also have different angles (χ) between the glycosidic bond and the average plane of the deoxyribose ring (see Section 1.1.8).

Different DNA conformations actually have different energetic preferences for their sugar conformations arising from both environmental conditions and DNA sequence. Thus, the conformation C3′-*endo* ($P \sim 0°$, ^{3}E, or *N*) predominates in A-form DNA, while that of C2′-*endo* ($P \sim 180°$, ^{2}E, or *S*) is the most common in B-DNA (see Section 1.1.8). Table 1.1 summarizes these comparisons.

1.1.5. Variations of B-DNA Structure

Considerable variations have been found among the structures of B-DNA studied by X-ray diffraction of single crystals of oligodeoxyribonucleotides. A structural study by Dickerson and Drew of the self-complementary dodecamer, d(CGCGAATTCGCG)$_2$, revealed that this oligomer contains a variety of deoxyribose conformations, although the overall structure is still in the category of B-DNA [17]; it still has a double-helix structure, with a pitch of 35.8 Å, which corresponds to 10.1 base pairs (bp) per turn, and, unlike A-form DNA, all the base-pair planes are nearly perpendicular to the helical axis. However, standard B-form DNA has its sugars in the C2′-*endo* (or "south") conformation, whereas the deoxyribose conformations found in this structure indicate that they are, in reality, greatly dependent on the nucleotide sequence (see Table 1.3). Higher-resolution studies of this dodecamer confirmed these results [18,19].

An extreme deviation from the B-DNA canonical structure was found by Nelson and others [20], in the "A-tract sequence" of the oligonucleotide duplex:

$$d(CGCAAAAAAGCG) \cdot d(GCGTTTTTCGC)$$

Here it was found that the extended string of contiguous (non-alternating) A-T base pairs results in an increased "propeller twist" within each base pair. In other words the bases within an A-T base pair are non-planar with respect to each other, and are rotated in opposite directions (contrarotated) along their longitudinal axis, as seen in Fig. 1.13.

TABLE 1.3. Deoxyribose Conformations in the Crystal of DNA Double-Helix Dodecamer, d(CGCGAATTCGCG)$_2$

Residue	Conformation	Residue	Conformation
C_1	$C2'$-endo	C_{13}	$C2'$-endo
G_2	$C1'$-exo	G_{14}	$C1'$-exo
C_3	$O1'$-endo	C_{15}	$O1'$-endo
G_4	$C2'$-endo	G_{16}	$C2'$-endo
A_5	$C1'$-exo	A_{17}	$C2'$-endo
A_6	$C1'$-exo	A_{18}	$C2'$-endo
T_7	$O1'$-endo	T_{19}	$C1'$-exo
T_8	$C1'$-exo	T_{20}	$C1'$-exo
C_9	$C1'$-exo	C_{21}	$C1'$-exo
G_{10}	$C2'$-endo	G_{22}	$C2'$-endo
C_{11}	$C2'$-endo	C_{23}	$C1'$-exo
G_{12}	$C1'$-exo	C_{24}	$C3'$-endo

Source: Ref. 17.

The average angle of rotation is 20°, which improves the overall stability of the helix by maximizing purine–purine base-stacking interactions. In this case, the helical repeat averages 10.0 bp per turn, and the helical structure is still similar to that of B-DNA. Nevertheless, A-tracts have unusual properties that prevent them from associating into nucleosomes, and short runs cause bending of the DNA.

Many other DNA oligomers have been crystallized and analyzed by X-ray diffraction, some of which are listed in Table 1.4. Those with duplex B-DNA structures are listed in the middle column [21–29]. A more complete review of single crystal X-ray diffraction studies of oligonucleotides is available in Ref. 30.

1.1.6. Participation of Water Molecules in B-DNA Structure

So far, we have considered three factors that stabilize the double-helix B-DNA structure: (1) hydrogen bonding between complementary base pairs, (2) π-electron interaction between the bases stacked vertically, and (3) electrostatic forces between the phosphate group (PO_2^-) and solvated cations (Na^+, K^+, Mg^{2+}, etc.). Another indispensible contribution to the stability of the DNA structure is hydration. This is demonstrated by the fact that the sodium salt of a DNA (Na^+-DNA) fiber produces different X-ray diffraction patterns depending on the relative humidity of its environment. These fibers produce a clear, standard, B-DNA X-ray diffraction pattern only if the fiber is maintained in an environment with 92% relative humidity. When the humidity drops lower than 75%, the X-ray diffraction pattern changes significantly, indicating that the conformation of the DNA fiber has changed to that of A-form DNA. When the humidity drops below 60%, it becomes difficult to obtain a well-defined X-ray diffraction pattern, suggesting that the DNA no longer has a well-defined structure [31].

DNA STRUCTURES

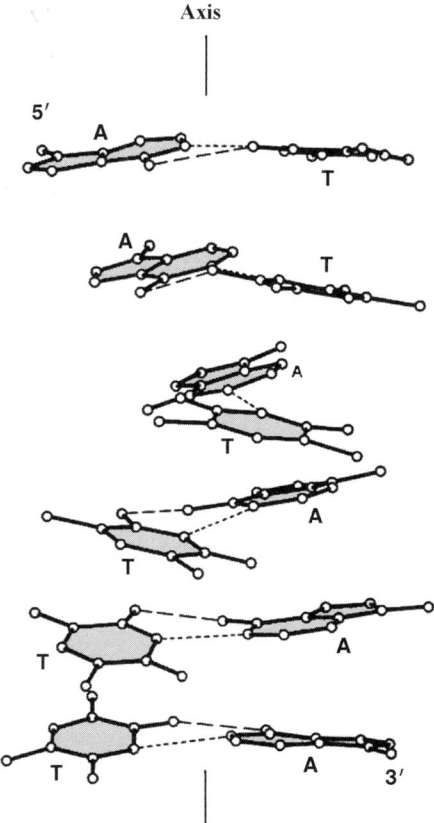

FIGURE 1.13. The central d(A)·d(T) tracts of the structure of double-helix DNA oligomer, d(CGCAAAAAAGCG)·d(GCGTTTTTT CGC) [20]. Only the bases are depicted.

High-resolution X-ray diffraction studies of single crystals of DNA oligomers have also revealed that DNA has a spine of water molecules along its minor groove [18,32]. As shown in Fig. 1.14 for B-DNA, this spine of water molecules has a zigzag shape running along the helix; they alternate between radial positions, where one H_2O molecule is located deep inside the minor groove while the neighboring H_2O molecule is located toward its outer edge. All of these water molecules stabilize the structure through hydrogen bonding to the O and N atoms of the duplex.

In most cases, groove-binding drug molecules such as netropsin [33] replace the spine of water molecules, thereby stabilizing both the duplex structure and the drug–DNA complex.

1.1.7. Partially Dehydrated Structure: A-DNA

In Section 1.1.6 we noted that the sodium salt of a DNA (Na-DNA) fiber adopts a conformation that is largely in the A form when the relative humidity is >75%,

TABLE 1.4. Oligodeoxyribonucleotides Subjected to X-Ray Crystallographic Studies[a]

A-DNA	B-DNA	Z-DNA
CCCCGGGG [35]	CGCGAATTCGCG [21]	CGCG [53]
GCCCGGGC [36]	CGCAAAAAAGCG [20]	CGCGCG [51]
GGCCGGCC [24]	CGCAAAAAATGCG [22]	CGCGCGCG [51]
GGGCGCCC [37]	CGCATATATGCG [23]	CGCATGCG [54]
CTCTAGAG [38]	CGTGAATTCACG [24,24]	CGTACGTACG [55]
GTACGTAC [39]	CGATCGATCG [26]	
ATGCATGC [40]	CCAACGTTGG [27]	
GGTATACC [25]	CCAGGCCTGG [28]	
GGGATCCC [41]	ACCGGCGCCACA [29]	
GGGTACCC [42]		
GGGGCCCC [43]		
GGATGGAG [44]		
GTGTACAC [45]		
ACCGGCCGGT [46]		

[a]Reference numbers appear in brackets.

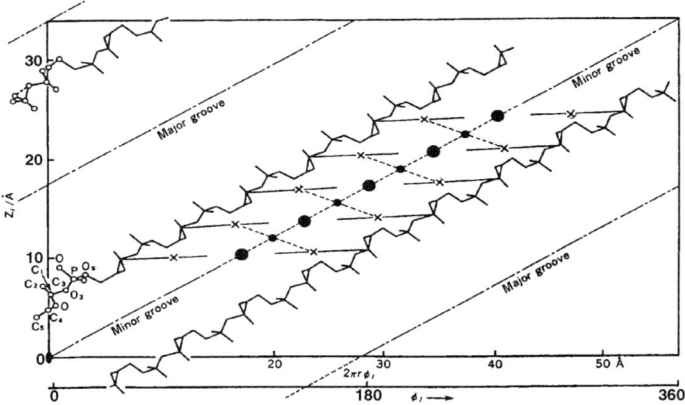

FIGURE 1.14. A spine of water molecules in the minor groove of B-DNA. This is a projection of the B-DNA structure on a plane that is formed by unrolling of a cylinder, that is, constructed by linking all P atoms in the duplex. As given in Table 1.2, every P atom is located at $r = 8.91$ Å from the helix axis, and therefore the P cylinder should have a circumference of $2\pi r = 56.0$ Å. Thus, the plane for projection should be a rectangle of 33.8×56.0 Å as shown here. The projections of polynucleotide mainchains are running from the lower left corner toward the upper right corner. A line of large and small circles is seen, along the minor groove that lies between two parallel polynucleotide chains. The small circles represent water molecules located deep inside the groove; the large circles represent water molecules located closer to the surface of the groove. The deeper water molecules form hydrogen bonds with O and N atoms (shown by "X") of the base residues (shown by horizontal lines projected out from C1' atom in the mainchain), whereas the shallow water molecules form hydrogen bonds with the O atoms in the diphosphoester and the deoxyribose ring [18, 52].

and adopts a B-form conformation when the humidity is ~92%. That the A and B forms are different conformations of the same DNA molecule is confirmed by the observation that changes in humidity induce reversible changes between A and B conformations [31].

As does B-form DNA, the A-form structure [34] (Fig. 1.9a) has a double-stranded structure consisting of two interwound right-handed helices, which run antiparallel to each other. In contrast to B-DNA, however, the midpoint between the base-pairs is located outside the center of the helical axis, and have an inclination angle of ~19°. The overall structure of A-DNA is shorter, fatter, and more compact than that of B-DNA. The helical pitch of A-DNA (~28 Å) is significantly less than that of B-DNA (~36 Å), having 11 nucleotide residues per turn instead of 10. The rise per base pair is much shorter (~2.6 Å) compared to B-form DNA (~3.4 Å). An important indicator of A-form DNA is that its deoxyribose conformation is in the C3'-*endo* (*N*) conformation, instead of the C2'-*endo* (*S*) conformation found in canonical B-DNA. These differences are summarized in Table 1.1.

Many relatively short DNA oligomers have crystallized into the A form, as determined by X-ray diffraction studies. Some of these A-DNA structures are listed in the first column of Table 1.4 [24,25,35–46].

In solution, a long stretch of A-DNA can be obtained by adding ethanol (85% in volume) to the aqueous solution of long B-DNA. The biological significance of A-DNA has not yet been clearly established, as so far there is no direct evidence for its presence in a living system. It should be pointed out, however, that double-helix RNA is structurally very similar to A-DNA, and that DNA–RNA hybrids have A-DNA-like structures even in aqueous solution without ethanol. In addition, the A form may be induced when DNA binds with certain drugs or proteins.

1.1.8. Internal Dihedral Angles

The bond lengths and bond angles in DNA remain nearly unchanged even when there is a significant conformational change caused by the binding of a drug or ligand. Instead, the conformational changes are attributable to changes in some of the dihedral angles within the DNA structure. As shown in the upper left corner of Fig. 1.15, every deoxyribonucleotide unit has seven single bonds, around which the atomic groups at both ends can rotate nearly freely (across lower potential energy barriers). These dihedral angles are defined as follows. If a system of four atoms A−B−C−D is projected onto a plane perpendicular to ("normal to") the B−C, then the angle between the projection of A−B and that of C−D is described as the dihedral angle about the B−C bond. The value of the dihedral angle is considered to be positive, if a clockwise turn of the A−B bond about the central bond is necessary to eclipse it on the C−D bond. The value is negative if a counterclockwise turn of the A−B bond is needed to eclipse it on the C−D bond (upper right corner of Fig. 1.15). It is convenient to remember that this definition can be rephrased as follows. If a clockwise turn of the bond containing the front atom about the central bond is needed for it to eclipse the bond to the rear

20 DNA STRUCTURES AND SPECTRA

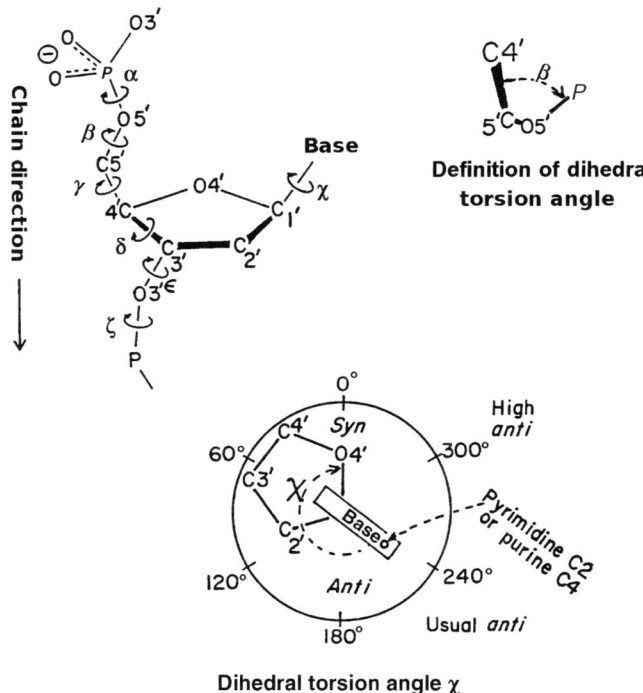

FIGURE 1.15. Internal torsion angles for expressing a local conformation of deoxyribonucleotide chain.

atom, then the angle is positive, regardless of the end from which the system is viewed [47].

A histogram of the internal dihedral angles within a DNA chain can be constructed [48] with reference to the crystallographic data from several oligodeoxyribonucleotides [24,25,49], and the results are shown in Fig. 1.16. The dihedral angles α, β, and γ are approximately $-60°$, $180°$, and $60°$, respectively, for both B- and A-DNA conformations. However, B-DNA and A-DNA are distinctly different conformers, because δ, ε, and ζ are different. Thus, $\delta = 120° \pm 40$ for B, but $\delta = 90°$ for A; $\varepsilon = -180°$ for B, but $-150°$ for A; and $\zeta = -90°$ for B and $-80°$ for A. The torsion angle δ is related to the pseudorotation phase-angle P of the five-membered ring C1'–C2'–C3'–C4'–O4', as shown in Fig. 1.17. Thus, the distribution of δ values in B-DNA (shown in Fig. 1.16, lower) indicates that its deoxyribose ring can adopt a wide range of P values, as pointed out by Dickerson et al. [49]. The ζ distribution has a tail toward $-60°$ in A, and toward $-180°$ in B. This unusual distribution of the dihedral angle ζ seems to be related to a sequence-dependent conformational variation. Similar, but much more detailed, histograms of the torsion angles have been obtained by high-resolution structural studies of B-DNA, A-DNA, and Z-DNA [50].

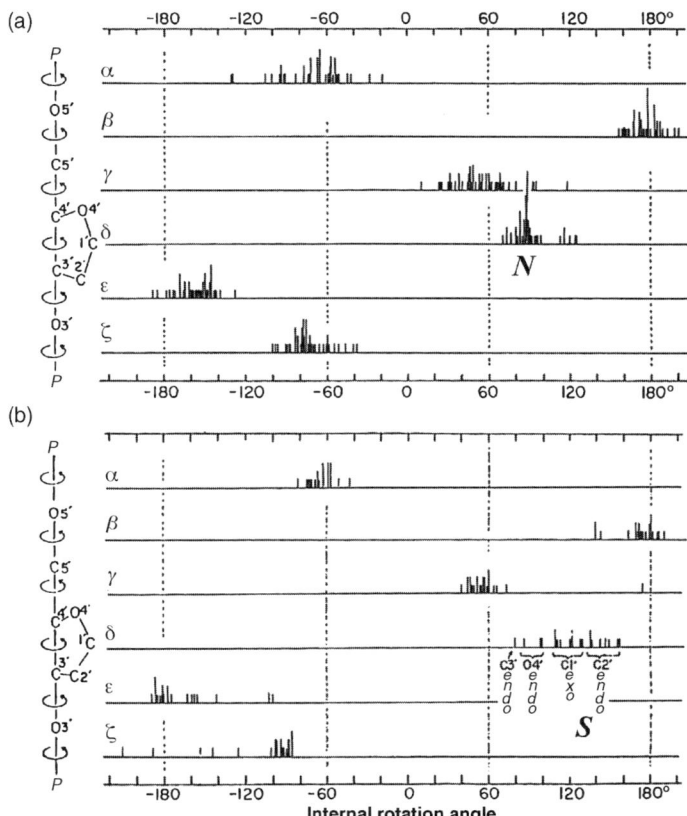

FIGURE 1.16. Histograms of the torsion angles in single crystals of double-helix DNA fragments: (a) d(GGCCGGCC)$_2$ [24] and d(GGTATACC)$_2$ [25]; (b) d(CGCGAATTCGCG)$_2$ [21].

The dihedral angle about the N-glycosidic bond (C1'−N), which links the base to the sugar, is denoted by the symbol χ. The definition of χ is illustrated in the lower right corner of Fig. 1.15. The sequence of atoms chosen to define this angle is O4'−C1'−N9−C4 for purine and O4'−C1'−N1−C2 for pyrimidine derivatives. Thus, when $\chi = 0°$, the O4'−C1' bond is eclipsed with the N9−C4 bond in purines and eclipsed with the N1−C2 bond in pyrimidines. This conformation is called *syn*. A conformation with $\chi \gg 0°$ is called *anti*. In general, the glycosidic bonds in B-DNA have an *anti* conformation.

1.1.9. Left-Handed Duplex: Z-DNA

Yet another type of duplex DNA conformation is that of Z-DNA. This duplex DNA structure is quite different from the preceding forms, A and B, in that it is a left-handed helix, rather than a right-handed one. This means that if one imagines

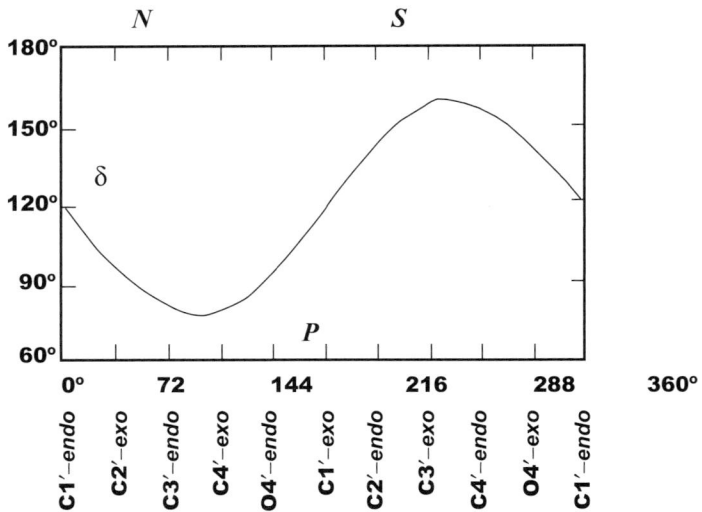

FIGURE 1.17. Relation between the C5′–C4′–C3′–O3′ internalrotation angle δ and the pseudorotation phase angle P of the deoxyribose ring.

grabbing a Z-DNA strand with one's left hand, with the thumb pointing in the direction of the helical axis, then the twist of the helix will be in the same direction as the curve of the fingers. One of the earliest reports on this conformation is the crystal structure study of the self-complementary oligo-DNA chain, d(CGCGCG)$_2$ [51,52]. This duplex structure (Fig. 1.9c) has the standard Watson–Crick, G-C base pairing, but the glycosidic bond angles (χ) of the guanine bases adopt the *syn* conformation, while the cytosine remain in the standard *anti* conformation. Since the sequence alternates between G and C, their attached sugars also alternate between *syn* and *anti*, as one moves along the strand. The deoxyguanosine sugar residues are also unusual in that they are in the C3′-*endo* (N) conformation, while the deoxycytidine residues are in the C2′-*endo* (S) conformation. (See Table 1.1 for a summary of these comparisons.) Some of other DNA oligomer sequences that have been found to have the Z-DNA structure by X-ray crystallographic investigations [53–55] are listed in the third column of Table 1.4.

Numerous studies have shown that poly(dG-dC)$_2$ can adopt a Z-DNA conformation in solution and that this conformation is stabilized by several chemical modifications of the base residues [56]. The Z-DNA conformation can also be adopted by the family of DNAs, poly[d(A-C)] • poly[d(G-U)], which bear at least one methyl or halogen (Br or I) at the pyrimidine C5 heterocyclic position [57]. Also, many G4-quadruplex structures (RNA and DNA) adopt Z-like conformations in which the guanines have alternating *syn*/*anti* glycosidic bond angles (χ) along the strands (Section 1.1.12).

Ha and others [58] constructed a synthetic DNA structure that contained a junction between B-DNA and Z-DNA. A 15-bp DNA duplex was designed such that it had an additional two nucleotides at the 5' end of each strand that were not base-paired, and which overhung the duplex formed by $5'-d(GTC_{-8}G_{-7}C_{-6}G_{-5}C_{-4}G_{-3}C_{-2}C_{-1}A_0T_1A_2A_3A_4C_5C_6) - 3'$ and $5'-d(ACG_{-6'}G_{-5'}T_{-4'}T_{-3'}T_{-2'}A_{-1'}T_{0'}G_{1'}G_{2'}C_{3'}G_{4'}C_{5'}G_{6'}C_{7'}G_{8'}) - 3'$.

The alternating dG-dC sequence composed the Z-DNA portion of the structure, and the other half of the duplex formed B-DNA. In this case, the bases were numbered from the center junction, with the two strands differentiated by the prime (') notation. The Z-DNA structure did not stabilize on its own; instead, these workers took advantage of the tight binding of a Z-DNA binding protein to induce this conformation in one half of the DNA duplex, while the remainder of the DNA duplex retained its B conformation. One B-Z junction was thus formed in the middle of the DNA duplex, connecting the Z- and B-DNA. Their X-ray crystallographic study indicated that the structure of the 8-bp region in the Z conformation is essentially the same as the Z-DNA structure described earlier (Fig. 1.9c), with a zigzag sugar phosphate backbone and alternating *anti–syn* base conformation. The other end of the DNA duplex adopts a standard B conformation (Fig. 1.9b). Between the B and Z conformations, an A-T base pair is disrupted and one base from the center of each strand (T0' and A0) is extended out from the double helix (Fig. 1.18).

The glycosyl bonds of A_0 and $T_{0'}$ are both in the *anti* conformation. The base of $T_{0'}$ is oriented almost parallel to the helix axis, whereas the A_0 base is fully extended out from the helix (Fig. 1.18). Thus, the handedness of the DNA duplex can be completely reversed merely by breaking one base pair and extruding the bases from the duplex. However, Z-DNA is inherently unstable, and to form Z-DNA under physiological conditions requires additional energy, such as can be supplied by negative supercoiling of the DNA (see Section 1.1.10). The base extrusion at the B-Z junction is accompanied by a sharp turn in the sugar–phosphate backbone, a slight bend (11°) at the junction, and displacement of the two helical axes by 5.2 Å. Thus, with only a single base pair broken at the junction, the structure affords a remarkable conservation of energy and maintenance of helical structure.

1.1.10. Superhelical Form of DNA

The tertiary structure of DNA includes superhelices of the double-stranded DNA chain. A *superhelix* is a helix formed by the curvature of the axis of the DNA duplex. The axis of B-DNA, for example, appears as a straight line in terms of its local structure, but in the cell, it is actually slightly curved, such that in a long strand of B-DNA duplex, it is no longer straight but forms another helix. Hence, in essence, a superhelix is a helix of a helix. A superhelix is often produced by a slight unwinding of the duplex, which may be induced by drug binding. In the canonical B-DNA, a base pair can be superimposed upon an adjacent one by translating it 35.5 Å along the helical axis, while simultaneously rotating it 36° around the helical axis (since there are 360° in

24 DNA STRUCTURES AND SPECTRA

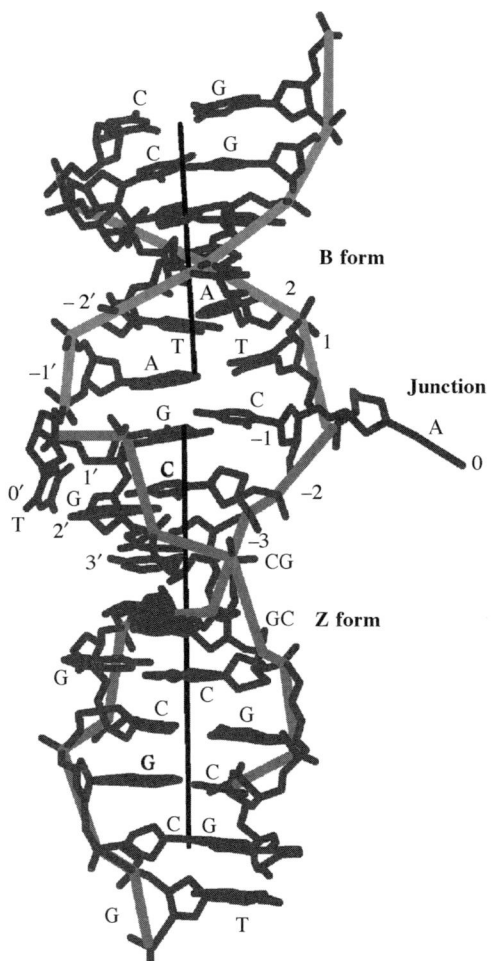

FIGURE 1.18. A 15-base-pair (15-bp) DNA structure containing a junction (black) between left-handed Z-DNA (magenta) and right-handed B-DNA (cyan) [58]. (See color insert.)

a complete turn and ~10 bases per turn; hence 360°/10). If, for example, the structure is deformed so that this angle is reduced from 36° to 31°, then the unwinding angle is 5°.

Superhelix formation often takes place in closed, circular DNA duplexes, as in *E. coli*, where the 3' end and 5' head of each strand are covalently connected through a phosphodiester linkage (Fig. 1.19). If a relaxed string of B-form DNA is closed into a circle, there will be no supercoiling (Fig. 1.19a). However, if the DNA double helix is unwound by 360° prior to joining its ends (Fig. 1.19b), it is harder for it to adopt the B form, even though it is energetically favorable to do so. The B form can be attained, however, by "winding up," or tightening, by 360°. Since the DNA is enclosed within a circle, this tightening introduces a strain, which is compensated by winding the circle of DNA into a superhelix with one turn (see Fig. 1.19c).

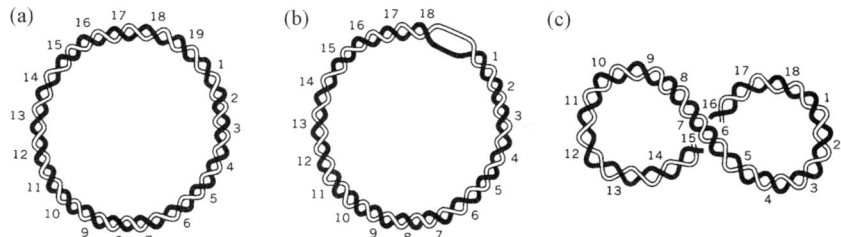

FIGURE 1.19. Schematic illustrations of closed circular DNA duplex. (a) Relaxed form, where linking number (α) = 19, twisting number (β) = 19, and writhing number (τ) = 0 (no supercoiling). The figures along the chain indicate the number of times one strand (black chain) goes around the other (white chain). (b) A hypothetical form that might be produced immediately after the circle is closed with the linking number (α) = 18, instead of 19. The closing portion does not have the B-DNA conformation and must be very unstable. (c) The unstable portion has been brought into a stable B-DNA structure, by twisting the whole circle by one turn, producing a negative superhelix.

In general, the topological properties of closed circular duplex DNA are defined by the relationship [59,60],

$$\alpha = \beta + \tau \tag{1.2}$$

where α is the "linking number" of the closed circular duplex, which is the number of times one strand goes around the other, and is a fixed integral number; β is the "twisting number," which is the number of helical turns of the DNA duplex; and τ is the "writhing number," which is the number of superhelical turns. In the model shown in Fig. 1.19a, $\alpha = 19$, $\beta = 19$, and $\tau = 0$. The conformation given in Fig. 1.19b does not actually exist. For the conformation shown in Fig. 1.19c, $\alpha = 18$, $\beta = 19$, and $\tau = -1$. The meaning of $\tau = -1$ is that the DNA has been unwound by one turn.

Natural, closed, circular DNAs are much larger than the model shown in Fig. 1.19. For example, consider the intact viral plasmid pBR322, which is purified from *Escherichia coli*, and is a closed circular DNA duplex consisting of 4361 bp. It is known to have a highly negative supercoiling ($\tau < -20$). By the action of topoisomerase I, however, it can be brought into the relaxed B form ($\alpha = 428$, $\beta = 428$, and $\tau = 0$). This was demonstrated by the preparation of moderately supercoiled pBR322 closed circles (e.g., $\alpha = 420$, $\beta = 428$, and $\tau = -8$) using the addition of appropriate amounts of aclacinomycin, followed by the addition of topoisomerase [61].

1.1.11. Other Duplex Structures

1.1.11.1. Mismatched Base Pairs Errors during replication or repair of DNA can lead to the presence of mismatched base pairs or unpaired bases within the double helix. For example, a mismatched base pair was found by X-ray crystallography [62] in the dodecamer duplex d(CGCAAATTCGCG)$_2$. Notice that this 12-base sequence is almost self-complementary—the string of Ts is one base shorter than the string of As—and it will prefer to exist in solution as a hydrogen-bonded duplex, even though the A-C base pairing at the ninth position is not correct. Figure 1.20 shows the A-C base

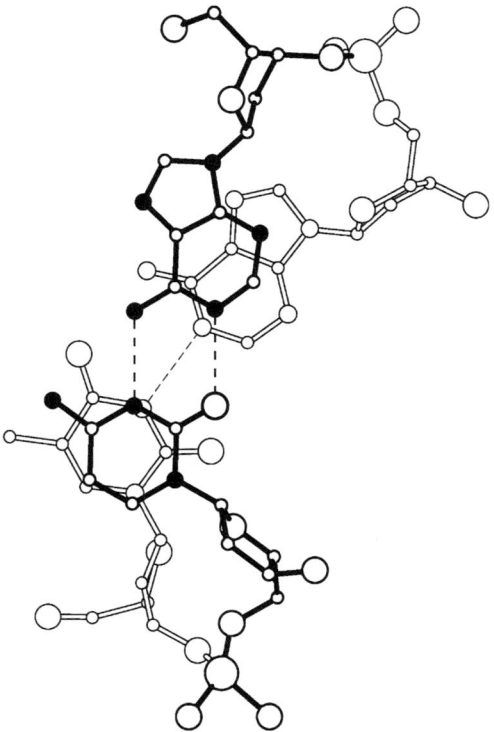

FIGURE 1.20. The A-C base pair, seen in bold lines, as established by the X-ray analysis [62]. This is proposed as a model of the A-C mismatch in double-helical DNA.

pair observed in the crystal structure of this dodecamer. The probability of incorporating an incorrect base during replication and its subsequent detection and excision by polymerase is thought to be related to the thermodynamic stability of the particular base-pair mismatch in the double helix. The relative energetic stabilities of different mismatches were studied by NMR (nuclear magnetic resonance) in a series of sequence-related dodecamers by determining their midpoints in the helix–coil transition. Relative to the fully hydrogen-bonded duplex form of the initial dodecamer, it was found that the G-T and A-G base pairs destabilize the double helix by 17–20°C, while the A-C base pair destabilizes it by 30°C [63]. Thus, even a single mismatch results in a significant lowering of the overall stability of a duplex, making it a looser structure. The stability of this short duplex is increased by the terminal G-C base pairs, each of which has three hydrogen bonds, and as such are said to be "stickier." As the length of the duplex portion of the DNA increases, this destabilization will contribute less on a percentage basis to the total energy. This becomes important in biological systems, where the native DNA is many thousands of bases long, and as such is capable of withstanding several mismatches.

1.1.11.2. Bulged Double Helix Sometimes two strands of DNA are complementary, with the exception of a stretch of extra bases in the middle of one strand.

These extra bases cannot form base pairs, and hence are bulged out of the duplex. The 13-base oligomer d(CGCAGAATTCGCG) is an example a nearly self-complementary sequence:

$$5'\text{-d}(C_1\text{-}G_2\text{-}C_3\text{-}A_4\text{-}G_5\text{-}A_6\text{-}A_7\text{-}T_8\text{-}T_9\text{-}C_{10}\text{-}G_{11}\text{-}C_{12}\text{-}G_{13})\text{-}3'$$

$$|\quad|\quad|\quad\quad/\quad/\quad/\quad/\quad/\quad/\quad\quad|\quad|\quad|$$

$$3'\text{-d}(G_{26}\text{-}C_{25}\text{-}G_{24}\text{-}C_{23}\text{-}T_{22}\text{-}T_{21}\text{-}A_{20}\text{-}A_{19}\text{-}G_{18}\text{-}A_{17}\text{-}C_{16}\text{-}G_{15}\text{-}C_{14})\text{-}5'.$$

When the two strands of this sequence come together to form a duplex, the A at the fourth position might be expected to form a mismatched base pair with the C on the other strand (e.g., $A_4\text{-}C_{23}$). This is not a Watson–Crick base pair, however, and instead the two additional adenosines (A_4, and its equivalent, A_{17}) are looped out from the double helix, forming a small bulge (Fig. 1.21) [64]. The looped-out bases cause little

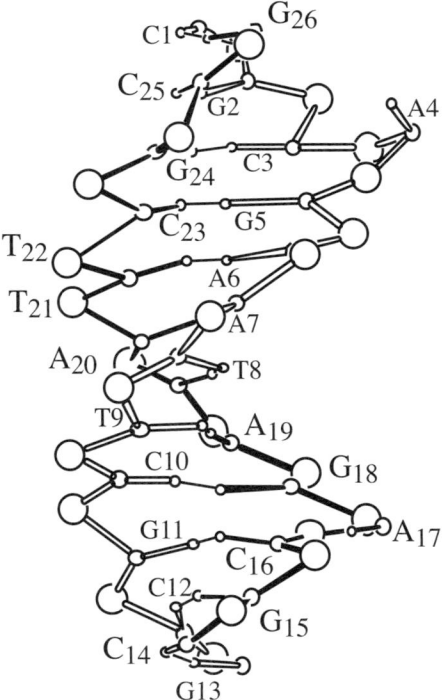

FIGURE 1.21. A schematic representation of the structure of the tridecamer d(CGCA-GAATTCGCG)$_2$ structure showing the extra adenosines (A_4 and A_{17}) bulging out [64]. The sugar–phosphate backbone is represented by phosphorus atoms (large spheres) linked to C1′ atoms of sugars (medium-sized spheres), and the bases by links from the C1′ atoms to the N9 atoms in purines or to the N1 atoms in pyrimidines (small spheres). The base-pair interactions are drawn as thin lines.

disruption of the rest of the structure, and the overall conformation is very close to that of B-form DNA.

1.1.11.3. Hairpin Double Helix   A DNA duplex does not need to be composed of two separate strands. When the DNA strand concentration is low, an otherwise self-complementary sequence can fold upon itself and adopt a monomeric hairpin conformation. This can also happen with a sequence that is only partially self-complementary. One example is the synthetic, 16-base DNA oligomer d(CGCGCGTTTTCGCGCG). A crystal structure of this sequence was solved at 2.1 Å resolution and was observed to form a monomeric hairpin having a Z-DNA hexamer stem and a loop consisting of four thymine bases (called a "T4" loop) (Fig. 1.22) [65]. The bases in the loop stack with one another, rather than with the base pairs of the hexamer stem. A similar stacked-base arrangement was proposed for the T4 loop of d(CGCGCGTTTTCGCGCG), on the basis of NMR measurements in solution and distance geometry calculations [66] (Section 1.7.2.1). In that case, however, the hairpin has a B-DNA stem, and the thymine bases in the loop were stacked parallel to the final base pair on the top of the stem. The difference in the T4 loop conformation between the X-ray and NMR studies probably reflects differences between crystal and solution environments, and illustrates the flexibility of a four-base loop. In general, it has been found that the details of the

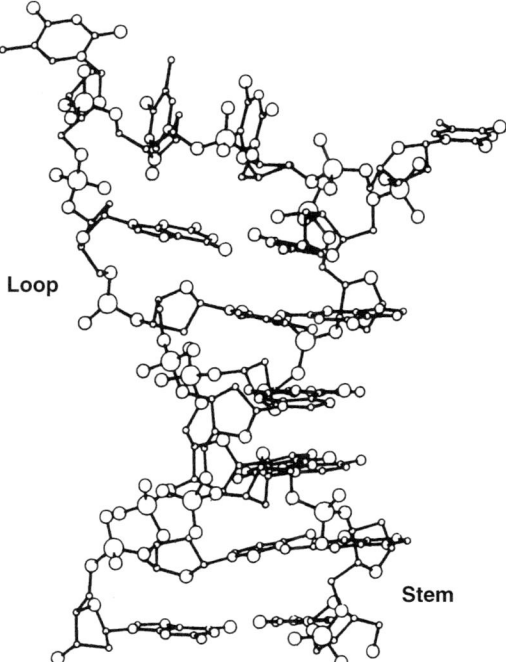

FIGURE 1.22. A structure of the hairpin of d(CGCGCGTTTTCGCGCG) [65]. Atoms in decreasing size are P, O, N, C. The stem of the hairpin is a normal left-handed Z-DNA hexamer.

stacking of bases within a loop is correlated to the adjacent base pairs at the top of the stem, with some stabilizing stacking more than others.

Two nearly self-complementary strands can also come together to form a duplex with an internal loop. In this case, there is a loop, or bulge, in the middle of both strands.

1.1.12. Higher-Order DNA Structures

The polymorphism of DNA has become a subject of considerable study in recent years, particularly because it has been suggested as a means of gene regulation and control of gene-based disorders and diseases. These non-B-DNA conformations include intra- and intermolecular triplex DNA, quadruplex DNA, cruciforms, and slipped mispaired structures. The first two of these are briefly discussed below, owing to their more unusual hydrogen-bonding patterns. For a variety of other DNA structures, see Ref. [67], for example.

1.1.12.1. Triple Helix Triple-helix formation between a homopyrimidine–homopurine duplex and a third homopyrimidine strand was discovered by Felsenfeld and Rich [68] in 1957, but remained a curiosity for many years. A possible biological role for triplex DNA has emerged in the area of gene regeneration with the discovery of H-DNA, which is a DNA structure formed by a pair of mirror repeat duplex sequences that contains a triple helix region, as seen in Fig. 1.23 (top) [69,70]. Triplex formation (including H-DNA formation) has received attention as a possible means to artificially control gene expression [71]. The idea here is that a third strand could be designed to bind in a sequence-specific manner to the duplex DNA of a particular gene and inhibit it; hence it is sometimes called an "antigene" approach. This approach, which involves the use of a piece of DNA as if it were a drug, could be ideal in terms of sequence specificity. It has faced two important limitations: (1) only some disease states have DNA target sequences that can form a triplex; and (2) it is difficult to deliver these short oligonucleotides to the correct gene in the cell without being digested by nucleases.

Triple-stranded DNA may be formed by laying a third strand into the major groove of a DNA duplex. Complementary hydrogen bonding interactions are responsible for the specificity of the third strand interaction. In these structures, the Watson–Crick base pairing within the duplex is unchanged, while the third strand is hydrogen-bonded to the duplex via a Hoogsteen base-pairing scheme. The hydrogen bonding patterns within in triple-stranded DNA are called "base-triples," and their schemes are *pur–pur–pyr* (GGC and AAT) and *pyr–pur–pyr* (TAT, CGC), the latter of which is shown in Fig. 1.23 (bottom). The central strand of the triplex must be purine-rich since pyrimidines do not have any additional hydrogen-bonding surfaces. Thus, triple-stranded DNA requires a homopurine–homopyrimidine sequence that is at least several bases long. If the third strand is purine-rich, it forms reverse Hoogsteen hydrogen bonds in an antiparallel orientation with the purine strand of the Watson–Crick duplex. If the third strand is pyrimidine-rich, it forms Hoogsteen bonds in a parallel orientation with the Watson–Crick-paired purine strand (Fig. 1.23, bottom).

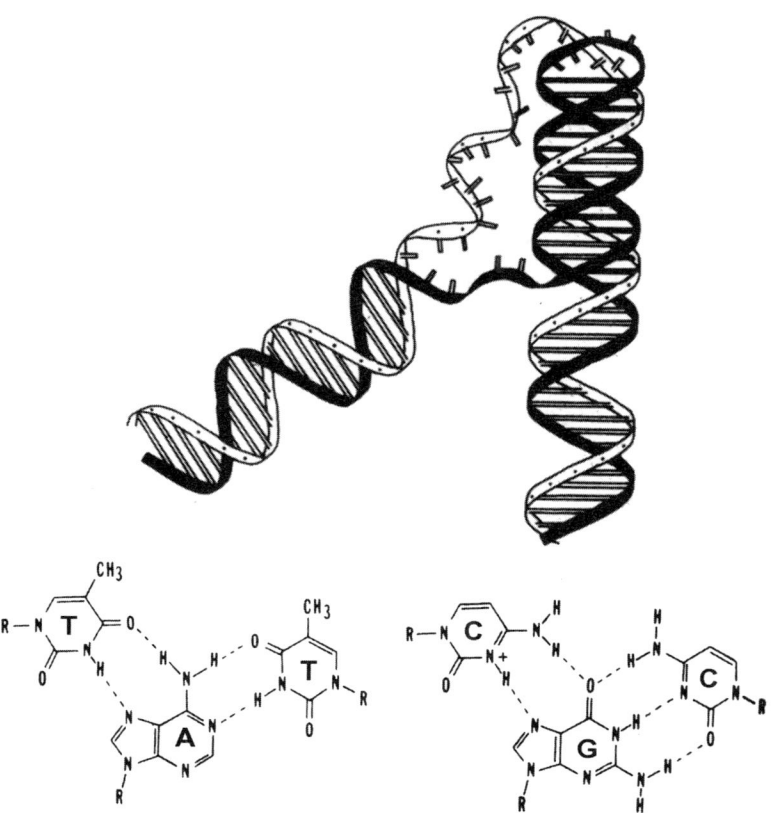

FIGURE 1.23. Structure of intramolecular triplex [70]. The strand containing the pyrimidine-rich sequence is shaded, whereas the strand containing the purine-rich sequences is unshaded. The 3′ half of the purine strand of the insert remains Watson–Crick-paired with the 5′ half of the pyrimidine strand. The 3′ half of the pyrimidine strand, after dissociation from its Watson–Crick complement, occupies the major groove of the duplex region described above and is specifically complexed with it through Hoogsteen base pairs. The mirror repeat and the oligopurine–oligopyrimidine nature of these sequences enable this specific association.

1.1.12.2. Quadruplex DNA   Four strands of DNA may interact to form a stable structure in a number of ways, and evidence for the importance of these structures *in vivo* is increasing. Similar sequences have been found, for example, in the telomeric DNA at the ends of chromosomes, suggesting that quadruplexes play a role in cell replication, and hence are likely to be important in both cancer and the aging process [72–76]. Involvement of quadruplexes in the promoter regions of oncogenes [77] and in the dimerization of the HIV-1 retrovirus [78] has been proposed. They also have been implicated in the regulation of the insulin gene [79], in the triplet repeat sequences associated with certain neurological disorders, such as the "Fragile X syndrome" [80–83], and in autoimmune diseases such as erthemaosus lupus [84]. Similarly, there is also a growing list of proteins that appear to bind specifically to quadruplex structures [85–90]. Some proteins seem to facilitate their formation, whereas the activity of other proteins appears to be suppressed by these structures.

Because of the prevalence of those quadruplex structures *in vivo*, they have been proposed as drug targets, and this has stimulated interest in the possibility of a whole new category of drug function. The primary action of most drugs is to bind to DNA, thereby interfering with transcription or replication. However, these proposed drugs would operate by altering the tertiary structure of the DNA or RNA, inducing gross structural changes through the enhanced formation of quadruplexes, or the destabilization of existing ones. For example, controlling the formation of G-quadruplex structures (discussed below) in telomeric DNA has been proposed as a means of treating cancer [91], while destabilizing them in triplet repeat sequences might aid in the treatment of other diseases. The binding mechanisms of these drugs normally follow the same general principles as has been determined for duplex DNA, and are not treated separately in this book.

The unique structures and properties of quadruplex DNA require some additional explanation. Instead of "base pairs," quadruplexes have "base-quartets" consisting of four bases that are hydrogen-bonded to each other. The most common form of quadruplex is based on the guanine quartet, and has been recognized since 1962. Optical spectroscopic studies using infrared (IR) and circular dichroism (CD) demonstrated that individual guanines, and its derivatives, can self-associate in the presence of cations to form Hoogsteen-type, hydrogen-bonded G-quartets that can then be stacked vertically (on top of each other) [92–95]. Under physiological conditions, guanine-rich sequences of DNA or RNA readily form these structures *in vitro*. Such guanine-rich structures have been variously named "G-quadruplex," "G4-quadruplex" or "G4-DNA." Quadruplexes may be composed of four separate strands that have associated with each other. In this case, the structure is called "linear," and all of the DNA strands are aligned in a parallel manner (Fig. 1.24a). Quadruplexes can also be formed when one or more DNA strands fold upon themselves. They can have as many as three folds per structure. Each fold results in a loop region (e.g., see Fig. 1.24b), and usually results in the reversal of the strand polarity in the core of the quadruplex. In other words, within the guanosine quartet region of a G-quadruplex, the polarity of one end of the folded strand is usually antiparallel to its neighboring strands. The arrangement of the strand polarities within these folded G-quadruplexes is sensitive to the sequence, as well as other experimental conditions.

Quadruplexes have considerable variation in their structures, depending, in part, on the length and constituents of any adjacent loop regions. When a G-quadruplex is formed with all of its strands aligned in the linear, all-parallel manner, then all of the guanosine glycosidic dihedral angles are typically in the *anti* conformation (Fig. 1.24a), just as they are in A- and B-form DNA. One notable exception to this was found in the early work by Henderson et al. [96]. However, when the strands are aligned in an antiparallel manner (as in a folded structure), a stable quartet can be formed only if a proportional number of guanosines have glycosidic bond angles that are in the *syn* conformation (e.g., see Fig. 1.24b). In most folded (and therefore, antiparallel) G-quadruplexes,

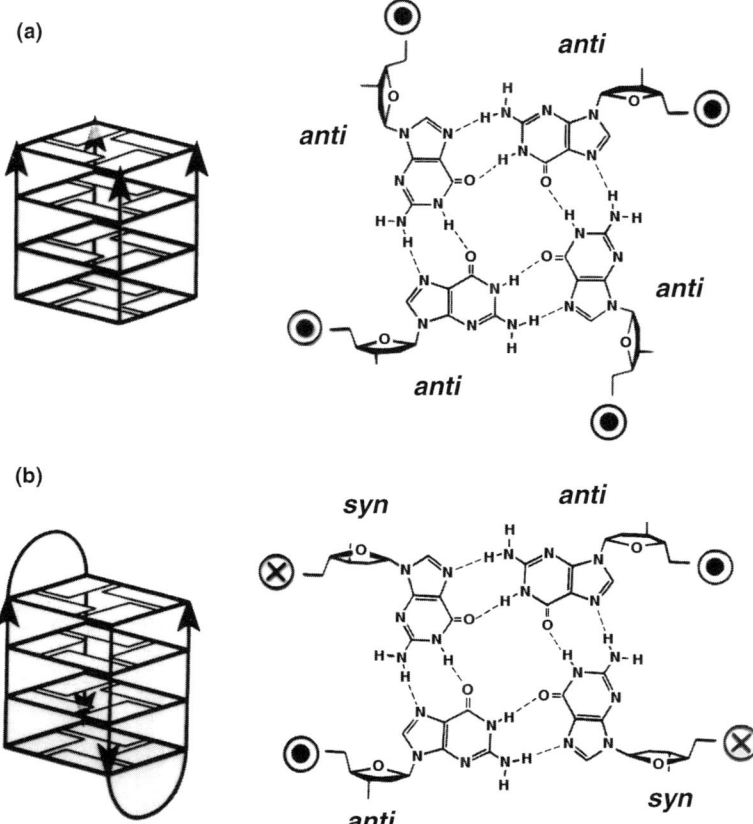

FIGURE 1.24. The core of the quadruplex consists of four guanosine bases arranged in a square, each participating in Hoogsteen base pairing with both of its neighbors. There are a total of eight hydrogen bonds in each "quartet" or "tetrad." The orientations of the strands is illustrated in the side-view (left side of figure) and by the symbols (circle and X) on the right side of the figure. (a) A linear quadruplex in which all strands are parallel and all *N*-glycosidic torsion angles are in the *anti* conformation. (b) A quadruplex arising from strands that have folded into the "edge"-type arrangement. In this arrangement, all of the adjacent strands are antiparallel to each other, and the glycosidic torsion angles alternate between *anti* and *syn*, as one proceeds clockwise around the quartet. These torsion angles also alternate as one moves along the DNA strand.

these glycosidic angles alternate between *anti* and *syn* along the strand (although this is not always the case). This alternation of glycosidic dihedral angles is similar to that of Z-DNA.

Within the G-quartets, many different arrangements are possible for *syn* and *anti* dihedral angles. They may be all *syn* [95], all *anti* [97–100], alternating *syn* and *anti* (Fig. 1.24) [101,102], or *syn–syn–anti–anti* [103–105]. The specific arrangements of *syn* and *anti* dihedral angles also affect the groove widths. Hence, although a base quartet may appear to have fourfold symmetry, the overall molecule may not have any symmetry at all.

As with duplex and hairpin loop DNA structures, base-stacking interactions are important to the stability of quadruplexes. This includes base stacking within the guanosine core [106,107], or between bases in the core and those in a loop or overhanging region [107–109]. Base-stacking interactions within the guanosine core are also affected by the glycosidic dihedral bond angles, as well as by the orientations of strands [104].

The DNA sequence can affect the folding pattern of a G-quadruplex structure. Sequences that prefer linear topologies are usually short oligomers containing a stretch of several guanines, such as d(G_n) or d(G_nT_m), where n and m are integers (n is typically 2–5, and m can be almost any number). Longer DNA sequences containing repeating strings of guanines, interspersed with loop segments (consisting mostly of thymines), generally prefer to form folded structures, but they can also form long linear structures. These latter sequences usually take the form of d($G_nX_mG_n$), or d[$(G_nX_m)_4$], where X represents the bases in the loops (mostly T), and both n and m are typically 2–5. These sequences are similar to those found in telomeric DNA. The concentration of the DNA also affects the topology through the principle of mass action. High DNA concentrations favor the four-stranded (linear) species, while lower concentrations favor uni- or bimolecular (folded) species. An equilibrium exists between these forms, and they are often found together in solution.

Critical to the stability of a G-quadruplex structure is the presence of internal, coordinated, cations. The most common cations are the physiologically relevant K^+ and Na^+. Generally, they are sandwiched in the cavity between adjacent quartets (Fig. 1.25). These cations stabilize the G-quadruplex by electrostatically coordinating with the O6 atoms of the eight guanosine bases in the quartet stack, and by contributing favorable free energies of dehydration [110–112]. The K^+ ion has a snug fit inside this

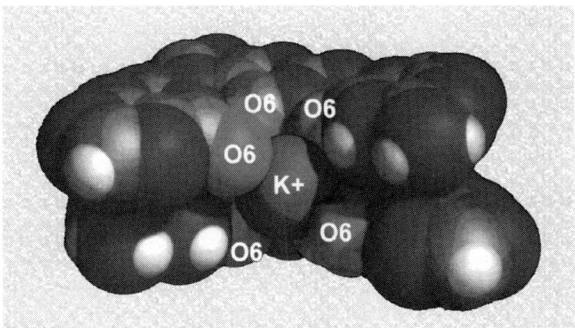

FIGURE 1.25. A side view of two G quartets stacked on top of each other, with an internal, coordinated cation in the center. The view of the stack is from one corner of the square, with the two closest guanosine bases removed to allow visualization of the core. The atom in the center is a K^+ ion, which can be seen coordinating with the O6 atoms in each of the surrounding bases. This figure is derived from experimental NMR structures determined by a combined approach using molecular modeling and experimental NMR data [108]. (See color insert.)

central core, where it touches all eight O6 atoms. The Na^+ ion is smaller and has been sometimes found within the quartet plane. In this case, the Na^+ ion coordinates with only four guanosine bases. The structures stabilized by K^+ are generally more stable than those stabilized by Na^+, as indicated by the fact that they have higher T_m values (Section 1.2.1). This is consistent with both the more optimal fit and the more favorable dehydration energy of K^+ [111]. Thus, the role of cations in quadruplex DNA is quite different from that in duplex DNA, where they neutralize the charges on the phosphodiester backbone along the outside of the double helix. However, several workers have also suggested that some G-quadruplex topologies may derive additional stability from either groove- or loop-bound cations [113,114].

The cations also preferentially stabilize certain G-quadruplex topologies [115–125]. It has been found that the presence of K^+ tends to stabilize linear quadruplex structures, whereas the presence of Na^+ tends to stabilize structures with folded topologies. This is likely related to the size and location of the cations, but is still a subject of controversy. An interesting observation was that the equilibrium and interconversion pathway between folded and linear topologies may be kinetically blocked for some Na^+-stabilized structures, preventing it from achieving a thermodynamically preferred configuration [108]. This same equilibrium appears to occur more easily for similar K^+-stabilized structures. The evidence suggests that the K^+ ion allows for a greater range of molecular motion within the quadruplex, thereby opening up more dissociation pathways for its folded structures, relative to those of Na^+. Such a kinetically controlled process could be associated with differing degrees of freedom, with the extent of molecular motion depending on the size of the internal cation. This also suggests that it may be possible to develop drugs that can control the specific quadruplex topology that is formed, or even act as another kind of conformational switch.

1.2. ELECTRONIC SPECTRA

1.2.1. Ultraviolet Absorption Spectra

The absorption spectra of five mononucleotides at two different pH values are shown in Fig. 1.26 [62]. These nucleotides have strong bands near 260 nm. At the maxima of their peaks (λ_{max}), the molar extinction coefficients (ε) are as great as 10,000–15,000 $M^{-1} cm^{-1}$ and their transition moments are in the base planes. These are attributed to $\pi \rightarrow \pi^*$ transitions of the base residues. In addition, another $\pi \rightarrow \pi^*$ transition is observed around 210 nm.

A typical absorption spectrum of a B-DNA solution at room temperature is shown in Fig. 1.27a [126]. It has an absorption maximum at 260 nm and a trough at 230 nm. The 260 nm peak is considered to be caused by $\pi \rightarrow \pi^*$ transitions of the base residues. Such a spectral feature, however, depends only slightly on the base composition and the base sequence of DNA.

The molar extinction coefficient for a DNA nucleotide, at the 260 nm absorption maximum, is always about 6000 $M^{-1} cm^{-1}$ (where M = mole/liter per nucleotide), as

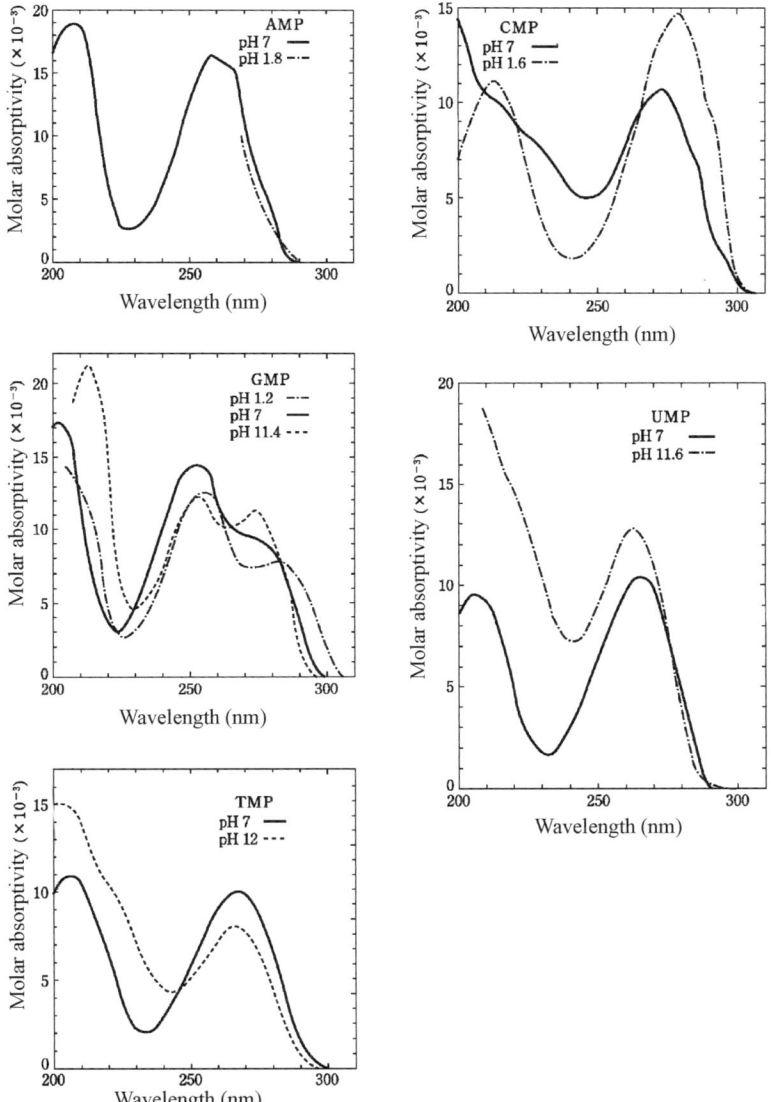

FIGURE 1.26. Absorption spectra of the nucleotides, at 77 K in ethylene glycol–water glass [126].

long as it is in B-DNA conformation [127]. When the duplex is separated into single strands of DNA, such as by heating or some other means, the absorption intensity at 260 nm is greatly increased (see Fig. 1.27a), as the molar extinction coefficient at 260 nm of any mononucleotide far exceeds that of a nucleotide in B-DNA duplex. The peculiarly smaller extinction coefficient of B-DNA results from the interaction of the π electrons between the base aromatic residues when they are stacked vertically.

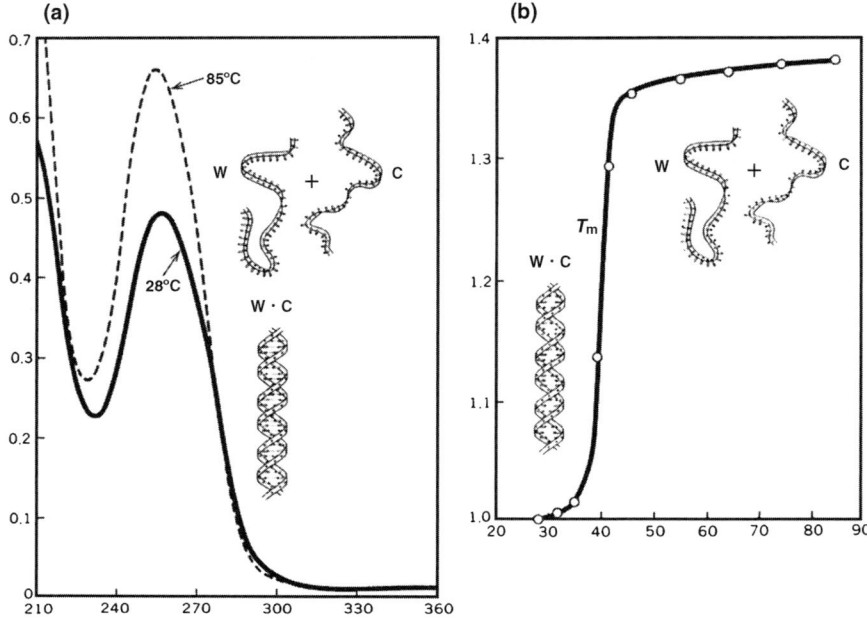

FIGURE 1.27. (a) Absorption spectra of DNA from a crab (*Cancer borealis*) in aqueous solution observed at two different temperatures. Concentration, 7×10^{-5} M; solvent, 0.005 M phosphate buffer, pH = 7. The guanine · cytosine content of this DNA is only 3% [128]. At 28°C, two polydeoxyribonucleotide chains (Watson–Crick chains) are interwound with each other to form a duplex, whereas at 85°C they are separated from each other. (b) A melting curve of the same DNA. In the course of an elevation of the temperature of the solution from 28°C to 82°C, a sharp rise in absorbance at 260 nm takes place at 40°C ($=T_m$, melting temperature).

This decrease in absorption intensity is called *hypochromism*, and its theoretical treatment was described by Tinoco [128].

The intensity of the absorption band at 260 nm can be used to measure DNA concentrations in solution as low as a few μg/mL. In addition, the degree of hypochromism can be a good indication of how closely the aromatic chromophores are packed together. By following the changes in the intensity measurement at 260 nm, one can monitor the course of heat denaturation, the action of a DNase, or the extent of drug binding (especially intercalation), because the base stacking in DNA should decrease as these events proceed. Figure 1.27b shows the result of tracing the change of the 260 nm intensity by raising the temperature of the solution [129]. It is seen that a transition occurs in a narrow temperature range, with a sharp ascent, instead of a gentle slope. This is called a "melting curve" of the DNA, as it refers to the dissociation of the duplex into two separate strands, resulting in an increased disorder in the base stacking. The temperature at the midpoint of the transition is called the *melting temperature*, and denoted by T_m. The melting temperature T_m of DNA in solution varies according to the base composition and the salt concentration. The T_m of natural B-DNA plotted against the GC content forms a straight line (see Fig. 1.28), from ($T_m = 81°C$, GC = 30%) to

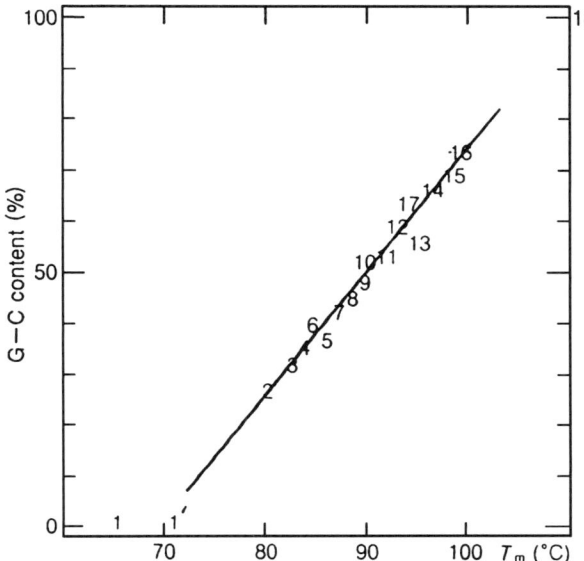

FIGURE 1.28. Relation between the melting temperature T_m of DNA duplex and its guanine + cytosine content (GC content) [63]. Solvent: 0.15 M NaCl + 0.015 M Na citrate. DNA samples are (1) poly(dA-dT)·poly(dA-dT), (1′) poly(dA)·poly(dT), (2) DNA from *Clostridium perfringens*, (3) from *Bacillus cerius*, (4) from *B. thurigiensis*, (5) from *Proteus vulgaris*, (6) from phage T6, (7) from *B. subtilis*, (8) from *Vibrio cholerae*, (9) from phage T7, (10) from *E. coli*, (11) from *Neisseria meningitidis*, (12) from *Aerobacter aerogenes*, (13) from *Azotobacter vinerandii*, (14) from *Mycobacterium plei*, (15) from *Sarcina lutia*, (16) from *Micrococcus lysodeikticus*, (17) from *Pseudomonas aeruginosa*, and (18) poly(dG)·poly(dC).

($T_m = 100°C$, GC = 74%), as long as the salt concentration of the solution is kept constant (0.15 M NaCl plus 0.015 M Na citrate) [129]. The transition moment of the 260 nm band is in the plane of the base, perpendicular to the helical axis in B-DNA, as shown by flow linear dichroism (FLD) experiments [130] (Section 1.2.3).

1.2.2. Fluorescence Spectra

The low-temperature fluorescence and phosphorescence spectra of the common mononucleotides [131] are shown in Fig. 1.29. At room temperature and in neutral aqueous solutions, the fluorescences of the common base residues are very weak, as they have quantum yields of only 3×10^{-6}, 2×10^{-6}, 4×10^{-5}, 3×10^{-5}, and 1×10^{-5}, respectively, for 5′-AMP, 5′-GMP, 5′-TMP, 5′-CMP, and 5′UMP [131]. In contrast, many of the modified and unusual bases involved in natural nucleic acids are strongly fluorescent. Four examples of such fluorescence spectra are shown in Fig. 1.30 [132–136]. Their strong fluorescence is useful in designing biochemical and biophysical experiments of various systems involving these base residues. For example, chloroacetaldehyde can react with the adenine bases in DNA to give the fluorescent product, ethoxyadenosine (Fig. 1.30a). This was used to study the

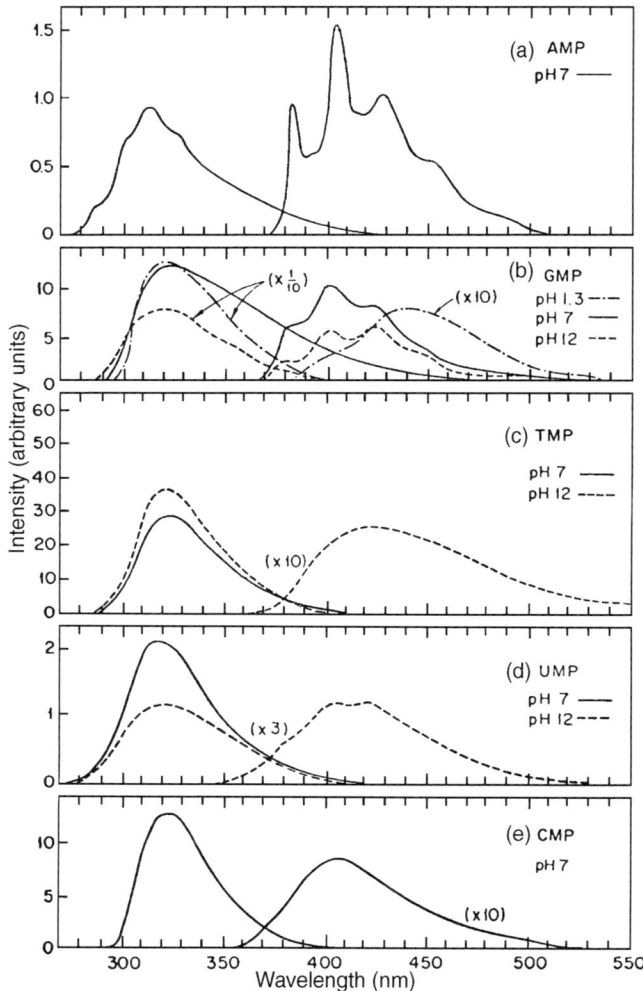

FIGURE 1.29. The fluorescence (near 310 nm) and phosphorescence (near 410 nm) spectra of nucleotides [126]. The solvent was an ethylene glycol : water (1 : 1) glass and spectra were obtained at 80 K.

availability of positions 1 and 6 of adenine in DNA for intermolecular reaction with the helix-destabilizing protein [132].

1.2.3. Linear and Circular Dichroism

1.2.3.1. Linear Dichroism

Linear dichroism (LD) measures the differences in the absorption by a molecule when it is irradiated by parallel ($A_\parallel$) versus perpendicular ($A_\perp$) plane-polarized light. As this technique is very general, arising from asymmetry in the molecular structure or environment, it has found many

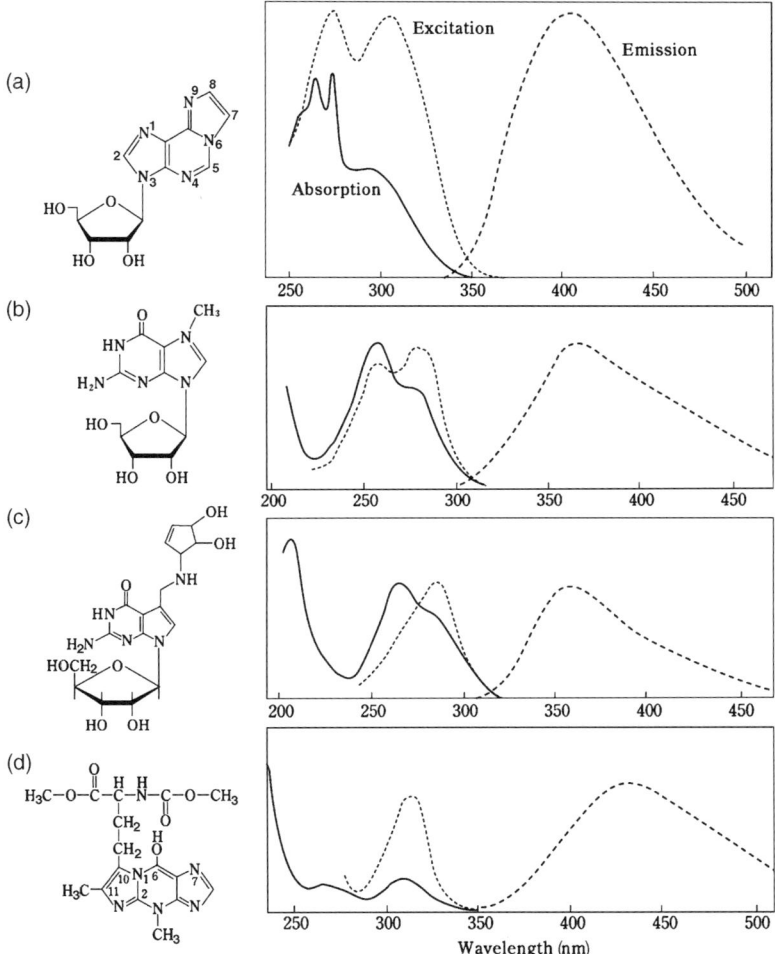

FIGURE 1.30. The fluorescence and fluorescence excitation spectra of four less common, naturally occurring bases of nucleotides [132–136] (——) absorption spectra; (- - - - -) emission spectra; (......) excitation spectra; (a) ethoxyadenosine; (b) 7-methylguanosine in aqueous solution at pH 4.5; (c) Q-base nucleoside-5′-phosphoric acid in aqueous solution at pH 6.2; (d) Yt base in aqueous solution at pH 6.2.

kinds of uses in biomolecular studies [137–139]. Linear dichroism is used most frequently to measure the average relative orientation of a molecule when it is in an aligned matrix, such as when a probe molecule is in a membrane, or when an aromatic molecule is in the presence of the ordered structure of the bases in a DNA duplex.

The measured absorption bands may be in the UV–visible region or in the infrared (IR) region. The former generally provides information based on an electronic–vibrational envelope, whereas the latter probes individual vibrational modes within the ground electronic state. The general equation for linear dichroism is

$$\text{LD} = A_\| - A_\perp \tag{1.3}$$

It is often helpful to express these results as the dichroic ratio, by dividing LD by the absorbance of a randomly aligned sample A_r or $(A_\| + A_\perp)$. This ratio allows one to calculate the order parameter S of the sample [140,141].

The sample can be a crystal, deposited as a film on a window, or oriented by an electric field (ELD; see Sections 3.5) or by a rotating flow cell. In the last case, it is called *flow-aligned LD* (FLD). When a sample rotates or flows through a narrow space while the spectrum is measured, the shearing forces have the effect of orienting the molecules along the direction of flow. Using flow to align a sample is common in most other spectroscopic techniques as well. In all cases, the sample is only partially aligned. Only in studies of crystals can full alignment be achieved.

Linear dichroism has been used to observe the transitions between different forms of DNA, as well as the effects of drug binding [142,143]. As stated in Section 1.1.1, the sugar moiety in a nucleotide has a D-stereoisomeric configuration, and must therefore have optical activity. This means that a nucleotide can rotate the plane of polarization of linearly polarized light. With duplex DNA, the helical axis may be aligned along the laboratory z axis (such as by flow), but the bases will be perpendicular to it. Light polarized perpendicular to the helical axis is absorbed by the DNA bases because the electronic transitions of the bases all involve electron movement within their planes. Light polarized parallel to the helical axis is not absorbed. The difference is therefore a negative signal. Single-stranded DNA, however, is too flexible to show any signal. Hence, this technique can distinguish differences in the length and rigidity of polymeric DNA, as each of these affects the hydrodynamic properties of the molecule.

This technique can also be used to study drug-binding to DNA, as a free ligand in solution has no average orientation, and hence no signal. When it is bound to DNA, however, its orientation is fixed with respect to the helical axis. If the molecule is intercalated between its bases, it will have a negative signal, as do the bases (see above). If the molecule is bound in either the major or minor grooves, then it can have a positive signal, depending on the orientation of the chromophore relative to the helix [142].

1.2.3.2. Circular Dichroism Circular dichroism (CD) measures the differences in the absorption of a molecule by left and right circularly polarized light. These differences arise from asymmetry in the structure of the molecule. It is particularly useful for assessing the secondary structure of proteins and nucleic acids, especially the extent to which it is helical. The CD spectrum describes the average molecular conformation in the bulk sample. As such, it cannot determine the orientation of any specific base or residue, but it is very helpful in determining whether a macroscopic structural change occurs when solution or temperature conditions are changed, or when a drug (or ligand) has bound to the biomacromolecule. Note that unlike linear dichroism (LD), CD does not require an oriented sample. For DNA, the spectrum is usually measured between 210 and 290 nm, where the aromatic bases absorb and the spectra are strongly affected by stacking interactions. Unstructured, single-stranded,

DNA and the different DNA helical forms (for duplex, triplex, and quadruplex DNA) can usually be distinguished on the basis of their unique CD profiles [144].

This optical activity can be detected as the difference in absorption intensities for left (ε_L) and right (ε_R) circularly polarized light [145]. The amount of circular dichroism (CD) is usually given by molar ellipticity (θ)

$$\theta = 3300\,\Delta\varepsilon \qquad (1.4)$$

where $\Delta\varepsilon$ is the difference in molar extinction coefficients ($\varepsilon_L - \varepsilon_R$). It is also possible to perform CD measurements on individual bands within a vibrational spectrum (e.g., an IR spectrum), in which case it is termed *vibrational circular dichroism* (VCD) (Section 6.1.1.3).

The CD spectra of nucleic acids, in general, reflect the relative locations and orientations of the chromophores (i.e., base residues) within the same molecule. As an example, CD spectra of four tetranucleotide sequences are compared in Fig. 1.31

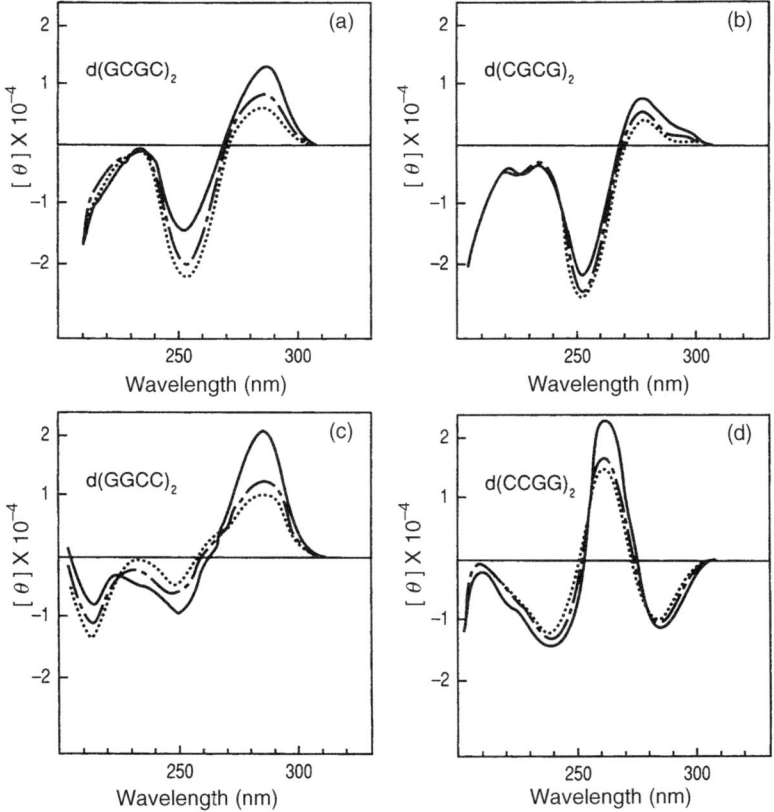

FIGURE 1.31. The circular dichroism (CD) spectra of four tetranucleoside triphospates with two cytidine and two guanosine residues [146]. Concentrations of the oligomer molecules are all in 130 OD units (absorbance/mole). Solvent: (———), no salt, at 5°C; (— - —) 0.5 M NaCl, at 5°C; (· · · · ·) 0.5 M, at −5°C.

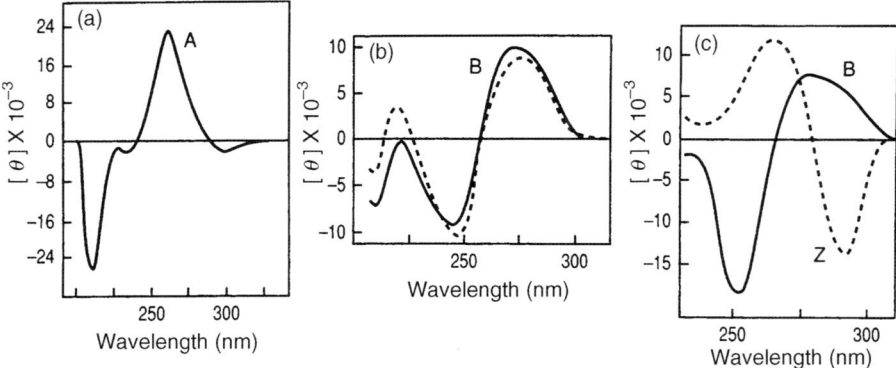

FIGURE 1.32. The circular dichroism (CD) spectra of (a) rice dwarf virus RNA (A-form duplex) [147], (b) (———) DNA from *Micobacterium plei* (B-form duplex) and (- - - -) DNA from *E. coli* (B-form duplex) [148], and (c) poly(dG–dC) • poly(dG–dC) in low ionic-strength solvent (B-form duplex, - - - -) and in high-ionic-strength solvent (Z-form duplex, ———) [149].

[146]. Here, all the molecules have the same number and the same types of chromophores; thus, every molecule has two cytosines and two guanines. Nevertheless, their CD spectra vary significantly. Even more pronounced differences are seen among the A, B, and Z duplexes of polynucleotides, as shown in Fig. 1.32 [147–149].

1.3. VIBRATIONAL SPECTRA

1.3.1. Infrared Spectra of Bases

In general, infrared (IR) absorptions are caused by the vibrations of atoms in a molecule. An IR spectrum is usually presented as a curve having the wavenumber scale (cm^{-1}) along the abscissa and the percent transmittance along the ordinate. It then consists of a number of minima, whose positions correspond to the characteristic vibrational frequencies of the molecule. These characteristic vibrational frequencies correspond to a unique set of ways in which a specific molecular structure can vibrate, and are called "normal modes" of vibration, or often, "normal vibrations." The number of normal modes for a given molecule is $3N - 6$, where N is the number of atoms in the molecule. As with all absorption processes, the depth of each minimum (or the absorption intensity of each band) is a measure of the amount of dipole oscillation caused by the vibration in question. The vibrational frequencies of a molecule depend on the geometrical structure and the interatomic forces of the molecule [150]. The atomic masses of the chemical groups participating in the vibration are a particularly important determinant of the vibrational frequencies. An easy and advantageous experimental approach is therefore to replace the exchangeable hydrogens (^{1}H) with the slightly heavier isotope, deuterium (^{2}H). This is called *deuteration*, and the resulting difference in mass, although small, can shift the frequency enough to allow determination of the atoms involved in the vibration.

Thymine. Thymine is a molecule consisting of 15 atoms, and therefore it has 39 normal vibrations. Approximate descriptions of these 39 vibrational modes are listed in the first column of Table 1.5, and the lower trace of Fig. 1.33 shows the infrared absorption spectrum of thymine [151,152]. Most of the observed bands assigned to these modes, including their frequencies and relative intensities, are given in the second column. These assignments were based on the results of the following studies:

(1) Comparison with the Raman spectrum (see discussion below) of thymine.
(2) Examination of the effect of deuteration, N1−H → N1−D and N3−H → N3−D, on the IR spectrum.
(3) Comparison with the results of *ab initio* self-consistent field molecular orbital (SCF-MO) calculations of the thymine molecule for the equilibrium geometry, harmonic force constants, vibrational frequencies, vibrational modes, and IR and Raman intensities [151].
(4) Comparison with the known correlation of the vibrational modes and the vibrational frequencies of the benzene molecule. These ring vibrational modes [150,153,154] are shown in Fig. 1.34, and their characterizations are given in the upper half of Table 1.6 (under "six-membered ring").

In the 1750–1650 cm^{-1} region, two strong infrared bands appear at 1738 and 1675 cm^{-1} (Fig. 1.33, lower trace). The former is a nearly pure C2=O stretching, and the latter is an in-phase stretching of the C4=O and C5=C6 bonds. In the 1500–1400 cm^{-1} region, *ab initio* calculations predict two stronger bands due to the v_b^6 ring vibration (see Fig. 1.34) and N3−H in-plane bending vibration (modes 14 and 15 of Table 1.5). The latter is expected to be shifted to a lower frequency on deuteration. Actually, two stronger IR absorptions are observed at 1450 and 1430 cm^{-1} for thymine and apparently only one at 1470 cm^{-1} for thymine-d$_2$. The stronger band at ∼1200 cm^{-1} is assignable to a N3−C4/N1−C6 in-phase stretching mode (mode 18 in Table 1.5). It is generally accepted that the N−H out-of-plane bending vibration of a nucleic acid base gives a broad and strong IR band, whereas the corresponding Raman band is extremely weak. In fact, thymine shows a broad and strong infrared band at 846 cm^{-1}, which is removed by deuteration, whereas practically no Raman band is found in this vicinity (see Fig. 1.33). The theoretical calculations predict that mode 24 (Table 1.5) should produce a strong infrared band at 867 cm^{-1}, and that the corresponding Raman band should be very weak. Therefore, mode 24 is unquestionably assigned as the 846 cm^{-1} band. The calculations also predict the N1−H out-of-plane bending vibration (mode 29) at 645 cm^{-1}, but a strong and broad IR band is actually observed at 559 cm^{-1} (not shown in Fig. 1.33).

Cytosine. The infrared absorption spectrum of cytosine is shown in the lower trace of Fig. 1.35 [155,156]. This molecule has 13 atoms, and therefore 33 normal vibrations.

TABLE 1.5. Vibrational Modes of Thymine Molecule with IR and Raman Bands Assigned to These Modes

Mode Number	Vibrational Mode	Observed Frequencies (cm^{-1}) and Intensities	
		IR	Raman
1	N1–H stretching	—	—
2	N3–H stretching	—	3304(w)[a]
3	C6 stretching	~3100(w)[a]	3089(w)[a]
4	Methyl deg stretching A'	2963(s)[a]	2966(s)[a]
5	Methyl deg stretching A''	2934(s)[a]	2934(s)[a]
6	Methyl sym stretching	2934(s)[a]	2934(s)[a]
7	C2=O stretching	1738(s)	1700(w)
8	C4=O, C5=C6 in-phase streching	1675(s)	1672(s)
9	C4=O, C5=C6 out-of-phase stretching	—	1652(w)
10	Methyl deg deformation A'	—	1489(w)
11	N1–H in-plane bending	1480(m)	1459(w)
12	Methyl deg deformation A''	—	—
13	Methyl sym deformation	—	1489(w)
14	ν_b^6 ring vibration	1450(m)	1434(w)
15	N3–H in-plane bending	1430(m)	1408(w)
16	C6–H in-plane bending	1365(w)	1368(s)
17	Kekulé ring vibration	1242(m)	1246(m)
18	ν_a^6 ring vibration + C5–Cm stretching	1200(s)	1214(w)
19	$\nu_{\delta B}^6$ ring vibration	1156(w)	1156(m)
20	Methyl rocking A'	1030(m)	1048(m)
21	Methyl rocking A''	—	—
22	C6–H out-of-plane bending	985(m)	983(m)
23	$\nu_{\delta a}^6$ ring vibration	937(m)	934(w)
24	N3–H out-of-plane bending	—	846(bs)
25	Tr ring deformation + C5–Cm stretching	814(s)	804(m)
26	Ring out-of-plane vibration	809(w)	—
27	Ring out-of-plane deformation	762(m)	—
28	Ring breathing	743(m)	741(s)
29	N1–H out-of-plane bending	700(bm)	—
30	Ring deformation	602(w)	616(s)
31	δ_a^6 ring vibration	559(s)	565(m)
32	δ_b^6 ring vibration	—	478(m)
33	Ring out-of-plane vibration	—	428(m)
34	C2=O, C4=O in plane bending	—	—
35	Ring out-of-plane vibration	—	320(w)
36	C5–Cm in-plane bending	—	286(w)
37	Ring out-of-plane vibration	—	—
38	Methyl torsion	—	—
39	Ring out-of-plane vibration	—	—

Key: A'—in-plane mode, A''—out-of-plane mode; sym.—symmetric, deg—degenerate; ν_a^6 etc.—see Fig. 1.33; Cm—carbon atom of CH_3 group; s—strong, w—weak, m—medium.
[a]Observed for thymidine.
Source: Refs. 151 and 152.

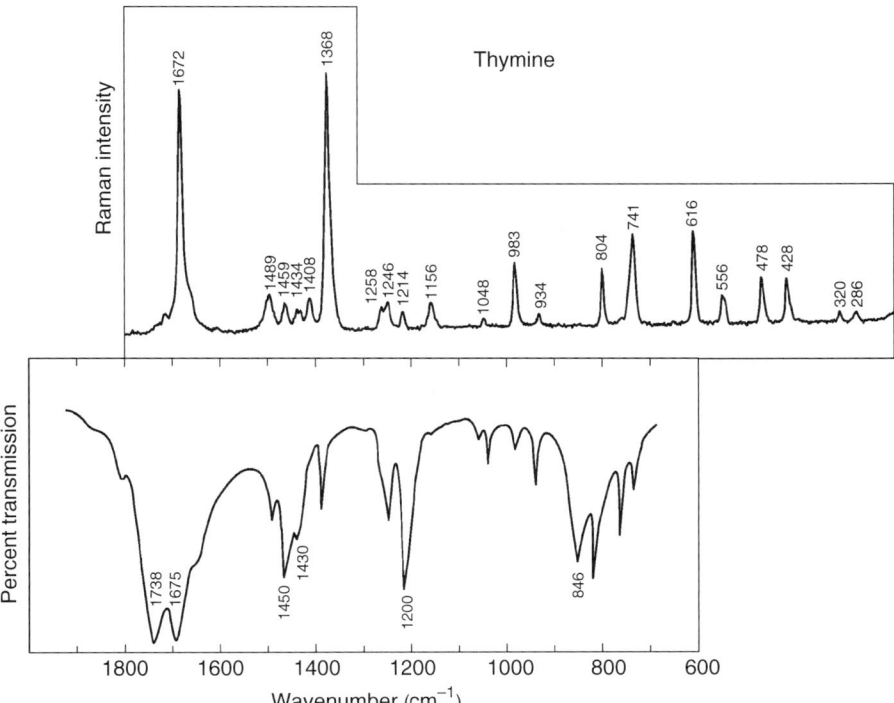

FIGURE 1.33. Infrared spectrum (lower trace) and Raman spectrum of thymine in crystalline powder [151]. The Raman spectrum was excited at 488.0 nm.

Of these, 23 ($= 2N - 3$) are in-plane and 10 ($= N - 3$) are out-of-plane vibrations. Most of the out-of-plane vibrations are weak in both IR and Raman spectra, and their vibrational modes are not always clear. Thus, they are less useful for the study of drug–DNA interactions. Therefore, only the vibrational modes of in-plane vibrations are given in Table 1.7 [156]. The assignments of these modes to the observed IR absorption bands were made in a manner similar to that for thymine, described above. The frequencies and intensities of the observed infrared bands are listed in the second column of Table 1.7.

The strong IR bands at 3380 and 3169 cm^{-1} of cytosine are assigned to the NH$_2$ antisymmetric and symmetric stretching vibrations, respectively, because they are shifted to 2545 and 2337 cm^{-1} when deuterated (cytosine-1, amino-d_3). It is known that the hydrogen-bonded imino group often gives a broad band at a much lower frequency. The broad IR band of cytosine around 2700 cm^{-1} and that of cytosine-1, amino-d_3 around 2150 cm^{-1} can be assigned to the N1–H and N1–D stretching vibrations, respectively.

In the 1800–850 cm^{-1} region, the highest two observed frequencies, 1703 and 1662 cm^{-1}, are respectively assigned as the C2=O and C5=C6 (or v_b^6) stretching vibrations. Molecular orbital calculations [156] indicate that the next highest frequency should be that of NH$_2$ scissoring motion, and the strong IR band at 1615 cm^{-1}

46 DNA STRUCTURES AND SPECTRA

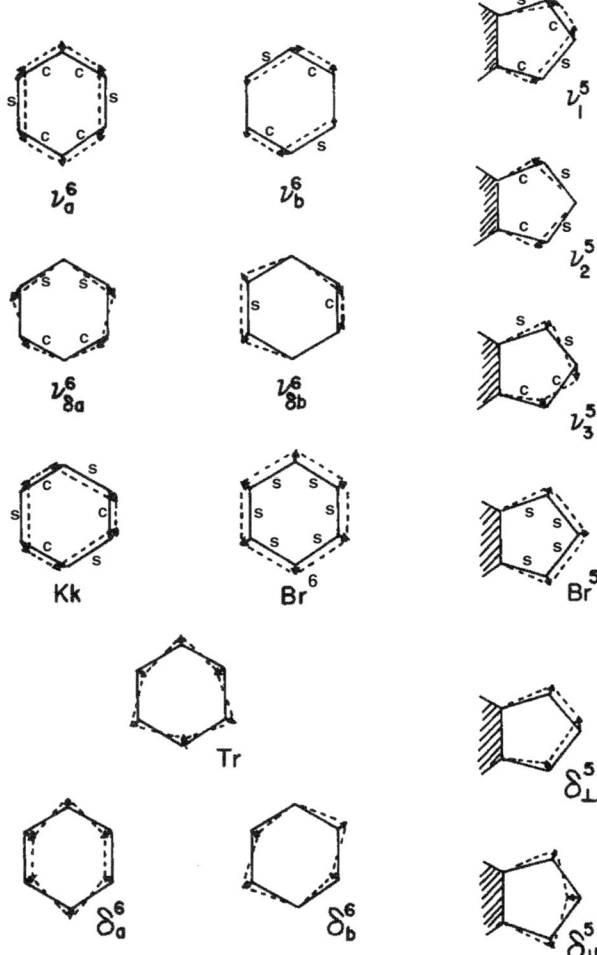

FIGURE 1.34. Standard modes of vibration of six-membered ring, and their abbreviated designation [153,154] (s—stretch; c—contract).

corresponds to it. As expected, it is shifted down to 1182 cm^{-1} on deuteration. The next three highest frequencies should be assigned to ν_a^6 of the hexagon, C4—(NH$_2$) stretching, and N1—H in-plane bending vibrations that correspond, respectively, to the observed 1538, 1505, and 1465 cm^{-1} IR bands. When deuterated, the first two remain nearly unchanged, while the last one (N1—H bending) shifts to 950 cm^{-1}. Molecular orbital calculations predict that the C5—H/C6—H in-phase bending vibration should come next, followed by the Kekulé vibration, and next the C5—H/C6—H 180° out-of-phase bending. Corresponding to these are three prominent IR bands, which are observed at 1364, 1277, and 1236 cm^{-1}, respectively. The NH$_2$ rocking vibration is expected to be the next highest one, and the 1150 cm^{-1} band can be assigned to it.

TABLE 1.6. Designation of Ring Vibrations

Origin of Vibration	Description
Six-membered ring	
ν_a^6 and ν_b^6	1595 cm^{-1} (e_{1g}) of benzene
$\nu_{a\delta}^6$ and $\nu_{b\delta}^6$	1484 cm^{-1} (e_{1u}) of benzene
Kk	Kekulé vibration, 1308 cm^{-1} (b_{2u}) of benzene
Br^6	Six-membered ring breathing vibration, 992 cm^{-1} (a_{1g}) of benzene
Tr	Triangle vibration, 1010 cm^{-1} (b_{1u}) of benzene
Da6 and db6	605 cm^{-1} (e_{2g}) of benzene
Five-membered ring (fused with another ring through the left vertical bond)	
ν_1^5	Antisymmetric stretching vibration with respect to twofold axis along horizontal direction
ν_2^5	Symmetric stretching vibration
ν_3^5	Another antisymmetric stretching vibration
Br^5	Five-membered ring stretching vibration
$\delta_\parallel^5$	Symmetric deformation vibration[a]
$\delta_\perp^5$	Antisymmetric deformation vibration[a]

[a] If the left (left-hand-side) vertical bond is fixed, the center of gravity of the five-membered ring moves either parallel ($\parallel$) or ($\perp$) to the horizontal line (molecular axis).
Sources: Refs. 150, 153 and 154; see also Fig. 1.34.

Markedly broad and strong infrared bands at 823 and 520 cm^{-1} of cytosine (Fig. 1.35) are not predicted to be in-plane vibrations; they must be N1–H out-of plane bending and NH$_2$ wagging vibrations. On deuteration (to cytosine-1,amino-d_3), they are shifted to 605 cm^{-1} (N1–D out-of-plane bending) and 388 cm^{-1} (ND$_2$ wagging), respectively.

Adenine. The infrared absorption spectrum of adenine is shown in the lower trace of Fig. 1.36 [157,158]. The adenine molecule consists of 15 atoms, and hence it should have 39 normal vibrations. Five of these are stretching vibrations involving N–H and C–H bonds, with frequencies higher than 2000 cm^{-1} (not shown in Fig. 1.36). In the region below 1800 cm^{-1}, 34 normal vibrations are expected, and their proposed assignments [158] are listed in the second column of Table 1.8. As indicated in the table, 12 out of these are out-of-plane vibrations, while the remaining 22 are in-plan vibrations. Thirteen (13) of the modes involve mostly hydrogen displacements, while 21 of the modes involve mostly skeletal motions of the purine ring. The modes of the latter category can be correlated to the known modes of benzene and imidazole, which are listed in Table 1.9, with a numbering scheme ranging from ν_1 to ν_{21}.

Guanine. The infrared spectrum of guanine [159] is shown in the lower trace of Fig. 1.37. This molecule has 29 in-plane normal vibrations. Table 1.10 lists the descriptions of these vibrational modes (column 1), along with the frequencies of the observed IR bands that have been assigned to them (column 2) [160].

TABLE 1.7. Vibrational Modes of Cytosine Molecule with the IR and Raman Bands that have been Assigned to These Modes

Mode Number	Vibrational Mode[a]	Observed Frequencies (cm^{-1}) and Intensities[b]	
		IR	Raman
1	NH_2 antisymmetric stretching	3380(s)	3354(m)
2	N1–H stretching	2700(m)	—
3	NH_2 symmetric stretching	3169(s)	3176(m)
4	C5–H stretching	—	3117(m)
5	C6–H stretching	—	3050(w)
6	C=O stretching	1703(w)	1694(m)
7	C5=C6 stretching	1662(s)	1653(m)
8	NH_2 scissoring	1615(s)	1612(w)
9	v_a^6	1538(m)	1533(s)
10	C4–N stretching	1505(m)	1498(m)
11	N1–H bending	1465(s)	1462(m)
12	C5–H and C6–H in-phase bending	1364(m)	1361(s)
13	Kekulé vibration	1277(m)	1276(s)
14	C5–H and C6–H out-of-phase bending	1236(m)	1247(s)
15	NH_2 rocking	1150(w)	1148(w)
16	$v_{\delta a}^6$	1100(w)	1108(m)
17	Triangle vibration	966(w)	971(m)
18	$v_{\delta b}^6$	—	894(w)
19	Ring breathing vibration	793(w)	792(s)
20	δ_a^6	600(m)	597(s)
21	δ_b^6	549(m)	546(s)
22	C=O in-plane bending	533(m)	533(s)
23	C–N in-plane bending	—	400(m)

[a]See Table 1.5.
[b]See Refs. 154–146.

In the frequency region of 4000–2000 cm^{-1}, special care must be taken in correlating the calculated and observed vibrations, because many of the IR bands here are severely perturbed by hydrogen bonding. Ab initio molecular orbital (MO) calculations [160] show that the vibration at 3686 cm^{-1} is the amino antisymmetric stretching and that at 3496 cm^{-1} is the amino symmetric stretching. In the IR spectrum, on the other hand, 2'3'-benzyliden-5'-tritilguanosine dissolved in deuterated chloroform shows two bands at 3515 and 3407 cm^{-1} due to the molecule free from any hydrogen bonding [161]. These two bands may correspond to the two vibrations described above:

In the 1200–800 cm^{-1} region, the contribution of the C8–H deformation mode is well established [160]. The vibrations calculated to be at 1174 cm^{-1} for guanine, at 885 cm^{-1} for guanine-8-d, at 1146 cm^{-1} for guanine-1,9-amino-d_4, and at 930 cm^{-1} for guanine-d_5, should involve a large contribution of C8–H or C8–D in-plane

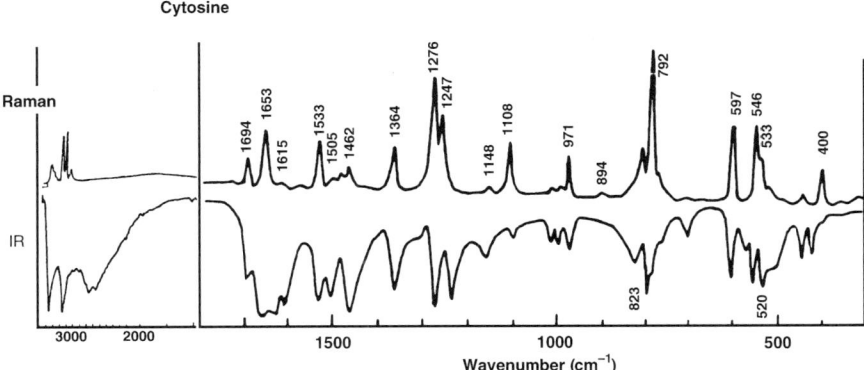

FIGURE 1.35. Raman and infrared spectra of crystalline cytosine (powder) [156].

bending mode. These vibrations are correlated with observed IR bands at 1174, 926, 1168, and 897 cm^{-1}, respectively.

1.3.2. Raman Spectra of Bases

The Raman spectra of thymine [151], cytosine [151], adenine [158], and guanine [162,163] are shown in the upper traces of Figures 1.33, 1.35, 1.36 and 1.37, respectively. Raman spectra are generally obtained by using excitation wavelengths in the visible region. When light having a frequency v_e strikes a molecule, an electronic polarization is induced within it. This polarization has the same frequency as the incident light (v_e), and causes light to be scattered with that same frequency, v_e. This is

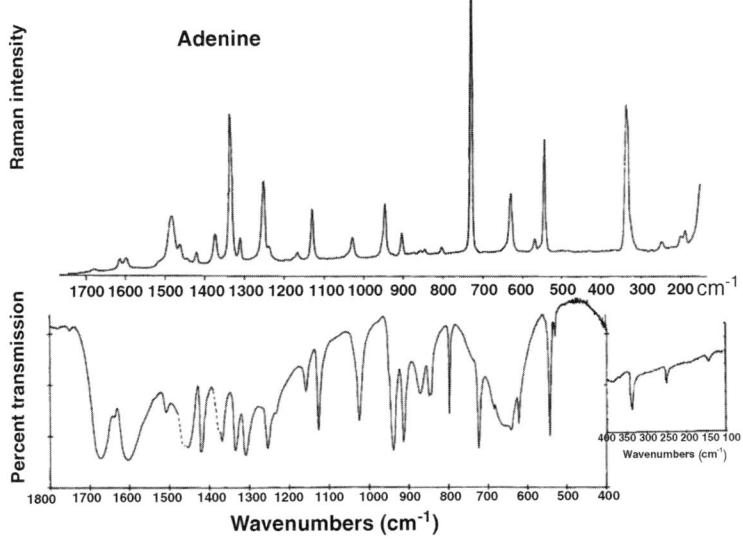

FIGURE 1.36. Vibrational spectra of adenine [158].

TABLE 1.8. Vibrational Modes of Adenine Molecule with IR and Raman Bands Assigned to These Modes

Mode Number	Vibrational Mode[a]	Observed frequencies (cm^{-1}) and Intensities	
		IR	Raman
1	NH_2 scissoring	1673.2(s)	1676.6(w)
2	v_1	1637.6(w)	—
3	v_2	1604(s)	1621.1(w), 1591.1(w)
4	v_3	1507.8(m)	1514.0(w), 1507.8(w)
5	v_4	—	1485.3(m), 1481.6(m)
6	N9–H bending	1451.4(m), 1441.8(w)	—
7	v_5	1421.1(s)	1418.9(w)
8	C2–H bending	1368.0(m)	1370.8(m)
9	v_6	1334.7(s)	1332.3(s)
10	v_7	1308.7(s)	1307.0(m)
11	C6–NH_2 stretching	1252.8(s)	1248.7(m)
12	v_8	1234.0(w)	1234.8(w)
13	C8–H bending	1156.8(m)	1163.3(w)
14	v_9	1125.9(s)	1126.0(m)
15	NH_2 rocking	1024.7(s)	1024.4(w)
16	v_{14}	951(w)	852(w)
17	C8–H out-of-plane bending	940.3(s)	941.4(m)
18	v_{11}	913.3(s)	898.8(w)
19	C9–H out-of-plane bending	872.1(m)	—
20	C2–H out-of-plane bending	849.2(m), 844.0(m)	845.5(w), 839.9(w)
21	v_{16}	797.1(m)	797.0(w)
22	v_{10} (ring breathing)	723.3(s)	723.6(s)
23	v_{19}	684.2(w)	—
24	NH_2 wagging	660(s)	—
25	v_{20}	640.8(m)	—
26	v_{12}	621.6(m)	622.8(m)
27	v_{17}	—	560.2(w)
28	v_{13}	543.0(s)	535.1(m)
29	v_{15}	—	530(w)
30	v_{18}	529.9(w)	—
31	NH_2 torsion	380(w)	—
32	C6–NH_2 out-of-plane bending	337.1(m)	—
33	C6–NH_2 in-plane bending	—	331.3(s), 327.2(s)
34	v_{21}	294.3(m)	242.0(w), 238.3(w)

[a]See Table 1.5.
Source: Ref. 158.

called *Rayleigh scattering*. In addition, when a molecule vibrates, its polarizability oscillates with the same frequency as the vibration (v). All of these processes occur simultaneously. Therefore, when light with a frequency v_e strikes a vibrating molecule, the light is scattered as a linear combination of the frequencies, $v_e \pm v$. This is called *Raman scattering*. A *Raman spectrum* is a plot of the scattering intensity as a

TABLE 1.9. Classification of the 21 Ring Vibrations of Adenine with Respect to Origin

Type of Motion: Bond Stretching or Angle Bending	Origin		Mode Number
	Six-Membered Ring	Five-Membered Ring	
In-plane stretching	1595 cm^{-1} (e_{2g}) of benzene		ν_1, ν_2
	1485 cm^{-1} (e_{1u}) of benzene		ν_3, ν_4
		Double-bond stretching, 1446 cm^{-1} of imidazole	ν_5
	Kekulé vibration, 1380 cm^{-1} (b_{2u}) of benzene		ν_6
		1324 cm^{-1} of imidazole	ν_7
		Single-bond stretching 1261 cm^{-1} of imidazole	ν_8
	Ring breathing, 992 cm^{-1} (a_{1g}) of benzene		ν_9
		1142 cm^{-1} of imidazole	ν_{10}
In-plane bending	Triangle vibration 1010 cm^{-1} (b_{1u}) of benzene		ν_{11}
	605 cm^{-1} (e_{2g}) of benzene		ν_{12}, ν_{13}
		Elongation ∥ to the long axis of the molecule	ν_{14}
		Rocking ⊥ to the long axis of the molecule	ν_{15}
Out-of-plane bending	Crown vibration		
	703 cm^{-1} (b_{2g}) of benzene		ν_{16}
	405 cm^{-1} (e_{2u}) of benzene		ν_{17}, ν_{18}
		Twisting 619 cm^{-1} of imidazole	ν_{19}
		Envelope formation, 660 cm^{-1} of imidazole	ν_{20}
		Butterfly motion	ν_{21}

Source: Ref. 158.

function of frequency, ν. The frequency of a Raman band corresponds to a vibrational frequency ν of the molecule, and its intensity is proportional to the square of the amplitude of the scattered light. Since a molecule vibrates in more than one way, as given by its normal modes, a Raman spectrum is composed of a complex mixture of all of those vibrations, ν.

The frequency of a Raman band, in general, is expected to be the same as that of an IR band for the same molecular vibration. In the case of thymine, all of the prominent

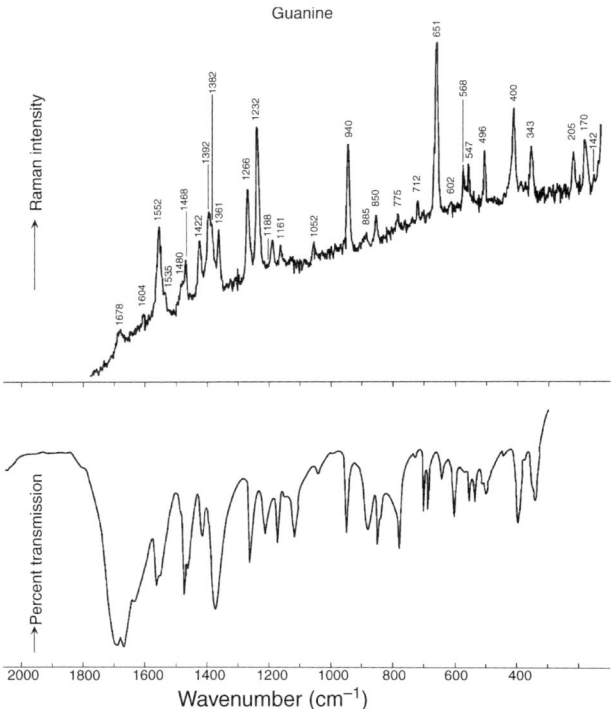

FIGURE 1.37. Vibrational spectra of guanine [159,162].

Raman bands have counterparts of nearly the same frequencies in the IR spectrum, and vice versa (see Fig. 1.33), however, the frequencies of each counterpart are not always exactly the same (Table 1.5). In this example, these differences arise because the sample used for these IR and Raman investigations was a crystalline powder of thymine, not an isolated thymine molecule. In the crystalline environment, some intermolecular vibrational couplings take place (space group $P2_1/c$), causing each molecular vibration to split into a group of vibrations with slightly different frequencies. Thus, the selection rules are different for Raman scattering and IR absorption in the crystalline state.

The relative intensities of Raman versus IR bands are seldom the same, as they arise from different selection rules. For example, the carbonyl (C=O) stretching vibrations always give strong IR bands (1738 cm^{-1} for thymine, 1703 cm^{-1} for cytosine, and 1702 cm^{-1} for guanine), whereas the corresponding Raman bands are weak. Ring-breathing vibrations always exhibit strong Raman bands (741, 792, 724, and 651 cm^{-1} for thymine, cytosine, adenine, and guanine, respectively), whereas IR counterparts are weak. In drug–DNA interaction studies, a difference in the intensity of a Raman band compared to its IR counterpart can be an interesting and useful indicator of binding.

As with other optical spectroscopic techniques, Raman spectroscopy can often yield useful information when the polarization of the signal is analyzed. This is

TABLE 1.10. Vibrational Modes of Guanine Molecule

Mode Number	Vibrational Mode[a]	Observed Frequency (cm^{-1}) IR	Raman
1	NH$_2$ antisymmetric stretching	3515	—
2	N9–H stretching	—	—
3	N1–H stretching	—	—
4	NH$_2$ symmetric stretching	3403	—
5	C8–H stretching	3125	—
6	C6=O stretching	1702	1678
7	C2–N stretching + NH$_2$ scissoring	1675	—
8	ν_a^6 ring vibration	1638	1604
9	NH$_2$ scissoring	1565	1552
10	ν_b^6 ring vibration	1550	1535
11	Kekulé + ν_1^5	1477	1480
12	Kekulé − ν_1^5	1464	1468
13	$\nu_{\delta\beta}^6$ ring vibration	1420	1422
14	ν_2^5 ring vibration	1375	1392
15	N1–H in-plane bending	1362	1361
16	N9–H in-plane bending	1261	1266
17	Br^5	1174	1188
18	C8–H in-plane bending	1216	1232
19	ν_3^5	1150	1161
20	NH$_2$ rocking	1118	—
21	$\nu_{\delta a}^6$	1042	1052
22	$\delta_\parallel^5$	950	940
23	Tr+$\delta_\perp^5$	850	850
24	$\delta_b^6 + \delta_\perp^5$	703	712
25	Br^6 + Br^5	645	651
26	δ_b^6	557	568
27	δ_a^6	501	496
28	C2–N in-plane bending	404	400
29	C6=O in-plane bending	348	343

[a]See Table 1.5 for ring vibrations.
Source: Ref. 160.

because the Raman scattering light is partially polarized even in the case of randomly oriented molecules in a solution sample. In such isotropic samples, the components of the induced electric dipole change as a function of the spatial coordinates of the molecule. The signal intensity is thus measured both parallel and perpendicular to the plane of the polarization of the incident light. This is called the *Raman depolarization ratio*:

$$\rho = \frac{I_\perp}{I_\parallel} \qquad (1.5)$$

When an isotropic sample (e.g., a solution) is excited with plane-polarized light, the totally symmetric vibrations will produce Raman bands that are strongly polarized in the plane parallel ($I_∥$) to that of the incident light. In this case, ρ is small or zero, and the band is referred to as being *polarized*. Nontotally symmetric vibrations, however, produce bands which are called *depolarized*. These have a ρ ≈ 0.75, meaning that the intensity perpendicular ($I_⊥$) to the incident light is 75% as strong as from the parallel orientation ($I_∥$).

The effects of polarized excitation is more complex when the sample is in the crystalline state, where its orientation is fixed with respect to the incident light and detector. Here, the components of the polarization depend on the orientation of the crystal axes with respect to the direction of the incident light, its plane of the polarization, and the placement and polarization of the detector. This requires the measurement of the different components of the polarized Raman intensities (I_{aa}, I_{bb}, I_{cc}, etc.) along the crystal axes (i.e., a, b, and c), where the orientations of incident light and the detector are indicated More details on depolarization studies may be found in Appendix A.1. A more quantitative discussion requires discussion of the polarizability tensor of an ellipsoid [164].

1.3.3. Vibrational Spectra of Nucleosides

Most of the molecular vibrations of a nucleoside are localized primarily in either the base residue or the sugar moiety. Thus, the Raman spectrum of thymidine shows Raman bands attributable to both the thymine residue and the deoxyribose moiety. The origins of these bands have been determined by examining the Raman spectrum of thymidine-1′,2′,3′,4′,5′-(^{13}C′)$_5$, in which the five carbon atoms in the deoxyribose moiety were uniformly replaced with the ^{13}C isotope [165]. This heavier isotope causes the vibrational frequencies within the sugar component of thymidine-^{13}C$_5$ to shift with respect to those of natural thymidine. Consequently, those Raman bands that shift may be attributed to deoxyribose motions. The medium-intensity band at 898 cm^{-1} showed the greatest frequency shift, -17 cm^{-1}. It is assigned to a ring-breathing vibration of the deoxyribose residue [165].

To understand the vibrational modes of the C2′ methylene group in deoxyribonucleosides, the thymidine stereoisotopomer, (2′S)-[2′-^{2}H]thymidine, was synthesized to incorporate deuterium in the S configuration at the furanosyl 2′ carbon [166]. The crystalline powder of this molecule was analyzed by IR and Raman spectroscopies (the latter using excitation at 1063 nm and 514.5 nm) and were compared with those of normal thymidine. On the basis of these experiments, detailed vibrational characterizations were made for many of its observed Raman and IR bands (shown in part in Table 1.11). For example, the strong Raman band at 898 cm^{-1}, which was assigned to a deoxyribose ring-breathing vibration (see above), turns out to be sensitive to C2′(S) deuteration. Thus, this band arises from a vibration that also has a significant amount of 2′CH$_2$ motion, probably methylene rocking. However, a vibration that consists entirely of a methylene rocking motion is expected to exhibit a ^{13}C shift larger than 20 cm^{-1}. Shifts of that magnitude were not observed in this region of the thymidine spectrum, indicating that thymidine has no purely methylene rocking mode. Instead,

TABLE 1.11. Vibrational Bands of the C2'-Methylene Group of Thymidine

Bands of C2'H2 Species			Bands of C2'H(S)D Species		
Infrared[a]	Raman[a]	Assignment[b]	Infrared[a]	Raman[a]	Assignment[b]
2994(m)	2996(s)	CH_2 antisym st	2984(w)	2984(s)	C–H st
2956(w)	2956(m)	CH_2 sym st	2187(w)	2187(m)	C–D st
1410(w)	1404(w)	CH_2 scissor	1078(s)		C2'DH def
1173(s)	1174(m)	CH_2 wag	990(s)	993(m)	C2'DH def
1069(s)	1103(w)	CH_2 twist	903(m)	905(m)	C2'DH def
896(w)	898(s)	CH_2 rock	799(m)	801(m)	C2'DH def

[a]Band frequencies are given in wavenumber units (cm^{-1}), and relative intensities are indicated in parentheses.
[b]Key: sym—symmetic, antisym—antisymmetric, st—stretch, def—deformation, scissor—scissoring, wag—wagging, twist—twisting, rock—rocking.
Source: Ref. 166.

the 2'CH_2 rocking modes are likely coupled with deoxyribose skeletal modes (such as breathing motions), and may contribute to several bands in the 700–1000 cm^{-1} region. It is suggested, however, that the 2'CH_2 contribution is especially large for the 898 cm^{-1} vibration [166]. By synthesizing [5'-^{13}C]thymidine, which has the C5' position selectively replaced with ^{13}C, it is possible to identify those bands in the Raman spectrum that arise from motions involving that atom [167]. Whereas most of the Raman bands remained the same as in natural thymidine, several were shifted significantly to lower frequencies. These are listed in Table 1.12, along with their assigned vibrational modes.

A comparison of the Raman spectra of thymine and thymidine reveals that 22 of the Raman bands from the nucleoside correlate directly with those of the base [167]. These are listed in Table 1.13 (for the assignments based on the thymine vibrational analysis, see Table 1.5). Similar frequency correlations can be tabulated for the IR spectra of

TABLE 1.12. Raman Bands Showing Prominent Frequency Shifts on Replacing 5'-^{12}C by ^{13}C in the Thymidine Molecule

Frequency (cm^{-1})			
5'-^{12}C Species	5'-^{13}C Species	Shift	Vibrational Mode
2934	2931	−3	5'CH_2 symmetric stretching
1481	1477	−4	5'CH_2 scissoring
1066	1060	−6	5'C–4'C–H bending
1029	1022	−7	5'CH_2 rocking
1016	1012	−4	O–5'C–4'C antisymmetric stretching
852	849	−3	O–5'C–4'C symmetric stretching
396	393	−3	O–5'C–4'C bending

Source: Ref. 167.

TABLE 1.13. Raman Bands Common to Thymidine and Thymine

Thymidine		Thymine	
Frequency (cm^{-1})	Intensity[a]	Frequency (cm^{-1})	Intensity[a]
1692	w	1700	w
1665	s	1672	s
1643	m	1652	w
1483	m	1489	m
1458	w	1459	w
1437	w	1434	w
1390	w	1408	m
1365	s	1368	s
1233, 1227	s	1258, 1246	m
1199	s	1214	m
1124	w	1156	m
1067	m	1048	w
1016	m	983	m
974	w	934	w
794	m	804	m
772	s	741	s
675	s	616	s
565	s	556	m
495	s	478	m
379, 397	m	320	w
297, 305	m	286	w

[a]Relative Raman intensity (s—strong, m—medium, w—weak) in the spectra of crystalline powder.
Source: Ref. 167.

thymidine and thymine. However, the IR intensities are more variable, presumably owing to the greater sensitivity of infrared transition moments to the effects of intermolecular interactions, which differ significantly in thymidine and thymine crystal lattices. In the Raman spectra, the observed differences in frequency and intensity between thymidine and thymine are attributable mostly to the base–sugar vibrational couplings. Thymine exhibits, for example, a strong band at 616 cm^{-1} that has been assigned [151] to the in-plane bending mode in which the six-membered ring undergoes a skeletal deformation, while the C2=O and C4=O groups simultaneously bend in phase with one another (a "windshield-wiper" motion). In thymidine, this vibration is upshifted to 675 cm^{-1} in the Raman spectrum, possibly due to coupling with a deoxyribose mode at 632 cm^{-1}. In thymidine-1',2',3',4',5'-^{13}C$_5$, this mode is observed at 671 cm^{-1} in the Raman spectrum [165]. For thymidine-C2'-*d*, the corresponding mode occurs at 663 cm^{-1} in the Raman spectrum [166]. The C2'-deuteration shift of 12 cm^{-1} (675 → 663 cm^{-1}) in the Raman spectrum demonstrates vibrational coupling with the deoxyribose moiety.

The strategy of site-specific furanose deuteration was initiated by Harada and coworkers [168], who investigated the coupling of the base and sugar vibrations in purine ribonucleisides. These workers synthesized and obtained vibrational spectra of adenosine and guanosine that are deuterated at the C1′ carbon, the site of attachment of the purine base.

Vibrational coupling between the base and sugar in a nucleoside varies according to the sugar conformation. For example, the ring-puckering vibration of guanine in guanosine produces a Raman band at 682 ± 3 cm^{-1} if the ribose has $S(C2'\text{-}endo)\text{-}anti$ conformation, at 671 ± 2 cm^{-1} for $S(C2'\text{-}endo)\text{-}syn$, at 664 ± 2 cm^{-1} for $N(C3'\text{-}endo)\text{-}anti$, and at 625 ± 2 cm^{-1} for $N(C3'\text{-}endo)\text{-}syn$ conformation. Such an empirical rule was derived from a series of Raman spectra of crystals containing the guanosine residues, whose structures are known through crystallographic analyses [169]. Other marker bands were also proposed (Section 6.1.1).

1.3.4. Vibrational Spectra of Nucleotides

In the vibrational spectra of nucleotides and polynucleotides, there appear some Raman and IR bands that are assigned to vibrations localized in their phosphate portions, in addition to those assigned to their nucleoside moieties. These bands, along with their properties, are listed in Table 1.14 [170,171]. Mononucleotides and terminal residues of polynucleotides have a phosphomonoester group. At pH higher than 6, this group takes the form of $-O-PO_3^{2-}$, which shows a strong and broad IR band at 1100 cm^{-1} and a strong and sharp band at 980 cm^{-1} in the IR spectrum. The Raman band corresponding to the latter is strong and polarized (see discussion below). The former, at 1100 cm^{-1}, is assigned to the PO_3^{2-} degenerate stretching vibration and the latter at 980 cm^{-1} to the PO_3^{2-} symmetric stretching vibration. These two characteristic bands of the PO_3^{2-} group are seen in the spectra of adenosine-5′-monophosphate, shown in Fig. 1.38 [171,172].

As examples of nucleotide Raman spectra in solution, those of 5′-CMP and 5′-dAMP are shown in Fig. 1.39 panels (a) and (b), respectively [173]. Two scans are seen in each of these, which were measured such that the polarization of the scattered light is selected according to whether it is parallel (∥) or perpendicular (⊥) to

TABLE 1.14. Characteristic Frequencies of the $>PO_2^-$ and $-PO_3^{2-}$ Groups

Group	Frequency (cm^{-1})	Infrared Band	Transition Moment	Raman Band	Assignment
$>PO_2^-$	1230	Strong	Along the O....O	Weak	PO_2^- antisymmetric stretching
	1090	Strong	Along the bisector of <OPO	Strong	PO_2^- symmetric stretching
$-PO_3^{2-}$	1100	Strong, broad	⊥ to the threefold symmetry axis	Weak	PO_3^{2-} degenerate stretching
	980	Strong, sharp	∥ to the threefold symmetry axis	Strong	PO_3^{2-} symmetric stretching

58 DNA STRUCTURES AND SPECTRA

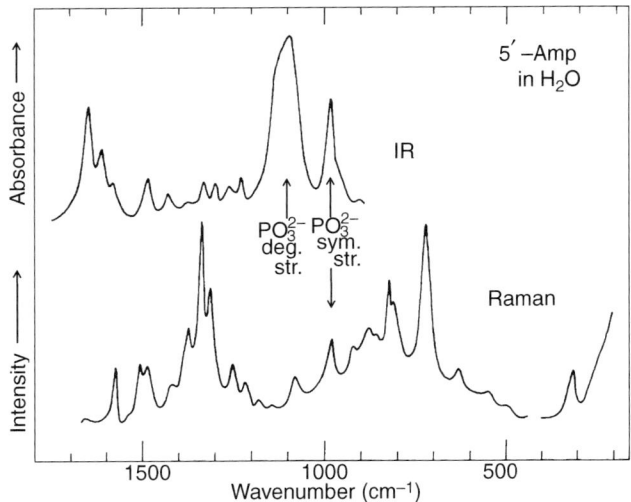

FIGURE 1.38. Infrared and Raman spectra of 5′-AMP in H_2O solution [159,160].

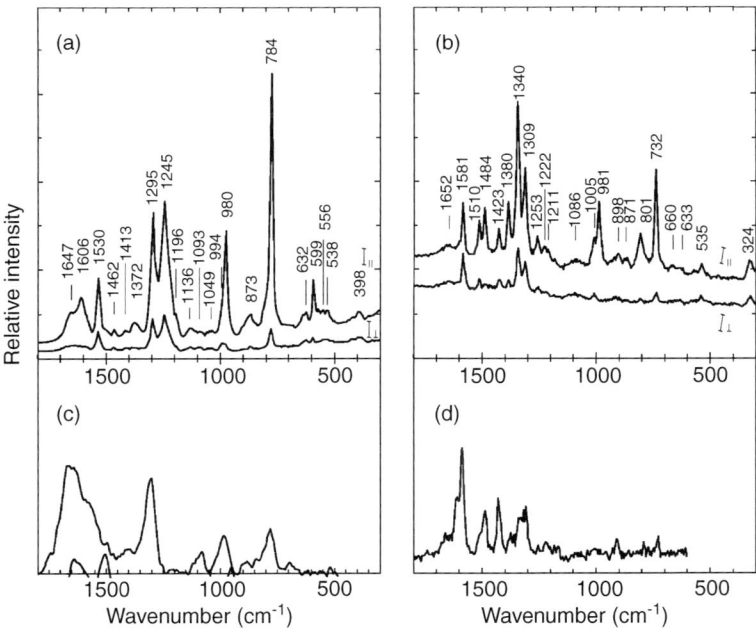

FIGURE 1.39. Raman spectra (1800–300 cm^{-1}) of some nucleotides in H_2O solutions: (a) polarized Raman spectra of 5′-CMP excited at 488.0 nm ($I_{\parallel}$—intensity of Raman scattering beam polarized in same direction as that of exciting beam polarization; $I_{\perp}$—intensity of Raman scattering beam polarized in direction perpendicular to that of exciting beam polarization [173]) of; (b) polarized Raman spectra of 5′-dAMP excited at 488.0 nm. [$I_{\parallel}$ and $I_{\perp}$ are of same width as in (a) [173]]; (c) UV resonance Raman spectrum of 5′-CMP excited at 200 nm [174]; (d) UV resonance Raman spectrum of 5′-dAMP excited at 200 nm [175].

the incident excitation. Together, they constitute the "depolarization ratio" of each Raman band, as discussed Section 1.3.2 and Appendix A.1.

1.3.5. Resonance Raman Spectra

Resonance Raman spectroscopy is essentially the same technique as Raman spectroscopy, except that the energy of the incident light is tuned to an absorbance maximum of the molecule in the UV–visible region. Under this condition, the input energy is the same as (or, is "in resonance with") the energy gap between the electronic ground and excited states that are associated with the absorbance feature. When the energy of excitation is properly tuned, it will specifically excite vibrations associated with that excited state, resulting in a significant enhancement of the Raman signal. An excited electronic state is a real energy state of the molecule that has energy higher than that of the ground state. This exited state has an electron distribution different from that of the ground state, and often, a somewhat different geometric structure. This allows the Raman transitions to satisfy certain special selection rules. This is in contrast to standard, or "nonresonant" Raman spectroscopy, in which the excitation energy corresponds to that of "*virtual* excited states," which consist of a large number of overlapping rotation–vibration energy levels within the electronic ground state.

Since each pyrimidine and purine base of nucleic acids has a few strong electronic absorption bands in the 200–280 nm range (Section 1.2.1), the resonance Raman spectroscopy in this wavelength region serves as a powerful tool for the conformational study of nucleic acids with a high sensitivity and selectivity. With the 200–280 nm excitation, the Raman scattering intensity is, in general, so high that the concentration of a nucleotide solution as low as 10^{-4} M is sufficient for observing Raman band and determining its frequency. In addition, the relative intensities of the Raman bands of the same molecule may change considerably as the excitation wavelength of the incident light is tuned to different absorption bands. It is thus often possible to choose the excitation wavelength to maximize the intensity of a particular Raman band of interest relative to other bands. This effectively increases the selectivity of the technique for conformational studies.

To illustrate the marked changes of the absolute intensities of the Raman bands of nucleotides, resonance Raman spectra of 5'-CMP [174] and 5'-dAMP [175] are shown in Fig. 1.39 panels (c) and (d), respectively, along with the off-resonance Raman spectra (Fig. 1.39a, b). When the Raman scattering intensity is plotted as a function the excitation wavelength, it is called an *excitation profile*. Excitation profiles were reported for many of the bands of 5'-CMP (1650, 1529, 1294, 1243, and 784 cm^{-1} bands). Similar excitation profiles were obtained for some Raman bands of 5'-AMP and 5'-dAMP. The intensities of most of the AMP Raman bands are maximized at 260 nm, where the absorption maximum of AMP is located.

1.3.6. Vibrations of the Phosphodiester Linkage

Nucleic acids have an intense and polarized Raman band between 1000 and 1100 cm^{-1}, which can be assigned to the symmetric P–O stretching mode within the phosphodioxy (PO^{2-}) group. This can be done on the basis of detailed analyses of the Raman and infrared spectra of orthophosphates and their simple alkyl esters [176,177]. Likewise, intense bands in the 750–850 cm^{-1} region can be assigned to vibrations that involve a large contribution from phosphorus–oxygen stretching of the phosphodiester linkages. As there is significant vibrational coupling between the sugar and phosphate groups, these bands are extraordinarily sensitive to the nucleic acid backbone conformation and can provide a firm basis for DNA and RNA conformational analysis. Figure 1.40 shows the Raman frequencies and depolarization ratios of dimethylphosphate, together with descriptions of the assigned vibrational modes, on the basis of a preliminary mathematical analysis.

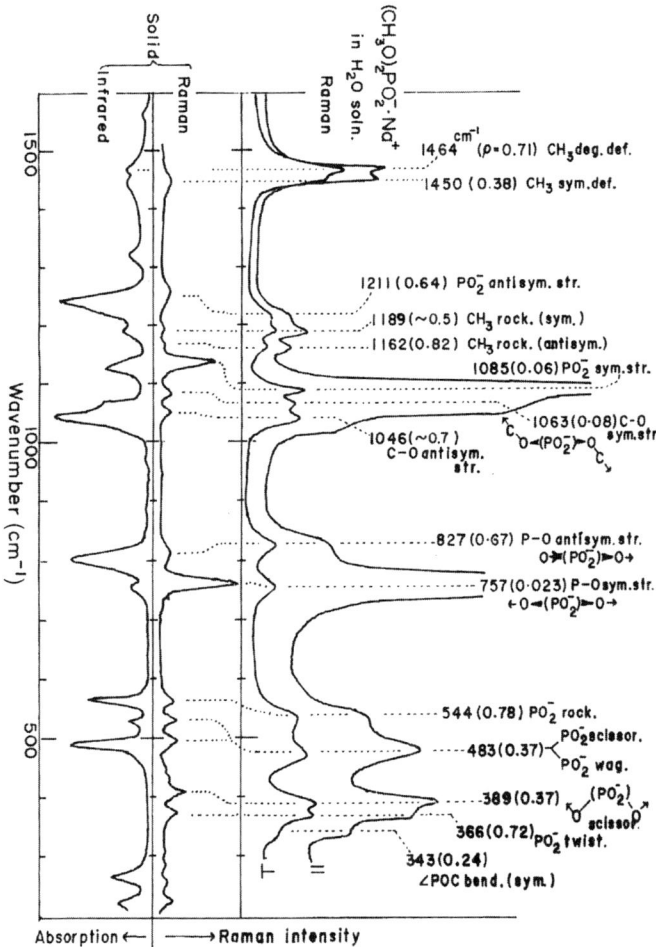

FIGURE 1.40. Raman and infrared spectra, Raman depolarization ratios, and preliminary vibrational assignments of dimethylphosphate anion [177].

Note that the symmetry of the dimethylphosphate anion leads to distinct symmetric (757 cm^{-1}) and antisymmetric (827 cm^{-1}) stretching modes with the expected relative intensities: The 757 cm^{-1} vibration is strong in Raman and weak in IR, whereas 827 cm^{-1} is weak in Raman and strong in IR spectra. However, in nucleic acids, the strict twofold local symmetry is lost in the 3′,5′-diester form, and it is not possible to rigorously define the symmetric and antisymmetric character of the phophodiester O−P−O stretching bands. This may account for the phophodiester vibrations of DNA and RNA (approximately 810–840 cm^{-1}) exhibiting both relatively high frequencies and relatively high Raman (and also IR) intensities.

1.3.7. Infrared Spectra of DNA

Infrared absorption spectra of typical DNA, as undeuterated, partially deuterated, and deuterated films, are shown in Fig. 1.41 [178]. Since the film was kept at 92% relative humidity, the DNA molecules are considered to be B-DNA. As seen in the figure, B-DNA gives a broad and strong band at 3400 cm^{-1} (or 2500 cm^{-1}), which is assigned to the O−H (or O−D) stretching vibration of the hydrated water molecules. The bands in the 1800–1500 cm^{-1} region are assigned to the C=O stretching, skeletal stretching, and N−H bending vibrations of the base residues [137]. They are partially overlapped by HOH (or HOD) bending vibrations of the water molecules. A strong band at 1220 cm^{-1} is assigned to the PO$_2$ antisymmetric stretching vibration, and another strong band at 1100 cm^{-1} is assignal to the PO$_2$ symmetric stretching vibration of the phosphodiester group [137,170]. There are a number of bands with strong, medium, and weak intensities in the region between 1000 and 700 cm^{-1}. These bands are due to the P−O stretching, C−O stretching, and N−H out-of-plane bending vibrations and also to liberations (incomplete rotation in a film or crystalline state) of adsorbed water molecules. The N−D out-of-plane bending liberation of D$_2$O should appear in the 600–400 cm^{-1} region. Assignments of some of these bands are indicated in Fig. 1.41.

Figure 1.42 shows the infrared spectra of the A and B forms of an oriented DNA film from *Micrococcus lysodeikticus* observed with polarized radiation [179]. Prominent spectral differences between the A and B forms of DNA are observed in the 1300–1000 cm^{-1} region [179,180]. In A-DNA, the PO$_2^-$ *antisymmetric* stretching band (located at 1230 cm^{-1} in parallel-polarized spectrum, and at 1243 cm^{-1} in perpendicular-polarized spectrum) shows a perpendicular dichroism (i.e., $I_\perp > I_\|$), whereas in B-DNA the antisymmetric stretching band (at 1220 cm^{-1} in ∥, and at 1230 cm^{-1} in ⊥) is almost nondichroic (i.e., $I_\perp \approx I_\|$).

In A-DNA, the PO$_2^-$ *symmetric* stretching band at 1092 cm^{-1} (both ∥ and ⊥ spectra) shows a parallel dichroism (i.e., $I_\perp < I_\|$), whereas in B-DNA the symmetric stretching band at 1088 cm^{-1} (both ∥ and ⊥ spectra) shows a perpendicular dichroism (i.e., $I_\perp > I_\|$). These differences in the IR dichroism are ascribed to the difference in the orientations of the PO$_2^-$ group between A-DNA and B-DNA. Thus, in A-DNA the O······O line of the PO$_2^-$ group is directed nearly perpendicular to the helical axis

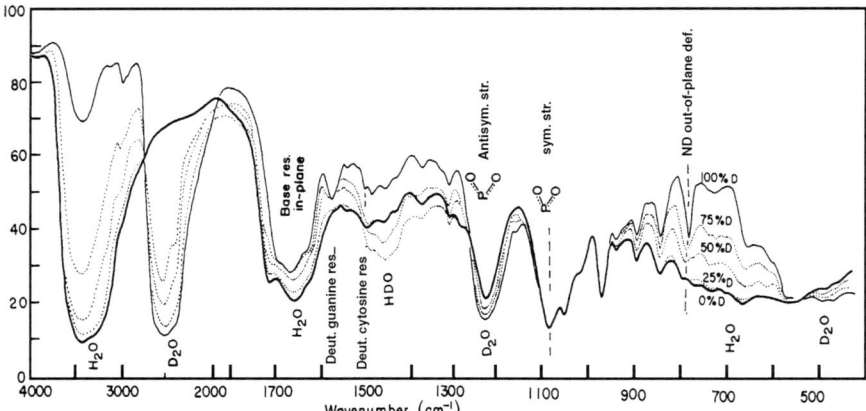

FIGURE 1.41. Infrared absorption spectra of calf thymus DNA film placed in air of 92% relative humidity [178]. The spectra were observed in undeuterated, partially deuterated, and nearly completely deuterated states. Deuteration is shown on the right side of the figure.

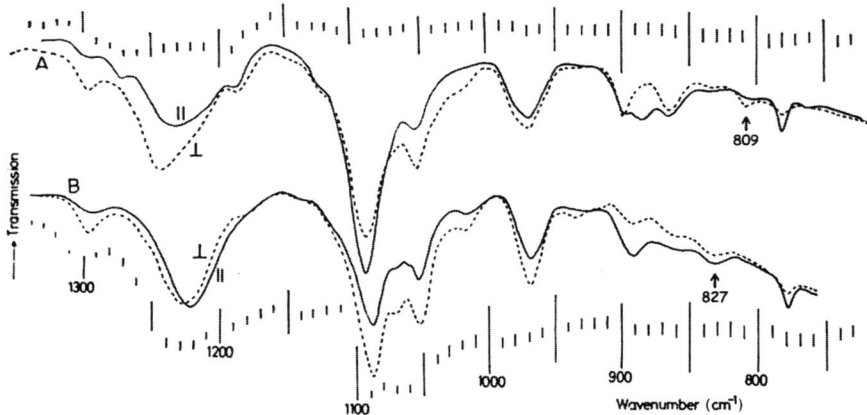

FIGURE 1.42. Infrared absorption curves of oriented films of DNA from *Micrococcus lysodeikticus* in its A and B forms, observed with polarized radiation [179]. The solid line represents the spectrum when the electric vector of the incident radiation is parallel to the fiber axis; the broken line, the spectrum when the electric vector is perpendicular to the fiber axis. The film was prepared by evaporating an aqueous solution of DNA with 10^{-5} M NaCl onto an AgCl plate. The spectrum of the A form was obtained by maintaining a prolonged environmental humidity of 75%. The B-form spectrum was easily obtained at 92% humidity. The conversion from A to B occurs readily when the humidity is changed from 75% to 92%, but the conversion of B form back to the A form in the film requires a constant environmental 75% humidity lasting for a week or so (Section 1.1.7).

TABLE 1.15. DNA Samples Used in Spectroscopic Studies of Fig. 1.43

Source	CG Content (%)
Tetrahymena pyrifromis GL	25[a]
Cloastridium perfringens	31[b]
Calf thymus	43[b]
Escherichia coli	51[b]
Pseudomonnas aeruginosa	65[a]
Micrococcus lysodeikticus	72[b]

[a]Estimated from the melting temperature (61.8°C in 0.015 M NaCl + 0.0015 M Na citrate) and buoyant density in CsCl density gradient centrifugation (1.686, as given by Flavell and Jones [182]).
[b]Ulizur [183].
[c]Sueoka and Cheng [184].
Source: Ref. 184.

and the bisector of its O—P—O angle is directed nearly parallel to the helical axis. In B-DNA, on the other hand, the O .... O line is inclined by around 54° from the helical axis and the bisector is nearly perpendicular to the helical axis.

In the 800–700 cm^{-1} region, A- and B-DNAs show a difference in the vibrational frequency that probably involves mainly the P—O single bond stretching mode (Fig. 1.40). Thus, in A-DNA a weak band is observed at 807 cm^{-1}, while in B-DNA a broad and weak band appears at 832 cm^{-1}. The former shows a perpendicular dichroism, whereas the latter is almost nondichroic. As will be described below, these two diagnostic frequencies, 807 cm^{-1} for A-DNA and 832 cm^{-1} for B-DNA, are observed more conspicuously in Raman spectra.

Nishimura et al. [181] examined the IR spectra (within 1750–1480 cm^{-1}) in D_2O of six different DNA samples, each with different guanine–cytosine (GC) content (Table 1.15) [182–184]. As may be seen in Fig. 1.43, the absorption peaks at 1621, 1642, 1668, and 1694 cm^{-1} become more prominent with decreasing GC content, while those at 1580, 1648, and 1682 cm^{-1} become more prominent as the GC content increases. In fact, after a proper adjustment of the absorbance scale in each absorption curve, these frequencies are found to have an almost linear intensity relationship with the GC content. This result indicates that, on the average, the A-T base pairs (or the G-C base pairs) in all of these DNA samples are in similar structural environments with respect to adjacent base pairs. The vertical stacking interactions between the base pairs in a double-helix polynucleotide structure should certainly affect the IR absorptions of the base pairs, but this was not clearly detected in this series of DNAs, even though their GC content spans the range of 25–72%.

1.3.8. Raman Spectra of B-DNA

Figure 1.44 shows the polarized Raman spectra of a B-DNA fiber obtained with a laser excitation of 514.5 nm [185], and the polarized Raman intensities I_{cc}, I_{bb}, I_{bc}, and I_{cb} were measured in the (cc), (bb), (bc), and (cb) configurations (see Section 1.3.2). Table 1.16 lists frequencies of the Raman bands observed for B-DNA fiber and for B-DNA in solution. The vibrational modes assigned to these bands and depolarization

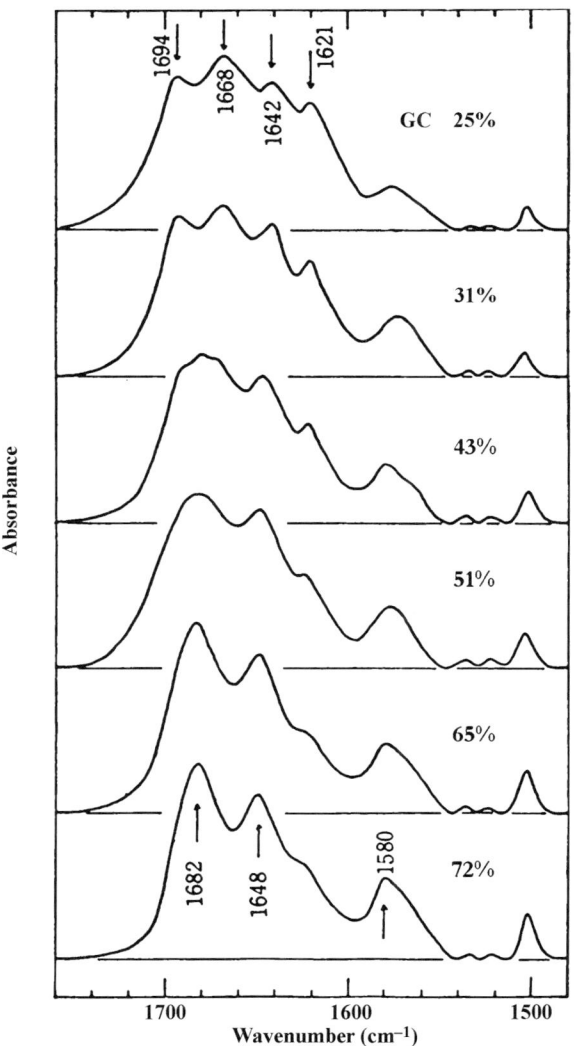

FIGURE 1.43. Infrared absorption curves of different DNA samples in D_2O solutions at about 40°C. Each sample had different GC content (see Table 1.15) [181].

ratios observed for the B-DNA in solution (with random, isotropic orientation) are also given [185]. The 1670 cm^{-1} band of B-DNA is assigned to the exocyclic carbonyl (C=O) bond stretching vibrations of the bases. Among the three base residues containing carbonyl groups, the thymine contribution to the band intensity is large, whereas that of guanine is small, and that of cytosine is negligible. The depolarization ratio ($\rho = 0.26$) observed for the 1670 cm^{-1} band of randomly oriented B-DNA can be explained as a weighted average of the ρ values of the contributing base residues [173], in accordance with their respective intensities.

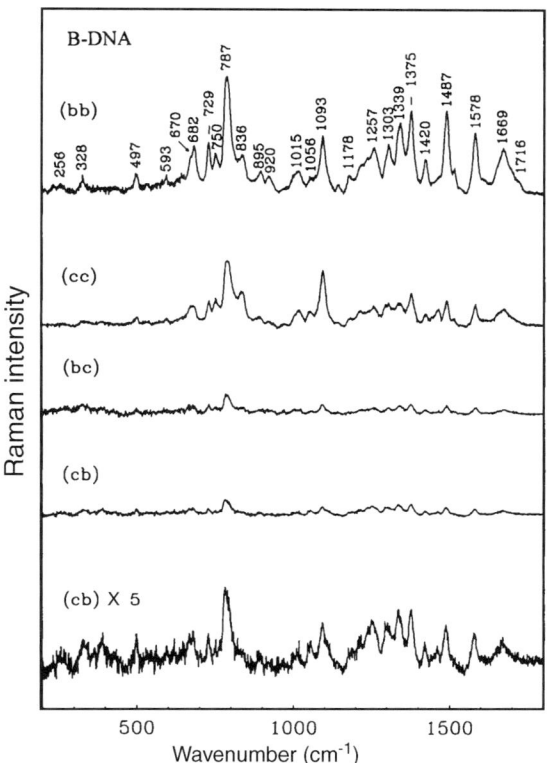

FIGURE 1.44. The Raman spectra (I_{bb}, I_{cc}, I_{bc}, and I_{cb}) of a B-DNA fiber corresponding to four tensor components (bb, cc, bc, and cb, respectively) using 514.5 nm excitation, and at 10°C and 92% humidity [185]. Here a, b, and c refer to the crystal axes. The DNA fibers were about 200 μm thick and were mounted in a thermostatically controlled hygrostatic chamber (microsample cell), which was designed specifically for the sample platform of a Raman microscope. This allowed the humidity to be carefully controlled, and the sample, wellequilibrated. The electric vector of the exciting radiation was directed along the laboratory x axis. A polarizer located before the entrance slit of the monochromator selects for Raman scattering along either the x or y axis that is perpendicular to it. The fiber axis of the sample was placed along either the x or y axis.

The 1578 cm^{-1} band of B-DNA comprises overlapping contributions from adenine and guanine. The adenine contributor is almost completely depolarized ($\rho = 0.64$), whereas that of guanine is significantly polarized ($\rho = 0.34$) [173]. The intermediate value of ρ (0.54) observed for the 1578 cm^{-1} band of B-DNA is thus consistent with values observed for purine mononucleotides. The 1489 cm^{-1} band of B-DNA, also a composite of contributions from two purines, is very strongly polarized in the nucleic acid ($\rho = 0.17$), as well as in the mononucleotides ($\rho = 0.11$ in GMP and 0.13 in dAMP); however, in this case the purine mononucleotide depolarizations are apparently not simply additive. Interestingly, the depolarization characteristics of the two purine ring modes at 1578 and 1489 cm^{-1} for isotropically oriented DNA in solution are dissimilar.

TABLE 1.16. Raman Bands of B-DNA in Fiber and in Solution

Frequency			
Fiber (cm^{-1})	Solution (cm^{-1})	Principal Assignment$_a$	Depolarization Ratio (ρ)
1669	1670	C=O stretching (T)	0.26 ± 0.05
1578	1578	Ring mode (A, G)	0.54 ± 0.05
1487	1489	Ring mode (G, A)	0.17 ± 0.02
1465	—	5'CH2 scissoring	—
1420	1420	2'CH2 scissoring	0.42 ± 0.02
—	1375	Ring mode (T)	0.27 ± 0.03
1339	1340	Ring mode (G, A)	0.43 ± 0.02
—	1302	Ring mode (A, C)	0.33 ± 0.05
1257	1254	Ring mode (C, T)	0.47 ± 0.05
1093	1092	PO$_2^-$ symmetric stretching	<0.04
920	923	Deoxyribose	0.11 ± 0.05
895	900	Deoxyribose	0.36 ± 0.05
836	832	O–P–O stretching	0.17 ± 0.10
792	792	O–P–O stretching	0.13 ± 0.05
784	782	Ring breathing (C)	0.17 ± 0.05
750	749	Ring breathing (T)	0.10 ± 0.05
729	728	Ring breathing (A)	0.13 ± 0.05
682	683	Ring breathing (G)	0.19 ± 0.05
—	670	Ring mode (T)	0.18 ± 0.05
497	497	PO$_2^-$ scissoring	0.15 ± 0.10

^{a}Key: T—thymine; A—adenine; G—guanine; C—cytosine.
Source: Ref. 185.

Depolarization ratios of CH$_2$ scissoring modes near 1420 cm^{-1} have been examined in Raman spectra of nucleotides [173]. In every case the ρ value is near 0.4. This is also the case for isotropically oriented B-DNA (see Table 1.16). The four intense Raman bands of B-DNA at 1254, 1302, 1340, and 1375 cm^{-1} can be assigned to in-plane ring vibrations of the base residues. A common characteristic of each of these Raman bands for isotropically oriented B-DNA in solution is that they will have a depolarization ratio with an intermediate magnitude ($0.25 < \rho < 0.50$). In accordance with the discussion in Section 1.3.2, such a ρ value is indicative of a relatively anisotropic Raman tensor. (Using the notation of Appendix A.1 and Fig. 1A.1, these would be given as $|r_1|, |r_2| > 1$.) Accordingly, these bands are potentially good candidates for probing DNA base orientations in complex biological assemblies by polarized Raman spectroscopy.

The PO$_2^-$ symmetric stretching band at 1092 cm^{-1} is highly polarized ($\rho < 0.04$), and the Raman tensor should be relatively spherical, or isotropic. (Using the notation of Appendix A.1, and Fig. A.1, we expect $r_1 \approx r_2 \approx 1$, i.e., $\alpha_{xx} \approx \alpha_{yy} \approx \alpha_{zz}$.) Accordingly, this band in the polarized Raman spectra is not expected to be a useful indicator of the residue orientation.

Raman bands at 895 and 920 cm^{-1} of B-DNA fiber and those at 900 and 923 cm^{-1} of B-DNA solution can be assigned to bond stretching vibrations of deoxyribose ring. Because of its highly Raman depolarization ($\rho = 0.11$), the 923 cm^{-1} band is considered to represent a more symmetric furanose ring stretching vibration than does the 900 cm^{-1} band ($\rho = 0.36$). These assignments are consistent with those proposed for the sugar group of thymidine [165].

Depolarization ratios for phosphodiester modes of the 750–850 cm^{-1} interval are shown in Table 1.16. It is seen that the bands at 792 cm^{-1} ($\rho = 0.13$) and 832 cm^{-1} ($\rho = 0.17$) are both highly polarized. In B-DNA, therefore, the depolarization ratios suggest that a distinction between symmetric and antisymmetric motions of the C–O–P–O–C bond network is inappropriate (see Section 1.1.8 and Fig. 1.39).

Of the bands assigned to ring-breathing-type motions of the base residues, 683 (G), 728 (A), 749 (T), and 782 (C) cm^{-1}, those of G and C exhibit significantly higher ρ values, that is, lower polarizations, in B-DNA than in the corresponding mononucleotide [173]. This may result from Raman tensor perturbations specific to pairing and stacking of these residues in B-DNA.

1.3.9. Comparison of Raman Spectra of A- and B-DNA

Figure 1.45 shows the polarized Raman spectra of A-DNA fibers [185]. These were obtained through exactly the same procedure with that used for the polarized Raman spectra of B-DNA (Section 1.3.9) except that the relative humidity in the sample chamber was maintained at 75%, instead of at 92%, which is the requirement for B-form DNA.

Raman bands originating from C–C and C–N bond stretching vibrations of the base residues are expected to dominate the 1100–1700 cm^{-1} spectral region. Normal coordinate calculations suggest that many of the modes here involve 180° phase differences in the stretching of adjacent bonds of the heterocycles [153,154]. For both A- and B-DNA fibers, seven strong peaks are detected in this region of the Raman spectrum. In the I_{bb} spectrum of the B-DNA fiber (Fig. 1.44, trace) the peaks are centered at 1257, 1304, 1339, 1375, 1487, 1578, and 1669 cm^{-1}; the corresponding bands of A DNA (Fig. 1.45, trace) occur at 1257, 1304, 1339, 1376, 1486, 1577, and 1677 cm^{-1}. In B-DNA, the band near 1339 cm^{-1}, due to adenine, is less intense than the thymine band near 1376 cm^{-1}. The relative intensities of these adenine and thymine markers are reversed in A-DNA. Figures 1.44 and 1.45 show that all of the strong I_{bb} Raman bands noted above are greatly reduced in I_{cc} spectra of both A-DNA and B-DNA. This fact may indicate that, for each of the corresponding normal modes of vibration, the out-of-plane tensor component (α_{zz}) is much smaller than at least one of the in-plane tensor components (α_{xx} or α_{yy}) (see Section 1.3.5).

Although in both A and B fibers the base planes are oriented close to the normal to the fiber axis (c axis), the base tilt angle in A-DNA (about 20°) is greater than that in B-DNA (nearly 0°) (Sections 1.1.3 and 1.1.7). Accordingly, it might be anticipated that Raman anisotropies for the above-noted in-plane base vibrations, as measured by the polarized Raman intensity ratio I_{bb}/I_{cc}, should be greater for B-DNA than for A-DNA. However, Figures 1.44 and 1.45 indicate that this is not the case for any of the bands in

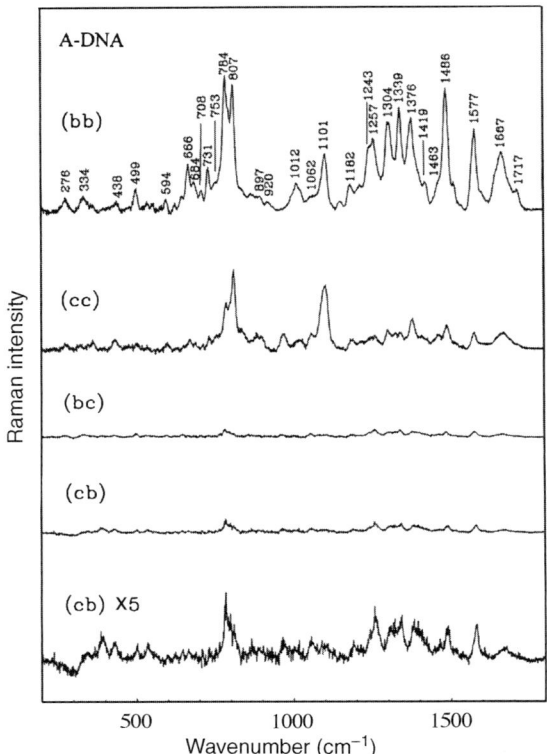

FIGURE 1.45. Raman spectra (I_{bb}, I_{cc}, I_{bc}, and I_{cb}) of an A-DNA fiber corresponding to four tensor components (dd, cc, bc, and cb, respectively). Here a, b, and c refer to the crystal axes. The spectra were acquired by exciting the sample at 514.5 nm and maintaining the temperature at 10°C and the humidity at 75% [185].

question. A possible cause for this is the higher crystallinity and greater degree of unidirectional orientation achievable for fibers of the A form than for fibers of the B form. This is consistent with fiber X-ray diffraction patterns reported for sodium salts of A- and B-DNA [12,13].

The I_{cc} spectrum of A-DNA reveals a very strong Raman band at 807 cm^{-1} for the sugar in the C3′-endo geometry (Fig. 1.45). B-DNA exhibits a much weaker band at 836 cm^{-1} for sugars in the C2′-endo geometry (Fig. 1.44). The 807 and 836 cm^{-1} frequencies correspond to the well-characterized "marker" bands of DNA, which are considered diagnostic of A and B conformations, respectively [177,187,188] (see Section 1.3.10). Extensive study of nucleic acid model compounds [176,177] indicates that these modes originate from vibrations associated with the 3′,5′-phosphodiester bond network (C−O−P−O−C) and not from sugar puckering (see also Section 1.3.9).

For B-DNA, an additional marker band near 792 cm^{-1} has been proposed on the basis of deuteration effects observed in fiber of calf thymus DNA [188]. A putative B backbone marker near 790 cm^{-1} would normally be masked by DNA of mixed-base composition due to the presence of overlapping Raman band near 784 cm^{-1} assignable to a ring mode of the cytosine residue. We see, from the I_{bb} and I_{cc} spectra of A-DNA (Fig. 1.45), that the out-of-plane Raman tensor component (α_{zz}) of the 784 cm^{-1} cytosine mode is relatively small. Accordingly, the strong band in the I_{cc} spectrum of B-DNA cannot be assigned to the cytosine mode. A reasonable assignment is to a phosphorus–oxygen single bond stretching vibration, which is expected in the 750–850 cm^{-1} region [176].

The two relatively weak bands of B-DNA at 895 and 920 cm^{-1} are assigned to bond stretching vibrations of the deoxyribose ring. This assignment is supported by the observation that similar bands, at 852 and 898 cm^{-1}, in the Raman spectrum of thymidine are shifted by 12 and 17 cm^{-1}, respectively, on ^{13}C substitution of C1', C2', C3', C4', and C5' positions of deoxyribose residue [165]. Further support is provided by significant band shifts of guanosine in the 850–1000 cm^{-1} region on deuterium substitution of the C1' position [168]. Similarly, weak bands are observed in A-DNA at 882 and 897 cm^{-1}. For the band of 882 cm^{-1} in A-DNA, $I_{bb} < I_{cc}$ is observed. For all other deoxyribose bands in this region, we find $I_{bb} > I_{cc}$.

The Raman band assigned to the symmetric O=P=O stretching vibration of the nucleic acid phosphodioxy (PO$_2^-$) group is expected near 1093 cm^{-1} in the structures of B form and near 1100 cm^{-1} in A-form structures [137,170]. These PO$_2^-$ markers are observed at the expected frequencies both B and A fibers (see Fig. 1.44 and 1.45). For both DNA conformations, we note also that the I_{bb} and I_{cc} intensities are nearly equal. This contrasts sharply with the results from polarized infrared spectroscopy, where $I_\parallel > I_\perp$ for A DNA, but $I_\parallel < I_\perp$ for B-DNA (see Section 1.3.10). The fact that $I_{bb} \approx I_{cc}$ for the Raman PO$_2^-$ mode can therefore be attributed to an intrinsically isotropic local Raman tensor.

The antisymmetric O=P=O stretching vibration of the phosphodioxy group is expected near 1200 cm^{-1}. Although this vibration generates intense infrared absorption, the Raman scattering is characteristically very weak. Actually, no band definitely assignable to the antisymmetric O=P=O stretching mode can be detected in spectra of either the A- and B-DNA fiber (see Figs. 1.44 and 1.45).

Relatively intense Raman bands that can be assigned to the "ring breathing" motions of guanine, adenine, thymine, and cytosine are observed, respectively, at 666, 731, 753, and 784 cm^{-1} in A-DNA and 682, 729, 750, and 787 cm^{-1} in B-DNA. As noted above, the 787 cm^{-1} band of B-DNA actually represents the unresolved center of two overlapping bands due to cytosine (near 790 cm^{-1}) and the B backbone (near 790 cm^{-1}). For all of the ring modes, $I_{bb} > I_{cc}$, which is consistent with the fact that the out-of-plane tensor component (α_{zz}) is smaller than those of in-plane components (α_{xx} and α_{yy}). It should be noted that the Raman anisotropies for the in-plane base modes are greater in A-DNA than in B-DNA.

In both A and B fibers, a weak but prominent band is observed near 498 cm^{-1}. Assignment of this band to a valence angle bending mode (scissoring) of the phophodioxy group is supported by normal coordinate analysis [189]. High Raman anisotropy ($I_{bb} > I_{cc}$) distinguishes this phosphate group bending vibration from the abovementioned phosphorus-oxygen stretching modes in the 750–1100 cm^{-1} interval.

All Raman bands in the I_{bc} and I_{cb} spectra (Figs. 1.44 and 1.45) are weak, and therefore their signal-to-noise ratios are low. Nevertheless, some interesting conclusions can be drawn from these observations:

(1) The results show that $I_{bc} = I_{cb}$ for a given DNA fiber; that is, the Raman spectrum of a fiber oriented along the c axis is invariant to an orthogonal interchange of the polarization directions of the incident and scattered light. This is consistent with precise optical alignment of the Raman microscope–spectrometer system, with the absence of polarization artifacts, and with a high degree of unidirectional orientation of DNA along the fiber axis.

(2) Comparisons between A- and B-DNAs indicate important tensor similarities and differences. For example, the bands near 1487, 1578, and 1669 cm^{-1} exhibit nearly equal intensities and band shapes in A and B fibers. On the other hand, the CH$_2$ scissoring mode near 1420 cm^{-1} gives a sharp band in B-DNA, but a more diffuse band in A-DNA. The bands at 1303, 1339, and 1375 cm^{-1} are prominent only in B-DNA, whereas the 1257 cm^{-1} band is prominent only in A-DNA.

(3) Finally, of particular interest is the observation that the PO$_2^-$ symmetric stretching band appears much more prominently in the I_{bc} (or I_{cb}) spectrum of B fiber than that of the A fiber. Thus, the different orientations of phosphodioxy groups of A- and B-DNA, which are not obvious in the I_{bb}/I_{cc} ratio, are clearly revealed in the I_{bb}/I_{bc} anisotropy.

1.3.10. Raman Spectra of Z-DNA

As stated in Section 1.1.9, the self-complementary hexamer DNA sequence, d(CGCGCG)$_2$, forms a left-handed helix (Z-DNA) in the crystalline state, as determined by X-ray diffraction [51,52]. Raman measurements on this crystal are shown in Fig. 1.46. For each intense band in the 300–1700 cm^{-1} region of the Raman spectrum, the relative scattering intensities, I_{aa}, I_{bb}, and I_{cc}, corresponding to the aa, bb, and cc components of the crystal Raman tensors, were determined [190]. In the crystal, the helical axis of the molecule is aligned nearly along the crystallographic c axis, and the base planes are in the crystallographic ab plane, which is exactly perpendicular to the c axis (see Fig. 1.9c).

As seen in Fig. 1.46, Raman bands of the I_{cc} spectrum at 1579 (G), 1486 (G), 1318 (G), 1264 (C), 784 (C), 670 (G), and 625 (G) cm^{-1} are much weaker than their counterparts in the I_{aa} and I_{bb} spectra. This observation is consistent with the crystal structure, which shows all base-planes arranged nearly in the ab plane, that is, nearly perpendicular to the c-axis. Thus, at least one in-plane component (α_{xx} or α_{yy}) must be much greater than the out-of-plane component (α_{zz}) for the Raman tensor corresponding to each band. The two ring breathing bands of guanine at 670 and 625 cm^{-1} are well known as deoxyguanosine conformation markers (see Section 1.3.2 and Fig. 1.37). The 670 cm^{-1} band is assigned to two terminal dG residues with C2'-*endo/syn* conformation, and the 625 cm^{-1} band is assigned to the four internal dG residues with C3'-*endo/syn* conformation.

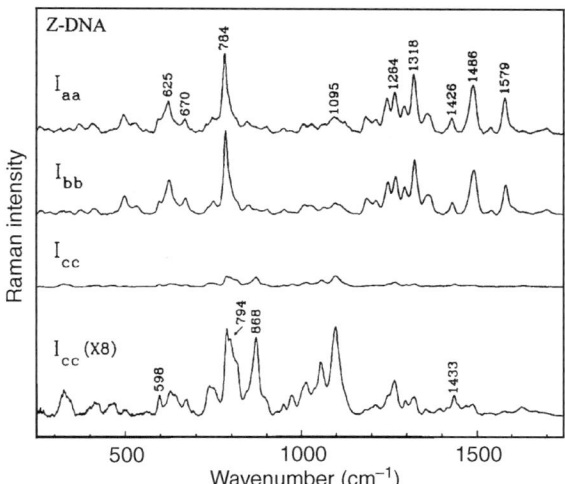

FIGURE 1.46. Raman spectra of the single crystal of d(CGCGCG)$_2$, which has a Z-DNA conformation. The three spectra correspond to the three different orientations of the unit cell axes with respect to the electric vector of exciting and scattered radiation [190].

The weak band at 1426 cm^{-1} in I_{aa} and I_{bb} spectra, assigned to CH$_2$ scissoring, is not evident in I_{cc} spectrum, where instead a weak band is observed at 1433 cm^{-1}. The crystal structure reveals that most C2'H$_2$ groups are oriented such that a line connecting the two hydrogen atoms is approximately perpendicular to the c axis. Since the H....H line is the direction along which the Raman tensor should have its greatest component [172], the corresponding Raman band should exhibit much greater intensity in the I_{aa} and I_{bb} spectra than in the I_{cc} spectrum. On this basis, the 1426 cm^{-1} band may be reasonably assigned to the C2'H$_2$ groups, and the 1433 cm^{-1} band to the 5'CH$_2$ groups.

The 1095 cm^{-1} band, assignable to the symmetric stretching vibration of the anionic phosphodioxy group (O=P=O)$^-$, gives similar intensities in all polarized spectra. This is expected because the 10 PO$_2^-$ groups per asymmetric unit of the duplex are oriented along markedly different directions. Additionally, the Raman tensor of this vibration is expected to be intrinsically isotropic.

The Raman band at 868 cm^{-1} draws special attention. This is the only band that is more intense in the I_{cc} spectrum than in either the I_{aa} or I_{bb} spectrum. The Raman tensor of the 868 cm^{-1} band, therefore, exhibits its greatest component along the crystallographic c axis. This band is tentatively assigned to a furanose ring-breathing vibration (see Section 1.3.3 and Ref. [165]). In the d(CGCGCG)$_2$ crystal, all furanose rings are arranged with their "planes" parallel to the c axis and with the C1'—O—C4' angle bisectors nearly parallel to the c axis. It is possible that one component of the Raman tensor is localized along the bisector, and that this Raman tensor component is much greater than the other two, thus accounting for the observed anisotropy.

1.4. NMR SPECTRA

Nuclear magnetic resonance (NMR) spectroscopy is a powerful method for studying the structures of drug–DNA interactions, as well as drug-binding affinities. It can be used in different ways to obtain many different types of information; hence, it can be regarded as a family of closely related methodologies, rather than as just one technique. As the details of NMR far exceed the scope of this book, we will consider only the most essential aspects of this subject, focusing primarily on its application to the most common elements in biochemical research, namely, H, C, N, and P. More details on selected aspects of NMR can be found in Appendix A.2. The interested reader is referred to the many excellent texts for information on the principles of NMR [191–196] and on its uses in nucleic acid research [197–203].

The term *nuclear magnetic resonance* (NMR) is derived from its reliance on the intrinsic quantum mechanical properties of the nuclei of atoms. Atoms contain the subatomic particles, electrons, neutrons, and protons, each of which has a property called "spin." NMR is based on the combined spin of the protons and neutrons in the atomic nucleus, which is described by I, the *nuclear spin*, or *nuclear angular momentum*. By contrast, the related technique, electron spin resonance (ESR), is based on the spin of electrons (see Section 1.5).

Both electrons and nuclei can be regarded as very small bar magnets, with a small *magnetic moment*, μ. The magnitude of μ is dependent on a constant, called the *gyromagnetic ratio*, or γ. Whereas all electrons are identical, the nuclei are not. The values of the nuclear spin and γ value depend on the atomic and mass numbers of a given isotope (see Table 1.17) [204]. Those isotopes having nuclei with large absolute values of γ are easier to detect, and are said to be "sensitive," whereas those with small γ values are "insensitive."

The most commonly used isotopes in biological NMR studies are ^{1}H, ^{13}C, ^{15}N, and ^{31}P (see Table 1.18). Of these, ^{1}H is the easiest to detect. This isotope is almost 100% abundant and is NMR active—its nucleus has a spin quantum number of $I = \frac{1}{2}$, as well as a large γ. The phosphorus isotope, ^{31}P, also exists in 100% abundance, and its nucleus has a spin of $I = \frac{1}{2}$, but its γ is much smaller. It also suffers from an efficient relaxation process called *chemical shift anisotropy* (CSA), which broadens its peaks (see Appendix A.2). The nuclei of the most abundant isotope of carbon, ^{12}C, is not NMR-active ($I = 0$), whereas that of the less abundant isotope (1.07%), ^{13}C, has $I = \frac{1}{2}$, and is active. Therefore, the ^{12}C carbons are often replaced with ^{13}C by isotopic

TABLE 1.17. Properties Determining Nuclear Spin[a]

Mass Number	Atomic Number	Nuclear Spin (I)
Even	Even	0
Even	Odd	Integral (1, 2, 3...)
Odd	Even or odd	Half-integral ($\frac{1}{2}, \frac{3}{2}, \frac{5}{2}, \ldots$)

[a] When $I = \frac{1}{1}$, the nucleus has a spherical electric field, resulting in sharp peaks in the NMR spectrum. When $I > \frac{1}{2}$, the nucleus has an ellipsoidal electric field, resulting in quadrupolar relaxation and hence broad NMR lines. All nuclei with $I \neq 0$ are observable by NMR (i.e., are "NMR-active").

TABLE 1.18. NMR Properties of Selected Nuclei[a]

Nucleus	Nuclear Spin (I)	Number of Spin States ($m = 2I+1$)	Gynomagnetic Ratio (γ) (T·s)$^{-1}$	Natural Abundance (%)
$^{1}_{1}\text{H}$	$\frac{1}{2}$	2	2.675×10^{8}	99.988
$^{2}_{1}\text{H}$	1	3	4.107×10^{7}	0.012
$^{12}_{6}\text{C}$	0	0	0	98.89
$^{13}_{6}\text{C}$	$\frac{1}{2}$	2	6.728×10^{7}	1.07
$^{14}_{7}\text{N}$	1	3	1.934×10^{7}	99.63
$^{15}_{7}\text{N}$	$\frac{1}{2}$	2	-2.712×10^{7}	0.37
$^{16}_{8}\text{O}$	0	0	0	99.96
$^{17}_{8}\text{O}$	$\frac{5}{2}$	6	-3.628×10^{7}	0.04
$^{31}_{15}\text{P}$	$\frac{1}{2}$	2	10.839×10^{7}	100.00

[a]The atomic number is given as a subscript and the atomic mass, as a superscript.
Source: Data taken from Ref. 204.

enrichment in biological samples. In the case of nitrogen, there are two NMR active isotopes, ^{14}N and ^{15}N, of which ^{14}N is the most abundant, but its nucleus has a quantum spin number of $I = 1$. Nuclei with $I = 1$ produce spectra that are often difficult to interpret, and ^{14}N has the added disadvantage of undergoing very fast quadrupolar relaxation, producing broad bands in the spectrum (see Appendix A.2.3). On the other hand, ^{15}N, which is not as abundant (0.37%), has a spin of $\frac{1}{2}$ and has better relaxation properties. Once again, the concentration of ^{15}N is usually enriched in biological samples to enhance its detection.

Bar magnets interact and align themselves with respect to each other. Similarly, when nuclei are placed in a strong, external magnetic field, the nuclear magnetic moments also align with that magnetic field. NMR spectroscopy is based on manipulating that alignment by irradiating the sample with radiofrequency pulses of appropriate energy. By pulsing the sample in different ways, a wide variety of experiments can be performed, each of which probes the molecule for different properties. The most common analyses made via NMR are chemical shift (indicative of the local chemical environment), resonance intensity (a measure of the relative numbers of a given type of atom), spin–spin coupling (a function of through-bond connectivity, and hence chemical structure), the nuclear Overhauser effect (a function of through-space proximity, and hence of molecular conformation), local and global dynamics as determined by relaxation properties (arising in part from the previous two), chemical exchange, and diffusion.

1.4.1. NMR Signals and Magnetization Relaxation

As noted above, the magnetic fields of nuclei align themselves in the presence of a strong magnetic field. The net magnetization of the nuclei can be expressed as a vector that points in the same direction as the magnetic field, which is defined as being along the z axis (Fig. 1.47). In a typical modern NMR experiment, the sample is briefly

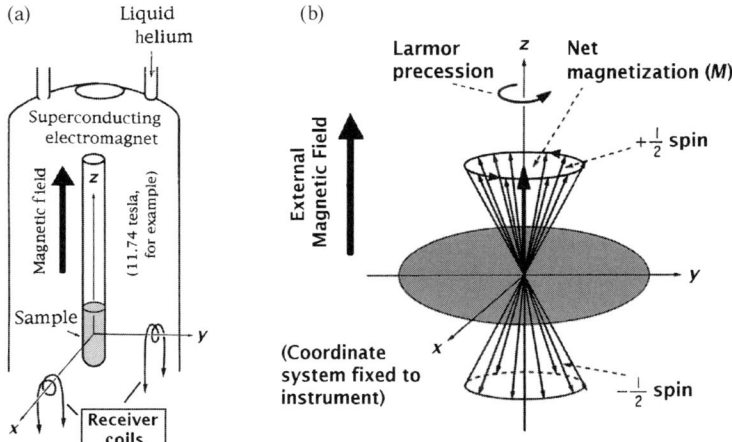

FIGURE 1.47. A comparison of the macroscopic and microscopic aspects of an NMR experiment with respect to the laboratory's physical coordinate system, in which the magnetic field defines the $+z$ axis. Diagram (a) depicts the placement of the sample and receiver coils within the instrument, and indicates that located along the x and y axes. Diagram (b) shows the orientations of the nuclear magnetic moments of the sample with respect to their spin states. There are more spins aligned with the external magnetic field (B_0) than opposed to it, hence the net magnetic vector is also aligned with it, and points in the $+z$ direction. The dipole moments of the nuclei rotate around the z axis with a periodicity proportional to the gynomagnetic ratio (γ). This is called the *Larmor precession frequency*. For a magnetic field strength of 11.74 T, a proton (^{1}H) will precess at a rate of 500 MHz.

irradiated with a radiofrequency (RF) pulse, which rotates the nuclei into the x,y plane (perpendicular to the z axis). The intensity of an NMR signal is determined by the amount of magnetization in the x,y plane as a function of time. When the RF pulse is turned off, the nuclei begin to return to the initial alignment via a variety of relaxation processes (Appendix A.2.3). These processes are depicted in Fig. 1.48, and fall into two general categories: longitudinal relaxation (*spin–lattice*) and transverse relaxation (*spin–spin*). Together, they give rise to all measurable NMR properties. For example, dipole–dipole relaxation, which is one of several longitudinal relaxation processes that involve the interaction of nuclei through space, gives rise to the nuclear Overhauser effect (NOE; see Section 1.4.4) as well as dipolar coupling (see Section 1.4.5). Similarly, transverse relaxation, which arises from spin–spin coupling (also called *scalar*, or *J coupling*), is the origin of the splitting patterns in one-dimensional (1D) spectra. It is this process that enables through-bond correlation experiments, which provide information on the connectivity of atoms within a molecule (e.g., COSY and HSQC; see Section1.4.3).

These relaxation properties are often affected by experimental conditions, such as temperature and viscosity, which can alter the NMR spectrum, especially the peak locations and their linewidths. The former is illustrated in Fig. 1.49, which shows the phosphorus (^{31}P) NMR spectrum of d(CCTAGG)$_2$ at 11°C, as well as a graph of how the positions of the peaks change with temperature [205,206].

NMR SPECTRA 75

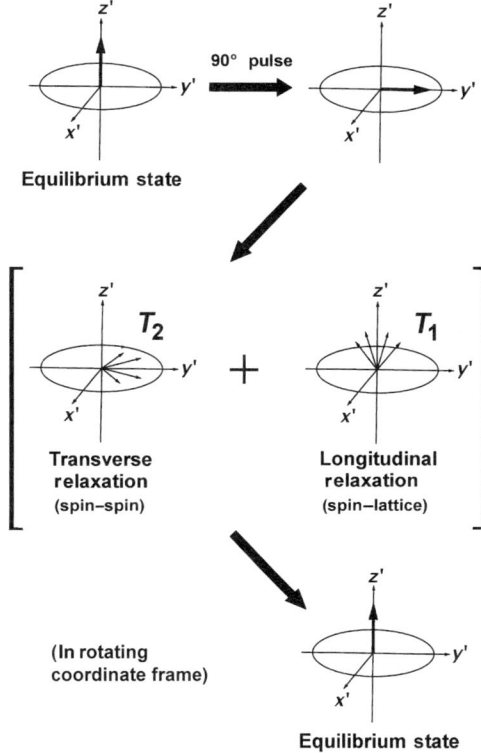

FIGURE 1.48. In their initial, equilibrium states, the nuclei are aligned such that their net magnetization is along the $+z$-axis ($M_z \neq 0$), with no net magnetization in the x, y plane ($M_{x,y} = 0$). Then, the nuclei are depicted as being irradiated with a 90° pulse radiofrequency along the x axis, causing the net magnetization (**M**) to rotate away from the z axis, and aligning it along the y axis ($M_y \neq 0$, $M_z = 0$). After the pulse is turned off, the magnetization relaxes back to its equilibrium alignment with the magnetic field via two relaxation processes. Longitudinal relaxation is the vertical realignment of **M** along the $+z$ axis ($M_z \neq 0$). Transverse relaxation is the restoration of zero net magnetization within the x, y plane ($M_{x,y} = 0$). The latter is achieved by the radial dispersion of the individual nuclei within the x, y plane, resulting in a net cancellation of the magnetization. These processes produce all measurable NMR phenomena and result in the free induction decay (FID) curve. The coordinate system ($x'\ y'\ z'$) is rotating with respect to the fixed coordinate system ($x\ y\ z$) around the z axis. See text and Appendix A.2, Section A.2.3, for more information.

In addition, these relaxation properties can provide useful information about the dynamical motion, or conformational fluctuation, within a molecule or its complex. Such motion is usually altered when a drug or other ligand binds to DNA, and can sometimes be correlated with the preferential binding of the drug at certain base sequences. For example, an early study of the flexibility of calf thymus DNA (CT-DNA) measured the transverse relaxation time of its ^{31}P NMR spectrum [207].

76 DNA STRUCTURES AND SPECTRA

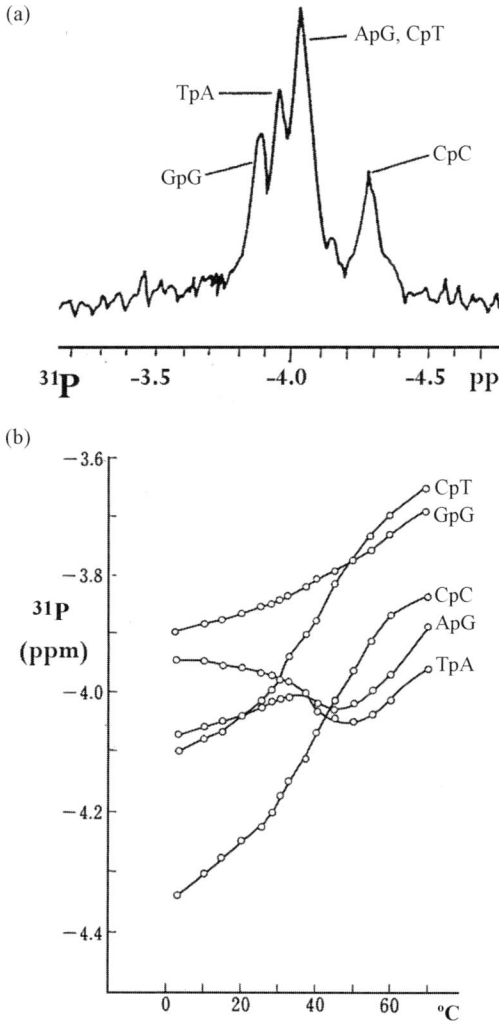

FIGURE 1.49. (a) The phosphorus NMR spectrum of d(CCTAGG)$_2$; (b) ^{31}P chemical shifts of d(CCTAGG)$_2$ as a function of temperature and their assignments [205].

More information about the mechanisms and measurement of relaxation properties can be found in Appendix A.2.3.

1.4.2. Proton NMR Spectra of Bases, Nucleosides, and Nucleotides

1.4.2.1. Chemical Shifts Figure 1.50 is an example of a proton (^{1}H) NMR spectrum of a nucleoside. This is a spectrum of cytidine that was obtained in a D$_2$O solution, using a spectrometer with a magnetic field strength of 5.87 T (teslas) [208]. Cytidine has 12 hydrogen atoms, each of which is observed in the NMR spectrum as a

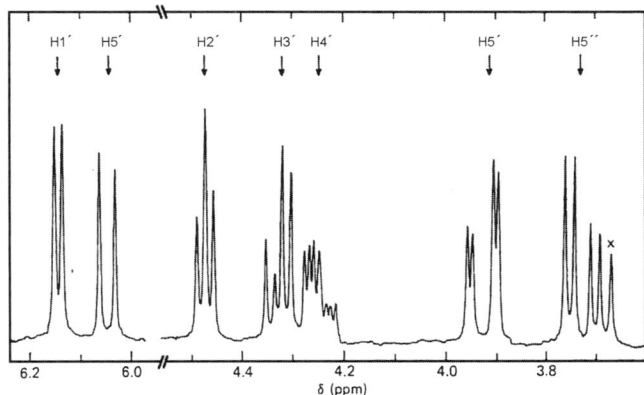

FIGURE 1.50. The 250-MHz ^{1}H NMR spectrum of cytidine in ^{2}H$_2$O [208].

peak occurring at a different resonance frequency, or "chemical shift" (δ), expressed in units of ppm (see Appendix A.2.2). These different frequencies arise from the fact that each proton (or ^{1}H hydrogen atom) is in a slightly different electronic environment, and hence "feels" the external magnetic field slightly differently. These resonances can be summarized as follows:

Base protons: H5: $\delta = 6.049$ ppm; H6: $\delta = 7.752$ ppm (not shown);
Sugar Protons: H1′: $\delta = 6.144$ ppm; H2′: $\delta = 4.478$ ppm;
               H3′: $\delta = 4.310$ ppm; H4′: $\delta = 4.254$ ppm;
               H5′: $\delta = 3.930$ ppm; H5″: $\delta = 3.731$ ppm.

Nuclei in similar environments always have comparable resonance frequencies and generally fall within a relatively small range of ppm values [198,209,210]. These vary systematically as a function of nucleotide sequence and global structure, arising from differences in the local chemical environments. For example, the H8 of deoxyguanosine in the second G (or G$_2$) of the sequence d(CGCAGAATTCGCG)$_2$ has $\delta = 7.95$ ppm, whereas the H8 of the 12th base (or G$_{12}$) has $\delta = 7.76$ ppm. Thus, the effect of DNA sequence and base-pairing influences the observed resonance frequencies, but the different DNA residues can still be distinguished. Table 1.19 lists the chemical shift values of hydrogen, carbon, and nitrogen of some nucleoside residues in B-form DNA.

1.4.2.2. Spin–Spin Coupling Constants The ^{1}H NMR spectrum of cytidine (Fig. 1.50) also demonstrates another important feature of NMR spectra—the resonance lines for most of the hydrogen atoms do not appear as single peaks, but rather are split into multiplets. The separation between these peaks gives the *coupling constant J*, and is measured in Hertz (Hz). The physical mechanism for this effect is the transfer of magnetization via the spin pairing of each atomic nucleus to the spins of the electrons in chemical bonds between them. It is thus called an *electron–coupled, spin–spin*

78 DNA STRUCTURES AND SPECTRA

TABLE 1.19. Normal Ranges of Proton (^{1}H) and "X"-Atom [Carbon (^{13}C) or Nitrogen (^{15}N)] Resonance Frequencies in a Standard DNA Duplex

Atom Position/Group	Resonances (ppm)	
	^{1}H	X
Deoxyribose		^{13}C
1'	5.2–6.7	84–91
2' (H', H")	0.9–3.9	37–44
3'	4.1–5.6	72–82
4'	3.8–5.0	84–89
5' (H', H")	3.3–4.6	62–72
Nucleoside[b]		^{13}C
2 (A)	7.3–8.4	153–155
5 (C)	4.5–6.5	92–99
6 (C, T)	6.5–8.1	139–143
7 (T)	0.8–1.6	13.5–15
8 (A, G)	7.4–8.5	137–142
		^{15}N
—NH$_2$ (A, C, G)	6.2–9.0	(A) 79–81
		(G) 74–76
		(C) 96–98
>NH (ring NH of G, and T)	10–15 m	(G) ~147
		(T) 158–160

[a]These numbers were compiled from multiple sources [198,209]. The ranges listed here are approximate, as many outlying examples exist. In particular, the resonances arising from cytidine may have a much wider range than those given here, as they may exist in both neutral and acid forms.
[b]Notice that the chemical shifts of aromatic atoms in the nucleosides are much higher than those in deoxyribose moiety.

interaction, which can be depicted for a C–H bond as

$$\uparrow H- \downarrow e^- - \uparrow e^- - \downarrow C$$

where the arrows refer to the orientations of the quantum spin states. Thus, each nucleus is sensitive to the possible orientations of the spin of its neighbor. This method of transferring magnetization is an important transverse relaxation process in NMR called T_2 (see Fig. 1.48, left, Section 1.4.3, and Appendix A.2, Section A.2.3). The strength of the coupling between two nuclei i and j is given by the magnitude of the interaction energy

$$E = \mathsf{h}J(\mathbf{I}_i \cdot \mathbf{I}_j), \tag{1.6}$$

where h is Planck's constant, $\mathbf{I}_k$ is the spin quantum number of nucleus k, and J is the *coupling constant*. This coupling is also called *J coupling* or *scalar coupling* because of the form of this mathematical equation.

Under normal circumstances, this spin–spin coupling extends out to about three bonds (e.g., H–C–C–H), although coupling through more bonds is possible. This *through-bond* coupling provides information on an atom's chemical environment, and it can be exploited to determine unambiguously the connectivity of a chain of atoms by using one of the spectroscopic, 2D correlation experiments, such as COSY (Section 1.4.3).

When a nucleus is coupled to a neighbor with a nuclear spin of $I = \frac{1}{2}$, its resonance will be split into a doublet; if the neighboring atom has a spin of $I = 1$, then it will be split into a triplet. Thus, the general rule is that the number of lines in a multiplet is $(2nI + 1)$, where n is the number of identical atoms in a neighboring chemical group and I is the spin of these nuclei. The intensity of each line in the multiplet is determined by the number of possible combinations of individual spins that can produce a given value of total spin, and follows a binomial distribution. That is, when a nuclei is coupled to a group of n identical protons, its resonance is split into a series of peaks whose number and intensity ratios are given in the nth row of Pascal's triangle (for couplings to nuclei with $I = \frac{1}{2}$):

n	Resonance Pattern					Splitting Pattern
0			1			Singlet
1		1		1		Doublet
2		1	2	1		Triplet
3	1	3		3	1	Quartet
			(etc.)			

The situation can become quite complex if a nucleus is coupled to several non-equivalent nuclei, or if there is longer-range coupling, such as through a double bond.

Returning to the cytidine spectrum as an example (Fig. 1.50), the H1' has an $I = \frac{1}{2}$ neighbor, H2', and hence has its resonance line split into a doublet with $J = 3.2$ Hz. The H2', however, has a deceptively simple triplet pattern. In fact, this is not a triplet at all, but a doublet of doublets in which the two inner lines overlap. This happens because H2' is coupled with both H1' and H3', and the coupling constant is nearly the same with each, $J = 3.7$ Hz. This results in the same intensity pattern as the $n = 2$ case, but really results from two $n = 1$ groups.

It is possible to predict these coupling constants by using the Karplus equation [211], which is an experimentally parameterized relationship between the magnitude of the coupling constant, J and the dihedral torsion angle ϕ about the central bond in a chain of four atoms. This equation has several forms depending on the number of dimensions in the spectrum being analyzed. For a 1D NMR spectrum, it is

$$J(\phi) = A\cos^2(\phi-60) - B\cos(\phi-60) + C \quad (1.7)$$

where A, B, and C are empirical parameters, which depend on the types of atoms involved in the coupling. This relationship between coupling constant and dihedral torsion angle (ϕ) is graphed in Fig. 1.51 for HCCH coupling. It clearly reveals that these coupling constants are largest when ϕ is either 0° or 180°, and smallest at 90°.

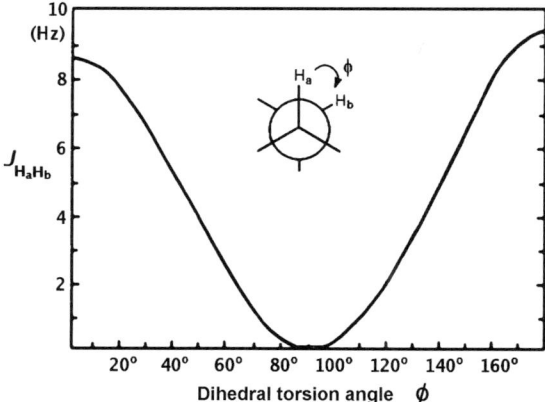

FIGURE 1.51. The Karplus curve; showing dependence of the coupling constant J (in Hz) on the dihedral angle ϕ, which is defined by the planes of $H-C_i-C_j$ and C_i-C_j-H bonds in an $H-C_i\text{-}C_j\text{-}H$ fragment.

This equation is very important when determining 3D structures of biological macromolecules from NMR data.

This mathematical function applies only to weakly coupled systems, where J is much smaller than the chemical shift difference between the coupled nuclei (and all frequencies are expressed in units of Hertz); that is, it applies when $10\,|J| < \Delta\delta$, where $\Delta\delta$ is the difference in the chemical shifts, for example $(\delta(H_i) - \delta(H_j))$ in Hertz. This defines a "first-order" spectrum. If J has at least the same order of magnitude as $\Delta\delta$, then more complex spectral patterns are observed, as in the 4.2–4.4 ppm region of the cytidine spectrum (Fig. 1.50). As the external magnetic field strength $\mathbf{B_0}$ increases, the chemical shift difference ($\Delta\delta$, in Hz) between the coupled peaks is also increased. Since almost all modern spectrometers operate above 400 MHz for the ^{1}H, $\Delta\delta$ is greater than $10\,|J|$. This means that all protons are weakly coupled, and their spectra can now be considered to be first order.

1.4.3. Correlation Analysis, or "Spin–Spin Coupling"

When determining the three dimensional structure of a molecule, it is necessary to first assign the NMR resonances to specific atoms within the molecule. An atom's chemical shift value (ppm) often provides a general classification of its chemical environment, but it gives no information regarding the atom's connectivity (bonding) to other specific atoms. As noted in the previous section, the nuclear magnetic dipole of one atom can interact with a chemically attached neighboring atom through their shared electrons in the chemical bond. This is a T_2 relaxation process (see Appendix A.2.3) and there are many NMR experiments that take advantage of this to determine chemical connectivity.

For example, chemical connectivity may be determined by a one-dimensional experiment in which a sample is irradiated with a radiofrequency equal to the resonance of a particular atomic nucleus. This "decouples" any other resonances that are spin-coupled to that irradiated nucleus, resulting in the collapse of the coupling patterns. Hence, irradiating the H2' of cytidine at 4.48 ppm will cause the H1' doublet at 6.14 ppm to collapse into a singlet located at its midpoint (compare with Fig. 1.50). This radiofrequency causes the irradiated nuclei to rapidly transit between their possible spin orientations, rendering the different spin states indistinguishable [193].

By using a variety of such pulse sequences, it is possible to take advantage of these T_2 relaxation processes in order to transmit magnetization information along a chain of atoms of two bonds, or more. This can unequivocally establish a series of atoms that are chemically connected to each other. Common two-dimensional experiments that measure this are COSY, TOCSY, HMBC, HSQC, HMQC, and HETCOR (or heteronuclear COSY). The first two of these are homonuclear experiments that establish through-bond connectivities between protons. The remaining four experiments establish through-bond connectivities between protons and heteroatoms (e.g., ^{13}C, ^{15}N, and ^{31}P). In addition to these experiments, many others exist that have been designed specifically for certain chains of atoms, especially those found in proteins. We will not consider all of these different experiments, but instead only discuss the uses of general types.

1.4.3.1. Homonuclear Correlations and Coupling Analysis The homonuclear COSY experiment establishes which protons (^{1}H) are on the same or adjacent carbon atoms, but it provides no direct information about the intervening carbon atoms. In nucleic acids, the ^{1}H-^{1}H coupling patterns within the sugar rings are particularly informative, as they can allow one to unequivocally distinguish most of their resonances. The TOCSY experiment provides similar information as the COSY, but it relays magnetization through an entire network of spin-coupled nuclei. So, the COSY shows interatom connectivity between adjacent protons on a pairwise basis, whereas the TOCSY reveals all protons in a group. Together, these experiments can be used to assign the signals of all the sugar protons in a nucleotide.

The phase-sensitive variant of the COSY (called DQF-COSY) contains additional information about the *J* coupling between protons. It is often used to determine the sugar pucker conformations of the nucleotides. This is possible because each proton in a sugar ring is coupled to its adjacent neighbor, and the coupling constant between protons is an indication of the dihedral torsion angle (ϕ), as given by the Karplus equation (Section 1.4.2). Thus, by carefully determining the various coupling constants, one is able to determine the pseudorotation angle *P* (Section 1.1.4) [212–216]. These angles can, in turn, be correlated to the backbone torsion angles. Although several programs are available for this type of detailed analysis, it remains fairly tedious. It is also often possible to make a more qualitative classification of the pseudorotation angle by simply comparing the experimental coupling patterns of the H1'-H2' and H1'-H2'' COSY cross-peaks to those found in the literature. If more than one conformation exists for a molecule, or if a sugar undergoes rapid dynamical motion, such as rapid flipping of its pseudorotational angle ("repuckering"), then these coupling patterns can become diffuse, or indistinct, or their different phases may result in cancellation of the peak intensities. There are other techniques for obtaining this information that do not rely on COSY information. For example, in RNA studies, it is often sufficient to ascertain the strength of a H1'-H2' cross-peak in a 2D-NOESY experiment (Section 1.4.4). Reliable information about sugar conformations can be also obtained from *residual dipolar coupling* analysis (Section 1.4.5).

1.4.3.2. Heteronuclear Correlations and Coupling Analysis Heteronuclear correlation NMR experiments are quite important when determining the sequential base connectivity. The prototype of these is the heteronuclear COSY experiment (also called "HETCOR" or [^{1}H,X]-COSY), which can determine which proton is coupled to a nearby heteroatom X, where $X = {}^{13}C$, ^{15}N, or ^{31}P. An example of a heteronuclear C,H-COSY experiment of the self-complementary DNA dedecamer, d(CGCGAA*T*TCGCG)$_2$ is shown in Fig. 1.52. In the seventh residue of this duplex, thymidine-7 (T$_7$), the carbon atoms of the deoxyribose sugar were uniformly, isotopically enriched with ^{13}C [217]. This enhanced the carbon signals from this residue. These maps show cross-peaks ("spots") for each C—H bond, giving the chemical shifts of each atom in the bonding pair, as well as allowing the measurement of their spin couplings. More sensitive experiments, such as HSQC or HMQC, give the same kind of information and have been used in more recent publications.

If a sample has been properly enriched, it is also possible to perform heteronuclear experiments of the type [X,Y]-COSY, where X and Y can be any NMR-active isotope (e.g., ^{13}C, ^{15}N, or ^{31}P). For example, all of the atoms within the aromatic rings of a nucleotide might be assigned by using an [^{13}C, ^{15}N]-COSY, or [^{13}C, ^{15}N]-HSQC experiment.

1.4.3.3. Detection of Hydrogen Bonding in Base Pairs Scalar coupling NMR experiments can directly detect the imino hydrogen bonds in DNA and RNA base pairs. These imino hydrogen bonds occur between two nitrogens, one of which is a hydrogen donor and the other is a hydrogen acceptor: N—H···N. For this experiment, the aromatic bases of the DNA are enriched with ^{15}N to enhance their detection, and the nitrogens are assigned by means of a heteronuclear correlation experiment, such as an HSQC (see discussion above). Then, by using another scalar coupling experiment (such as HNN-COSY), one can measure the couplings between the nitrogens *through* the hydrogen bond of the base pair, $^{15}N-{}^1H\cdots{}^{15}N$ [218,219].

These types of scalar experiments provided the first direct proof of such hydrogen bonds in nucleic acids. Prior to this, the presence of hydrogen bonds could be inferred only from the geometries of donor and acceptor groups found in 3D models of nucleic acid structures, or from NOE measurements (Section 1.4.4.2). These scalar couplings are also particularly important as they allow the direct observation of unique base-pairing patterns, such as Hoogsteen base pairs [220, 221]. The N—N couplings appear to be very sensitive to the atomic level details of the hydrogen bonds, as there is evidence of a correlation between the size of the scalar coupling and the strength of the hydrogen bond [222].

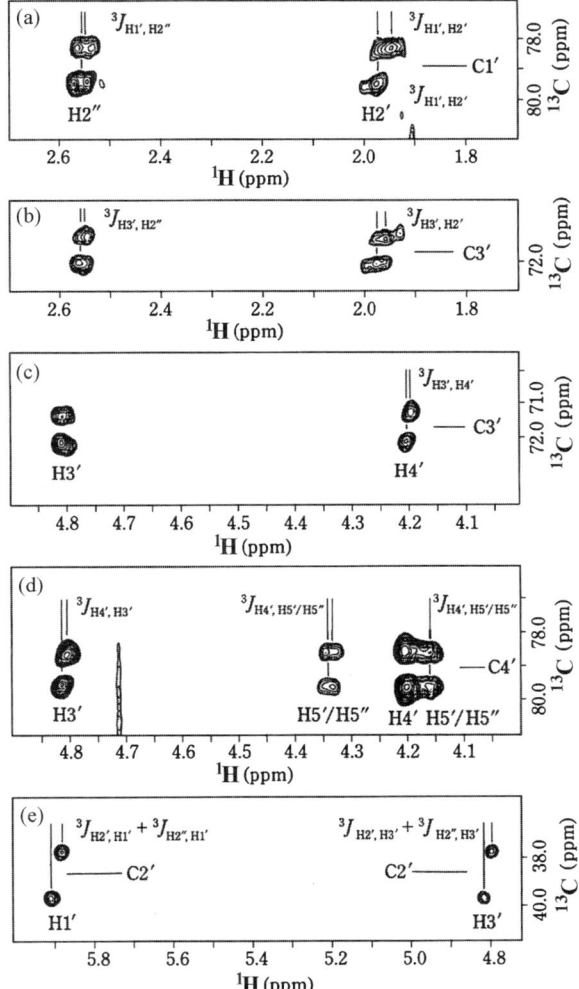

FIGURE 1.52. Two-dimensional maps of spin–spin couplings between ^{13}C and ^{1}H in d(CGCGAATTCGCG)$_2$ from a C,H–COSY. The seventh base, T_7, is a thymidine in which the deoxyribose carbons (C1′, C2′, C3′, C4′, C5′) were uniformly enriched with the isotope ^{13}C [217]. The chemical shifts were determined to be 81.0 (C1′), 38.6(C2′); 71.6(C3′), and 79.1 (C4′) ppm, and their couplings measured: (a) C1′–H2′/H2″ couplings; (b) C3′–H2′/H2″ couplings; (c) C3′–H3′/H4′ couplings; (d) C4′–H5′/H5″/H3′ couplings; (e) C2′–H1′/H3′ couplings.

1.4.3.4. Phosphorus–Proton Couplings and Backbone Dihedral Angles

For nucleic acids, a particularly useful heteronuclear experiment is the [^{1}H, ^{31}P]-COSY (HETCOR), as it can establish the sequential connectivities between sugar moieties. It does this by identifying the coupling between the H3′ protons of base n, through to the ^{31}P in the phosphodiester linkage, and then from the ^{31}P to the H4′ or

H5'/H5" of the next base, $(n+1)$. The H5'/H5" peaks are usually weaker and more difficult to resolve; hence H4' is more commonly utilized.

As an example of the utility of ^{31}P NMR in the study of short DNA oligomers, the 2D [^{1}H,^{31}P]-COSY spectrum of d(CCTAGG)$_2$ is shown in Fig. 1.53 [206]. By comparing this spectrum with the resonance assignments for H5' and H4' (determined from ^{1}H experiments), the assignments of the ^{31}P signals were established. This approach is often used to confirm, or determine, the sequential assignment of the DNA resonances. The different chemical shift values of these phosphorus atoms reflect their different environments within the polynucleotide chain. It was noted that the ^{31}P chemical shift has a nearly linear correlation to the local twist angle t_g of the polynucleotide chain [207]. Here, t_g is the change in the orientation of the C1' − C1' vectors in two successive base-pairs when viewed down the helical axis (the standard B structure is assumed to have a value of $t_g = 35.9°$; see Table 1.1 and Section 1.1.8).

These **P**−(O−C)−**H** couplings are also dependent on the dihedral torsion angle and can be used to determine the backbone conformation of an oligonucleotide by an equation similar to the Karplus equation [223–231]. For example, $J_{PH} = 28$ Hz corresponds to the *trans* conformation, while $J_{PH} = 1.5$ Hz corresponds to the *gauche* conformation. Unfortunately, this is not a very sensitive technique, making such analyses difficult. In addition, the chemical shift dispersion of ^{31}P is usually small (only a couple of ppm units), which limits the utility of this approach to smaller oligonucleotides (less than ~20 residues).

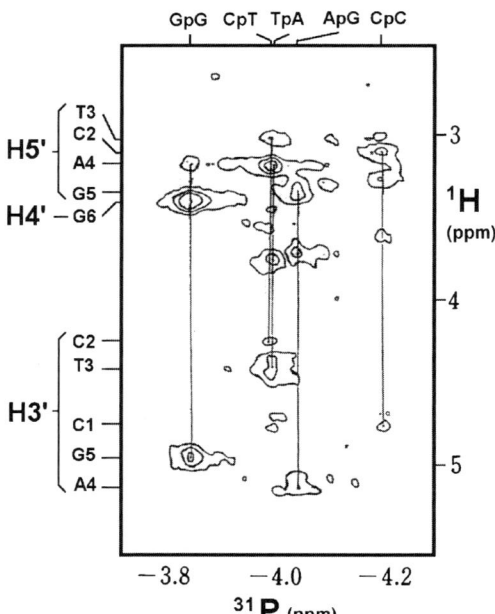

FIGURE 1.53. A contour plot of a ^{31}P-^{1}H shift correlated two-dimensional NMR map ([^{31}P, ^{1}H]- COSY) of d(CCTAGG)$_2$ (1400 OD/mL) at 31°C [206].

In some cases one can obtain reasonable ranges of the backbone torsion angles based on the combined use of linewidth analysis and the sums of the J couplings for different sugar protons [232]. The sum of the H4′ J couplings

$$\Sigma J_{H4'} = (J_{H4'-H3'} + J_{H4'-H5'} + J_{H4'-H5''}) \tag{1.8}$$

is dependent on the values of the sugar pseudorotation angle P and the backbone angle γ (see Figs. 1.2 and 1.15). $J_{H4'-P}$ depends on both γ and β, but its dependence on γ is ambiguous. Similarly, the sum of the H3′ J couplings

$$\Sigma J_{H3'} = (J_{H3'-H2'} + J_{H3'-H2''} + J_{H3'-H4'}) \tag{1.9}$$

is dependent on the backbone angle ε. Some of the other backbone angles are better estimated from NOE distance information (Section 1.4.4).

In the discussion above, only the heteronuclear COSY experiment (HETCOR) was mentioned, but there are a number of more sensitive pulse sequences that give similar information, such as the HSQC and HMQC experiments. These experiments can provide the same sequential connectivity information as the HETCOR. However, even with these experiments it can be difficult to obtain such information because of spectral overlap and the fast relaxation properties of ^{31}P nuclei. As it is not always possible to obtain all this information, spectroscopists sometimes use general estimates of the backbone conformation, which serve to keep the molecular models physically reasonable, without unduly biasing the final result.

A number of synthetic modifications have been suggested as a means for learning about the nucleic acid backbone connectivity and conformation, without relying on ^{31}P. For example, oligonucleotides can be synthesized with the phosphodiester linkage replaced with a phophoramidate linkage. Here, one of the phosphoester oxygen atoms is replaced by ^{15}NH. Although this substitution can be done with the oxygen atom on either the 3′ or 5′ side of the P atom, only the -C3′-NH-P5′-O-C5′ variant forms a stable duplex (compare with Fig. 1.2). This is the N3′ → P5′ derivative, and it prefers to stabilize the A form over the B form in both DNA and RNA [233]. Since RNA prefers the A form, this modification is generally more effective in RNA studies.

1.4.4. Nuclear Overhauser Effect

The *nuclear Overhauser effect* (NOE) enables the measurement of proton–proton interatomic distances through space. This makes it extremely important when studying biomolecules, and is especially critical in the determination of nucleic acid structures. This effect occurs when there is a transfer of magnetization between two nuclei through space via the dipole–dipole relaxation process (see Appendix A.2.3), and it is observed as a change in the intensity of an NMR signal. An NMR signal can be enhanced by a positive NOE, or decreased by a negative one. Three primary factors affect the sign and intensity of an NOE: (1) the gyromagnetic ratios of the interacting nuclei, (2) the interatomic distance between the nuclei, and (3) the mobility of the molecule (or of a substructure within the molecule). This can be expressed by the relationship [192,193]:

$$I_{\text{NOE}} \propto \frac{\gamma_j}{\gamma_i} \frac{1}{r_{ij}^6} \cdot f(\tau_c) \tag{1.10}$$

where i is the observed nucleus, j is the interacting nucleus, r_{ij} is the distance between the two nuclei, γ is the gyromagnetic ratio of a nucleus, and $f(\tau_c)$ is a complex function of the correlation time of the molecule τ_c. The correlation time is a measure of the tumbling rate of the molecule in solution (Appendix A.2, Section A.2.3), and is, itself, a function of many factors, including the size and shape of the molecule, especially its molecular weight. The correlation time is therefore a description of molecular mobility.

Since the NOE is dependent on (γ_j/γ_i), it is clear that the relative values of γ_i and γ_j affect the intensity of the NOE. It is also clear that the signs of these gyromagnetic ratios can determine the sign of the NOE. When both nuclei have positive values of γ (e.g., ^{1}H or ^{13}C), then the NOE is also positive and the NMR signal is enhanced. When one of the nuclei has a negative value of γ (e.g., ^{15}N), then the NOE is negative, and the NMR signal is decreased.

The correlation time τ_c also affects the sign and magnitude of NOEs. This influence can be seen by plotting NOE intensity as a function of molecular weight (a primary determinant of τ_c; see discussion above). Small molecules, such as many drugs, tumble rapidly in solution, resulting in small τ_c values and hence small, *positive*, NOEs. Large macromolecules, such as DNA, tumble slowly. These have larger τ_c values and large, *negative* NOEs. In the middle are "medium-sized" molecules, with molecular weights corresponding to intermediate values of τ_c. Such molecular weights fall near the point where the NOEs change from positive to negative. As a result, medium-sized molecules have NOEs that are small or nearly zero. This can be a problem when trying to determine 3D structures of such molecules. Consequently, experiments were developed that measure the *rotating-frame nuclear Overhauser effect* (ROE), instead of the NOE. ROEs provide the same kind of information as NOEs, but they have the same sign regardless of the molecular weight. Whereas the NOESY experiment measures NOEs, the ROESY measures ROEs. The ROESY experiment can also give information about molecules that are in exchange between different conformations in the sample. The description of the ROESY experiment is more complicated than the NOESY, and it will therefore not be discussed in detail here.

Perhaps the most important aspect of Eq. (1.10) is that the magnitude of NOE (and ROE) enhancement is inversely proportional to the distance between the atoms (r_{ij}), raised to the sixth power: $I_{\text{NOE}} \propto (1/r_{ij})^6$. This allows the calculation of interatomic distances based on NOE intensities. The inverse, sixth-power relationship means that NOE intensity diminishes quickly as the distance between the atoms increases and indicates that less magnetization is transferred at greater separations. Thus, NOEs are quite strong when atoms are <2 Å apart, but they become very small when the atoms are separated by more than ~5 Å. This means that NOEs can be used only to determine short distances.

As with the scalar coupling experiments, NOEs may be measured by either a series of 1D experiments, or by a single 2D experiment. In the standard 1D difference experiment, the magnitude of an NOE is determined one resonance at a time. A steady-state protocol is used in which the sample is selectively irradiated with a radio-frequency that is equal to the energy of the resonance signal of interest. This saturates the irradiated spin—that is, it equalizes the populations in all the possible spin states. This, in turn, perturbs the spin populations of any neighboring nuclei that are dipole-coupled to the irradiated nucleus. A control spectrum is then acquired in which the irradiation is set to a frequency far away from any resonances. Any intensity differences between these two spectra (either positive or negative) may be considered to be the result of an NOE between the two nuclei. Using this information, one can identify the resonances of nuclei that are within 5 Å of each other. As the approach described here uses a steady-state protocol, it is able to give only relative internuclear distances.

The 2D NOESY experiment uses a transient protocol that starts by moving the net magnetization of the nuclei into the $-z$ axis. This is a simultaneous inversion of all the spin-state populations, which allows all possible pairs of nuclei to interact. Then, the buildup of the NOE signals is measured as the nuclei return to the $+z$ axis via *longitudinal relaxation* (see Fig. 1.48 and Appendix A.2.3). This procedure provides accurate internuclear distances [193]. Its spectral resolution in either of its dimensions is not as great as the 1D variant, but this is usually more than compensated by the dispersion into a second direction. The 2D NOESY experiment is generally more visually intuitive to interpret, as it can be regarded as a "contact map" in which only protons that are close to each other show a cross-peak (a spot). Although NOESY data are not acquired as a series of 1D NOE experiments, the analysis can be performed as if the NOEs were measured one resonance at a time.

1.4.4.1. The NOESY–Walk The NOESY experiment provides the basis for the "NOESY–walk" analysis of nucleic acids. The NOESY-walk takes advantage of the fact that in DNA, certain pairs of inter-residue protons are almost always within 5 Å of each other. The most important "walk" is between the aromatic protons of the bases and the H1′ of the sugar moieties. The intra-residue H_{aro}–H1′ distance is typically ∼3.6–3.9 Å (for the *anti* glycosidic dihedral torsion angle), resulting in a medium-intensity cross-peak. The inter-residue NOESY cross-peaks between the aromatic protons on base n, and the H1′ of base $(n-1)$, are usually weaker, being separated by ∼4–4.9 Å (see discussion below and Fig. 1.54c.). Together these peaks normally enable one to draw a connectivity chain from one end of the DNA strand through to the other end. The other sugar protons are also used for NOESY–walk analysis, as are the distinctive thymine methyl protons. Each of these "walks" contributes additional structural details, and provide informational redundancy that can help resolve ambiguous overlaps of NMR resonances in other parts of the spectrum. Often, the most revealing information comes from an unexpected "break" in a walk sequence, or when additional cross-peaks are observed between two adjacent aromatic bases or sugar residues. All of this information is compiled into a large table of short-range (≤ 5 Å) interproton distances, which can then be used with computational structure building programs to develop a high-resolution model.

As an example of a simple NOESY–walk, the H1'-(aromatic base) portion of such a spectrum is shown in Fig. 1.54 [234] for the sequence d(TGGGGT). This DNA hexamer most likely forms a quadruplex with four equivalent, and indistinguishable, strands. In the one-dimensional ^{1}H NMR spectrum (Fig. 1.54a), four downfield H8 resonance lines are observed, corresponding to the four guanosines in each strand. There are also two upfield H6 resonance lines from the two thymidines. The two-dimensional spectrum (Fig. 1.54b) is displayed with the aromatic region plotted

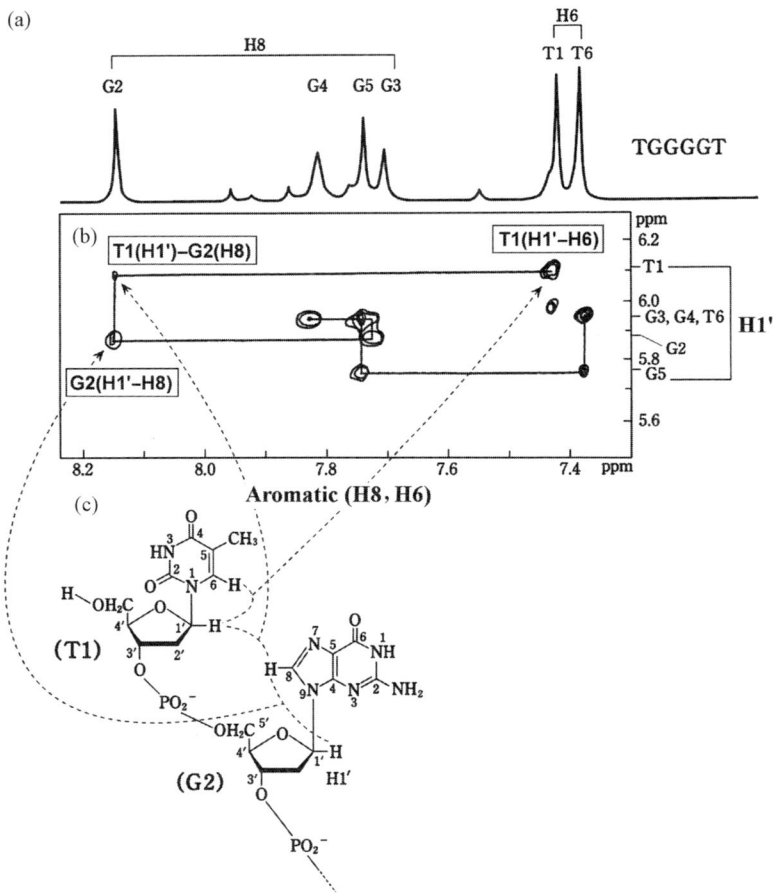

FIGURE 1.54. The one- and two-dimensional proton NMR spectra of the hexamer oligonucleotide, d(TGGGGT) [234]. This sequence mostly likely forms the linear DNA quadruplex, d(TGGGGT)$_4$, consisting of four strands in identical conformations (Section 1.1.12): (a) one-dimensional NMR spectrum in the resonance region of aromatic protons (8.2–7.4 ppm); (b) two-dimensional NOESY spectrum showing a map of the nuclear Overhauser effects between the aromatic protons (G-H8 and T-H6) of the bases and the H1' of the deoxyribose rings; (c) a part of the molecular structure. The protons giving rise to each of the NOESY cross-peaks are indicated by dotted lines, and the associated peaks are shown with arrows.

across the horizontal axis and the H1′ region plotted along the vertical axis. It is clear from Fig. 1.54(b, c) that the *intra*-residue distances between the H1′ of residue T_1 and its own aromatic H6 are very short, as the intensity of the cross-peak is relatively strong. Likewise, the H1′ of *intra*-residue G_2 is near its own aromatic base, H8. What makes this experiment particularly useful, however, is the fact that one also observes an *inter*-residue cross-peak between the H1′ of T_1 and the H8 of G_2. This is not a strong peak, indicating that these two protons are separated by ~4–5 Å. Thus, by using alternating intra- and inter-residue H1′-base cross-peaks, one can "walk" from the first residue of the strand (at the 5′ end), through to the last base (3′ end). This "walk" can be confirmed by the other possible "walks" mentioned earlier. Not only do these NOE cross-peaks allow the assignment of most of the aromatic and sugar resonances, but also, their intensities can be used to determine their approximate interproton distances, which are a measure of the proximity of the bases themselves.

An important indicator of an *intercalating* drug is that it often causes a conspicuous break in the NOESY walk, as the drug causes the DNA base step to expand at the site of the intercalation. If the drug molecule is tightly bound to the DNA, there also will be NOE cross-peaks between the drug and the DNA, especially between the aromatic or exchangeable protons of the base and the drug. Depending on how tight is the binding, it may even be possible to observe two sets of peaks, arising from both bound and unbound molecules. These intermolecular NOEs arise from the saturation transfer of magnetization between the molecules (see discussion below).

If two protons are not near each other, they do not give rise to a cross-peak in a NOESY spectrum. This "negative" information can be important, and it is often used to disprove certain molecular conformations. However, the absence of a peak in an NOESY spectrum is not always conclusive proof, as several factors can give false negatives. If a portion of the molecule is in dynamical motion (with respect to the NMR timescale), or if multiple conformations exist, the resulting peak may be too broad and diffuse to be observed. Likewise, if a drug is in rapid equilibrium (again, with respect to the NMR timescale), then it will also not produce a peak in a standard NOESY spectrum, although other approaches can be used (see text below).

1.4.4.2. Base-Pair Determination and Exchangeable Protons It is often necessary to determine the hydrogen bonding pattern in a DNA or a DNA–drug complex. In most NMR studies of nucleic acids, the solvent chosen is D_2O. This has the advantage of being an invisible solvent in proton NMR experiments, allowing the dense region of sugar protons to be observed near the water peak. However, hydrogen bonds can not be observed in this way because the hydrogen atoms are labile—they are in rapid exchange with the solvent and are quickly replaced with the ^2H (D) atoms that have no signals in the proton spectrum. In order to observe these hydrogen bonds, the sample is dissolved in a solution of 10% D_2O and 90% H_2O. This mixed solvent provides a strong proton signal (^1H), enabling the observation of the hydrogen bonds. All of the standard 2D NMR experiments can then be performed, especially the 2D NOE. This allows one to observe the proximity of an N–H to other protons in nearby bases and sugar groups, and provides important interbase structural information, especially for nonstandard base pairing. Imino protons are usually sharper, more easy to analyze and are more likely to give the desired structural information. Solvent suppression is required for these experiments, as the solvent HOD peak would otherwise overwhelm the digitized signal. The standard presaturation approach for suppression of the solvent signal cannot be used because the rapid exchange between the exchangeable protons and the water peak results in the suppression of both peaks [235].

Only those exchangeable protons that are protected from exchange with the solvent are observed. Protected protons are usually involved in hydrogen bonding, although some exchangeable protons are protected by virtue of being buried within a hydrophobic portion of a molecule, slowing their exchange rate with the solvent. Hydrogen-bonded protons are typically more downfield (have higher ppm values) than expected, as they are deshielded by the electronegative atoms (e.g., N and O) between which they are bound. These heavy atoms enhance the positive charge on the proton by pulling some of its electron density away: $N^{\delta-}-H^{\delta+}\cdots O^{\delta-}$, where δ denotes a partial electron charge.

Several examples of base-pairing schemes are depicted in Fig. 1.55, along with potentially observable interproton distances (represented as arrows in the figure) that are distinctive for that particular base pair. As shown in Fig. 1.55b, a NOESY cross-peak is expected between the N3H of thymine (or uracil) and the C2H of adenine. For the G-C base pair (Fig. 1.55a), on the other hand, an NOE signal is expected between the N1H of guanine and the NH_2 of cytosine, although in practice it is difficult to observe. It is also possible to observe sequential NOE contacts between adjacent DNA base pairs [198,236]. This can be used to confirm sequential assignments, which is especially important in nonstandard DNA structures (see, e.g., Ref. 237). Also, it is quite common to observe intermolecular NOE cross-peaks between these labile DNA–base protons and a drug molecule. These may arise with both *intercalating* drugs and *groove-binding* drugs. In the latter case, the drugs usually form hydrogen bonds with the exposed hydrogen bond acceptors and donors of the base pairs (for examples, see Chapter 3).

Besides the Watson–Crick A-T and G-C base pairs (Fig. 1.10), other special base pairings, called *Hoogsteen* and *reversed Hoogsteen base pairs*, [see the A-U variants in Fig. 1.55, parts (c) and (d), respectively] have been found [238]. The NOE patterns of these base pairs differ from those of the Watson–Crick pairings, as the N3H of uracil has a NOE signal to the C8H of adenine, instead of C2H. A standard Hoogsteen pairing is expected to have a weak (~4.4 Å) NOE between the adenine amino protons (N6H) and the uracil C5H. A reversed Hoogsteen pairing might also have weak NOE contacts between the adenine amino protons and some of the sugar protons of the uracil. Confirmation from other information is also helpful, such as long-range NOE contacts with neighboring bases, or other knowledge of the structure (which could come from NMR data), or the use of scalar coupling experiments (see Section 1.4.3.3). Evidence for a reversed Hoogsteen base pair was found in natural transfer RNA (yeast $tRNA^{Phe}$), where the N3H of a uracil showed an NOE cross-peak with the C8H of an adenine [239]. The presence

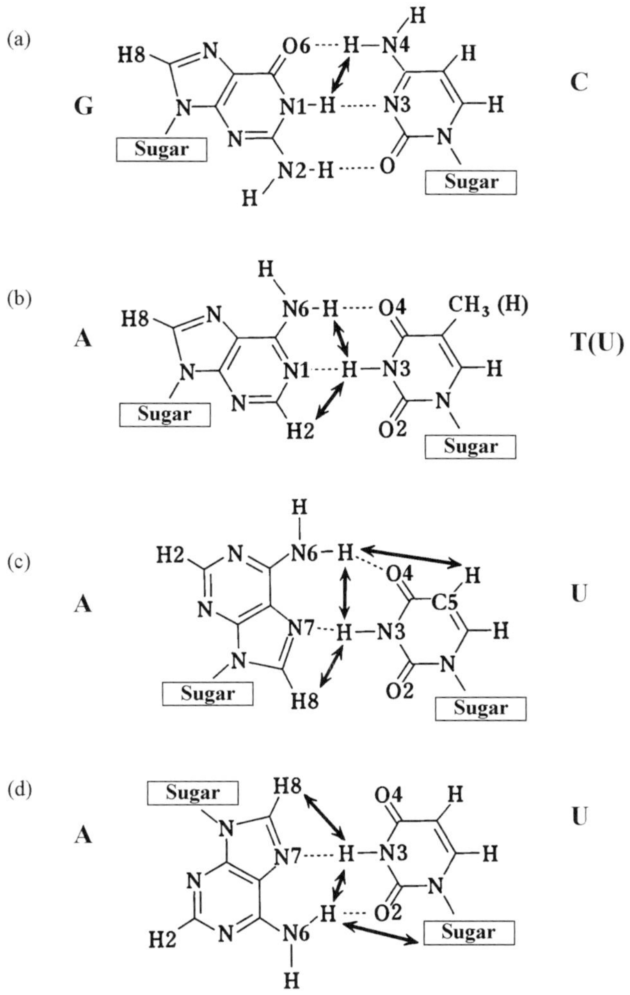

FIGURE 1.55. Examples of common base-pairing schemes, with hydrogen bonding indicated by dotted lines and potential interbase NOEs (dipole couplings) indicated by arrows. The longer arrows [in (c) and (d)] represent NOEs that are generally expected to be weak (4–5 Å), or perhaps not observable. However, the interresidue distances are shown for illustrative purposes only. (a) Watson–Crick-type G-C base pair; (b) Watson–Crick A-T (A-U) base pair; (c) Hoogsteen A-U base pair; (d) reverse Hoogsteen A-U base pair. Many other base-pairing schemes are also possible.

of this reversed Hoogsteen base pair was confirmed by labeling the C8H of adenine with ^{2}H (which is not observed in ^{1}H NMR), resulting in the disappearance of the NOE.

The presence or absence of Hoogsteen and reversed Hoogsteen base pairs may also be determined indirectly. One can sometimes examine the conformations of the glycosidic torsion angles, as the *syn* or *anti* conformations are significantly different (see Section 1.1.8 and Fig. 1.15). If a nucleoside is in a Hoogsteen base pair, its sugar will be in the *syn* conformation, where the interproton distance between C1'H and C8H is very short (2.2 Å), causing a particularly strong NOE cross-peak. Interestingly, this method has been used more commonly to disprove the presence of a Hoogsteen base pair [240].

1.4.4.3. Structure Elucidation from NOE Data
There are several ways to generate interproton distances from NOE data, but it is critical to do it accurately. Since the final molecular structure is constructed from the sum of many short distances, even a small systematic error can become amplified when calculating long distances, resulting in a distorted structure. In addition, the protons in nucleic acids are close together, especially in their sugar rings, resulting in interproton NOE cross-peaks with intensities that can be significantly different from those expected. This happens because the magnetization is transferred through spin diffusion networks to other nearby protons. Nucleic acids analysis therefore requires the careful integration of the intensity of each NOE cross-peak.

One of the more accurate methods for calculating interproton NOE distances in nucleic acids is the *complete relaxation matrix analysis*, which determines interproton distances that are independent of an initially proposed structure [241]. This method does not calculate a simple NOE distance from each separate cross-peak, but instead uses general atomic connectivity information within the molecule to determine and analyze the spin diffusion network. A set of model-independent interproton distances (and error estimates) are then iteratively calculated. For this type of analysis, the NOE signal arising from each proton pair is individually integrated and used as input data. Although preliminary structural information is used in this calculation, the final set of interproton distances is independent of the conformations of that structure.

Another approach is the *distance geometry* (DG) algorithm, in which atom placement is based on the geometric network of competing pairs of distances [242]. This method is quite common in protein structure determination, but its use with nucleic acids requires special attention, as nucleic acids (especially DNA) have fewer long-range interresidue NOE peaks. In addition, DG does not normally consider the effects of spin diffusion on the NOE intensities in nucleic acids. This method can be easier to apply than the complete relaxation method, and is particularly useful for the structure determination of DNA and RNA complexes with proteins, drugs, or other ligands, such as in analysis of the complex formed by d(CGCGAATTCGCG)$_2$ and the ligand, berenil [=1,3-bis(4'-amidinophenyl)triazene] [243].

Regardless of the method by which the set of interproton distances is calculated, molecular modeling is then used to turn that information into an experimentally reasonable structure. The final structures are then verified by comparing them to the experimental data. Details on molecular modeling may be found in Section 1.7.

As it is often difficult to avoid systematic errors when using NOE intensities for model-building purposes, NMR spectroscopists do not rely solely on NOE information, but also use *J*-coupling analyses to calculate pseudorotation and backbone dihedral torsion angles by means of equations similar to the Karplus relationship (see the preceding section). Since NOE information is limited to interproton distances shorter than 5 Å, important longer-range structural information can sometimes be missed. The technique of residual dipolar coupling analysis (Section 1.4.5.) is complementary to NOE analysis as it can determine interproton directionality vectors (Section 1.4.5). It has a distance cutoff of ~8 Å, and hence can provide other essential structural information for DNA.

1.4.4.4. Transfer NOE, Saturation Transfer, and Related Methods The use of NOE data is not restricted to interproton distances within the same molecule, as any proton within 5 Å will be detected, allowing the determination of intermolecular distances between bound ligand molecules (drugs) and DNA. Direct observation of such intermolecular NOEs is usually the first approach attempted, but is often difficult to observe them. It has the disadvantage of working only with tightly bound ligands, for which one often observes two very clearly defined sets of resonances for the bound and unbound states. (The bound ligands in such a system are said to be in "slow exchange" with those that are unbound.) Another disadvantage is that the decreased tumbling rate of macromolecules often leads to broad peaks for both the DNA and the drug (which "goes along for the ride"). To circumvent these problems, several techniques have been designed to utilize the various effects of NOE on a spectrum. These are especially designed for use in drug discovery programs for screening a large pool of ligands to find those that bind to a target receptor molecule, but can be also used to obtain some structural information [244,245].

One important technique is the transfer NOE (trNOE) experiment, which is used to determine whether small ligands can bind to large molecules (>10,000 Da), but it works only for drugs that bind weakly ($k_{dissociation} \sim 10^{-6}$–$10^{-3}$ M). Small, unbound ligand molecules tumble rapidly in solution and have small, positive, intra-ligand NOEs. Large macromolecules, like DNA, tumble slowly and have large, negative NOEs. However, when ligands bind to a large molecule, they acquire the tumbling characteristics (defined by the *correlation time* τ_c; see Appendix A.2.3.1) of that larger receptor molecule, resulting in large, negative NOEs. Thus, bound and unbound ligands can be distinguished by examining the sign and magnitude of the NOE peaks. Typically, the NOEs of an unbound ligand require more than 4 times as much time to reach their maxima, compared to a bound ligand. Hence, by monitoring the signs of *intra-ligand* NOEs, one can determine whether a ligand is bound to a macromolecule. These trNOEs are indicative of the conformation of the bound ligand. The structural conformation of the bound ligand can be determined via experiments such as the 3D TOCSY-trNOE. Intramolecular trNOEs determine the ligand conformation in the bound state, while intermolecular trNOEs can indicate the orientation of the ligands in the binding site [246].

A similar technique is the saturation transfer difference (STD-NOE) experiment. In this method it is not the sign of the intra-ligand NOEs that is monitored, but the transference of magnetization between the molecules. When a macromolecular resonance is selectively irradiated, the spin diffusion networks enable the magnetization to be rapidly dispersed throughout the macromolecule and eventually transferred to any bound ligands (drugs). Ligands that are not tightly bound to the DNA can dissociate while still carrying the transferred NOE magnetization. Since large molecules, and any bound ligands, have short T_2 relaxation rates (see Fig. 1.48 and Section 1A.2.3.2), they typically have broad, poorly resolved peaks. However, small, unbound, molecules have longer T_2 relaxation rates and narrower peaks. Thus, one can use a relaxation-time "filter" to selectively remove signals from the large complex on the basis of their shorter T_2 values. The STD analysis is performed by acquiring two spectra, one in which the macromolecule has been irradiated, and one in which it is irradiated a great distance from any resonances (effectively, it is unirradiated). The two spectra are subtracted from each other, and the enhanced magnetization is then detected through the unbound ligand molecule, which has a much longer relaxation time [247]. In these experiments a greater than 10-fold excess of the drug is used, which ensures that nearly all of the large receptor molecules (DNA) are involved in binding to the ligand. As this dissociation constant can be in the range of nM to mM, STD-NOE can cover a wider range that trNOE. This technique may be combined with any of the other standard 2D NMR experiments.

1.4.5. Residual Dipolar Coupling

Owing to its short-range potential, NOE information is capable of accurately describing local conformational parameters such as sugar pucker and glycosidic torsion angles. However, NOE is less useful for determining the more macroscopic parameters, such as backbone torsion angles and helical parameters, and detecting an overall bend in the molecule. Even with the addition of *J*-coupling constraints, reliable global conformations are difficult to obtain.

As noted in Sections 1.4.1 and A.2.3.1), dipolar relaxation arises from the through-space interaction (coupling) of nuclear magnetic dipoles. It is the major source of longitudinal relaxation (Fig. 1.48) measured by T_1 and is the origin of the NOE. Dipolar couplings, such as those of ^{1}H-^{1}H, ^{1}H-^{13}C, and ^{1}H-^{15}N, are a function of both distance and angular orientation. In (non-spinning) solids the angular orientations are fixed with respect to the direction of the magnetic field $\mathbf{B_0}$, producing very strong dipolar couplings. The resulting relaxation rate is so fast that the NMR spectrum is dominated by very broad and diffuse lines. The effect of this dipolar interaction can extend to many angstroms. In solution, the rapid isotropic tumbling of the molecules causes these dipolar couplings to average out to zero, producing no observable splittings, and the resulting spectrum is comparatively simple, with typically sharp and narrow lines. By adding a liquid crystalline cosolute (see discussion below) it is possible to inhibit this free tumbling and induce a slight orientational preference and partial alignment (∼0.3%) of the molecule. This anisotropic rotational diffusion and partial alignment of the molecule results in *residual* dipolar couplings (RDCs) that are weaker than in solids, but stronger than in liquids. There is a growing list of cosolutes that have been used as alignment media. Some alignment media form lipid bicelles, which are disk-shaped particles that typically have a diameter of several hundred angstroms and a thickness of ∼40 Å. As these lipids are diamagnetic, the bicelles orient with their normal orthogonal to the magnetic field. In addition to bicelles, some other cosolutes suitable for aqueous solutions are filamentous phage and purple membranes. By varying the concentrations of these media, these residual couplings can be tuned to 10–50 Hz. This is optimal for measuring single bond dipolar couplings, especially those of ^{1}H-^{13}C and ^{1}H-^{15}N. These bonds are the most commonly studied as they are easily interpreted. Since the interatomic bond distances are essentially fixed, the strength of the single bond residual dipolar coupling will be dependent only on their angular orientations. This allows the construction of a matrix consisting of the orientations of the bonds (the bond vectors). It has the important advantage of enabling the determination of long-range orientational restraints. These constraints are fundamentally different from those obtained by the local NOE and spin coupling constraints, and provide additional data for the structural analysis. From this, in combination with connectivity information, a high-resolution, stereospecific, 3D structure can be created [249].

As an example of the utility of this approach, the solution structure of Dickerson's dodecamer (Section 1.1.5) d(CGCGAATTCGCG)$_2$ was determined in an aqueous liquid crystalline medium containing 5% w/v (mg/mL) bicelles [249]. This allowed the determination of 198 single-bond dipolar coupling restraints (^{13}C-^{1}H and 10 ^{15}N-^{1}H), which were in addition to the more standard restraints of 200 approximate ^{1}H-^{1}H dipolar coupling restraints and 162 structurally meaningful NOE restraints. The use of more restraints improves the final structure, producing a DNA duplex that is a highly regular B form, with no significant bends or kinks, and having the expected C2$'$-endo/C1$'$-exo sugar pseudorotation angles P. These angles, as well as the glycosyl torsion angles and the base-pair propeller twists, were all tightly determined.

When this same dodecamer was later reexamined, an even larger set of residual dipolar couplings were used [250]. This study used the same heteronuclear ^{13}C-^{1}H and ^{15}N-^{1}H and qualitative homonuclear ^{1}H-^{1}H dipolar couplings that were previously measured in the bicelle medium. Then, an additional 300 quantitative ^{1}H-^{1}H and 22 ^{31}P-^{1}H dipolar restrains, together with 22 ^{31}P chemical shift anisotropy restraints, were obtained in a liquid crystalline Pf1 (bacteriophage) medium. The ^{31}P-^{1}H dipolar couplings, H3$'$-C3$'$-O-^{31}P spin couplings, and ^{31}P chemical shift anisotropy data were used to restrain the phosphodiester backbone dihedral angles. The final structure represents a quite regular B-form helix with a modest bend of $\sim$10°, which is essentially independent of whether electrostatic terms are used in the calculation. It was indicated, however, that the dipolar coupling data cannot be fit by a single structure, but are compatible with the presence of a rapid equilibrium between C2$'$-endo and C3$'$-endo deoxyribose puckers. The C2$'$-H2$'$/H2$''$ dipolar couplings in B-form DNA were particularly sensitive to sugar pucker and yielded the largest discrepancies when fit to a single structure.

1.5. ELECTRON SPIN RESONANCE SPECTRA

Electron spin resonance (ESR), or electron paramagnetic resonance (EPR), has been the subject of many reviews and books, covering both its theory and its application to biomolecular systems [251–253]. It is a technique that relies on the same fundamental equations of physics as NMR, except that the spins of electrons (given by the quantum number, $S = \frac{1}{2}$) are excited, rather than the spins of the nuclei. In both NMR and ESR, the sample is placed in a high magnetic field, although the magnetic field strength needed for ESR is considerably less than that for NMR. Since the resonance frequency for an unpaired electron is ~1000 times greater than that used for nuclei, ESR requires microwave irradiation to alter the spin populations of electrons, whereas NMR uses radiofrequency.

As with NMR, the ESR signal is influenced by factors such as longitudinal (T_1) and transverse (T_2) relaxation. When an unpaired electron interacts with a nucleus having a spin quantum number of I, it produces a "nuclear hyperfine splitting" of the ESR signal with $(2I + 1)$ lines. Hence, the interaction of an electron with an ^{14}N nucleus ($I = 1$) results in a spectrum with three lines. Unlike NMR spectra, ESR spectra are typically displayed as the first derivative of the absorption spectrum. This enables careful measurement of both the hyperfine splitting constants and the lineshape [254].

The spins of electrons can be detected as an ESR signal only for molecules, ions, or atoms that have unpaired electrons—especially free radicals or transition metal complexes. Since the majority of compounds do not have unpaired electrons under normal conditions, they do not produce an ESR signal. As a result, ESR applications have not been as widespread as those of NMR. However, many proteins and some drug-ligand complexes require the presence of paramagnetic metal ions to be biologically active. In such cases, ESR is a particularly effective way to obtain mechanistic and structural information (Section 5.1.2).

It is also common to modify a compound of interest by adding one or more radicals to it. This is called a *spin label*, such as the nitroxyl radical (see Section 6.2.2, for example). The hyperfine splitting of the ESR signal is sensitive to both the orientation and polarity of the environment of the label; hence it can provide important information about the structure and dynamics of the biomolecular system. More recently, it has become possible to spin-label DNA and RNA, which has greatly aided the study of nucleic acids by ESR. These labels must be used with care, however, as their presence may alter the structure of the complex in such a way that it no longer represents the one that is biological relevant.

1.6. X-RAY CRYSTALLOGRAPHY

This technique examines the transmission and scattering (diffraction) of X rays by a crystal and is used to determine the three-dimensional atomic coordinates of a molecular structure. Crystals can be considered to be infinitely repeating array of identical copies of the molecule of interest, and hence they can produce a scattering signal that is strong enough to be detected. This is not true of single molecules, which lack sufficient scattering power. The photons of X-ray radiation do not interact with the atomic nuclei themselves, but instead interact with the electrons surrounding the atoms. When the photons strike the electron cloud, they are absorbed and instanteously re-emited via a Rayleigh scattering process (Section 1.3.2). Diffraction results from the summation of all the scattered waves of a given wavelength. The resulting diffraction pattern is then analyzed using Bragg's law to give the spacing between atoms or ions in a crystal lattice. This enables the generation of a 3D electron density map, indicating the positions of the heavy atoms (hydrogens are not observed). Using computer modeling approaches, the macromolecular structure is fit to the map.

This brief overview is intended only to provide the most essential background information on this important and specialized technique. The reader is refered to many excellent references on X-ray diffraction methods [255,256] and their applications to DNA, nucleic acids, and drug interactions [10,17,49,257,258].

1.6.1. Bragg Diffraction

The wavelengths associated with X-ray electromagnetic radiation are very short and are of the same order of magnitude as the distance between atoms in crystals ($\sim$1.5 Å). In a crystal, the atoms are arranged in planes, with each plane stacked on top of another. Bragg determined that the diffraction of X ray by atoms in a crystal can be treated mathematically as if the crystal were a reflection grating [259]. When a monochromatic beam of X ray strikes the stacked planes of a crystal, each plane reflects some of the radiation, while the rest is transmitted through to the next plane, which will also reflect and transmit some of the rays (Fig. 1.56a). The waves of these reflections can be either in-phase with each other, resulting in constructive interference (additive enhancement of the signal), or out-of-phase, resulting in destructive interference (cancellation of the signal). The Bragg equation describes the necessary conditions for constructive interference by relating the wavelength of the X ray (λ) to the distance between the atomic planes in the crystal (d; the lattice spacing), and the angle between the incident beam and the scattering planes (θ) called the "grazing" or "glancing" angle. The one dimensional form of the Bragg equation is given by

$$n\lambda = 2d \sin \theta \qquad (1.11)$$

where n is an integer representing the number of the plane from which the X ray is reflected. The rays striking the lower planes must necessarily travel a greater distance than those striking the top most plane (Fig. 1.56a). When the difference in the distance travelled is equal to an integral number of X-ray wavelengths ($n\lambda$), the emerging beams will be in phase and will reinforce each other. A reflection is called "first-order" when $n = 1$; when $n = 2$, it is called "second-order," and so on. If the X ray strike the crystal at angles other than the grazing angle θ, then the extra distance will not be an integral number of wavelengths, resulting in destructive interference, not reinforcement. In this case, the intensity of the reflected X ray is weakened. Thus, constructive interference occurs only at certain geometries, and when it does, it produces a set of spots (reflections) in the diffraction pattern. Bragg's equation allows the determination of the distance d between the planes of the atoms, if the wavelength of the irradiation is known. The crystal is slowly rotated through a range of θ in order to obtain

98 DNA STRUCTURES AND SPECTRA

(a)

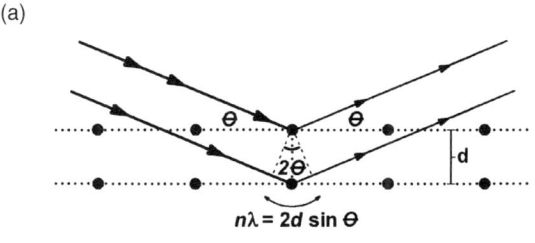

$n\lambda = 2d \sin \Theta$

(b)

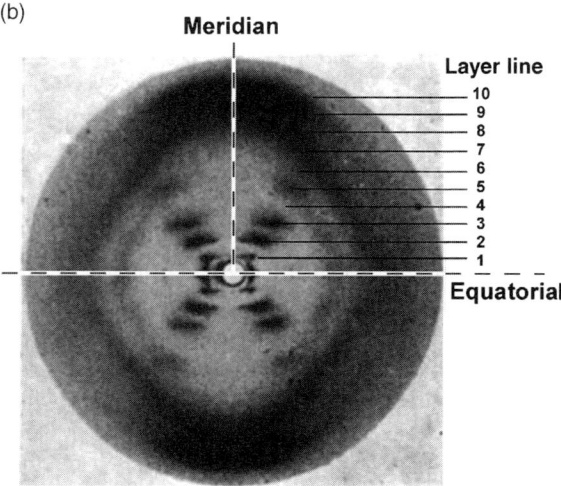

FIGURE 1.56. (a) The scattering of monochromatic light from an array of atoms in a crystal, as described by Bragg's equation. The angles and distances displayed were arbitrarily chosen for illustrative purposes only. (b) Rosalind Franklin's famous photograph of the X-ray fiber diffraction of B-DNA [3], annotated to indicate the layer lines and necessary salient features. The diffraction pattern is a reciprocal lattice image of the DNA fiber, with the distances inversely proportional to those in the fiber.

reflections from as many Bragg planes as possible, and hence increase the number of observed interference spots.

From this information, an electron density map can be generated. The resolution of this map is determined by the quality of the crystal, the X-ray wavelengths used, and the scattering angles observed. The resolution is usually expressed in terms of the interplanar spacings d. At a resolution of 2.5 Å the DNA bases begin to be discernable, at 1.5 Å they are clearly seen, and by 0.8 Å the electron densities of individual atoms are well defined. The highest resolution is usually possible only in crystals of small molecules.

1.6.2. Diffraction of Helical DNA

In order to obtain an accurate three-dimensional model of a macromolecule, it is necessary to obtain atomic resolution diffraction images. Unfortunately, these are often difficult to obtain because of imperfections in their crystal structure, such as those that arise from impurities, multiple stable conformations, or an inherent assymmetry of the crystal. This is particularly true of polymeric, double-helix DNA, which is more difficult to crystallize than more globular biomolecules, such as RNA, or DNA bound to proteins. (For comparision, consider the great strides that have been made in the X-ray crystal analysis of the ribosome [260–262].) Instead, polymeric DNA has been generally studied as fibers which are drawn from viscous gels, and is referred to as *fiber diffraction*.

A DNA fiber is a continuous helix that can be considered as kind of a one-dimensional crystal. Figure 1.56b shows the now-famous fiber diffraction photograph of B-DNA acquired by Franklin [3]. The difraction from a helix is scattered perpendicular to the helical axis in a series of planes that appear as lines (layer lines). Bragg spots are not seen because the sample consists of a bundle of helices that are randomly oriented around the helical axis. There is a reciprocal relationship between the distances on the photograph and the distances in real space. The positions of the first and strongest peaks in each layer line describe an X pattern, called the *helical cross*. The angle between the arms of the X is inversely related to the helical radius (~ 10 Å), and the separation of the layer lines is the reciprocal of the pitch of the helix (P), indicating that a complete helical turn is ~ 34 Å. The symmetric placement of the spots along the layer lines demonstrates that the helix has a regular twist. The missing spot at the fourth layer line arises because it is here that the two strands of the double helix cross and cancel each other out. The intense reflection on the meridian at the 10th layer line occurs at a distance of ~ 3.4 Å, indicating that the helix contains ~ 10 nucleotides per turn, and that the bases are stacked perpendicular to the helical axis with a separation of ~ 3.4 Å. The intensity of this reflection results from the fact that the sample was composed of many DNA strands with imperfect alignment and signficant rotational disorder. When the fibers are tilted, the reflection from the base pairs is increased, thereby producing the strong meridian reflection on the 10th layer line.

In order to obtain a high-resolution X-ray structure, significant data analysis is required. The intensity of each diffraction spot is proportional to the square of the *structure factor* amplitude, which is a complex number containing information related to both the amplitude and phase of the reflected wave. Once this "phase problem" is solved, an electron density map may be constructed, and from this a molecular model may be built.

1.7. MOLECULAR MODELING AND MOLECULAR MECHANICS

1.7.1. Molecular Modeling

Molecular modeling is a collective term referring to many different theoretical and computational approaches used to calculate molecular structure, molecular dynamics, and molecular properties such as reaction intermediates and thermodynamic analyses. Molecular modeling has become ubiquitous in biomacromolecular structure determination as there are very few ways to coalesce a large amount of structural data into a 3D structure. It can be a purely theoretical pursuit, but it is also extensively used as a tool in both NMR and X-ray crystallography, where the results are driven by the experimental data. X-ray structure determination is generally more straightforward than it is for NMR because of the intrinsic natures of their experimental data. For this reason, much of what follows pertains to methods that are more often used for NMR.

All the methods used in molecular modeling rely on a set of equations and a large set of parameters, which together compose a *semiempirical forcefield*. The equations attempt to describe mathematically the general motion and interaction energies of atoms. The parameters define the specific characteristics of the atoms and their bonds, and their values are derived from a combination of the results of quantum mechanical (QM) calculations and experimental studies. Much of the underlying experimental and QM data on which these parameters are based are determined from gas-phase or *in vacuo* studies of much smaller, more simple molecules. A brief description of the basic theories of molecular modeling and mechanics can be found in Appendix A.3. For detailed descriptions of the methods and fundamentals of molecular modeling, the interested reader should consult Refs. 263–267.

1.7.2. Modeling of Experimental Data

1.7.2.1. Distance Geometry   Unlike most modeling techniques, the distance geometry approach, which is often used in NMR structure refinement (Section 1.4.4), does not use energy functions. Instead, it describes atoms as rigid-sphere potentials (i.e., "solid balls") and determines their relative three-dimensional positions in a molecule by using connectivity information (bonds) combined with NOE interproton distances. There are also many programs used for crystallographic structure refinement that do not include energy functions. This makes model building more simple, and much faster. Since they are guided by experimental information, the resulting structures are usually very nearly the same as those that have been derived using a combination of energy functions and experimental information. This is not always true, however, and the comparisons can be enlightening.

1.7.2.2. Molecular Dynamics   Molecular modeling can also be used to simulate the movement of a molecule by a process called *molecular dynamics* (MD). This technique is often used to model the movement and flexibility of a molecule. It is also frequently used in NMR structure determination as a means of molding an initial structure into one that is experimentally realistic. This is done by subjecting the initial molecular structure to a process called *simulated annealing*, which is very similar to DNA melting curve analysis (Section 1.2.1). The kinetic energy of the atoms in the model structure is increased, such as by raising the temperature (often to 400–600 K), and then, after a short time, it is slowly lowered back to 275 K, where it is allowed to equilibrate for further period of time. This kinetic energy increase allows the molecule to explore more conformations by allowing bond vibrations and rotations to move more freely. As it "cools back down," the potential energy surface of the molecule guides it toward a more realistic structure with a relatively lower total energy. The structure is usually then subjected to an *energy minimization* procedure to remove any structural anomalies and produce a lowest-energy structure. During this process an additional component to the potential energy may be imposed on the molecule to represent experimental information. The experimental information usually takes the form of distance or dihedral angle constraints, and functions like tethers or springs that pull or twist the atoms and bonds toward a better fit with the experimental data. These are often derived from NMR NOE and *J*-coupling analyses. When used in this way, this procedure is called *restrained molecular dynamics* (rMD).

Unlike energy minimization, rMD does not result in only one lowest-energy structure, but rather in a family of related structures that together fit the data better than does any single structure alone. This family is usually represented by those structures that are most similar. This is determined by calculating the RMSD (root mean square deviation) between the models. Low RMSD values indicate that the structure calculations have nearly converged to a single structure. The best structures have RMSD values of about <1 Å for the heavy atoms (i.e., excluding hydrogen). If there are large motions in the molecule, such as in loop regions (Section 1.1.11), or if several different conformations exist in solution, the overall RMSD values will be larger. When possible, these are subdivided and analyzed separately. Such situations are often better handled by imposing the experimental NMR information in a time-averaged manner over the course of the simulation (called "MD-tar") [268], or by performing an *ensemble-average* MD simulation (using multiple copies of the molecule) [269–275] and calculating an averaged RMSD. An alternative approach, which can be combined with the previous methods, is to rigorously analyze the simulations using statistical methods to identify families of structurally related conformers and calculate their relative populations according to the percentages needed to best fit the experimental data [216,275]. This has the advantage of reducing the number of representative structures to an experimentally relevant minimum, and provides a more easily visualized picture of the equilibrium mixture.

NMR is usually used to study macromolecules in solution, which often contain localized regions that undergo greater dynamical motion. When a portion of a molecule is moving, well-defined interatomic distances cannot (and perhaps should not) be imposed on that part of the final structure, resulting in regions that have less stringent structural constraints. This is not the case with X-ray crystallography, where molecular motions are rarely substantial. Such internal motions result in a family of multiple structures that are usually overlaid and displayed in a manner that reveals those regions having greater and lesser structural information. When MD and NMR relaxation experiments are used in combination, one can often obtain a clear picture of the true range of motions existing in solution. This is important, as it is often correlated to changes in molecular conformation upon drug binding.

REFERENCES

1. Watson, J. D. and Crick, F. H. C., "Molecular structure for nucleic acids: A structure for deoxyribose nucleic acid," *Nature* **171**:737–738 (1953).
2. Wilkins, M. H. F., Stokes, A. R., and Wilson, H. R., "Structure of nucleic acids: Molecular structure of deoxypentose nucleic acids," *Nature* **171**:738–740 (1953).
3. Franklin, R. E. and Gosling, R., "Molecular configuration in sodium thymonucleate," *Nature* **171**:740–741 (1953).
4. Rigandy, J. and Klesney, S. P. eds., *International Union of Pure and Applied Chemistry. Nomenclature of Organic Chemistry*, Pergamon Press, Oxford, 1979.
5. Taylor, R. and Kennard, O., "The molecular structures of nucleosides and nucleotides. 1. The influence of protonation on the geometries of nucleic acid constituents," *J. Mol. Struct.* **78**:1–28 (1982).
6. Geise, H. J., Adams, W. J., and Bartell, L. S., "Electron diffraction study of gaseous tetrahydrofurane," *Tetrahedron* **25**:3045–3052 (1969).
7. Almenningen, A., Seip, H. M., and Willardsen, T., "Studies on molecules with five-membered rings. II. An electron diffraction investigation of gaseous tetrahydrofurane," *Acta Chem. Scand.* **23**:2748–2754 (1969).
8. Engelholm, G. G., Gwinn, W. D., and Harris, D. O., "Ring puckering in five-membered rings," *J. Chem. Phys.* **50**:2446–2457 (1969).
9. Cremer, D. and Pople, J. A., "Molecular orbital theory of the electronic structure of organic compounds XXIII. Pseudorotation in saturated five-membered ring compounds," *J. Am. Chem. Soc.* **97**:1358–1367 (1975).
10. Saenger, W., *Principles of Nucleic Acid Structure*, Springer-Verlag, Heidelberg, 1984.
11. Langridge, R., Wilson, H. R., Hooper, C. W., Wilkins, M. H. F., and Hamilton, L. D., "The molecular configuration of deoxyribonucleic acid I. X-ray diffraction study of a crystalline form of the lithium salt," *J. Mol. Biol.* **2**:19–37 (1960).
12. Langridge, R., Marvin, D. A., Seeds, W. E., Wilson, H. R., Hooper, C. W., Wilkins, M. H. F., and Hamilton, L. D., "The molecular configuration of deoxyribonucleic acid. II. Molecular models and their fourier transforms," *J. Mol. Biol.* **2**:38–64 (1960).
13. Fuller, W., Wilkins, M. H. F., Wilson, H. R., and Hamilton, L. D., "The molecular configuration of deoxyribonucleic acid. IV. X-ray diffraction study of the A form," *J. Mol. Biol.* **12**:60–80 (1965).
14. Arnott, S., Dover, S. D., and Wonacott, A. J., "Least-squares refinement of the crystal and molecular structures of DNA and RNA from X-ray data and standard bond lengths and angles," *Acta. Cryst.*, **B25**:2192–2206 (1969).
15. Arnott, S. and Hukins, D. W., "Optimized parameters for A DNA and B-DNA," *Biochem. Biophys. Res. Commun.* **47**:1504–1510 (1972).
16. Arnott, S., Chandrasekaran, R., Puigjamer, L. C., and Walker, J. K., in eds., Pullman, B. and Jortner, J., Reiter, D., *Nucleic Acids: The Vector of Life*, 1983, pp. 17–31.
17. Dickerson, R. E. and Drew, H. R., "Structure of a B-DNA dodecamer. II. Influence of base sequence on helix structure," *J. Mol. Biol.* **149**:761–786 (1981).
18. Shui, X., McFail-Isom, L., Hu, G. G., and Williams, L. D., "The B-DNA dodecamer at high resolution reveals a spine of water on sodium," *Biochemistry* **37**:8341–8355 (1998).

19. Tereshko, V., Minasov, G., and Egli, M., "The Dickerson-Drew B-DNA dodecamer revised at atomic resolution," *J. Am. Chem. Soc.* **121**:470–471 (1999).
20. Nelson, H. C. M., Finch, J. T., Luisi, F., and Klug, A., "The structure of an oligo(dA)·oligo (dT) tract and its biological implications," *Nature* **330**:221–226 (1987).
21. Wing, R. M., Drew, H. R., Takano, T., Broke, C., Tanaka, S., Itakura, K., and Dickerson, R. E., "Crystal structure analysis of a complete turn of B-DNA," *Nature* **287**:755–758 (1980).
22. DiGabriele, A. D., Sanderson, M. R., and Steitz, T. A., "Crystal lattice packing isimportant in determining the bend of a DNA dodecamer containing a adenine tract," *Proc. Natl. Acad. Sci. USA* **86**:1816–1820 (1989).
23. Yoon, C., Prive, G. G., Goodsell, D. S., and Dickerson, R. E., " Structure of an alternating B-DNA helix and its relationship to A-tract DNA," *Proc. Natl. Acad. Sci. USA* **85**:6332–6336 (1988).
24. Larsen, T. A., Kopka, M. L., and Dickerson, R. E., "Crystal Structure Analysis of the B-DNA Dodecamer CGTGAATTCACG," *Biochemistry* **30**:4443–4449 (1991).
25. Narayana, N., Ginell, S. L., Russu, I. M., and Berman, H. M., "Crystal and molecular structure of a DNA fragment: d(CGTGAATTCACG)," *Biochemistry* **30**:4449–4455 (1991).
26. Grzeskowiak, K., Yanagi, K., Prive, G. G., and Dickerson, R. E., "The structure of B-helical C-G-A-T-C-G-A-T-C-G and comparison with C-C-A-A-C-G-T-T-G-G," *J. Biol. Chem.* **266**:8861–8889 (1991).
27. Prive, G. G., Yanagi, K., and Dickerson, R. E., "Structure of the B-DNA decamer C-C-A-A-C-G-T-T-G-G and comparison with isomorphous decamers," *J. Mol. Biol.* **217**:177–199 (1991).
28. Heinemann, U., and Alings, C., "Crystallographic study of one twin of G/C rich B-DNA," *J. Mol. Biol.* **210**:369–381 (1989).
29. Timsit, Y., Westhof, E., Fuchs, R. P. P., and Moras, D., "Unusual helical packing in crystals of DNA bearing a mutation hot spot," *Nature* **341**:459–462 (1989).
30. Kennard, O. and Hunter, W. N., "Oligonucleotide structure: A decade of results from single crystal X-ray diffraction studies," *Quart. Rev. Biophys.* **22**:327–379 (1989).
31. Franklin, R. E. and Gosling, R. G., "The structure of sodium thymonucleate fibers. I. The influence of water content," *Acta. Cryst.* **6**:673–677 (1953).
32. Drew, H. R. and Dickerson, R. E., "Structure of B-DNA dodecamer. III. Geometry of hydration," *J. Mol. Biol.* **151**:535–556 (1981).
33. Kopka, M. L., Yoon, C., Goodsell, D., Pjura, P., and Dickerson, R. E., "The molecular origin of DNA-drug specificity in netropsin and distamycin," *Proc. Natl. Acad. Sci. USA* **82**:1376–1380 (1985).
34. Arnott, S., "Polynucleotide secondary structures: an historical perspective," in Neidle, S., ed., *Oxford Handbook of Nucleic Acid Structure*, Oxford Univ. Press, 1999, p. 7.
35. Haran, T. E., Shakked, Z., Wang, A. H.-J., and Rich, A., "The crystal structure of d(CCCCGGGG): A new A-form variant with an extended backbone conformation," *J. Biomol. Struct. Dyn.* **5**:199–217 (1987).
36. Heinemann, U., Lauble, R., Frank, R., and Bloekel, H., "Crystal structure analysis of an A-DNA fragment at 1.8 Å resolution: d(GCCCGGGC)," *Nucl. Acids Res.* **15**:9531–9550 (1987).

37. Robinovich, D., Harqawn, T., Eisenstein, M., and Shakked, Z., "Structures of the mismatched duplex d(GGGTGCCC) and one of its Watson-Crick Analo[gs], d(GGGCGCCC)," *J. Mol. Biol.* **200**:151–161 (1988).
38. Hunter, W. N., D'Estaintot, B. L., and Kennard, O., "Structural variation in d(CTCTAGAG). Implications for protein-DNA interactions," *Biochemistry* **28**:2444–2451 (1989).
39. Courseille, C., Dautant, A., Hospital, M., D'Estaintot, B. L., Precigoux, G., Molko, D., and Teoule, R., "Crystal structure analysis of an A(DNA) octamer," *Acta Cryst.*, **A46**: FC9–FC12 (1990).
40. Clark, G. R., Brown, D. G., Sanderson, M. R., Chwalinski, T., Meidle, S., Veal, J. M., Jones, R. L., Wilson, W. D., Zon, G., Garman, E., and Stuart, D. I., "Crystal and solution structures of the oligonucleotide d(ATGCGCAT)$_2$: A combined X-ray and NMR study," *Nucl. Acids Res.* **18**:5521–5528 (1990).
41. Lauble, H., Frank, R., Bloeker, H., and Heinemann, U., "Three-dimensional structure of d(GGGATCCC) in the crystalline state," *Nucl. Acids Res.*, **16**:7799–7816 (1988).
42. Eisenstein, M., Frolow, F., Shakked, Z., and Rabinovich, D., "The structure and hydration of the A-DNA fragment d(GGGTACCC) at room temperature and low temperature," *Nucl. Acids Res.* **18**:3185–3144 (1990).
43. McCall, M., Brown, T., Hunter, W. N., and Kennard, O., "The crystal structure of d(G-G-G-G-C-C-C-C): A model for poly(dG)·poly(dC)," *J. Mol. Biol.* **183**:385–395 (1985).
44. McCall, M., Brown, T., Hunter, W. N., and Kennard, O., "The crystal structure of d(GGATGGGAG) forms an essential part of the binding site for transcription factor IIIA," *Nature* **322**:661–664 (1986).
45. Jain, S., and Sundaralingam, M., "Effect of crystal packing environment on conformationof DNA duplex: Molecular structure of the A-DNA octamer d(G-T-G-T-A-C-A-C) in two crystalline forms," *J. Biol. Chem.* **264**:12780–12784 (1989).
46. Frederick, C. A., Quigley, G. J., Teng, M.-K., Coll, M., van der Marel, G. A., van Boom, J. H., Rich, A., and Wang, A. H. J., "Molecular structure of an A-DNA decamer d(ACCGGCCGGT)," *Eur. J. Biochem.* **181**:295–307 (1989).
47. IUPAC-IUB Joint Commission on Biochemical Nomenclature (JCBN), "Abbreviations and symbols for the description of conformations of polynucleotide chains," *Eur. J. Biochem.* **131**:9–15 (1983).
48. Tsuboi, M., Hirakawa, A. Y., and Nakagawa, M., "A Few comments on the internal rotations in DNA," *J. Mol. Struct.* **126**:525–528 (1985).
49. Dickerson, R. E., Drew, H. R., Conner, B. N., Wing, R. M., Fratini, A. V. and Kopka, M. L., "The anatomy of A-, B-, and Z-DNA," *Science* **216**:475–485 (1982).
50. Schneider, B., Neidle, S., and Berman, H. M., "Conformations of the sugar-phosphate backbone in helical DNA crystal structures," *Biopolymers* **42**:113–124 (1997).
51. Wang, A. H.-J., Quigley, G. J., Kolpak, F. J., Crawford, J. L., van Boom, J. H., van der Marel, G., and Rich, A., "Molecular structure of a left-handed double helical DNA fragment at atomic resolution," *Nature* **282**:680–686 (1979).
52. Gessner, R. V., Frederick, G. A., Quigley, G. J., Rich, A., and Wang, A. H.-J., "The molecular structure of left-handed Z-DNA double helix at 1.0 A atomic resolution," *J. Biol. Chem.* **264**:7921–7935 (1989).

53. Drew, H. R., Takano, T., Tanaka, S., Itakura, K., and Dickerson, R. E., "High-salt d(CpGpCpG), a left-handed Z' DNA double helix," *Nature* **286**:567–573 (1980).
54. Fujii, S., Wang, A. H.-J., Quigley, G. J., Westerink, H., van der Marel, G., van Boom, J. H., and Rich, A., "The octamers d(CGCGCGCG) and d(CGTATGCG) both crystallize as Z-DNA in the same hexagonal lattice," *Biopolymers* **24**:243–250 (1985).
55. Brennan, R. G., and Sundaralingam, M., "Crystallization and preliminary crystallographic studies of the decadeoxyoligonucleotide d(CpGpTpApCpGpTpApCpG)," *J. Mol. Biol.* **181**:561–563 (1985).
56. Leng, M., in Helene, C., ed., *Structure, Dynamics, Interactions and Evolution of Biological Macromolecules*, Reidel, Dordrecht, 1983, pp. 45–53.
57. Jovin, T. M., McIntosh, L. P., Zarling, D. A., Arndt-Jovin, D. J., Robert-Nicoud, M., and van de Sande, J. H., "Probing for and with left-handed DNA: Poly[d(A-br5C) d(G-T)], a member of a new family of Z-forming DNAs," in Pullman, B. and Tortner., J., eds., *Nucleic Acids: Vector of Life*, Reidel, Dordrecht, 1983, pp. 89–99.
58. Ha, S. C., Lowenhaupt, K., Rich, A., Kim, Y.-G., and Kim, K. K., "Crystal structure of a junction between B-DNA and Z-DNA reveals two extruded bases," *Nature* **437**:1183–1186 (2005).
59. Crick, F. H. C., "Linking numbers and nucleosomes," *Proc. Natl. Acad. Sci. USA* **73**:2639–2643 (1976).
60. Fisher, L. M., Kuroda, R., and Sakai, T. T., "Interaction of bleomycin A_2 with deoxyribonucleic acid: DNA unwinding and inhibition of bleomycin-induced DNA breakage by cationic thiazole amides related to bleomycin A2," *Biochemistry* **24**:3199–3207 (1985).
61. Utsuno, K., and Tsuboi, M., "Degree of DNA unwinding caused by the binding of aclacinomycin A," *Chem. Pharm. Bull.* **45**:1551–1557 (1997).
62. Hunter, W. N., Brown, T., Anand, N. N., and Kennard, O., "Structure of an adenine-cytosine base pair in DNA and its implication for mismatch repair," *Nature* **320**:552–555 (1986).
63. Patel, D. L., Koslowski, S. A., Ikuta, S., and Itakura, K., "Dynamics of DNA duplexes containing internal G:T, G:A, A:C and T:C pairs: Hydrogen exchange at and adjacent to mismatch sites," *Fed. Proc.* **43**:2665–2670 (1984).
64. Leemor, J.-T., Rabinovich, D., Hope, H., Frolow, F., Appella, E., and Sussman, J. L., "The three dimensional structure of a DNA duplex containing looped-out bases," *Nature* **334**:82–85 (1988).
65. Chattopadhyaya, R., Ikuta, S., Grezeskowiak, K., and Dickerson, R. E., "X-ray structure a DNA hairpin molecule," *Nature* **334**:175–179 (1988).
66. Hare, D. R., and Reid, B. R., "Three-dimensional structure of a DNA hairpin in solution: Two-dimensional NMR studies and distant geometry calculations on d(CGCGTTTTCGCG)," *Biochemistry* **25**:5341–5350 (1986).
67. Sinden, R. R., *DNA Structure and Function*, Academic Press, 1994, p. 135.
68. Felsenfeld, G., and Rich, A., "Studies of the formation of two and three-stranded polyribonucleotides," *Biochim. Biophys. Acta* **26**:457–468 (1957).

69. Mirkin, S. M., Lyamichev, V. I., Drushlyak, K. N., Dobtynin, V. N., Filippov, S. A., and Frank-Kameneskii, M. D., "DNA H form requires a homopurine-homopyrimidine mirror repeat," *Nature* **330**:495–497 (1987).
70. Hanvey, J. C., Shimizu, M., and Wells, R. D., "Intramolecular DNA triplexes in supercoiled plasmids," *Proc. Natl. Acad. Sci. USA* **85**:6292–6296 (1988).
71. Duval-Valentin, G., de Bizemont, T., Takasugi, M., Mergny, J. L., Bisagni, E., and Helene, C., "Triple-helix specific ligands stabilize H-DNA," *J. Mol. Biol.* **247**:847–858 (1995).
72. Sen, D. and Gilbert, W., "Formation of parallel 4-stranded complexes by guanine-rich motifs in DNA and its implications for meiosis," *Nature* **334**:364–366 (1988).
73. Cech, T. R., "DNA biochemistry—G-strings at chromosome ends," *Nature* **322**:777–778 (1988).
74. Sundquist, W. I., and Klug, A., "Telomeric DNA dimerizes by formation of guanine tetrads between hairpin loops," *Nature* **342**:825–829 (1989).
75. Williamson, J. R., Raghuraman, M. K., and Cech, T. R., "Mono-valent cation induced structure of telomeric DNA—the G-quartet model," *Cell* **59**:871–880 (1989).
76. Blackburn, E. H., "Structure and function of telomeres," *Nature* **350**:569 (1991).
77. Murchie, A. I. H. and Lilley, D. M. J., "Retinoblastoma susceptibility genes contain 5' sequences with a high propensity to form guanine-tetrad structures," *Nucl. Acids Res.* **20**:49–53 (1992).
78. Sundquist, W. I. and Heaphy, S., "Evidence for intrastrand quadruplex formation in the dimerization of human immunodeficiency virus-1 genomic RNA," *Proc. Natl. Acad. Sci. USA* **90**:3393–3397 (1990).
79. Hammond-Kosack, M. C., Kilpatrick, M. W., and Docherty, K., "Analysis of DNA-structure in the human insulin gene-linked polymorphic region *in vivo*," *J. Mol. Endocrinol.* **9**:221–225 (1992).
80. Fry, M. and Loeb, L. A., "The fragile-X syndrome D(Cgg)(N) nucleotide repeats form a stable tetrahelical structure," *Proc. Natl. Acad. Sci. USA* **91**:4950–4954 (1994).
81. Kettani, A., Kumar, R. A., and Patel, D. J., "Solution structure of a DNA quadruplex containing the fragile-X syndrome triplet repeat," *J. Mol. Biol.* **254**:638–656 (1995).
82. Darnell, J. C., Jensen, K. B., Jin, P., Brown, V., Warren, S. T., and Darnell, R. B., "Fragile X mental retardation protein targets G quartet mRNAs important for neuronal function," *Cell* **107**:489–499 (2001).
83. Schaeffer, C., Bardoni, B., Mandel, J. L., Ehresmann, B., Ehresmann, C., and Moine, H., The fragile X mental retardation protein binds specifically to its mRNA *via* a purine quartet motif," *EMBO J.* **20**:4803–4813 (2001).
84. Brown, B. A., Li, Y. Q., Brown, J. C., Hardin, C. C., Roberts, J. F., Pelsue, S. C., and Shultz, L. D., "Isolation and characterization of a monoclonal anti-quadruplex DNA antibody from autoimmune 'viable motheaten' mice," *Biochemistry* **37**:16325–16337 (1998).
85. Pearson, A. M., Rich, A., and Krieger, M., "Polynucleotide binding to macrophage scavenger receptors depends on the formation of base-quartet-stabilized 4-stranded helices," *J. Biol. Chem.* **268**:3546–3554 (1993).
86. Walsh, K. and Gualberto, A., "Myod binds to the guanine tetrad nucleic-acid structure," *Biol. Chem.* **267**:13714–13718 (1992).

87. Fang, G. and Cech, T. R., "Characterization of a G-quartet formation reaction promoted by the beta-subunit of the oxytricha telomere-binding protein," *Biochemistry* **32**:11646–11657 (1993).
88. Wang, K. Y., Krawczyk, S. H., Bischoftberger, N., Swaminathan, S., and Bolton, P. H., "The tertiary structure of a DNA aptamer which binds to and inhibits thrombin determines activity," *Biochemistry* **32**:11285–11292 (1993).
89. Girald, R. and Rhodes, D., "The yeast telomere-binding protein Rap1 binds to and promotes the formation of DNA quadruplexes in telomeric DNA," *EMBO J.* **13**:2411–2420 (1994).
90. Marchand, C., Pourquier, P., Laco, G. S., Jing, N. J., and Pommier, Y., "Interaction of human nuclear topoisomerase I with guanosine quartet-forming and guanosine-rich single-stranded DNA and RNA oligonucleotides," *J. Biol. Chem.* **277**:8906–8911 (2002).
91. Hurley, L. H., "Secondary DNA structures as molecular targets for cancer therapeutics," *Biochem. Soc. Trans.* **29**:692–696 (2001).
92. Gellert, M., Lipsett, M. S., and Davies, D. R., "Helix formation by gyanylic acid," *Proc. Natl. Acad. Sci. USA* **48**:2013–2018 (1962).
93. Arnott, S. and Selsing, E., "Structure of polydeoxyguanylic acic and polydeoxycytidulic acid," *J. Mol. Biol.* **108**:551–552 (1974).
94. Thiele, D. and Guschlbauer, W., "Protonated polynucleotide structures. IX. Disproportionation of poly(G)·poly (C) in acid medium," *Biopolymers* **10**:143–157 (1971).
95. Howard, F. B., Frazier, J., and Miles, H. T., "Stable and metastable forms of poly(G).," *Biopolymers* **16**:791–809 (1977).
96. Henderson, E., Hardin, C. C., Walk, S. K., Tinoco, Jr., K., and Blackburn, E. H., "Telomeric DNA oligonucleotides form novel intramolecular structures containing guanine guanine base-pairs," *Cell* **51**:899–908 (1987).
97. Wang, Y. and Patel, D. J., "Solution structure of the oxytricha telomeric repeat d[$(G_4T_4G_4)_3$] G-tetraplex," *J. Mol. Biol.* **251**:76–94 (1995).
98. Aboul-ela, F., Murchie, A. I., and Lilley, D. M., "NMR-study of parallel-stranded tetraplex formation by the hexadeoxynucleotide d(TG_4T)," *Nature* **360**:280–282(1992).
99. Aboul-ela, F., Murchie, A. I., Norman, D. G., and Lilley, D. M., "Solution structure of a parallel-stranded tetraplex formed by d(TG_4T) in the presence of sodium-ions by nuclear-magnetic-resonance," *J. Mol. Biol.* **243**:458–471 (1994).
100. Wang, Y. and Patel, D. J., "Guanine residues in d(T_2AG_3) and d(T_2G_4) form parallel-stranded potassium cation stabilized G-quadruplexes with antiglycosidic torsion angles in solution," *Biochemistry* **31**:8112–8119 (1992).
101. Wang, Y. and Patel, D. J., "Solution structure of the human telomeric repeat d[1.1.12: $AG_3(T_2AG_3)_3$] G-tetraplex," *Structure* **1**:263–282 (1993).
102. Smith, F. W. and Feigon, J., "Quadruplex structure of oxytricha telomeric DNA oligonucleotides," *Nature* **356**:164–168 (1992).
103. Smith, F. W. and Feigon, J., "Strand orientation in the DNA quadruplex formed from the oxytricha telomere repeat oligonucleotide d($G_4T_4G_4$) in solution," *Biochemistry* **32**:8682–8692 (1992).
104. Strahan, G. D., Shafer, R. H., and Keniry, M. A., "Structural-properties of the d($G_3T_4G_3)_2$ quadruplex—evidence for sequential *syn-syn* deoxyguanosines," *Nucl. Acids Res.* **22**:5447–5455 (1994).

105. Smith, F. W., Lau, F. W., and Feigon, J., "d($G_3T_4G_3$) forms an asymmetric diagonally looped dimeric quadruplex with guanosine $5'$-*syn-syn-anti* and $5'$-*syn-anti-anti* N-glycosidic conformations," *Proc. Natl. Acad. Sci. USA* **91**:10546–10550 (1994).

106. Ross, W. S. and Hardin, C. C., "Ion-induced stabilization of the G-DNA quadruplex—free-energy perturbation studies," *J. Am. Chem. Soc.* **119**:6070–6080 (1994).

107. Guo, Q., Lu, M., and Kallenbach, N. R., "Adenine affects the structure and stability of telomeric sequences," *J. Biol. Chem.* **267**:15293–15300 (1992).

108. Strahan, G. D., Keniry, M. A., and Shafer, R. H., "NMR structure refinement and dynamics of the K^+-[d($G_3T_4G_3$)]$_2$ quadruplex *via* particle mesh Ewald molecular dynamics simulations," *Biophys. J.* **75**:968–981 (1998).

109. Smirnov, I. and Shafer, R. H., "Effect of loop sequence and size on DNA aptamer stability," *Biochemistry* **39**:1462–1468 (2000).

110. Gu, J. D. and Leszczynski, J., "Structures and properties of mixed DNA bases tetrads: Nonempirical ab inito HF and DFT studies," *J. Phys. Chem. A* **104**:1898–1904 (2000).

111. Hud, N. V., Smith, F. W., Anet, F. A. L., and Feigon, J., "The selectivity for K^+ versus Na^+ in DNA quadruplexes is dominated by relative free energies of hydration: A thermodynamic analysis by H-1 NMR," *Biochemistry* **35**:15383–15390 (1996).

112. Miyoshi, D., Karimata, H., and Sugimoto, N., "Hydration regulates thermodynamics of G-quadruplex formation under molecular crowding conditions," *J. Am. Chem. Soc.* **128**:7957–7963 (2006).

113. Marathias, V. M., Wang, K. Y., Kumar, S., Pham, T. Q., Swaminathan, S., and Bolton, P. H., "Determination of the number and location of the manganese binding sites of DNA quadruplexes in solution by EPR and NMR in the presence and absence of thrombin," *J. Mol. Biol.* **260**:378–394 (1996).

114. Marathias, V. M. and Bolton, P. H., "Determinants of DNA quadruplex structural type: Sequence and potassium binding," *Biochemistry* **38**:4355–4364 (1999).

115. Lee, J. S., "The stability of polypurine tetraplexes in the presence of monovalent and divalent-cations," *Nucl. Acids Res.* **20**:6057–6060 (1990).

116. Hardin, C. C., Watson, T., Corregan, M., and Bailey, C., "Cation-dependent transition between the quadruplex and Watson-Crick hairpin forms of d(CGCG$_3$GCG)," *Biochemistry* **31**:833–841 (1992).

117. Jin, R., Gaffney, B. L., Wang, C., Jones, R. A., and Breslauer, K., "Thermodynamics and structure of a DNA tetraplex: A spectroscopic and calorimetric study of the tetramolecular complexes of d(TG$_3$T) and d(TG$_3$T$_2$G$_3$T)," *J. Proc. Natl. Acad. Sci. USA* **89**:8832–8836 (1992).

118. Lu, M., Guo, Q., and Kallenbach, N. R., "Thermodynamics of G-tetraplex formation by telomeric DNAs," *Biochemistry* **31**:2455–2459 (1993).

119. Chen, F.-M., "Sr^{2+} facilitates intermolecular G-quadruplex formation of telomeric sequences," *Biochemistry* **31**:3769–3776 (1992).

120. Xu, Q., Deng, H., and Braunlin, W. H., "Selective localization and rotational immobilization of univalent cations in quadruplex DNA," *Biochemistry* **32**:13130–13137 (1993).

121. Miura, T. and Thomas, G. J., Jr., "Structural polymorphism of telomere DNA—interquadruplex and duplex-quadruplex conversions probed by Raman-spectroscopy," *Biochemistry* **33**:7848–7456 (1994).

122. Miura, T., Benevides, J. M., and Thomas, G. J., Jr., "A phase-diagram for sodium and potassium-ion control of polymorphism in telomeric DNA," *J. Mol. Biol.* **248**:233–238 (1995).

123. Deng, H. and Braunlin, W. H., "Kinetics of sodium ion binding to DNA quadruplexes," *J. Mol. Biol.* **255**:476–483 (1996).

124. Töhl, J. and Eimer, W., "Interaction energies and dynamics of alkali and alkaline-earth cations in quadruplex-DNA structures," *J. Mol. Model.* **2**:327–329 (1996).

125. Hardin, C. C., Corregan, M. J., Lieberman, D. V., and Brown, B. A., "Allosteric interactions between DNA strands and monovalent cations in DNA quadruplex assembly: Thermodynamic evidence for three linked association pathways," *Biochemistry* **35**:15383–15390 (1997).

126. Gueron, M., Eisinger, J., and Shulman, R. G., "Excited states of nucleotides and singlet energy transfer in polynucleotides," *J. Chem. Phys.* **47**:4077–4091 (1967).

127. Sueoka, N. and Cheng, T. Y., "Natural occurrence of a deoxyribonucleic acid resembling the deoxyadenylate-deoxythymidylate polymer," *Proc. Natl. Acad. Sci. USA* **48**: 1851–1856 (1962).

128. Tinoco, I., Jr., "Hypochromism in polynucleotides," *J. Am. Chem. Soc.* **82**:4785–4790 (1960) and "Additions and corrections—hypochromism in polynucleotides," *J. Am. Chem. Soc.* **83**:5047 (1961).

129. Marmur, J. and Doty, P., "Determination of the base composition of deoxyribonucleic acid from its thermal denaturation temperature," *J. Mol. Biol.* **5**:109–118 (1962).

130. Honnas, P. I. and Steen, H. B., "X-ray- and U.V.-induced excitation of adenine, thymine and related nucleosides and nucleotides in solution at 77°K," *Photochem. Photobiol.* **11**:67–76 (1970).

131. Vigny, P., "Fluorescence des nucleosides et nucleotides a temperature ambiante," *C. R. Acad. Sci.*, **D272**:3206–3209 (1971).

132. Takahashi, S., Nishimura, Y., and Tsuboi, M., "Rate of photochemical protonation and electronic relaxation of excited 1,N6-ethenoadenosine in its aqueous solution," *J. Chem. Phys.* **15**:3831–3837 (1981).

133. Nishimura, Y., Yamamoto, T., and Tsuboi, M., "The nature of fluorescence spectra of some minor bases obtained from tRNA and mRNA," *Nucl. Acid Res.* (Spec. Publ. (3)) 85–88 (1977).

134. Kasai, H., Ohashi, Z., Harada, F., Nishimura, S., Oppenheimer, N. J., Crain, P. F., Liehr, J. G., von Minden, D. L., and McClosky, J. A., "Structure of the modified nucleoside O isolated from *Escherichia coli* transfer ribonucleic acid. 7-(4,5-cis-Dihydroxy-1-cyclopentene-3-ylaminomethyl)-7-deazaguanosine," *Biochemistry* **14**: 4198–4208 (1975).

135. Thiebe, R., Zachau, G., Baczynskyj, L., Biemann, K., and Sonnenbichler, J., "Study on the properties and structure of the modified base Y^+ of yeast tRNA^{Phe}," *Biochim. Biophys. Acta* **240**:163–169 (1971).

136. Kohwi-Shigematsu, T., Enomoto, T., Yamada, M., Nakanishi, M., and Tsuboi, M., "Exposure of DNA bases induced by the interaction of DNA and calf thymus DNA helix-destabilizing protein," *Proc. Natl. Acad. Sci. USA* **75**:4689–4693 (1978).

137. Sutherland, G. B. B. M., and Tsuboi, M., "The infrared spectrum and molecular configuration of sodium deoxyribonucleate," *Proc. Roy. Soc. A* (London) **239**:446–463 (1957).

138. Thulstrup, E. W. and Michl, J., *Elementary Polarization Spectroscopy*, Wiley, New York, 1989.
139. Rodger, A. and Norden, B., *Circular Dichroism and Linear Dichroism*, Oxford Univ. Press, 1997.
140. Kinosita, K., Jr., Kawato, S., and Ikegami, A., "A theory of fluorescence polarization decay in membranes," *Biophys. J.* **20**:289–305 (1977).
141. Lipari, G. and Szabo, A., "Effect of libration motion on fluorescence depolarization and nuclear magnetic resonance relaxation in macromolecules and membranes," *Biophys. J.* **30**:489–506 (1980).
142. Norden, B., and Kurucsev, T., "Analyzing DNA complexes by circular and linear dichroism," *J. Mol. Recogn.* **7**:141–155 (1994).
143. Fritzsch, H., "Infrared linear dichroism studies of DNA-drug complexes: Quantitative determination of the drug-induced restriction of the B-A transition," *Nucl. Acids Res.* **22**:787–791 (1994).
144. Fasman, G. D., ed., *Circular Dichroism and the Conformational Analysis of Biomolecules*, Plenum Press, New York, 1996.
145. Bush, C. A. and Brahma, J., "Optical activity of single-stranded oligonucleotides," *J. Chem. Phys.* **46**:79–88 (1967).
146. Matsuzaki, J., Hotoda, H., Sekine, M., Hata, T., Higuchi, S., Nishimura, Y., and Tsuboi, M., "Self-complementary tetradeoxyribonucleoside triphosphates: convenient chemical preparation and spectroscopic studies in solution," *Tetrahedron* **42**:501–531(1986).
147. Samejima, T., Hashizume, H., Imahori, K., Fujii, I., and Miura, K., "Optical rotatory dispersion and circular dichroism of rice dwarf virus ribonucleic acid," *J. Mol. Biol.* **34**:39–48 (1968).
148. Allen, F. S., Gray, D. M., Roberts, G. P., and Tinoco, I., Jr., "The ultraviolet circular dichroism of some natural DNAs and an analysis of the spectra for sequence information," *Biopolymers* **11**:853–879 (1972).
149. Pohl, F. M., and Jovin, T. M., "Salt-induced co-operative conformational change of sythetic DNA: Equilibrium and kinetic studies with poly(dG-dC)," *J. Mol. Biol.* **67**:375–396 (1972).
150. Wilson, E. B., Decius, J. C., and Cross, P. C., *Molecular Vibrations*, McGraw-Hill, New York, 1955, Chap. 10.
151. Aida, M., Kaneko, M., Dupuis, M., Ueda, T., Ushizawa, K., Ito, G., Kumakura, A., and Tsuboi, M., "Vibrational modes in thymine molecule from an ab initio MO calculation," *Spectrochim. Acta* **A53**:293–407 (1997).
152. Lee, A., Adamowicz, L., Nowak, M. J., and Lapinski, L., "The infrared spectra of matrix isolated uracil and thymine: An assignment based on new theoretical calculations," *Spectrochim. Acta* **48A**:1385–1395 (1992).
153. Tsuboi, M., Nishimura, Y., Hirakawa, A. Y., and Peticolas, W. L., "Resonance Raman spectroscopy and normal modes of the nucleic acid bases," in Spiro, T. G., ed., *Biological Applications of Raman Spectroscopy, Vol. 2, Resonance Raman Spectra of Polymers and Aromatics*, Wiley, New York, 1987, pp. 109–179.
154. Tsuboi, M., "New approaches to the analysis of vibrations of nucleic acids and their components," in Stepanek, J., Anzenbacher, P., and Sedlacek, B., eds., *Laser Scattering Spectroscopy of Biological Objects*, Vol. 45 in Studies in Physical and Theoretical Chemistry (series), Elsevier, 1987, pp. 351–368.

155. Susi, H., Ard, J. S., and Purcell, J. M., "Vibrational spectraof nucleic acid constituents II. Planar vibrations of cytosine," *Spectrochim. Acta* **29A**:725–733 (1973).

156. Nishimura, Y., and Tsuboi, M., "In-plane vibrational modes of cytosine from ab initio MO calculation," *Chem. Phys.* **98**:71–80 (1985).

157. Angell, C. L., "An infrared spectroscopic investigation of nucleic acid constituents," *J. Chem. Soc.* (Pt. 1) 504–515 (1961).

158. Hirakawa, A. Y., Okada, H., Sasagawa, S., and Tsuboi, M., "Infrared and Raman spectra of adenine and its ^{15}N and ^{13}C substitution products," *Spectrochim. Acta* **41A**:209–216 (1985).

159. Tsuboi, M. and Kyogoku, Y., "Infrared spectroscopy of nucleic acid components," in Zorbach, W. W. and Tipson, S., eds., *Synthetic Procedures in Nucleic Acid Chemistry*, Vol. II, Wiley, 1973, Chap. 6.

160. Nishimura, Y., Tsuboi, M., Kato, S., and Morokuma, K., "In-plane vibrational modes of guanine from an *ab initio* MO calculation," *Bull. Chem. Soc. Jpn.* **58**:638–645 (1985).

161. Kyogoku, Y., Lord, R. C., and Rich, A., "Hydrogen bonding specificity of nucleic acid purines and pyrimidines in solution," *Science* **154**:518–520 (1966).

162. Delabar, J. M. and Majoube, M., "Infrared and Raman spectroscopic study of ^{15}N and D-substituted guanines," *Spetrochim. Acta* **34A**:129–140 (1978).

163. Lane, M. and Thomas, G. J., "Kinetics of hydrogen-deuterium exchange in guanosine determined by laser-Raman spectroscopy," *Biochemistry* **18**:3839–3849 (1979).

164. Herzberg, G., *Molecular Spectra and Molecular Structure; Infrared and Raman of Polyatomic Molecules*, Krieger, (reprint edition), New York, 1991.

165. Tsuboi, M., Ueda, T., Ushizawa, K., Sasatake, Y., Ono, A., Kainosho, M., and Ishido, Y., "Assigments of the deoxyribose vibrations: isotopic thymidine," *Bull. Chem. Soc. Jpn.* **67**:1483–1484 (1994).

166. Tsuboi, M., Komatsu, M., Hoshi, J., Kawashima, E., Sekine, T., Ishido, Y., Russel, M. P., Benevides, J. M., and Thomas, G. J., Jr., "Raman and infrared spectra of (2'S)-[2-2H] thymidine: Vibrational coupling between deoxyribosyl and thymine moieties and structural implications," *J. Am. Chem. Soc.* **119**:2025–2032 (1997).

167. Tsuboi, M., Takeuchi, Y., Kawashima, E., Ishido, Y., and Aida, M., "Raman spectrum of [5'/12C]thymidine: Vibrations of its 5'-end atomic group," *Spectrochim. Acta* (Pt. A) **55**:1887–1896 (1999).

168. Toyama, A., Takino, Y., Takeuchi, H., and Harada, I., "Ultraviolet resonance Raman spectra of ribosyl C(1')-deuterated purine nucleosides; evidence of vibrational coupling between purine and ribose rings," *J. Am. Chem. Soc.* **115**:11092–11098 (1993).

169. Nishimura, Y., Tsuboi, M., Sato, T., and Aoki, K., "Conformation-sensitive Raman lines of mononucleotides and their use in a structure analysis of polynucleotides: Guanine and cytosine nucleotides," *J. Mol. Struct.* **146**:123–153 (1986).

170. Tsuboi, M., "Vibrational spectra of phosphite and hypophosphie anions, and the characteristic frequencies of PO_3^{2-} and PO_2^- groups," *J. Am. Chem, Soc.*, **79**:1351–1353 (1957).

171. Tsuboi, M., "Infrared and Raman spectroscopy," in Ts'o, P. O. P., ed., *Basic Principles of Nucleic Acid Chemistry*, Vol. 1, Academic Press, New York, 1974, p. 427.

172. Tsuboi, M., Takahashi, S., and Harada, I., "Infrared and Raman spectra of nucleic acids—vibrations in the base-residues," in Duchesne, J., ed., *Physico-Chemical Properties of Nucleic Acids*, Academic Press, New York, 1973, pp. 91–145.

173. Ueda, T., Ushizawa, K., and Tsuboi, M., "Depolarization of Raman scattering from some nucleotides of RNA and DNA," *Biopolymers* **33**:1791–1803 (1990).

174. Kubasek, W. L., Hudson, B., and Peticolas, W. L., "Ultraviolet resonance Raman excitation profiles of nucleic acid bases with excitation from 200 to 300 nanometers," *Proc. Natl. Acad. Sci. USA* **82**:2369–2373 (1985).

175. Fodor, S. P. A., Rava, R. P., Hays, T. R., and Spiro, T. G, "Ultraviolet resonance spectroscopy of the nucleotides with 266, 240, 218 and 200 nm pulsed laser excitation," *J. Am. Chem. Soc.* **107**:1520–1529 (1985).

176. Shimanouchi, T., Tsuboi, M., and Kyogoku, Y., "Infrared spectra of nucleic acids and related compounds," *Adv. Chem. Phys.* **7**:435–498 (1964).

177. Thomas, G. J., Jr., and Tsuboi, M., "Raman spectroscopy of nucleic acids and their complexes," *Adv. in Biophys. Chem.* **3**:1–70 (1993).

178. Tsuboi, M., "Application of infrared specrtroscopy to structure study of nucleic acids," *Appl. Spectrosc. Rev.* **3**:45–90 (1969).

179. Nishimura, Y., Morikawa, K., and Tsuboi, M., "Spectral difference of the A and B forms of deoxyribonucleic acid," *Bull. Chem. Soc. Jpn.* **47**:1043–1044 (1974).

180. Pilet, J. and Brahms, J., "Investigation of DNA structural changes by infrared spectroscopy," *Biopolymers* **12**:387–403 (1973).

181. Nishimura, Y., Morikawa, K., and Tsuboi, M., "Infrared spectra of deoxyribonucleic acids with different base compositions in their D_2O solutions," *Bull. Chem. Soc. Jpn.* **46**:3891–3892 (1973).

182. Flavell, R. A. and Jones, I. G., "Mitochondrial deoxyribonucleic acid from *Tetrahymena pyriformis* and its kinetic complexity," *Biochem. J.* **116**:811–817 (1970).

183. Ulitzur, S., "Rapid determination of DNA base composition by ultraviolet spectroscopy," *Biochim. Biophys. Acta* **272**:1–11 (1972).

184. Sueoka, N. and Cheng, T. Y., "Fractionation of nucleic acids with the methylated albumin column," *J. Mol. Biol.* **4**:161–172 (1962).

185. Thomas, G. J. Jr., Benevides, J. M., Overman, S. A., Ueda, T., Ushizawa, K., Saitoh, M., and Tsuboi, M., "Polarized Raman spectra of oriented fibers of A DNA and B DNA: Anisotropic and isotropic local Raman tensors of base and backbone vibrations," *Biophys. J.* **68**:1073–1088 (1995).

186. Tsuboi, M. and Thomas, G. J., Jr., "Raman scattering tensors in biological molecules and their assembly," *Appl. Spectrosc. Rev.* **32**:263–299 (1997).

187. Erfurth, S. C., Kiser, E. J., and Peticolas, W. J., "Determination of the backbone structure of nucleic acidsand nucleic acid oligomersby laser Raman scattering," *Proc. Natl. Acad. Sci. USA* **69**:938–941 (1972).

188. Prescott, B., Steinmetz, W., and Thomas, G. J., Jr., "Characterization of DNA secondary structure by Raman spectroscopy," *Biopolymers* **23**:235–256 (1984).

189. Guan, Y., Eurrey, C. J., and Thomas, G. J., Jr., " Vibrational analysis of nucleic acids I. The phosphdiester group in dimethyl phosphate model compounds: $(CH_3O)_2PO_2^-$, $(CD_3O)_2PO_2^-$, and $(^{13}CH_3O)_2PO_2^-$," *Biophys. J.* **66**:225–235 (1994).

190. Benevides, J. M., Tsuboi, M., Wang, A. H.-J., and Thomas, G. J., Jr., "Local Raman tensors of double-helical DNA in the crystal: A basis for determining DNA residue orientations," *J. Am. Chem. Soc.* **115**:5351–5359 (1993).

191. Bovey, F. A., Jelinski, L., and Mirau, P. A., *Nuclear Magnetic Resonance Spectroscopy*, Academic Press, San Diego, 1988.

192. Claridge, T. D. W., *High Resolution NMR Techniques in Organic Chemistry*, Pergamon Press, Elsevier Science, Amsterdam, 1999.

193. Wehrli, F. W. and Wirthlin, T., *Interpretaion of Carbon-13 NMR Spectra*, Heyden, London, 1976.

194. Cavanagh, J., Fairbrother, W. J., Palmer, A. G., III, and Skelton, N. J., *Protein NMR Spectroscopy: Principles and Practice*, Academic Press, San Diego, 1996.

195. Friebolin, H., *Basic One- and Two-Dimensional NMR Spectroscopy* (transl. J. K. Becconsall), Wiley-VCH, Weinheim, Germany, 2005.

196. Berger, S. and Braun, S., 200 and More NMR Experiments. A Practical Course, Wiley-VCH, Weinheim, Germany, 1998.

197. Van de Ven, F. J. and Hilbers, C. W., "Nucleic acids and nuclear magnetic resonance," *Eur. J. Biochem.* **178**:1–38 (1988).

198. Wüthrich, K., *NMR of Proteins and Nucleic Acids*, Wiley-Interscience, New York, 1986.

199. Shafer, R. H. and Brown, S. C., "Nuclear magnetic resonance studies of drug-nucleic acid interactions," in Kallenbach, N. R., ed., *Chemistry & Physics of DNA-Ligand Interactions*, Adenine Press, Schenectady, NY, 1990, pp. 109–142.

200. Zidek, L., Stefl, R., and Sklenar, V., "NMR methodology for the study of nucleic acids," *Curr. Opin. Struct. Biol.* **11**:275–281 (2001).

201. Driscoll, P. C., Esposito, D., and Pfuhl, M., " NMR of nucleic acids and proteins," *Nucl. Magn. Reson.* **29**:340–405 (2000).

202. Varani, G., Aboul-ela, F., and Allain, F. H.-T., "NMR investigation of RNA structure," *Prog. NMR Spectrosc.* **29**:51–127 (1996).

203. Flinders, J. and Dieckmann, T., "NMR spectroscopy of ribonucleic acids," *Prog. NMR Spectrosc.* **48**:137–159 (2006).

204. Harris, R. K., Becker, E. D., Cabral de Menezes, S. M., Goodfellow, R., and Granger, P., "Nuclear spin properties and conventions for chemical shifts (IUPAC Recommendations, 2001)," in *Encyclopedia of Nuclear Magnetic Resonance*, Vol. 9, Wiley, Chichester, UK, 2002, p. 5.

205. Takahashi, S., Nagashima, N., Nishimura, Y., and Tsuboi, M., "Anomalous temperature dependence of the phosphorus-31 nuclear magnetic resonance chemical shift in d(CCGG) and d(CCTAGG) at the junction of the pyrimidine stack followed by the purine stack," *Chem. Pharm. Bull.* **34**:3987–3993 (1986).

206. Takahashi, S., Nagashima, N., Nishimura, Y., and Tsuboi, M., "Nuclear magnetic resonance study on the interaction of aclacinomycin-A with a deoxyribo-hexanucleotide Pentaphosphate d(CCTAGG)$_2$ in Aqueous Solution," *Chem. Pharm. Bull.* **34**:4494–4499 (1986).

207. Bolton, P. H. and James, T. L., "Fast and slow conformational fluctuations of RNA and DNA. Subnanosecond internal motion correlation times determined by ^{31}P NMR," *J. Am. Chem. Soc.* **102**:25–31 (1980).

208. Neumann, J. M., Borrel, J., Thiery, J. M., Guschlbauer, W., and Tran-Dinh, S., "PMR-relaxation and steric computations give unequivocal nucleoside conformations," *Biochim. Biophys. Acta* **479**:427–440 (1977).
209. Biological Magnetic Resonance Bank, Univ. Wisconsin-Madison, Madison, 2006 (http://www.bmrb.wisc.edu).
210. Jardetzky, O. and Roberts, G. C. K., *NMR in Molecular Biology*, Academic Press, 1981, p. 188.
211. Karplus, M., "Contact electron-spin coupling of nuclear magnetic moments," *J. Chem. Phys.* **30**:11–15 (1959).
212. Haasnoot, C. A. G., de Leeuw, F. A. A. M., Leeuw, H. P. M., and Altona, C., "The relationship between proton-proton NMR coupling constants and substituent electronegativities. II. Conformational analysis of the suagr ring in nucleosides and nucleotides in solution using a generalized Karplus equation," *Org. Magn. Reson.* **15**:43–52 (1981).
213. Rinkel, L. J. and Altona, C., "Conformational analysis of the deoxyribofuranose ring in DNA by means of sums of proton-proton coupling constants: A graphical method," *J. Biomol. Struct. Dyn.* **4**:621–649 (1987).
214. Gonzalez, C., Stec, W., Kobylanska, A., Hogrefe, R. I., Reynolds, M., and James, T. L., "Structural study of a DNA/RNA hybrid duplex with a chiral phosphorothioate moiety by NMR: Extracton of distance and torsion angle constraints and imino proton exchange rates," *Biochemistry* **33**:11062–11072 (1994).
215. Schmitz, U., Zon, G., and James, T. L., "Deoxyribose conformation in [d(GTATATA)]2: Evaluationn of sugar pucker by simulation of double-quantum-filtered COSY crosspeaks," *Biochemistry* **29**:2357–2368 (1990).
216. Ulyanov, N. B., Schmitz, U., Kumar, A., and James, T. L., "Probability assessment of conformational ensembles: Sugar repuckering in a DNA duplex solution," *Biophys. J.* **68**:13–24 (1995).
217. Ono, A., Tate, S., and Kainosho, M., "Preparation and heteronuclear 2D NMR spectroscopy of a DNA dodecamer containing a thymidine residue with a uniformly ^{13}C-labeled deoxyribose ring," *J. Biomol. NMR* **4**:581–586 (1994).
218. Pervushin, K., Ono, A., Fernandez, C., Szyperski, T., Kainosho, M., and Wüthrich, K., "NMR scalar couplings across Watson-Crick base pair hydrogen bonds in DNA observed by transverse relaxation-optimized spectroscopy," *Proc. Natl. Acad. Sci. USA* **95**:14147–14151 (1998).
219. Dingley, A. J. and Grzesiek, S., "Direct observation of hydrogen bonds in nucleic acid base pairs by internucleotide $^{2}J_{NN_{NN}}$ couplings," *J. Am. Chem. Soc.* **120**:8293–8297 (1998).
220. Dingley, A. J., Masse, J. E., Peterson, R. D., Barfield, M., Feigon, J., and Grzesiek, S., "Internucleotide scalare couplings across hydrogen bonds in Watson-Crick and Hoogsteen base pairs of a DNA triplex," *J. Am. Chem. Soc.* **121**:6019–6027 (1999).
221. Wöhnert, J., Dingley, A. J., Stoldt, M., Görlach, M., Grzesiek, S., and Brown, L., "Direct identification of NH···N hydrogen bonds in non-canonical base pairs of RNA by NMR spectroscopy," *Nucl. Acids Res.* **27**:3104–3110 (1999).
222. Grzesiek, S., Cordier, F., Jaravine, V., and Barfield, M., "Insights into biomolecular hydrogen bonds from hydrogen bond scalar couplings," *Prog. NMR Spectrosc.* **45**:275–300 (2004).

223. Tsuboi, M., Kuriyagawa, F., Matso, K., and Kyogoku, Y., "Nuclear spin coupling and conformation of the P−O−C−H group," *Bull. Chem. Soc. Jpn.* **40**:1813–1818 (1967).
224. Kainosho, M., Nakamura, A., and Tsuboi, M., "Phosphorus−proton spin-spin coupling in the P−O−C−H group: A comparison of cyclic and acyclic systems," *Bull. Chem. Soc. Jpn.* **42**:1713–1718 (1969).
225. Wood, D. J., Mynott, R. J., Hruska, F. E., and Sarma, R. H., "^{1}H NMR study of the molecular conformation of the exocyclic linkage of 5′-B-nucleotides in solution–correlation between C(4′)-C(5′) and C(5′)-O(5′) conformer population and the influence of the base," *FEBS Lett.* **34**:323–326 (1973).
226. Lankhorst, P. P., Haasnoot, C. A. G., Erhelens, C., and Altona, C., "Carbon-13 NMR in conformational analysis of nucleic acid fragments 2. A reparameterization of the Karplus equation for vicinal NMR coupling constants in CCOP and HCOP fragments," *J. Biomol. Struct. Dyn.* **1**:1387–1405 (1984).
227. Roongta, V. A., Jones, C. R., and Gorenstein, D. G., "Effect of distortions in the deoxyribose phosphate backbone conformation of the duplex oligonucleotide dodecamers GT, GG, GA, AC and GU base-pair mis-matches on ^{31}P NMR spectra," *Biochemistry* **29**:5245–5258 (1990).
228. Majumdar, A. and Hosur, R. V., "Simulation of 2D NMR spectra for determination of solution conformations of nucleic acids," *Prog. NMR Spectrosc.* **24**:109–158 (1992).
229. Wijmenga, S. S., Mooren, M. W., and Hilbers, C. W., "NMR of nucleic acids: From spectrum to structure," in Roberts, G. C. K., ed., *NMR of Macromolecules: A Practical Approach*, Oxford Univ. Press, 1993, pp. 217–283.
230. Chary, K. V. R., Rastogi, V. K., Govil, G., and Miles, T., "Estimation of ^{31}P-^{1}H and ^{1}H-^{1}H vicinal coupling constants along the DNA backbone by 2D HELCO measurements," *J. Chem. Soc. Chem. Commun.* **1994**:241–243 (1994).
231. Ojha, R. P., Dhingra, M. M., Sarma, M. H., Shibata, M., Farrar, M., Turner, C. J., and Sarma, R. H., "DNA bending and sequence-dependent backbone conformation: NMR and computer experiments," *Eur. J. Biochem.* **265**:35–53 (1999).
232. Kim, S. G., Lin, L. J., and Reid, B. R., "Determination of nucleic acid backbone conformation by 1H NMR," *Biochemistry* **31**:3566–3574 (1992).
233. Ding, D., Gryazonv, S. M., Lloyd, D. H., Chandrasekaran, S., Yao, S., Ratmeyer, J., Pan, Y., and Wilson, W. D., "An oligonucleotide N3′-P5′ phosphoramidate duplex forms an A-type helix in solution," *Nucl. Acids Res.* **24**:354–360 (1996).
234. Onoye, H., Suzuki, E., Nishimura, Y., and Tsuboi, M., "Proton magnetic resonance studies of d(TGGGGT) and other oligodeoxyribonucleotides," (in press).
235. Haasnott, C. A. G. and Hilbers, C. W., "Effective water resonance suppression in 1D and 2D-FT-1H nmr spectroscopy of biopolymers in aqueous solution," *Biopolymers* **22**:1259–1266 (1983).
236. Otting, G., Grutter, R., Leupin, W., Minganti, C., Ganesh, K. N., Sproat, B. S., Gait, M. J., and Wüthrich, K., "Sequential NMR assignments of labile protons in DNA using two-dimensional nuclear-Overhauser-enhancement spectroscopy with three jump-and-return pulse sequences," *Eur. J. Biochem.* **166**:215–220 (1987).
237. Barbault, F., Huynh-Dinh, T., Paoletti, J., and Lancelot, G., "A peculiar DNA structure: NMR structure of a DNA kissing complex," *J. Biomol. Struct. Dyn.* **19**:649–658 (2002).

238. Hoogsteen, K., "The structure of crystals containing a hydrogen-bonded complex of 1-methylthymine and 9-methyladenine," *Acta Cryst.* **12**:822–823 (1959).

239. Roy, S., Papastavros, M. Z., and Redfield, A. G., "Procedure for C2 deuteration of nucleic acids and determination of Aψ31 pseudouridine conformation by nuclear Overhauser effect in yeast tRNA^{Phe}," *Nucl. Acids Res.* **10**:8341–8349 (1982).

240. Abrescia, N. G. C., Gonzáles, C., Gouyette, C., and Subirana, J. A., "X-ray and NMR studies of the DNA oligomer d(ATATAT): Hoogsteen base pairing in duplex DNA," *Biochemistry* **43**:4092–4100 (2004).

241. Schmitz, U. and James, T. L. "Nuclear magnetic resonance and nucleic acids," in James, T. L., ed., Methods in Enzymology, Academic Press, San Diego, 1995, Vol. 261, p. 3–44.

242. Crippen, G. M. and Havel, T. F., *Distance Geometry and Molecular Conformation*, Research Studies Press, Taunton, MA, 1988.

243. Lane, A. N., Jenkins, T. C., Brown, T., and Neidle, S., "Interaction of berenil with the EcoRI dodecamer d(CGCGAATTCGCG)$_2$ in solution studied by NMR," *Biochemistry* **30**:1372–1385 (1991).

244. Hajduk, P. J., Meadows, R. P., and Fesik, S. W., "NMR-based screening in drug discovery," *Quart. Rev. Biophys.* **32**:211–240 (1999).

245. Meyer, B. and Peters, T., "NMR spectroscopy techniques for screening and identifying liagand binding to protein receptors," *Angew. Chem. Int. Ed.* **42**:864–890 (2003).

246. Meyer, B., Weimar, T., and Peters, T., "Screening mixtures for biological activity by NMR," *Eur. J. Biochem.* **246**:705–709 (1997).

247. Ni, F., "Recent developments in transferred NOE methods," *Prog. NMR Spectrosc.* **26**:517 (1994).

248. Tjandra, N. and Bax, A., "Direct measurement of distances and angles in biomolecules by NMR in a dilute liquid crystalline medium," *Science* **278**:1111–1114 (1997).

249. Tjandra, N., Tate, S., Ono, A., Kainosho, M., and Bax, A., "The NMR structure of a DNA dodecamer in an aqueous dilute liquid crystalline phase," *J. Am. Chem. Soc.* **122**:6190–6200 (2000).

250. Wu, Z., Delaglio, F., Tjandra, N., Zhurkin, V. B, and Bax, A., "Overall structure and sugar dynamics of a DNA dodecamer from homo- and heteronuclear dipolar couplings and ^{31}P chemical shift anisotropy," *J. Biomol. NMR* **26**:297–315 (2003).

251. Wertz, J. E. and Bolton, J. R., *Electron Spin Resonance*, McGraw-Hill, New York, 1972.

252. Gaffney, B. J. "Electron spin resonance of biomolecules," in Meyers, R. A., ed., *Encyclopedia of Molecular Cell Biology and Molecular Medicine*, Wiley-VCH, Weinheim, Germany, 2004, Vol. 4, pp. 115–133.

253. Biwas, R., Kühne, H., Brudvig, G. W., and Gopalan, V., "Use of EPR spectroscopy to study macromolecular structure and function," *Sci. Prog.* **84**:45–68 (2001).

254. Cantor, C. R. and Schimmel, P. R., *Biophysical Chemistry. Part II: Techniques for the Study of Biological Structure and Function*, Freeman, San Francisco, 1980.

255. Lipson, H. S., *Crystals and X-rays*, Wykeham Publications, London, 1970.

256. Rhodes, G., *Crystallography Made Crystal Clear*, Academic Press, San Diego, 2006.

257. Egli, M., "Nucleic acid crystallography: Current progress," *Curr. Opin. Chem. Biol.* **8**:580–591 (2004).

258. Peek, M. E., Williams, L. D. "X-ray crystallography of DNA-drug complexes," in Claires, J. B. and Waring, M. J., eds., *Methods in Enzymology*, Academic Press, San Diego, 2001, Vol. 340, pp. 282–289.
259. Bragg, W. L., "The diffraction of short electromagnetic waves by a crystal," *Proc. Cambridge Phil. Soc.* **17**:43–57 (1914).
260. Ban, N., Nissen, P., Hansen, J., Moore, P. B., and Steitz, T. A., "The complete atomic structure of the large ribosomal subunit at 2.4 Å resolution," *Science* **289**:905–920 (2000).
261. Nissen, P., Hansen, J., Ban, N., Moore, P. B., and Steitz, T. A., "The structural basis of ribosome activity in peptide bond synthesis," *Science* **289**:920–930 (2000).
262. Schuwirth, B. S., Borovinskaya, M. A., Hau, C. W., Zhang, W., Vila-Sanjurjo, A., Holton, J. M., and Cate, J. H. D., "Structures of the bacterial ribosome at 3.5 Å resolution," *Science* **310**:827 (2005).
263. Allen, M. P. and Tildesley, D. J., *Computer Simulation of Liquids*, Oxford Univ. Press, New York, 1987.
264. Frenkel, D. and Smit, B., *Understanding Molecular Simulations: From Algorithms to Applications*, Academic Press, San Diego, 1996.
265. McCammon, J. A. and Harvey, S. C., *Dynamics of Proteins and Nucleic Acids*, Cambridge Univ. Press, Cambridge, 1987.
266. Burkert, U. and Allinger, N. L., *Molecular Mechanics*, American Chemical Society, Washington, DC, 1980.
267. Rapaport, D. C., *The Art of Molecular Dynamics Simulation*, Cambridge Univ. Press, Cambridge, UK, 1995.
268. Torda, A. E., Scheek, R. M., and van Gunsteren, W. F., "Time-averaged nuclear Overhauser effect distance restraints applied to tendamistat," *J. Mol. Biol.* **214**:223–235 (1990).
269. Fennen, J., Torda, A. E., and van Gunsteren, W. F., " Structure refinement with molecular dynamcis and a Boltzmann weighted ensemble," *J. Biomol. NMR* **6**:163–170 (1995).
270. Bonvin, A. M. J. J., and Brunger, A. T., "Conformational variability of solution nuclear magnetic resonance structures," *J. Mol. Biol.* **250**:80–93 (1995).
271. Kemmink, J. and Scheek, R. M., "Dynamic modelling of a helical peptide in solution using NMR data: Multiple conformations and multi-spin effects," *J. Biomol. NMR* **5**:33–40 (1995).
272. Pearlman, D. A. and Kollman, P. A., "Are time-averaged restraints necessary for NMR refinement? A model study for DNA," *J. Mol. Biol.* **220**:457–479 (1991).
273. Pearlman, D. A., "FINGAR: A new genetic algorithm-based method for fitting NMR data," *J. Biomol. NMR* **8**:49–66 (1996).
274. Gyi, J. I., Lane, A. N., Conn, G. L., and Brown, T., "Solution structures of DNA-RNA hybrids with purine-rich and pyrimidine-rich strands: Comparison with the homologous DNA and RNA duplexes," *Biochemistry* **37**:73–80 (1998).
275. Görler, A., Ulyanov, N. B., and James, T. L., "Determination of the populations and structures of multiple conformers in an ensemble from NMR data: Multiple-copy refinement of nucleic acid structures using floating weights," *J. Biomol. NMR* **16**:147–164 (2000).

276. This image was produced using the Chimera package from the Computer Graphics Laboratory, University of California, San Francisco: Huang, C. C., Couch, G. S., Petersen, E. F., and Ferrin, T. E., "Chimera: An extensible molecular modeling application constructed using standard components," *Proc. Pacific Symp. on Biocomputing*, 1996, Vol. 1, p. 724.
277. Berg, J. M., Tymoczko, J. L., and Stryer, L., *Biochemistry*, Freeman, New York, 2002.

CHAPTER 2

Intercalating Drugs

The concept of *intercalation* was pioneered in 1961 by Lerman [1], and modified in 1966 by Pritchard et al. [2]. These workers postulated that flat polycyclic aromatic ring molecules such as acridine dyes (Fig. 2.1) can be sandwiched (intercalated) between two base pairs of the DNA double helix, where the main binding force is van der Waals interaction between the π-electron systems of the dye molecule and the heterocyclic rings of the base pairs, and is reinforced by ionic interactions between the positively charged nitrogen atom of an acridine ring and the oxygen atom of the phosphodiester group of DNA (Fig. 2.2) [3]. Later, other workers confirmed their hypothesis by X-ray crystallographic studies on a series of acridine dyes complexed with dinucleoside monophosphates.

Consider the drug proflavine (also called *proflavin* and *diaminoacridine*), which is an acridine derivative (Fig. 2.1). It has been used as a disinfectant bacteriostatic against many gram-positive bacteria, as well as a topical antiseptic in its salt form. We now know that this drug functions primarily by intercalating between nucleic acid base pairs, causing base-pair deletions or insertions during replication. For this reason it is a mutagen, but it is an interesting one as most mutagens result in base-pair substitutions. X-ray structure analyses were performed on the complex of proflavine with the duplex formed by the dinucleoside monophosphate, r(CpG) [4]. These studies (at a drug: dinucleotide ratio of 3 : 2) provide a good picture of how this drug can bind. In Fig. 2.3 one observes two proflavine molecules stacked above and below the duplex, while the third molecule is *intercalated* between the C–G and G–C base pairs. The distance between the base pairs has increased in order to allow the insertion of the drug, and the five aromatic units form planes that are nearly equidistant in vertical separation. The amino groups of the intercalated molecule form two hydrogen bonds (dotted lines in the Figure) to the phophodiester oxygens and two more hydrogen bonds to the O(2′) hydroxyl oxygens of the cytosine riboses.

As the example above clearly illustrates, the DNA or RNA base step at the binding site is stretched when a drug molecule is intercalated between two base pairs. This causes the overall length of the DNA strand to increase by unwinding the double helix.

Drug–DNA Interactions: Structures and Spectra by Kazuo Nakamoto, Masamichi Tsuboi, and Gary D. Strahan
Copyright © 2008 John Wiley & Sons, Inc.

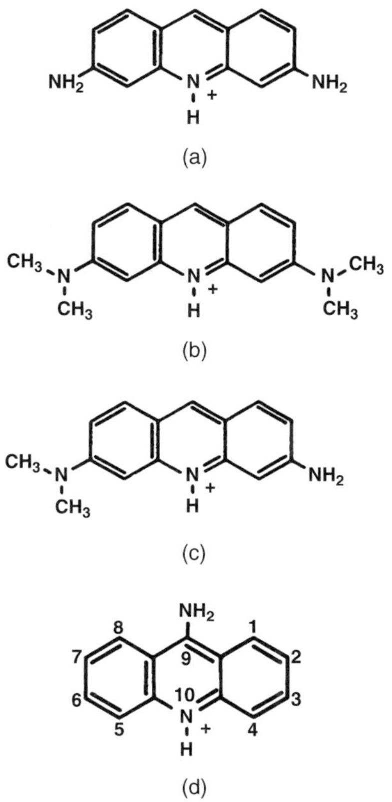

FIGURE 2.1. Chemical formula of (a) proflavine, (b) acridine orange, (c) *N,N*-dimethylproflavine, and (d) 9-aminoacridine. The numbering of the positions in the acridine ring is given in (d).

Since the vertical separation of successive base pairs is normally about 3.4 Å in B-form DNA, this distance must be increased markedly to accommodate a large aromatic ring molecule. The helical expansion can vary over a wide range (1.8–4.5 Å per intercalation), as can the unwinding angle (11–48°), depending on the structures of the drug and the nucleotides involved [5]. In the case of natural DNA, the degree of these distortions can be measured by the increase in viscosity [6], the decrease in sedimentation coefficient [7], the changes in circular dichroism (CD) [6–8], and electrophoresis of circularly closed DNA duplexes [9].

Intercalation is the most common mode of interaction by which small molecules may bind to DNA, although each drug has unique binding site preferences. Some prefer to intercalate between a 5′-purine–pyrimidine-3′ base step rather than a 5′-pyrimidine–purine-3′ base step (or vice versa), whereas others have no preference of order, but prefer certain base-pair sequences. For example, aclacinomycin intercalates at CpG or GpC sites, whereas adriamycin binds at TpA or ApT sites in DNA (Fig. 2.4). Slight differences in binding mechanisms determine the sequence preference of the drug, and usually result in differences in cytotoxic and pharmacological activity. These will be discussed whenever possible.

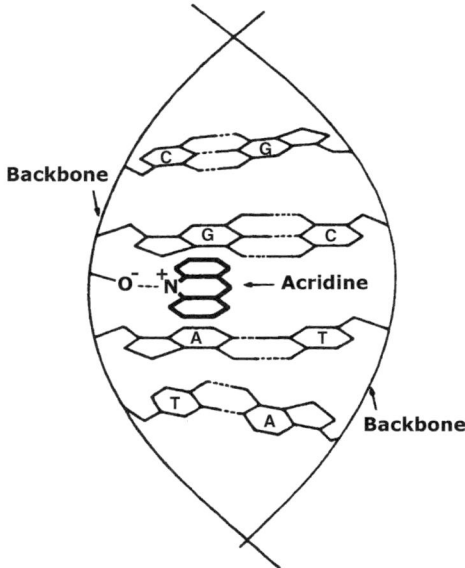

FIGURE 2.2. A conceptual drawing of the structure of, and interactions within, an acridine–DNA complex [3].

The intercalation of drug molecules is limited by the *neighbor exclusion rule*, which states that drug molecules cannot occupy two successive base steps. This rule is based on the observation that the drug/phosphate ratio is always smaller than 0.2 (which is the expected value if binding occurred at every other base step). This ratio would be 1 if all the possible base steps were occupied by the drug molecules [3]. The neighbor exclusion rule applies only to longer oligonucleotides and DNA. Short oligonucleotides, such as dinucleotides, can accommodate higher ratios of a drug.

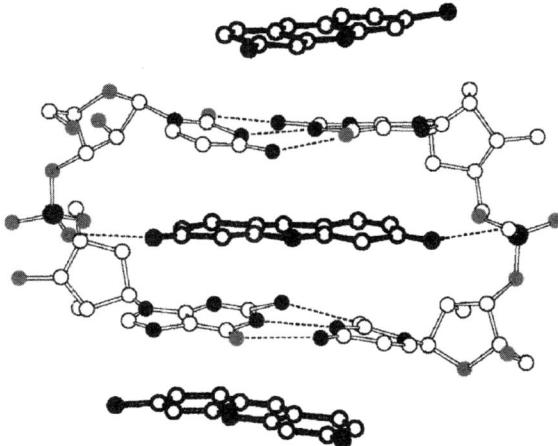

FIGURE 2.3. Crystal structure of proflavine–d(CpG) (3 : 2 ratio). The nitrogen and phosphorus atoms are black, and oxygen atoms are stippled. The dotted lines represent hydrogen bonds [4].

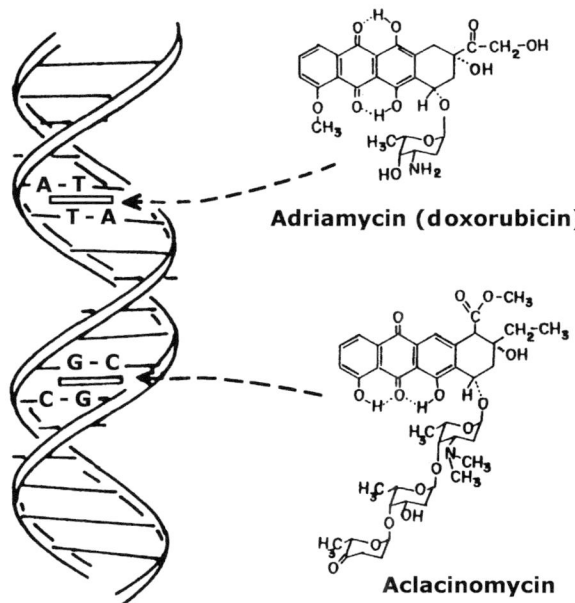

FIGURE 2.4. Different drugs go into different positions of DNA.

The intercalation of a drug alters the $\pi-\pi^*$ electron transition of the drug chromophore. This is observed in the visible region of the absorption spectrum as a red-shift of ~ 10 nm or more. In the fluorescence spectrum, the same band is blue-shifted by the same magnitude [3]. As discussed in Section 2.1.5, the phenomenon of fluorescence quenching is useful in distinguishing intercalation sites such as GC and AT. In the case of groove-binding drugs (Chapter 3), this phenomenon is less obvious since the distortion of the DNA structure is much smaller.

The precise stacking geometries of drug molecules relative to base pairs are different in each case, and can be determined only by X-ray crystallography, or NMR spectroscopy. Other techniques can provide important and detailed information on the relative orientation of the drug when it binds to nucleic acids (e.g., polarized Raman microphotometry). As this is our first chapter specifically dealing with drug–DNA interactions, this chapter elaborates on the different experimental methods used and how conclusions are reached. In this way, we hope that the reader will come to a deeper appreciation of the approaches that are possible, and thus better understand the later chapters.

2.1. ACRIDINE DYES

2.1.1. X-Ray Diffraction Studies of Acridine–Dinucleoside Monophosphate Crystals

The first detailed structure of an intercalation complex was determined in 1975 by Seemen et al. [10], who found that 9-aminoacridine (a mutagen; see Fig. 2.1d) and r(ApU) (adenyl-3′,5′-uridine) crystallize in a heavily hydrated monoclinic lattice. The adenine residues were found to be hydrogen-bonded to uracil residues, and the base pairs were stacked parallel to each other in the crystal with 9-aminoacridine sandwiched directly between them. Figure 2.5 shows the crystal structure of the 9-aminoacridine–ApU complex. The adenine–uracil residues are hydrogen-bonded in pairs between the N6 of adenine and the O4 of uracil, and between the N3 of uracil and the N7 of adenine. This hydrogen-bonding arrangement is a Hoogsteen base pair (Sections 1.1.12 and 1.4.4). As shown in Figure 2.5, the 9-aminoacridine molecules were sandwiched by the hydrogen-bonded base pairs. In addition, the adenine residues were hydrogen-bonded to the ribose hydroxyl groups of an adjoining uracil (rU), where the interactions were between the N6 amino group of adenine and O3′ of the ribose (2.95 Å), and between N1 of adenine and O2′ of the same uridine ribose (2.77 Å). Perturbation of the dinucleoside phosphate structure by this nitrogen results in some unusual features for the nucleosides. The adenosine ribose has a C2′-*endo* conformation with a glycosidic dihedral torsion angle χ of 77°, which is in a region between the *syn* and *anti* conformations (Section 1.1.8).

In contrast to the unusual nucleoside conformations in the crystalline 9-aminoacridine-r(ApU) complex, the complex of 9-aminoacridine with r(CpG) was found to have a structure that is more similar to that of B-DNA [11–13]. The crystalline complex of 9-aminoacridine with 5-iodocytidylyl-3′,5′-guanosine (5I-CpG) formed an asymmetric unit containing four 9-aminoacridine molecules, four 5I-CpG molecules, and 21 water molecules. In these crystals, 9-aminoacridine was observed to bind

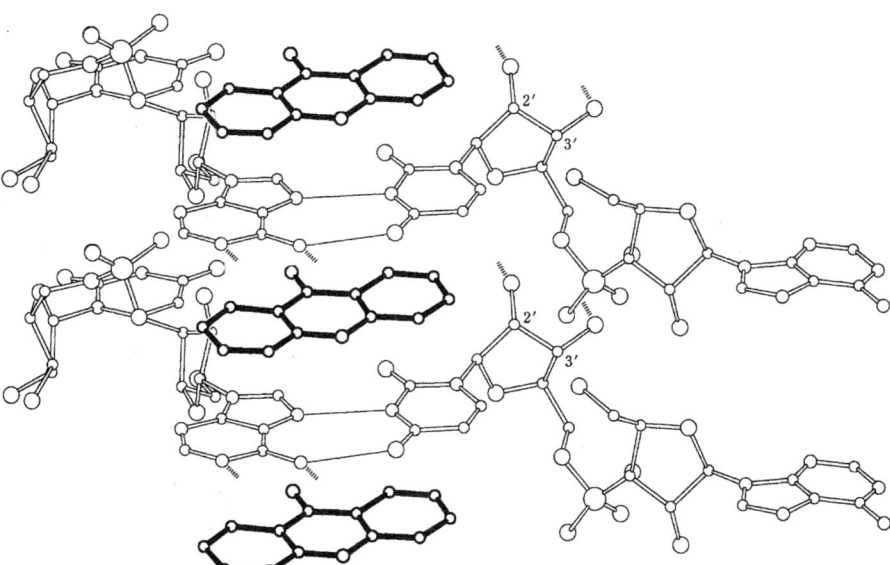

FIGURE 2.5. A view of the intercalative stacking interaction at an oblique angle. The nature of the vertical hydrophobic column of alternating base pairs and 9-aminoacridine are seen here [10]. The interstrand hydrogen bonds (between 6-NH$_2$ of adenine and 3′-O of uridine, and between 1-NH of adenine and 2′-O of uridine) are shown by the thin straight lines.

via two different intercalative modes and geometries. The first of these binding modes results in a nearly symmetric (or "pseudosymmetric") structure, with the 9-amino group of the drug lying in the minor groove of the intercalated base pairs (Fig. 2.6a). The second binding mode results in a grossly asymmetric intercalated structure in which the 9-amino group lies in the major groove of the dinucleotides (Fig. 2.6b). Both structures are similar in the following respects: (1) The 5′-cytidine residues have a C3′-endo sugar conformation, while the 3′-guanosine residues have a C2′-endo sugar conformation, and (2) the twist angles between the these base pairs is $\sim 10°$. However, these structures differ in the magnitude of the dislocation of the helical screw axis that accompanies intercalation (Fig. 2.6). In the pseudosymmetric intercalated structure, this value is about +0.5 Å, whereas in the asymmetric intercalated structure, this value is about +2.7 Å. These conformational differences can be described as a "sliding" of base pairs on the intercalated acridine molecule. Additional analysis using molecular modeling indicated that the pseudosymmetric intercalated structure of 9-aminoacridine is possible, but the asymmetric intercalated structure is unlikely, as there are stereochemical difficulties when connecting neighboring sugar–phosphate chains to the intercalated dinucleotides. This comparison of the results from X-ray analysis and modeled structures demonstrates the utility of obtaining information from several methodologies when evaluating the structures of drug-DNA complexes.

X-ray crystallographic analyses of the complexes of a variety of acridine dyes (Fig. 2.7) with dinucleoside monophosphate sequences are listed in Table 2.1. One observation that can be derived from these studies is that the complexes have very similar backbone conformations in the immediate vicinity of the intercalation site. When an intercalator slips between adjacent base pairs, the value of the torsion angle β (around the C5′–O5′ axis; see Section 1.1.8) at the base step changes from $-175°$ (A-DNA), or $-160°$ (B-DNA), to $-230°$. This change in the torsion angle is coincident with an increase in the separation of the two base-pair planes at the intercalation site, from 3.4 to 6.8 Å. As a whole, each complex has a pseudo-twofold symmetry about a line midway between the base pairs and parallel to their mean planes. The intercalated acridine molecule lies approximately on this twofold axis, although the symmetry is not exact.

One difference of these intercalator complexes is their base pair–base pair twist angle, also called the *base-turn angle*. This is the angle of rotation between adjacent nucleotide residues, and is defined as the twist around the common axis perpendicular to the base-pair plane. It is instructive to compare the structural characteristics of a series of drug–dinucleotide complexes consisting of a set of related acridine intercalators. Table 2.1 illustrates that the value of the base-turn angle depends on both the substituents of the acridine ring, and the sequence of the dinucleoside monophosphate (column 5). As an example, the crystal structure of the proflavine 5I-CpG complex has a base-turn angle of 36°, which is the same as in normal B-DNA. This stereochemistry, therefore, does not lead to significant unwinding in double-helix DNA (or RNA) at the immediate drug intercalation site. On the other hand, when the drug is the related molecule, acridine orange, the twist angle relating the base pairs above and below the intercalated molecule is $\sim 10°$ (Table 2.1, lines 7 and 8). This stereochemistry gives rise to significant unwinding of the double helix (26°) at the immediate site of drug

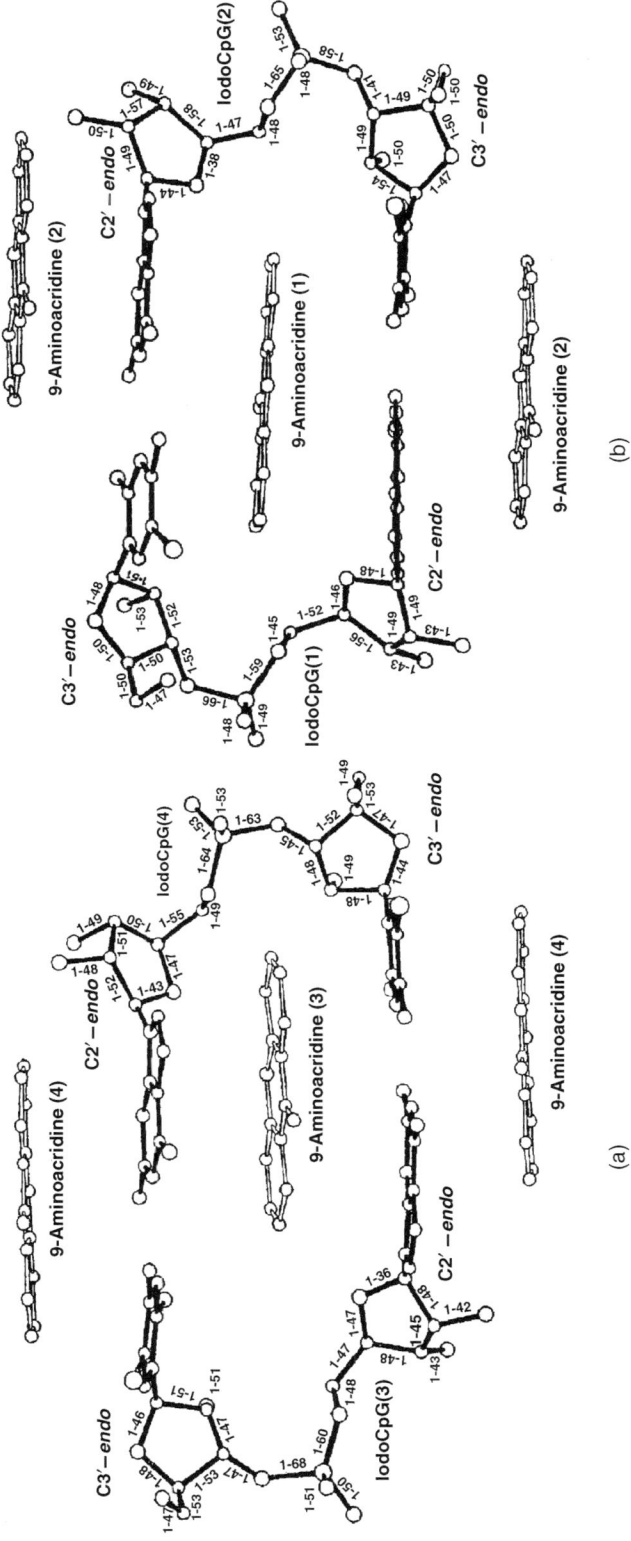

FIGURE 2.6. Two slightly different intercalative geometries found in a 9-aminoacridine-iodoCpG crystal structure, viewed approximately parallel to the plane of the guanine–cytosine base pairs and 9-aminoacridine molecule. The crystals were monoclinic, belonging to the space group $P2_1$, with a = 13.98 Å, b = 30.58 Å, c = 22.47 Å, and b = 113.9 Å. The structure was solved to atomic resolution by Patterson and Fourier methods, and refined by a combination of Fourier and sum function Fourier methods. (a) Pseudosymmetric intercalating binding mode; (b) asymmetric intercalating mode [11–13].

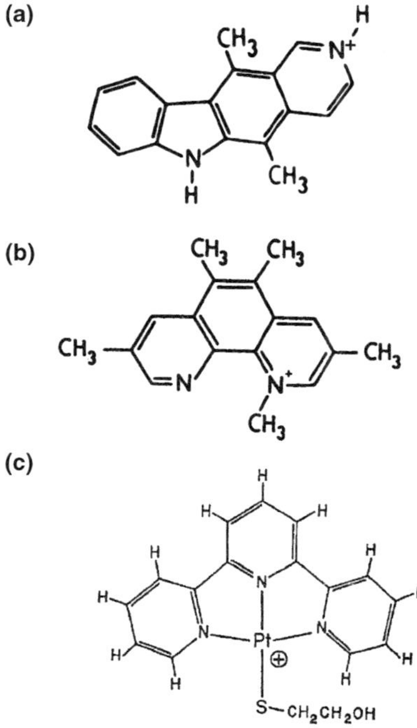

FIGURE 2.7. Structures of (a) Ellipticine, (b) 3,5,6,8-Tetramethyl-*N*-methylphenanthrolinium, and (c) 2-Hydroxyethanethiolato-2,2′,2″-terpyridineplatinum.

intercalation. These changes in twist angles appear to be related to the sugar pucker conformations within the complexes, which, in turn, are dependent on both the base sequence and the drug. The exact nature of these relationships is difficult to discern, and so it has been suggested that the sugar pucker angles are a "soft parameter" in the context of intercalation into RNA or DNA dimeric subunits. For example, all of those complexes with base-turn angles of 8–12° also have sugar pucker angles that "alternate" between the C2′-*endo* and C3′-*endo* conformations. Double-helix RNA has only C3′-*endo* puckers, whereas B-DNA has only C2′-*endo* puckers in B-DNA (for conformations of sugar rings, see Section 1.1.4). On the other hand, the proflavine-CpG and proflavine-5I-CpG complexes have base-turn angles of 34–36°, and all the sugars are C3′-*endo*. It was found [14] that proflavine-d(CpG) has a base-turn angle of 17°, which differs from the twist angle determined from studies of supercoiled DNA (Section 2.1.3). One of the two d(CpG) strands has the 5′ sugar in the C3′-*endo* conformation, whereas the 3′ sugar is in the C2′-*endo* conformation; this is written as C2′-*endo*-(3′,5′)-C3′-*endo*. The other d(CpG) strand has both sugars in the C3′-*endo* conformation. Thus, in this respect, these strands are more RNA-like.

Aggarwal et al. [15] examined the crystal structure of a complex involving proflavine with the dinucleotide duplex r(CpA) · r(UpG), neither strand of which is self-complementary. The strands of this duplex were arranged in the normal antiparallel manner, and were hydrogen-bonded to each other via standard Watson–Crick base pairs. This complex contained two proflavine molecules, rather than

TABLE 2.1. Crystallographic Studies of Complexes of Acridine Derivatves with Dinucleoside Monophosphate

Acridine Dye	Nucleic Acid Sequence	Base Pairs[a]	Separation[b] (Å)	Twist Angle[c]	Ref.
9-Aminoacridine	ApU	Hoogsteen	6.9	0°	10
9-Aminoacridine	5I-CpG	WC	6.8	4°	11
9-Aminoacridine	5I-CpG	WC	6.8	8°	12
2-Hydroxyethane thiolato-2,2′, 2″-terpyridine-Pt[d]	dCpG	WC	6.8	13°	12
Ellipticine[e]	5I-CpG	WC	6.7	10–12°	13
3,5,6,8-Tetramethyl-N-methyl phenanthrolinium[f]	5I-CpG	WC	6.7	10–12°	17
Acridine orange	CpG	WC	6.8	10°	18
Acridine orange	5I-CpG	WC	6.8	10°	19
Proflavine	CpG	WC	6.8	34°	4,20
Proflavine	CpG	WC	6.8	34°	21
Proflavine	5I-CpG	WC	6.7	36°	19
Proflavine	CpA	C·C, A·A (noncomplementary)	6.8	—	22
Proflavine	dCpG	WC	6.8	17°	14
Proflavine	(CpA)·(UpG)	WC	6.8	16°	15
N,N-Dimethylproflavine	5-ICpG	WC	6.7	8–10°	23
N,N-Dimethylproflavine	dCpdG	—	6.7	12°	24

[a]WC—Watson–Crick base pairs.
[b]Separation of the two base pairs intercalated by the drug molecule.
[c]The angle of clockwise rotation viewed along the local duplex axis in order to reach a base pair from the adjacent base pair located closer to the viewer.
[d]2-Hydroxyethanethiolato-2,2′,2″-terpyridineplatinum (see Fig. 2.7c).
[e]Ellipticine (Fig. 2.7a).
[f]3,5,6,8-Tetramethyl-N-methylphenanthrolinium (see Fig. 2.7b).

just one. The first of these was intercalated between the base pairs, while the other was stacked external to the double-helix fragment. The structure also has a relatively complex pattern of sugar puckers. The C and U riboses, at the 5′ends of the two strands, have sugar pucker angles that are both roughly C3′-*endo*, whereas the adenosine sugar has a conformation of C2′-*endo* pucker, thus giving the CpA strand the same kind of "alternation" of sugar pucker conformations as was found in one strand of the proflavine-d(CpG) structure [14]. Similarly, the UpG strand had both residues in the C3′-*endo* conformation, as was also found in proflavine-d(CpG).

Westhof et al. [16] crystallized a 1:1 complex between the non-self-complementary dinucleoside monophosphate, CpA, and proflavine. In the crystal, CpA forms a self-paired parallel chain dimer with a proflavine molecule intercalated between the protonated cytosine–cytosine (C) pair and the neutral adenine–adenine (A-A) base pair. The dimer complex exhibited a right-handed helical twist and an irregular girth. Through such a structural analysis, one can obtain insight into the binding of acridine dyes with denatured and single-stranded nucleic acids.

2.1.2. Complexes with Longer DNAs: X-Ray and Modeling Analyses

Although the dinucleotide studies described above revealed many details of acridine binding to nucleic acids in the immediate vacinity of the intercalation site, it is difficult to extrapolate from these structures to an understanding of how an intercalated acridine affects the DNA conformation at locations more distant from the site of binding.

There are not, as yet, any reports on a crystal structure of a complex between a simple acridine dye and a longer oligonucleotide sequence. However, in 1988, a nonexperimental molecular model was derived for the intercalation of proflavine at the CpG site in the duplex d(GATACGATAC) · d(CTATGCTATG) [17]. The initial structure consisted of two B-form pentanucleotides (Section 1.1.3), which were joined together at the CpG dinucleotide intercalation site. The conformation of the CpG site was based on the crystallographic studies of the d(CpG)-proflavine complex [13]. Energy minimization was then used to remove stereochemical clashes from this artificial structure, creating one with a lower energy (Fig. 2.8). The resulting structure was smoothly deformed from that of standard B-form DNA, deviating significantly in its backbone torsion angles and greatly widened major and minor grooves. Since this model was created, there have been significant improvements in both molecular modeling techniques and computational power. For example, in the year that this work was published, it was difficult to perform molecular dynamics simulations with explicit solvent and realistic salt concentrations. More modern approaches would be expected to lead to a more correct, low-energy structure.

Acridinecarboxamides are structurally related to acridine. They are known to intercalate into DNA, forming ternary complexes with mammalian topoisomerases that disrupt their cleavage and rejoining activities. For example, 9-amino-[*N*-(2-dimethylamino) ethyl]acridine-4-carboxamide (abbreviated as 9-amino-DACA; Fig. 2.9a) intercalates into DNA and the poisons topoisomerase II. At neutral pH it has two positive charges, as its pK_a values are 8.3 and 10 [27], and it has an affinity of $4 \times 10^5 \, M^{-1}$ for calf thymus DNA (CT-DNA) in 0.1 M NaCl [28]. This drug

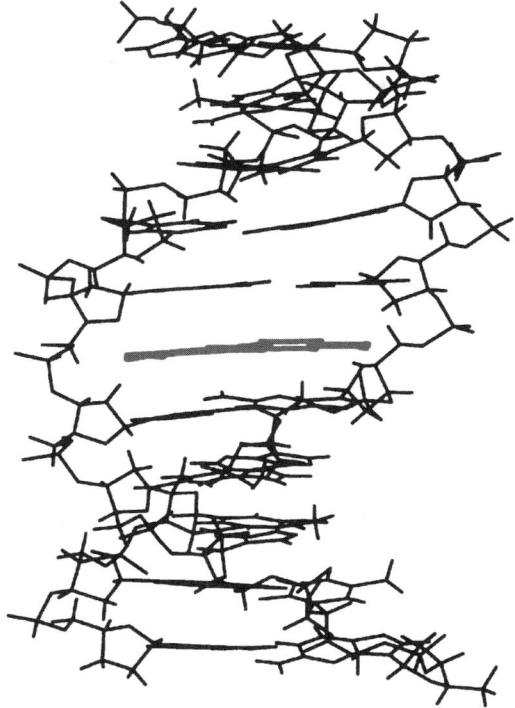

FIGURE 2.8. A molecular model of the complex of proflavine with the decamer d(GATAC-GATAC)$_2$ [17], viewed from the major groove of decamer. The proflavin molecule (red) is shown with its long axis (ordinate) aligned to the horizontal axis(abscissa). The structure was assembled from previously determined experimental pieces and energy minimized. The resulting structure deviated significantly from B-DNA. The major groove width increased from 11.4 Å in B-DNA, to a maximum of 17.6 Å, and the minor groove width was increased from 6.0 Å in B-DNA to a maximum of 9.1 Å. Helical twist angles of the base pairs are 28.5°, 35.6°, 57.1°, 33.0°, 20.9°, 28.9°, 25.3°, 44.3°, and 34.8°. The rises per base pair along the helical axis are 2.1, 3.4, 3.3, 3.1, 5.3, 2.4, 2.1, 2.2, and 0.7 Å. (See color insert.)

preferentially binds to sequences that are GC-rich [29], and induces a 17° unwinding of the DNA [28]. An X-ray crystal structure was determined for the complex of 9-amino-DACA with d(CGTACG)$_2$. This provided detailed empirical information about the effect of intercalation on the DNA structure at the location of the binding site, as well as insight into the longer-range consequences for those base pairs located farther from the binding site [25,26]. The 9-amino-DACA-d(CGTACG)$_2$ complex formed a crystal with an asymmetric unit cell containing a single strand of DNA, 1.5 drug molecules, and 29 water molecules. The two strands of the DNA have the standard conformation of a right-handed B-DNA-like duplex with Watson–Crick base pairs (Fig. 2.9b). The drug molecule intercalates between each of the CpG steps with its side chain lying in the major groove, while the protonated dimethylamino group partially occupies a position close to the N7 and O6 atoms of the guanine G$_2$, at ~3 Å (Fig. 2.9c). The X-ray crystal structure of the complex formed by 9-amino-DACA with d(CG(5-BrU)ACG)$_2$ was also

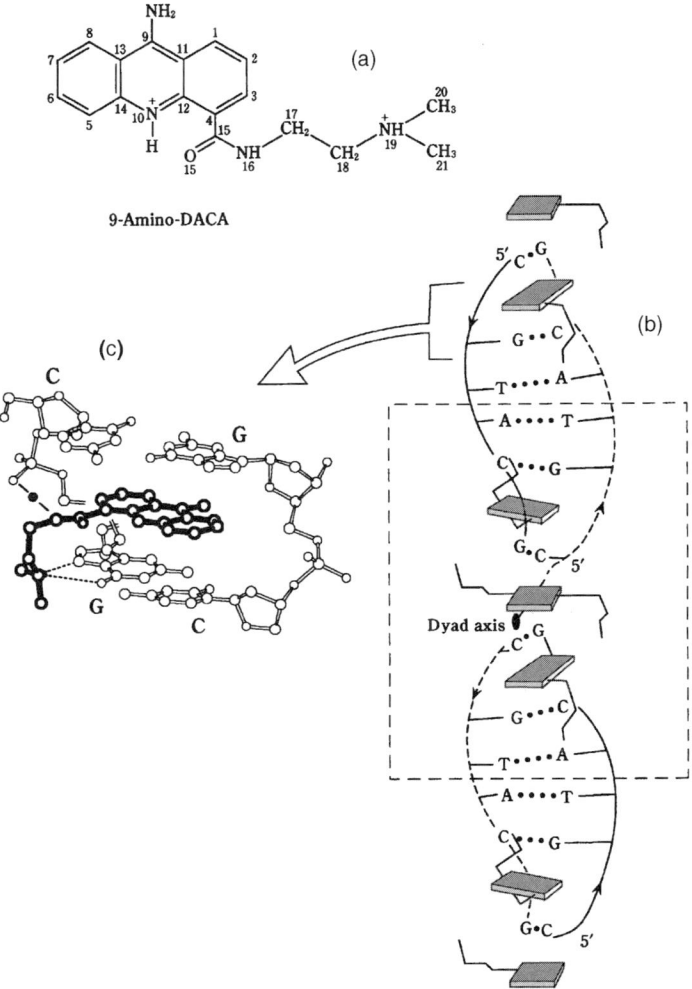

FIGURE 2.9. (a) Chemical structure of 9-amino-[N-(2-dimethylamino)ethyl]acridine-4-carboxamide (abbreviated as 9-amino-DACA); (b) schematic drawing of the structure of a complex formed between 9-amino-DACA and d(CGTACG)$_2$ as determined by X-ray crystallography [28]. (the rectangle formed by dashed lines involves two asymmetric units related by dyad of the space group $P6_4$); (c) 9-amino-DACA intercalated into d(CpG) · d(CpG) portion of d(CGTACG)$_2$. Dashed lines indicate hydrogen bonds.

studied, instead of d(CGTACG)$_2$, and was found to have a very similar intercalated conformation [29]. These studies provide an opportunity for interpreting the structure–activity relationships for biological activity among 9-amino-DACA in terms of a defined molecular structure for the DNA complex of the parent ligand. Previous studies with analogs of 9-amino-DACA revealed the importance of both the position and nature of the carboxamide side-chain for cytotoxicity, antitumor activity, and the kinetic stability of the DNA complexes formed [30, 31]. Only compounds bearing a

4-CONH(CH$_2$)$_2$NRR side-chain show significant cytotoxicity and antitumor activity, whereas moving the side-chain to position 2 or 3 produces inactive analogs. These findings are consistent with the crystal structures; when substituted in positions 2 or 3, the acridine ring cannot intercalate fully so as to maximize its stacking interactions with the base pairs and present the sidechain in the correct position for interacting with the O6/N7 atoms of guanine.

As discussed in Section 1.1.12, at the ends of chromosomes there exists a stretch of DNA, called *telomeres*, which usually consists of specialized tandem-repeating G-rich sequences. Telomeres are believed to protect the ends of the genetic code from degradation during recombination and end-to-end fusion. Under physiological conditions of monovalent cations (e.g., Na$^+$ and K$^+$), these G-rich sequences can fold, or associate, into a variety of inter- or intramolecular, four-stranded, quadruplex structures via a characteristic Hoogsteen hydrogen-bonded quartet of guanine bases. It has been shown that the presence of quadruplex folds in telomeric DNA can effectively inhibit the DNA from acting as a primer for telomerase, thereby inhibiting the synthesis of further telomeric repeats. This is of considerable interest, as the extent of telomerase activity has been linked to both cancer and the longevity of cell lines. One type of telomerase inhibitor consists of derivatives of amidoanthraquinone, which have been disubstituted at positions 1,4, 1,5, 1,8, 2,6, and 2,7 with amidoalkylamino substituents. Studies of the structures of the intercalation complexes of these drugs with G4-quadruplexes have been the subject of several X-ray structural studies [32–34]. It may be that these drugs can either stabilize quadruplexes, or lower the kinetic barriers involved in their formation, thereby inducing their presence and inhibiting telomerase.

The interaction of highly polymerized calf thymus DNA with proflavine was investigated by X-ray diffraction [35]. When this dye was at a concentration ratio of 0.17 ("dye-to-DNA phosphate ratio," or D/p) and at 100% relative humidity, the wet fiber of sodium salt of proflavine-DNA gave the diffraction pattern shown in the right half of Fig. 2.10. This figure illustrates how this diffraction pattern differs from that of B-DNA (shown in the left half of Fig. 2.10): (1) the layer line spacing (Section 1.6.2) is 48 Å, as compared to 34 Å for sodium-DNA; (2) as the hydration is lowered to 92% relative humidity, the layer line spacing for the complex decreases to 37 Å, whereas it remains at 34 Å for B-DNA fiber, at the same relative humidity; (3) for the complex, the first layer line is fairly sharp, but the second and third layer lines become progressively more diffuse. Conversely, for B-DNA layer lines 1–4 remain sharp. These results are qualitatively consistent with intercalation; the diffraction pattern arising from a "randomly intercalated DNA helix," with $D/p = 0.10$, was shown to have a layer line spacing of nearly 49 Å, as well as progressively diffuse second and third layer lines. Similar conclusions were reached from a fiber diffraction study of the binding of acridine orange with DNA [35].

2.1.3. Unwinding Angle Measurements

Intercalation models for the binding of acridines to DNA involve the postulate of drug-induced local uncoiling of the double helix. This uncoiling results in the removal and

132 INTERCALATING DRUGS

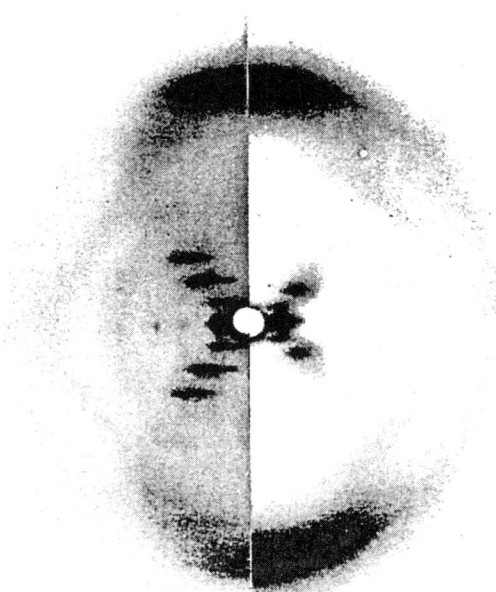

FIGURE 2.10. Comparison of X-ray diffractions of (left) B-form DNA and (right) proflavine-DNA fibers [35]. The ratio of dye to phosphate in the proflavine-DNA fiber was 0.17. The fibers were allowed to dry at 98% relative humidity in a closed vessel.

reversal of the supercoils in circular DNA, and can be followed by sedimentation coefficient measurements [36]. Ultracentrifugation analysis measures the differing rates at which the components of a sample sediment, or "sink," to the bottom of the tube. This is dependent on the relative molecular weights, charges, sizes, and shapes of the molecules in the sample. When DNA unwinds, its shape changes, becoming either more, or less, compact, thereby altering this rate of sedimentation. Curve (a) of Fig 2.11 shows the effect of proflavine on the sedimentation coefficient (S_{20}) of ϕX174 replicative-form DNA. The behavior of the closed circular DNA found here is interpreted in terms of a progressive loss of right-handed supercoils followed by the introduction of an increasing number of left-handed supercoils. Also included in the sample were "nicked circles" of DNA, in which one of the two strands were broken. When a drug intercalates into these DNA circles, the unwinding of the duplex can be compensated by rotations of the bonds in the unbroken strand, rather than by a change in the supercoiling. Thus, the inclusion of nicked circles in the sample preparation [Fig. 2.11, curve (b)] provided a useful internal control, and revealed how drug binding alone altered the sedimentation coefficient S_{20}, rather than effects other than changes in supercoil density. On binding proflavine the nicked circles show a monotonic decrease in S_{20} characteristic of intercalation (the same effect shown by linear DNA molecules).

The minimum in curve (a) in Fig. 2.11 for the closed circles, called the *equivalence point*, represents the level of drug binding at which the accumulated

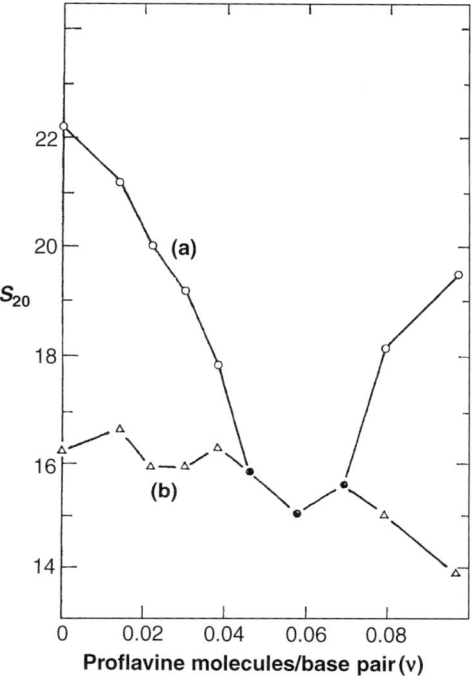

FIGURE 2.11. Effect of proflavine on the sedimentation coefficient (S_{20}) of φX174 replicative form DNA. Rotor speed was 32,000 rev/min (rpm). The symbols (O) and (Δ) show results from resolved boundaries of closed circles and nicked circles, respectively. The symbols (●) is used to indicate weight-average sedimentation coefficients from nonresolved boundaries [36].

uncoiling due to intercalation just balances the initial number of superhelical turns in the DNA. Thus

$$\tau = \frac{-N\phi v}{360} \qquad (2.1)$$

where τ is the initial number of superhelical turns in the absence of drug, N is the number of base pairs per replicative form of φX174 DNA, and φ is the unwinding angle per bound drug molecule. For this proflavine system, $\tau = -13$, $N = 10000$, and $v = 0.057$ (Fig. 2.11). This leads to the conclusion that the unwinding angle (φ) is 8.2° per bound proflavine molecule. Since the twist angle of *un*intercalated B-DNA is 36°, one can determine the twist angle between the two base pairs at the proflavine intercalation site to be (36° − 8° =)28°. Note that this is different from the twist angle (17°) found by X-ray analysis of the proflavine-d(CpG) complex (Section 2.1.1) [14]. Clearly, the intercalation geometry found in a single crystal of acridine–dinucleoside monophosphate is not, in general, applicable to the intercalation into a longer DNA duplex. The difference probably arises from the competing physical/structural forces that are present only in longer DNAs.

2.1.4. NMR Studies

NMR spectroscopy has been used extensively to study the intercalation of many clinical and experimental antitumor drugs. Many comparative studies in the earlier years used ^{1}H NMR to examine the intercalation of drugs into calf thymus DNA (e.g., see Ref. [37]). Among the most studied were derivatives of 9-acridinylmethanesulfonanilide (AMSA), which is derived from 9-aminoacridine (Fig. 2.1d) by addition of an aniline sidechain at position 9. The wide variety of derivatives that have been studied has allowed an evaluation of the effects of substituents on antileukemic activity [38,39].

The hydrogen-bonded imino protons of Watson–Crick base pairs in calf thymus DNA gave rise to two groups of resonances centered at $\sim$13.72 and $\sim$12.75 ppm for the A-T and G-C base pairs, respectively (Fig. 2.12). The half-widths of these two envelopes were rather broad ($\sim$190 Hz), and each envelope consisted of many overlapping lines with different chemical shifts. These arose primarily from "ring current effects" on the imino protons by neighboring base pairs. Ring current effects arise from the flow of electrons within the rings of aromatic groups; this generates a

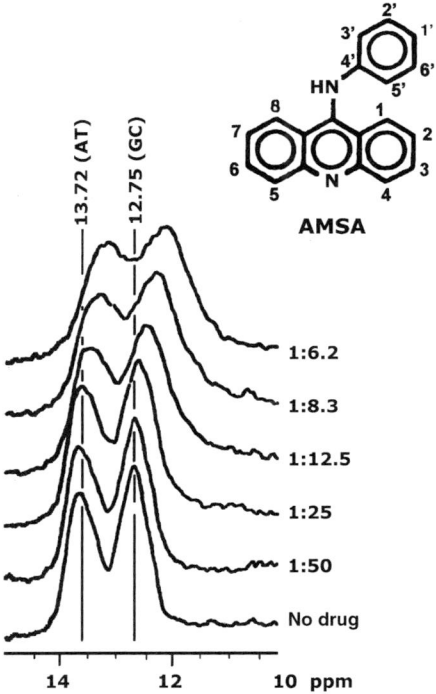

FIGURE 2.12. ^{1}H NMR (300-MHz) spectra of the low-field resonances of DNA as a function of AMSA level at 35°C [37]. The drug/DNA base-pair ratios are given by the spectra. Spectra were obtained in the correlation mode and required 23 min of signal averaging; 4000 transients were collected with a sweep time of 0.3 s and a predelay time of 50 ms. Spectra are linebroadened (exponential multiplication) by 2.5 Hz.

small, localized, magnetic field that can alter the resonance frequencies and lineshapes of nearby nuclei, especially protons. As AMSA was titrated into CT-DNA (at 35°C), the imino resonances of both the A-T and G-C base pairs were observed to shift upfield (i.e., they moved to smaller ppm values), and their resonance envelopes broadened even more (Fig. 2.12). These observations are often indicators of intercalative binding. In general, when a drug intercalates into the DNA, the ring current shifts exerted on each base pair at the intercalation site by the adjacent base pair are replaced by the larger ring current shifts from the drug. Thus, for each drug bound, the resonances from the two neighboring base pairs will be affected.

As described in Section 1.4.4, if a drug binds weakly to the DNA, then it will reside in the binding site for a short time, and it is said to be in "fast exchange" between bound and unbound states. Therefore, the observed chemical shift will be the weighted average of the chemical shifts from the bound and unbound ligands. As the drug concentration is increased, the equilibrium is shifted, increasing the percentage of drug molecules that are bound. This causes the low-field resonance (at higher ppm values, and associated with the unbound ligand) to gradually shift upfield (smaller ppm values) toward the chemical shift of the bound ligand. This is what was observed for the AMSA-DNA system, as shown in Fig. 2.12. Very different spectral changes were observed when drugs bind via modes other than intercalation, such as groove binding (Chapter 3) or nonspecific outside binding [41]. Thus, on the basis of the changes in the chemical shifts, NMR indicates that proflavine and its dimethyl, diethyl, and diisopropyl derivatives, all intercalate into DNA, to greater or lesser extents. In addition, one can conclude that at this temperature, the kinetics of their binding is intermediate to fast, relative to the NMR time-scale [38]. At temperatures lower than 35°C, however, the lifetime of this drug in the binding site becomes longer, and the chemical exchange between bound and unbound drugs becomes slower. This often enables the observation of separate peaks for the bound and unbound drugs (Section 1.4.4). In this case, intercalation may cause a new, upfield-shifted, resonance or group of resonances to appear, while the intensity of the original peak will be correspondingly decreased.

Proton NMR was used in this way to investigate the intercalation of 9-aminoacridine into the duplex formed by complementary decamer, $d(AT)_5 \cdot d(AT)_5$ [40]. The thymine imino protons of this duplex oligomer can be clearly seen in the NMR spectrum (at 6°C) between 13 and 14 ppm (Fig. 2.13a). These imino protons are involved in the hydrogen bonding of the base pairs and are shielded from the solvent by the surrounding DNA. This protection slows their exchange with the aqueous solvent and allows them to be observed by NMR [41]. In the absence of any bound drugs, the NMR spectrum of this oligonucleotide contains five resonances arising from these hydrogen-bonded T residues, and demonstrates that this is a symmetric piece of DNA in a duplex form, with all 10 T bases involved in stable hydrogen bonds. The assignments of these five resonances were determined on the basis of their temperature dependence [45]. Since the terminal base pairs of DNA have fewer base-stacking interactions to stabilize them, they tend to fray even at low temperatures. Hence, the "melting" (or disassociation) of a duplex DNA starts from its terminal ends and moves toward the middle. As the temperature of this oligomer was increased from 6°C to 23°C, peaks E, A, and B broadened and disappeared. Resonance E broadened first, and it was therefore attributed to the terminal base pairs at position 1. Considering the order in which they are broadened, the resonances were assigned to the following base pairs: E = 1, A = 2, B = 3, C = 4, and D = 5.

As 9-aminoacridine was titrated into d(AT)$_5$ · d(AT)$_5$, a new signal appeared in the ^{1}H NMR spectrum in the upfield proton region at 12.3 ppm, with a concomitant decrease in the intensity of resonance at 13.2 ppm (Fig. 2.13b) [42]. Moreover, at the relatively high drug : base pair ratio of 1 : 2, the original 13.2 ppm peak shifts slightly upfield (13.2 → 12.9 ppm). As noted above, the generation of an upfield-shifted peaks often indicates intercalative binding, in which the moderate shielding effect of an

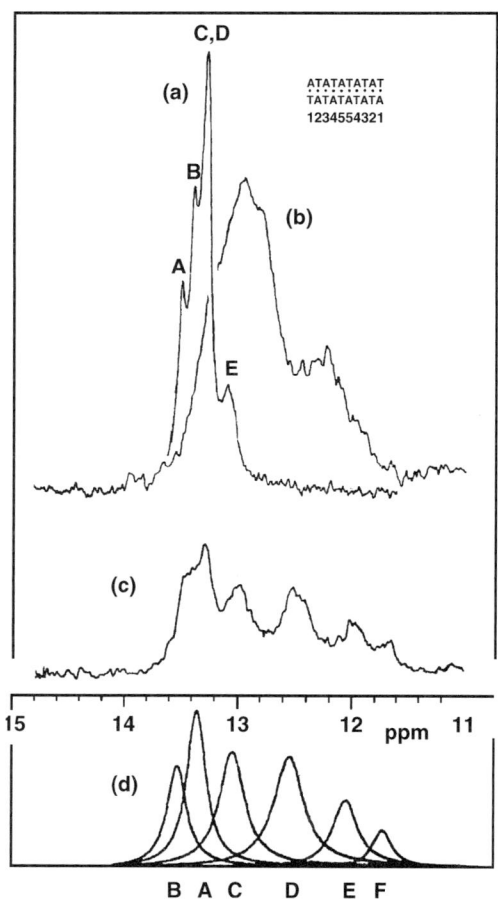

FIGURE 2.13. (a) Low-field spectrum of d(AT)$_5$·d(AT)$_5$ at 300 MHz and at 5°C, with assignment of the resonances indicated [40]; (b) low field spectrum of 2 : 1 9-aminoacridine [9-AA]-d(AT)$_5$·d(AT)$_5$ complex at 300 MHz and at 6°C; (c) spectrum of a 1 : 1 complex of pyrazole bis(acridine) [BAPY] with d(AT)$_5$·d(AT)$_5$. In (d), the observed spectrum (c) was simulated by superimposing a sum of resonances with positions and intensities shown here. The relative intensity of the peaks (D + E + F) corresponds to 41.4% of the total intensity.

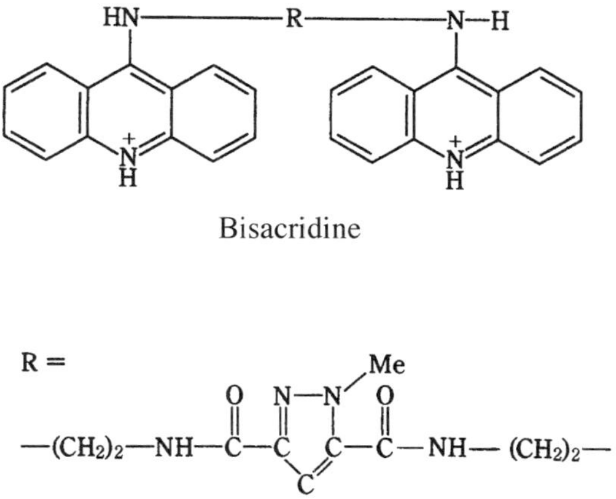

FIGURE 2.14. Two acridine chromophores joined by a molecular linker [40,42].

adjacent base pair is replaced by the somewhat greater shielding from the ligand chromophore. The intensity of the shifted peak relative to the unshifted peak is an important diagnostic monitor of the drug-binding properties. When one 9- aminoacridine chromophore is bound to this duplex (each strand of which contains five thymines), 2 of the 10 possible thymine resonances ($=20\%$) are shifted upfield together; this is consistent with a mono-intercalated, symmetric structure. This is observed in Fig. 2.13b. Similarly, when two chromophores were intercalated, 40% of the resonance intensity was shifted upfield.

It is also possible to create drugs containing two acridine chromophores joined by a molecular linking chain (Fig. 2.14) [40,42]. Figure 2.13c depicts the imino proton region of the spectrum of the complex of bis(acridine)pyrazole with the same oligomer duplex, $d(AT)_5 \cdot d(AT)_5$, at a molar ratio of 1 : 1, and at a temperature of 6°C. The interpretation of this spectrum was aided by simulating it with the sum of the curves shown in Fig. 2.13d. According to the positions and intensities of these simulated peaks, resonances A, B, and C appear to be nearly unshifted by the binding of the drug, while the resonances D, E, and F are shifted upfield (lower ppm); and the relative intensity of the peaks (D + E + F) is 41.4% of the total intensity. This, and other data, indicate that bis(acridine)pyrazole is a bis- intercalator. Similar NMR studies were made for other bis(acridine) derivatives, and the results for some of them are summarized in Table 2.2. For more examples of bis- and tris- intercalators, see Section 2.5.

2.1.5. Fluorescence Spectra

Fluorescence spectroscopy can be used to determine the orientation and tightness of a ligand when it is bound to DNA. In a simple solution, molecules tumble freely and are randomly oriented. In such solutions, a fluorescent molecule, like acridine, does not produce a polarized fluorescence spectrum. However, if these randomly oriented molecules are made stationary, and are illuminated with appropriately polarized

TABLE 2.2. Structures of Bis(acridine) Compounds and NMR Spectral Characteristics Observed for Their 1 : 1 Complexes with $d(AT)_5 \cdot d(AT)_5$

Compound Number	Nomenclature	Atoms in Linker	Linker R	Length of R (Å)	Change in Drug-Shifted Peaks	Intercalation Type	Intensity Shifted (%)
0	9AA				1 peak, broad	Mono-	20
1	C4	4	$-(CH_2)_4-$	6.3	1 peak, broad	Mono-	24
2	C6	6	$-(CH_2)_6-$	8.8	1 peak, broad	Mono-	26
3	C7	7	$-(CH_2)_7-$	10.0	1 peak, broad	Mono-	39
4	C8	8	$-(CH_2)_8-$	11.3	1 peak, broad	Bis-	38
5	C10	10	$-(CH_2)_{10}-$	13.8	2 sharp peaks	Bis-	41
6	Amd-6	6	$-(CH_2)NHC(=O)(CH_2)_2-$	8.6	1 peak, broad	Mono-	23
7	Amd-7a	7	$-(CH_2)_2NHC=O(CH_2)_3-$	9.8	1 peak, broad	Bis-	38
8	Amd-7b	7	$-(CH_2)_3NHC=O(CH_2)_2-$	9.9	1 peak, broad	Bis-	40
9	Amd-8	8	$-CH_2)_3NHC=O(CH_2)_3-$	11.1	1 peak, broad	Bis-	35
10	Spermidine	8	$-(CH_2)_3NH(CH_2)_4-$	11.2	1 peak, broad	Bis-	45
11	Spermidine (BAS)	12	$-(CH_2)_3NH(CH_2)_4NH(CH_2)_3-$	16.1	2 sharp peaks	Bis-	41
12	Pyrazol (BAPY)	11		13.7	3 sharp peaks	Bis-	42

Source: Ref. 40.

radiation, then the emitted light will be polarized. The intensity of the emitted light, when observed through a polarizing prism, will depend on the orientation of the prism. The polarized spectrum for a simple, dilute solution of an acridine derivative, quinacrine (atebrin), is shown in Fig. 2.15 [43]. Although almost no polarization is observed in the simple aqueous solution, substantial polarization is observed when quinacrine is in a solution containing DNA. This is because nearly all the quinacrine molecules are tightly bound to DNA, and since DNA tumbles extremely slowly in solution (i.e., it has an exceedingly small rotational diffusion constant), the bound drug is effectively immobilized.

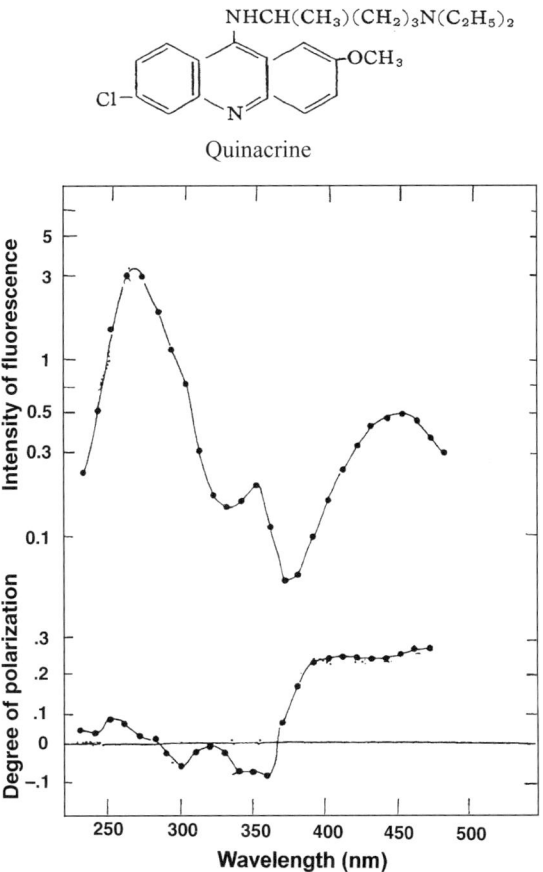

FIGURE 2.15. Excitation and polarization spectra for quinacrine (or atebrin, or acrinamine; see the chemical formula given above the plot) in DNA solution [44]. The upper curve shows the relative fluorescence intensity per unit concentration as a function of the wavelength of excitation, uncorrected for the energy distribution in the exciting radiation, for dilute solution of quinacrine in DNA. The lower curve shows the degree of polarization of the 90° emission at each wavelength. The DNA concentration was about 3.6×10^{-4} M in nucleotides. The quinacrine was 2.65×10^{-6} M in the DNA solution, corresponding to one dye molecule per 68 nucleotide pairs.

More information can be obtained if the DNA molecules in the sample are oriented in a uniform manner and excited with the polarized light. For example, when a solution of DNA is flowed, the length of the DNA (the molecular z axis) tends to align along the direction of the flow. So, if the excitation beam enters along the x axis (perpendicular to the flow), then the emission can be observed along the y axis. The degree of polarization is given by the difference between the two intensities measured with the plane of polarization of the analyzing prism respectively parallel to z and to x axes. This technique is called *flow-aligned linear dichroism fluorescence* (FLDF), and it is exactly analogous to FLD, which was discussed in Section 1.2.3. The data obtained for the binding of quinacrine to DNA are compatible with a model in which its $\pi^* \leftarrow \pi$ transition moment (at 450 nm) is perpendicular to the DNA chain. This is consistent with an intercalation model [44]. In related work, it has been found that the fluorescence of acridine is quenched on stacking with guanine. This property was used to establish that certain acridine derivatives can stabilize a triple-helix structure in polynucleotides [45].

2.1.6. Flow Linear Dichroism Spectra/Studies

Similarly, flow linear dichroism (FLD) experiments (Section 1.2.3) also support the intercalation model. The FLD measurements on DNA and complexes with quinacrine are presented in Table 2.3 [44]. Here, dichroism is given as the fractional change in absorbance due to flow. The flow linear dichroism of the quinacrine–DNA complex observed at 432.5 and 455.0 nm, is due entirely to the dye, and directly indicates its contribution to the dichroism at 260 nm, since both transitions are in the plane of the rings. It is known that the dye accounts for only 34% of the absorbance of the complex

TABLE 2.3. Flow Linear Dichroism[a] (FLD) of DNA and DNA–Quinacrine

	Wavelength (nm)		
	260.0	432.5	455.0
DNA only			
Dichroism, Z polarization	−0.19	—	—
Dichroism, Y polarization	+0.086	—	—
DNA–quinacrine			
Dichroism, Z polarization	−0.24	−0.25	−0.24
Dichroism, Y polarization	+0.11	+0.12	+0.11

[a]Dichroism is tabulated as the fractional change in absorbance of the DNA or the DNA–quinacrine complex when the solution is subjected to flow, with the polarizing prism either parallel or perpendicular to the flow axis. The shear rate at the windows was $1.18 \times 10^4 \, \text{s}^{-1}$. The absorbances in the stationary solution show the distribution of absorption between DNA, bound quinacrine, and the unattached dye. The DNA concentration (as nucleotides) when measured alone was 3.2×10^{-3} M, and in the complex, 3.6×10^{-3} M. The quinacrine bound to DNA was 0.661×10^{-3} M, involving an oligonucleotide as the third strand. The acridine was found to be incorporated in that oligonucleotide.

Source: Ref. 44.

at 260 nm. Nevertheless, the observed dichroism at 260 nm is very similar to those found at longer wavelengths. Thus, the dichroism due to DNA in the complex must be nearly the same as that of the dye. Also, the close similarity of the dichroism values for the quinacrine and DNA itself demonstrates that the plane of the purine–pyrimidine pairs must be parallel to the direction of the transition moment of quinacrine at long wavelengths. Similar results have also been obtained with proflavine.

2.1.7. Circular Dichroism Studies

Acridine dyes do not possess an intrinsic optical activity; rather, they acquire optical activity when bound to DNA—that is, when they are free to tumble in solution, they do not present a circular dichroism (CD) spectrum. However, when they are bound to DNA, they obtain an induced CD spectrum. The CD spectra for the complex of proflavine with calf thymus DNA (CT-DNA) complex are shown in Fig. 2.16a [46]. Here, $\Delta\varepsilon_p$ is the measured reading of the molar extinction coefficient $\Delta\varepsilon = (\varepsilon_l - \varepsilon_r)$, expressed per mole of DNA phosphate. The spectra show $\Delta\varepsilon_p$ for both high and low values of r, where r is defined as the total bound ligand concentration divided by the total DNA phosphate concentration. As shown in Fig. 2.16b, there is a regular increase

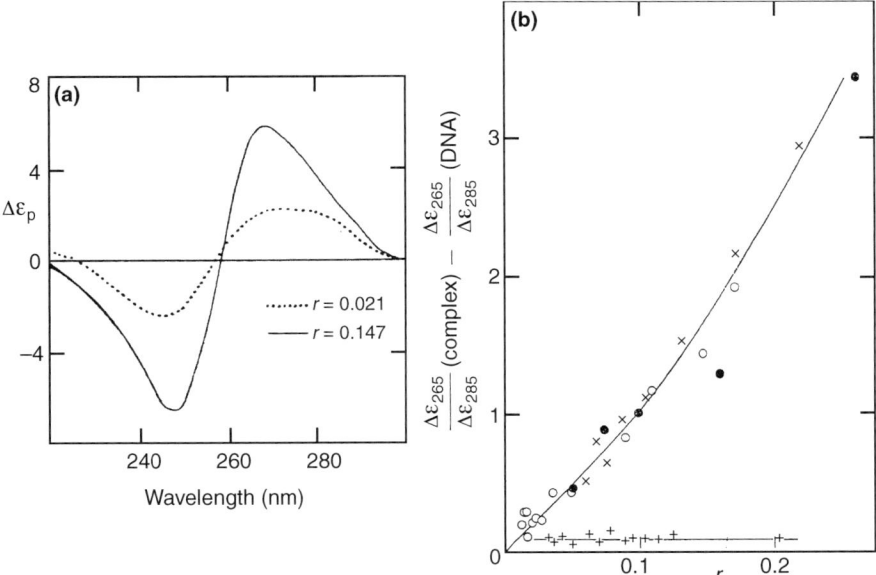

FIGURE 2.16. (a) Ultraviolet CD spectra of complexes of proflavine with calf thymus DNA [46], in a buffer of 0.1 M Na-phosphate, at pH 6.9. Here, r is defined as the total bound ligand concentration divided by the total DNA phosphate concentration. (b) The variation with r of the positive peak in the ultraviolet CD spectra of calf thymus DNA–ligand complexes; this illustrates the different shapes of the ultraviolet CD for complexes of aminoacridines and ethidium bromide with calf thymus DNA ○ = proflavine; ● = 4-ethyl-9-aminoacridine; × = 9-aminoacridine; + = ethidium bromide.

in the magnitude of a DNA-like doublet centered at ~265 nm, which increases as r increases; that is, r increases almost linearly with respect to $(\Delta\varepsilon_{265}/\Delta\varepsilon_{285})$[complex] − $(\Delta\varepsilon_{265}/\Delta\varepsilon_{285})$[DNA]. Interestingly, this is different from the CD spectra of the ethidium bromide–CT-DNA complex, the shape of which varies very little with respect to r in the region 265–285 nm. This may arise from the presence in ethidium of an additional transition in this region.

2.1.8. Biochemical Implications—a Few Examples

The ability to precisely cut a strand of DNA at a specific target sequence, called *site-selective scission*, is an important and challenging biochemical goal. Its application to the long strands of DNA from a higher organisms could lead to much more precise drug activity, as well as many other new biotechnologies. A common method of enhancing the sequence specificity of a drug is to attach a short strand of DNA that is complementary to the desired target sequence. Alternatively, one may attach a peptide, or peptide mimic, that has the desired sequence sensitivity. A variant of this approach was used by Ren et al. [28], for example, to restrict the digestion of single-stranded DNA by S1 to a predetermined position. Since S1 is an enzyme that specifically hydrolyzes single-stranded DNA, it is blocked in regions where the DNA exists as a duplex. In this case, they used a synthetic derivative of DNA, called a *peptide nucleic acid* (PNA), in which the sugars and phosphate backbone are replaced with repeating N-(2-aminoethyl)-glycine units linked by peptide bonds. Additionally, an acridine moiety was attached to the N terminus of the PNA to act as an anchor. Using a PNA chain that is complementary to the target sequence, they showed that H2NOC-(Gly)-TCTTCGCGGA-NHCO-(CH$_2$)-CO-acridine can effectively block the digestion of the single-stranded DNA sequence -AGAAGCGCCTGG-. The same PNA sequence without acridine was found to be less effective in blocking S1 activity, as it helps to hold the PNA in place.

In another example, the selectivity of aminoglycoside antibiotics was increased by attaching derivatives of the acridine molecule. In their unmodified form, aminoglycosides are highly selective in their preferential binding of RNA over DNA, but they exhibit much less discrimination between different RNA molecules. Several modified aminoglycosides were therefore prepared and tested for their affinity and selectivity in targeting the viral RNA site, the HIV-1 Rev response element (RRE) [47]. The attached acridine moieties in these experiments were conjugates of 9-aminoacridine (based on the drugs neomycin B, kanamycin A, and tobramycin). These derivatives were also compared to a series of dimeric aminoglycosides. All of the modified aminoglycosides exhibited a high affinity for the HIV-1 Rev binding site on RRE, but only a few compounds had a high specificity for RRE. Compared to the acridine conjugates, the dimeric and unmodified aminoglycosides exhibit selectivity for RNA over DNA, but show little ability to differentiate between different RNA molecules. To optimize these derivatives for RRE specificity, a series of neomycin–acridine conjugates with variable linker lengths were synthesized and evaluated; those that had the shortest linkers were observed to have the best RRE specificity.

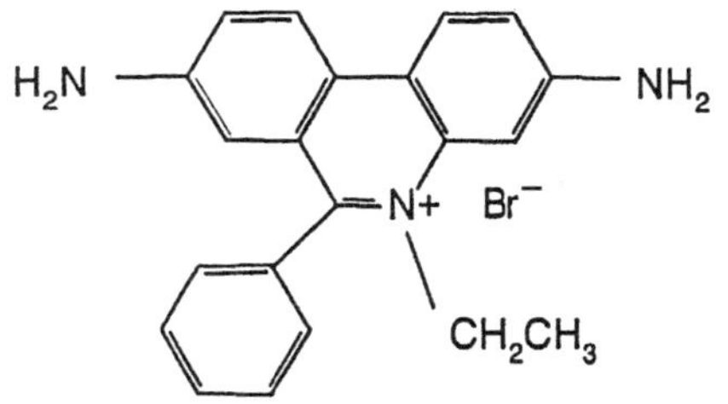

FIGURE 2.17. Chemical structure of ethidium bromide.

2.2. ETHIDIUM BROMIDE

Ethidium bromide (Fig. 2.17) is a trypanocidal drug [48], or an antimicrobial agent, which inhibits DNA-dependent nucleic acid synthesis [49,50]. It does this by very efficiently intercalating into the DNA, disrupting cell processes. As such, it is also a powerful mutagen and carcinogen. Its complexes with DNA have lower sedimentation constants, increased viscosity, and higher melting temperature compared to uncomplexed DNA [50]. The isolated drug absorbs in the visible region with a maximum at 480 nm. After complexation with DNA, the absorption maximum shifts to 518 nm and the intensity at the maximum decreases.

2.2.1. X-Ray and Modeling Analyses of Ethidium Bromide–Dinucleoside Monophosphate Crystals

Early X-ray analyses of the DNA intercalation structure of ethidium bromide included examinations of the complexes with two self-comlementary dinucleoside monophosphates: 5-iodouracyl(3'-5')adenosine and 5-iodocytidylyl(3'-5')guanosine [51,52]. Although each complex crystallized into a different lattice environment, both resulted in similar intercalated structures having a Watson–Crick-type duplex. The intercalative geometries of both complexes contain the sugar-puckering pattern C3'-*endo* (3'-5')C2'-*endo*—that is, the 5' sugar is in the C2'-*endo* conformation, and the 3' sugar is in the C3'-*endo* conformation. This, along with other conformational features in the sugar–phosphate backbone, allows the base pairs to separate by 6.7 Å and gives rise to a twist angle of 8–10° between base pairs above and below the phenanthridinium ring.

Later, an X-ray analysis of the complex of ethidium bromide with uridylyl(3'-5')-adenosine was performed at higher resolution and containing no iodine, or other heavy atoms [53]. Although this allowed the atomic positions to be determined with greater accuracy than in the previous work, it resulted in a very similar structure (Fig. 2.18); the detailed conformations of the sugar-phosphate backbone, and the ethidium intercalative geometry, were almost identical to that seen in the earlier, heavy-atom structures. This indicates that the early studies are valid, and that the presence of iodine (covalently bound to the C5 position on uridine and cytosine), did not alter the basic sugar-phosphate geometry, nor the mode of ethidium intercalation in these studies.

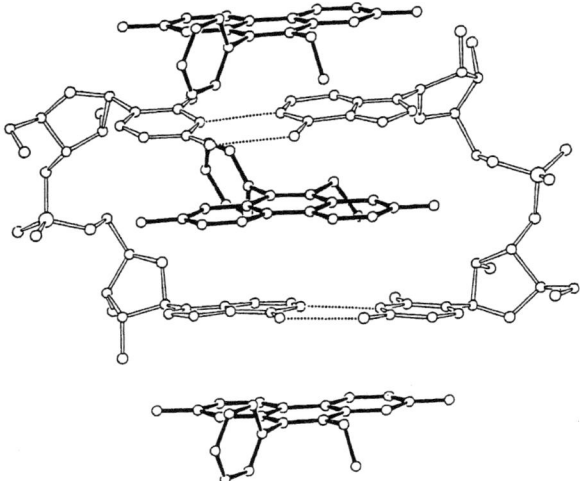

FIGURE 2.18. The ethidium–UpA complex viewed from an oblique angle to base pairs and ethidium molecules [53]. The top and bottom drug molecules are sandwiched between the dinucleotide stack, whereas the central one is intercalated.

2.2.1.1. Early Computational Modeling Studies Based on Dinucleoside Structures

A very early attempt (in 1977) was made to create a purely computational model of the intercalation structure of ethidium bromide into longer DNA, as experimental data were limited [54]. This was done by adding idealized B-DNA to either side of an ethidium:dinucleoside monophosphate complex, which resulted in a structure that had no steric clashes (Fig. 2.19). This early approach was even more limited by the computing power and techniques available than the work discussed for acridines (Section 2.1.2). Hence it was not possible to generate a fully relaxed structure in a realistic solvent environment. Nevertheless, several conclusions are worth mentioning: (1) When ethidium bromide intercalates, the helical axis of B-DNA is displaced by 1.0 Å; (2) base pairs in the immediate region of intercalation are twisted by 10°, indicating that ethidium bromide induces an angular unwinding of -26° at the binding site; (3) base pairs at the intercalation site are tilted relative to one another by 8°; and (4) the normal C2′-*endo* deoxyribose ring puckering in B-DNA is altered into a mixed puckering pattern where the 3′ base is in the C3′-*endo* conformation, and the 5′ base is in the C2′-*endo* [written as C3′-*endo*(3′-5′)C2′-*endo*). These conformational changes in the DNA produced a "kinked" structure. Additional structures were computed in which kinks were positioned at periodic intervals, with varying numbers of base-pair separation. For example, the superhelical DNA structure obtained when kinks were repeated every 10 bp was found to have a diameter of ~100 Å and about 1.5 turns per 140 handed base-pairs. It was suggested that such a structure might be utilized in the organization of DNA within the nucleosome in chromatin [54].

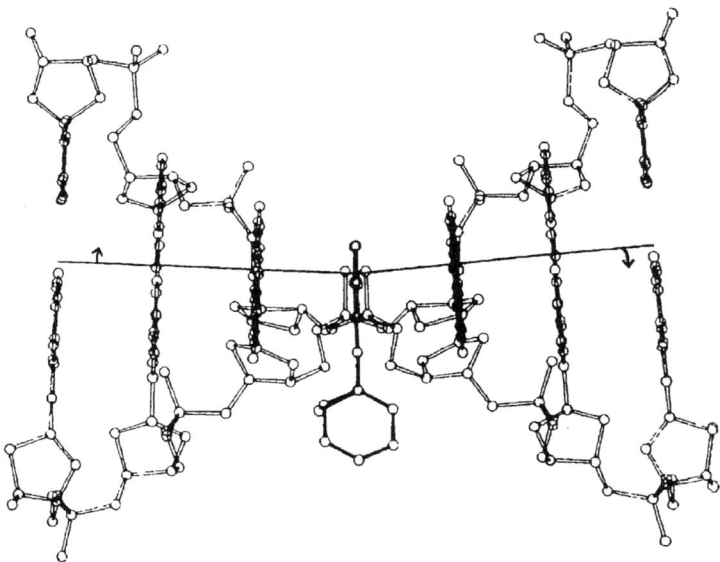

FIGURE 2.19. An ethidium–DNA-binding model constructed by utilizing structural information obtained by X-ray crystallography of ethidium–dinucleoside monophosphate complexes [54].

2.2.2. Fiber X-Ray Diffraction Studies

The X-ray diffraction pattern obtained from fibers of the ethidium complex with calf thymus (CT) DNA differs from those of B-DNA (Fig. 2.20a, c). In the former, there are no well-defined layer lines, and the first equatorial reflection line is further away from the center of the pattern [57]. The change in the position of this reflection indicates that the molecule has a smaller average diameter. The intermolecular separation (at 92% relative humidity) was calculated for sodium DNA as 26.6 Å and for DNA–ethidium as 23.9 Å. The overall intensity distribution in the diffraction pattern of the DNA–ethidium complex changed very little with the relative humidity of the environment, except for variations in the position of the equatorial reflection, which corresponds to the increased separation of the DNA molecules caused by that increased relative humidity. Intercalation appeared to be a reversible denaturation of the DNA structure, as the X-ray data indicated that removal of the drug molecule allowed the DNA to return to its native conformation [55].

X-ray diffraction patterns of better quality were obtained by studying the polycrystalline fibers of CT-DNA complexed with 2-hydroxyethane thiolato(2,2′,2″-terpyridine)platinum(II) (PtTS, Fig. 2.20b). This molecule is spatially homologous to ethidium, and contains a heavy atom, Pt [56]. When polycrystalline fibers containing PtTS were bound to calf thymus DNA by intercalation (Fig. 2.20b), the presence of the heavy metal in this molecular cation enabled an investigation of the distribution of intercalated molecules along the DNA backbone. The most striking feature is the presence of three layer lines in addition to the equator, with spacings of 10.2, 5.1, and

146 INTERCALATING DRUGS

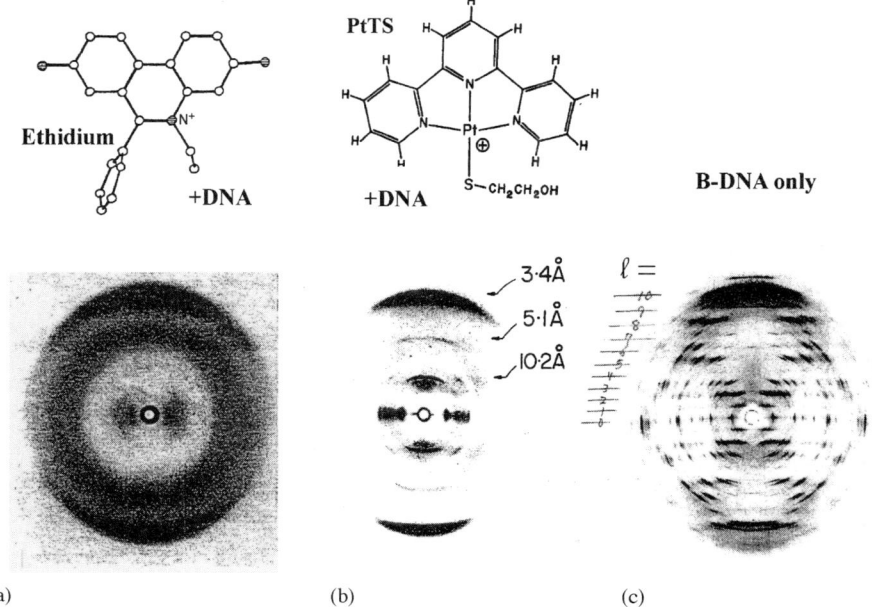

FIGURE 2.20. (a) X-ray diffraction from fiber of the sodium deoxyribonucleic acid–ethidium complex at 92% relative humidity, and with one drug molecule for every 5.5 base pairs [55]; (b) X-ray fiber diffraction diagram of fiber pulled from a solution having r(PtTS molecule/nucleotide residue) = 0.19 in 1 mM sodium phosphate, pH = 6.8, 3 mM sodium chloride the fiber was maintained at 75% relative humidity [56]; here, PtTS is the 2-hydroxyethan thiolate tetrapyridine platinum(II) cation]; (c) X-ray diffraction fiber diagram of B-DNA [57].

3.4 Å. These do not appear to vary with relative humidity, and within the range of 66–98%, the lowest angle equatorial reflection is very intense. At 66% relative humidity it can be indexed together with three other equatorial reflections on a trigonal lattice having a = b = 31.2 Å. Increased humidity expanded the lattice until at 85% relative humidity, when another intense but diffuse reflection (or "arc") appears on the equator. Higher humidity improves the crystal structure, producing the most reflections at 98% relative humidity. The effective molecular dimensions ($d_{\text{eff}} \approx a/3^{1/2}$) are a function of relative humidity, as shown in Fig. 2.21. At 66% relative humidity, the d_{eff} is 18 Å, which is comparable to the value of 19.3 Å reported for B-DNA. The values for d_{eff} at high humidities represent molecular center-to-center distances that are swollen by the hydration of the lattice. The d_{eff} of 24 Å for DNA containing intercalated PtTS (Fig. 2.20b) at 92% relative humidity may be compared to the values for DNA with ethidium bromide (23.9 Å) and for B-DNA itself (25 Å). The similarity of the effective molecular diameters of DNA containing intercalated PtTS and ethidium arises from their structural similarity.

The above mentioned fiber diffraction photographs of DNA containing intercalated ethidium bromide (Fig. 2.20a) do not show any interpretable meridional reflections apart from the usual base stacking one at 3.4 Å. If the ethidium molecules were

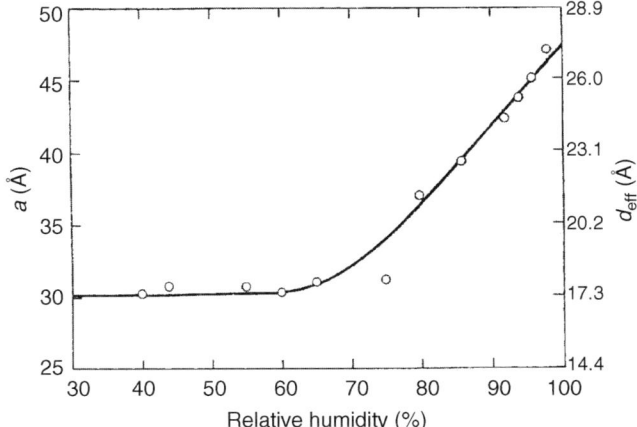

FIGURE 2.21. Dependence of the lattice constant a and the effective molecular diameter ($d_{\text{eff}} = a/3^{1/2}$) on relative humidity observed for PtTS/DNA complex fibers [56].

complexed at irregular intervals along the DNA molecule this would result in a molecule with an irregular pitch and resulting in the absence of well-defined layer lines in the diffraction pattern. If these intercalated cations bind to DNA with two base pairs per drug, then a repeating unit is created (Fig. 2.22), and the vertical separation between the two base pairs is expected to be 10.2 Å (3 × 3.4 Å). The presence of platinum in the intercalator of this chemical repeat should dramatically enhance the 10.2 Å periodicity in electron density, as PtTS contains almost twice as many electrons as either ethidium, or an average base pair. As anticipated, fibers pulled from PtTS-DNA solutions having a PtTS/nucleotide ratio of ~0.2 indeed give new layer

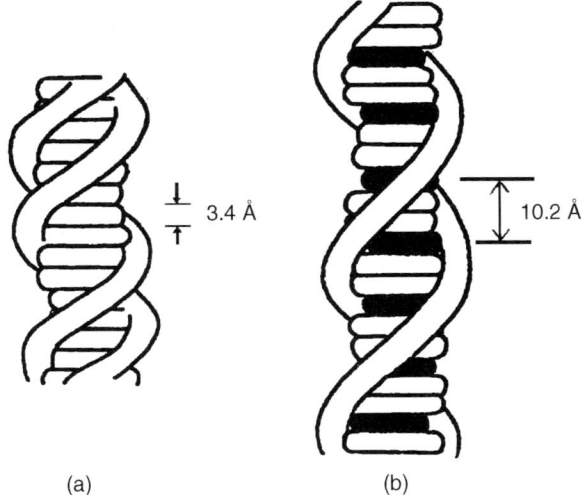

FIGURE 2.22. Schematic representations of a normal helix (a) and the neighbor exclusion binding (b) of an intercalator (shaded area) to a double-stranded polynucleotide [56].

lines with maxima near or on the meridian at 10.2, 5.1, and 3.4 Å (Fig. 2.20b). Thus, the fiber diffraction patterns were interpreted to mean that the helix requires 10.2 Å repeating units consisting of one PtTS molecular cation intercalated between two base pairs. The fact that the intercalators are not more closely spaced provides strong support for the neighbor exclusion model discussed earlier [57].

2.2.3. Visible Absorption Spectra

Both DNA and RNA are able to cause a bathochromic shift from 479 nm to 516–518 nm in the absorption spectrum of ethidium bromide. This effect is evident as a change in color of the solution from yellow-orange to bright pink. Figure 2.23 shows the effect of increasing concentrations of DNA from T2 bacteriophage ("T2-DNA") on the absorption spectrum of ethidium bromide [58]. The peak progressively shifts toward a limit (curve E) that represents the spectrum of the drug in a fully complexed form. All the curves pass through an isosbestic point at 510 nm, indicating that they result from the contribution of two forms of ethidium, free and bound, each form having a characteristic absorption spectrum. The proportions of the free and bound drug can be estimated from measurements of the absorbance in the region where the difference between curves A and E is great.

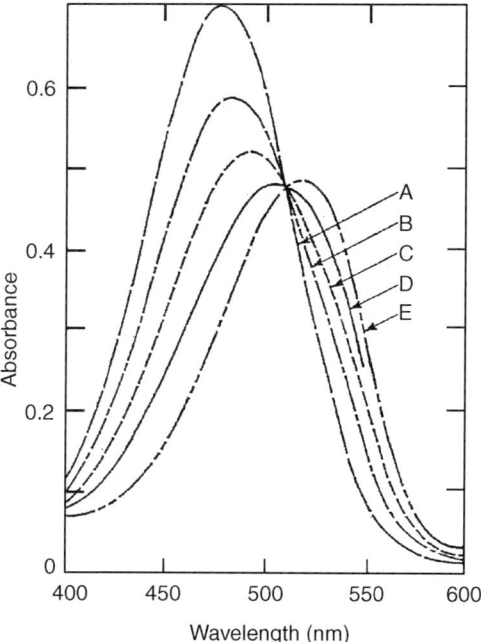

FIGURE 2.23. Effect of DNA on the absorption spectrum of ethidium bromide [58]. The solution contained 1.25×10^{-4} M ethidium bromide, and the aborbance was measured using a 1-cm light path. T2 DNA was present at the following concentrations: curve A, zero; B, 1.5×10^{-4} M; C, 3×10^{-4} M; D, 5×10^{-4} M; E, 1.2×10^{-3} M.

When all of the DNA binding sites of ethidium are the same and are independent of each other, the equilibrium binding process can be described by simple mass action terms, giving

$$k = \frac{c(n-r)}{r} \quad \text{or} \quad r/c = \frac{n}{k} - \frac{r}{k} \tag{2.2}$$

where k is the dissociation constant of the complex, c is the molar concentration of free drug, r is the number of drug molecules bound per nucleotide, and n is the number of total binding sites per nucleotide. According to this treatment, a plot of r/c versus r would yield a straight line with a gradient $-1/k$, and an intercept on the r axis equal to n. Such a plot (called a *Scatchard plot*) for T2 DNA is shown in Fig. 2.24. As seen here, a straight line could be drawn through most of the points and values of k and n were calculated to be $k = 5 \times 10^{-7}$ M and $n = 0.19$ [58].

This type of stoichiometric analysis of drug–DNA interaction is simple and attractive. However, as was pointed out by McGhee and von Hippel [59], Eq. (2.2) involves a serious defect in enumerating the number of free sites along the DNA duplex, as the number of free sites is assumed to be always proportional to the number of unoccupied base pairs. To solve this problem, they derived the following equation, to replace Eq. (2.2):

$$\frac{r}{c} = \left(\frac{1}{k}\right)(1-n^*r)\left[\frac{(1-n^*r)}{\{1-(n^*-1)r\}}\right]^{n^*-1} \tag{2.3}$$

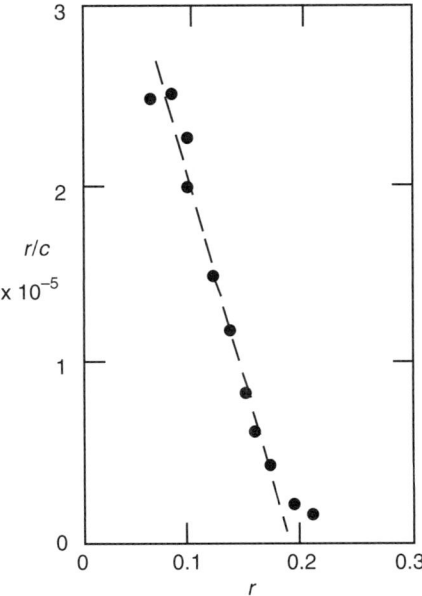

FIGURE 2.24. A Scatchard plot of ethidium bound to T2 DNA, in 0.04 M Tris-HCl buffer (pH = 7.9). Dotted line indicates what is expected from Eq. (2.2).

150 INTERCALATING DRUGS

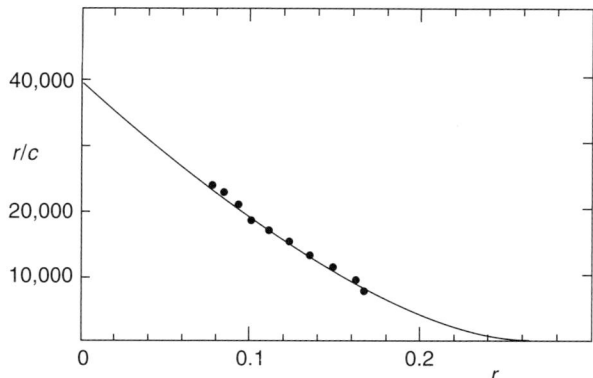

FIGURE 2.25. A *Scatchard plot* (●) of ethidium bromide bound to plasmid pBR322 DNA. Solvent: buffer-T (35 mM Tris-HCl, pH = 8.0, 72 mM KCl, 5 mM MgCl$_2$, 5 mM dithiothreitol, 5 mM spermidine and 0.01% bovine serum albumin. Solid line shows a theoretical curve on the basis of the equation developed by McGhee and von Hippel [Eq. (2.3)] in which it was assumed that $k = 2.5 \times 10^{-5}$ M (or $K = 40000$ M^{-1}) and $n^* = 3.3$ [9].

where n^* is the number of base pairs occupied by one drug molecule bound to DNA duplex. When r/c is plotted against r (*Scatchard plot*) using this equation, it is a curved, rather than straight, line. An example of the interaction of ethidium with the DNA plasmid, pBR322, is shown in Fig. 2.25. It should also be pointed out here that the dissociation constant k, or the equilibrium constant K ($=1/k$), of the ethidium–DNA interaction depends greatly on the constituents of the solvent. Table 2.4 demonstrates how the contents of buffer affect the ethidium–DNA interaction.

2.2.4. Circular Dichroism Spectra

The ends of eukaryotic chromosomes contain specialized structures called *telomeres*, which protect the natural termini from degradation or fusion processes. Telomeres contain DNA with variable lengths of simple tandemly repeated sequences, in which

TABLE 2.4. Equilibrium Constant (K) of Ethidium Bromide–Salmon Sperm DNA Interaction Determined in Various Solvents

Buffer: 35 mM Tris-HCl, pH = 8.0+	Equilibrium Constant (K_{eq})
	1.1×10^6 M^{-1}
+72 mM KCl	3.0×10^5 M^{-1}
+72 mM KCl + 0.5 mM MgCl$_2$	6.0×10^4 M^{-1}
+72 mM KCl + 0.5 mM MgCl$_2$ + 5 mM DTT	5.6×10^4 M^{-1}
+72 mM KCl + 0.5 mM MgCl$_2$ + 5 mM DTT + 5 mM spermidine	5.0×10^4 M^{-1}
+72 mM KCl + 0.5 mM MgCl$_2$ + 5 mM DTT + 5 mM spermidine +0.01% bovine serum albumin	4.0×10^4 M^{-1}

Source: Ref. 9.

there are generally clusters of G in one of the strands. Such sequences can adopt interesting conformations, which has focused attention on the *in vitro* properties of oligonucleotide models containing single or tandemly repeated G clusters (Section 1.1.12). In an elecrtrophoresis experiment, the octanucleotide, dT_4G_4 shows a mobility in the presence of Na^+ that is equivalent to a four-stranded, tetramer quadruplex structure. On the other hand, a longer strand, with the tandem repeat sequence, $d(T_4 G_4)_4$, exhibits anomalously high mobility in the presence of Na^+ or K^+. This was attributed to the formation of a more compact, folded quadruplex structure. Guo et al. [60] examined the interaction of these two molecules with ethidium bromide by circular dichroism spectroscopy. As is evident from a comparison of their CD spectra, the structure of the tetrameric complex with dT_4G_4 (Fig. 2.26a) differs fundamentally from that of the tandem repeat $d(T_4G_4)_4$ (Fig. 2.26b) in the presence of Na^+. These spectra were considered to be consistent with the idea that the former forms a fourfold G-quartet structure, while the repeat has a folded-back structure in which only two layers can form [61]. Ethidium associated tightly with the former structure by a mechanism with some degree of intercalation of the phenanthridinium ring between the $5'$-distal G bases; the dye shows no strong binding to the repeat structure in the presence of Na^+. The presence of K^+, however, appears to allow partial formation of four-layer G4 structures that now interact with ethidium (Fig. 2.26c,d). Such differences between the behavior of quadruplexes stabilized by Na^+ and K^+ have been seen many times, and have been attributed to the sizes of the cations and the possible presence of kinetic barriers during formation (Section 1.1.12). Thus,

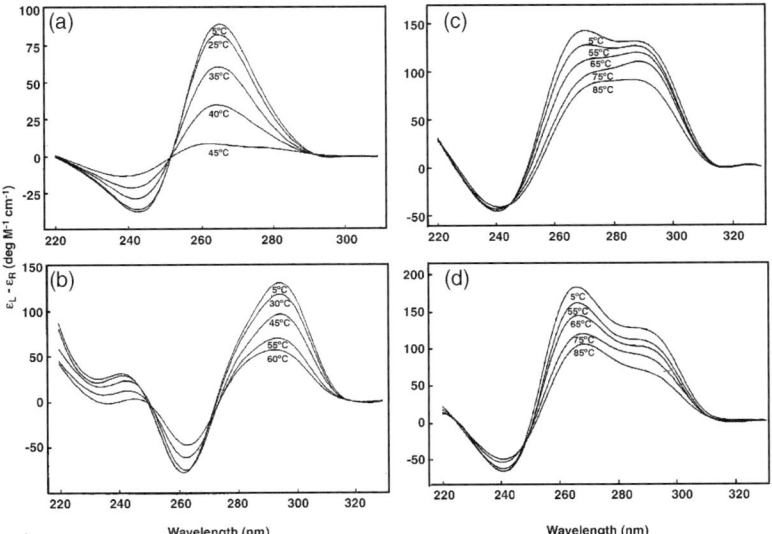

FIGURE 2.26. Temperature-dependent circular dichroism spectra: (a) $dT_4 G_4$ 100 μM; (b) d$(T_4 G_4)_4$ 25 μM [solvent for (a) and (b): 10 mM sodium phosphate buffer (pH 7), 0.1 mM EDTA and 200 mM NaCl; incubated at 5°C for 1 h]; (c) $d(T_4G_4)_4$ (25 μM) without ethidium; (d) $d(T_4 G_4)_4$ (25 μM) with ethidium (25 μM) [solvent for (c) and (d): 10 mM potassium phosphate butter (pH 7), 0.1 mM EDTA, and 200 mM KCl; incubated at 5°C for 1 h] [60].

152 INTERCALATING DRUGS

ethidium provided one of the first examples of agents other than metal ions that differentially bind to DNA *in vitro*, and thus may potentially affect telomerase activity.

2.2.5. Fluorescence Spectra

When a solution of DNA or RNA is mixed with a solution of ethidium, there is a very marked increase in the ethidium fluorescence. Figure 2.27 shows the fluorescence excitation spectra for free ethidium and ethidium in the presence of a large excess of DNA (a) and RNA (b) [62]. The width and the maxima of the fluorescence emission spectra ($\lambda_{em} = 590$ nm) are identical for the free and bound ethidium. When the excitation energy is in the region of the spectrum where there is no polynucleotide absorption, the ratio Q between the quantum efficiency of bound and free ethidium is independent of wavelength, and is 21 for DNA and 26 for RNA. In the ultraviolet region, where the polynucleotide has absorption features, the quantity Q is not independent of wavelength. Figure 2.28 displays the ratio Q_λ/Q_{vis} as a function of λ in the region of DNA absorption [62]. This ratio should be compared with the ratio $(A_{DNA} + A_{dye})/A_{dye}$, where A denotes the absorbance. If there is any energy transfer from DNA to ethidium, then Q_λ/Q_{vis}

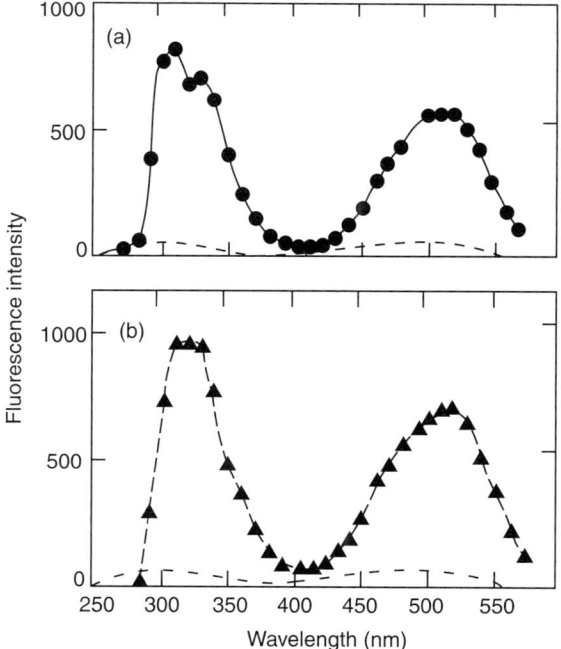

FIGURE 2.27. Apparent excitation spectra of pure solution of ethidium bromide and a solution of ethidium bromide with a large excess of native calf thymus DNA or RNA. Solutions were in 0.1 M NaCl, 0.1 M Tris-HCl (pH 7.5). DNA and RNA concentrations were 350 and 500 µg/mL, respectively. Emission monochromator was set at 500 nm, and no corrections were made. (a) --.-- .--, ethidium; –●—●–, ethidium + DNA; (b) --.--.--, ethidium; –▲—▲–, ethidium + DNA [62].

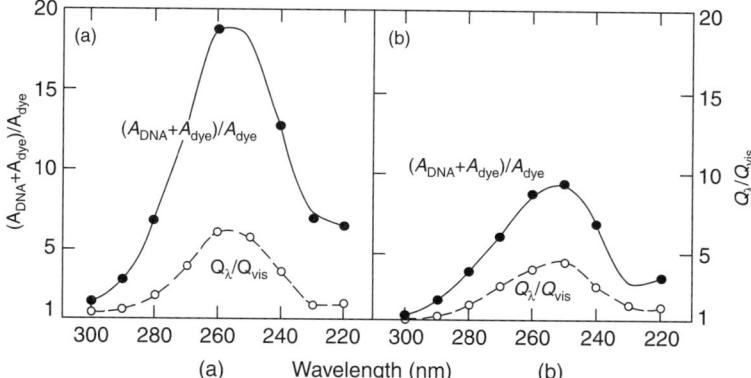

FIGURE 2.28. Ratios Q_λ/Q_{vis} (--o—o--) and $(A_{DNA} + A_{dye})/A_{dye}$ (-●—●-) were measured in the region of DNA absorption as a function of wavelength for two values of nucleotide/ethidium ratio (P/D). (a) P/D = 28; (b) P/D = 14. Ethidium concentration was in all experiments 4 μg/mL; saline concentration was 0.01 M NaCl, 10^{-3} M EDTA. Calf thymus DNA was used [63].

must be greater than unity; that is, the ratio

$$\frac{(Q_\lambda/Q_{vis}) - 1}{\{(A_{DNA} + A_{dye})/A_{dye}\} - 1}$$

measures the fraction of the light absorbed by DNA that is transferred to dye. This ratio, directly calculated from the results of Fig. 2.28, shows that, for a nucleotide/ethidium ratio (P/D) of 14, about half the amount of energy absorbed by DNA is transferred to ethidium. The variation in Q_λ/Q_{vis} relative to P/D is illustrated in Fig. 2.29 [62]. As no further increases in Q_λ/Q_{vis} occur after P/D > 20, energy transferred to ethidium does not originate from a base that is more than ~5 bp away on either side of the ethidium molecule.

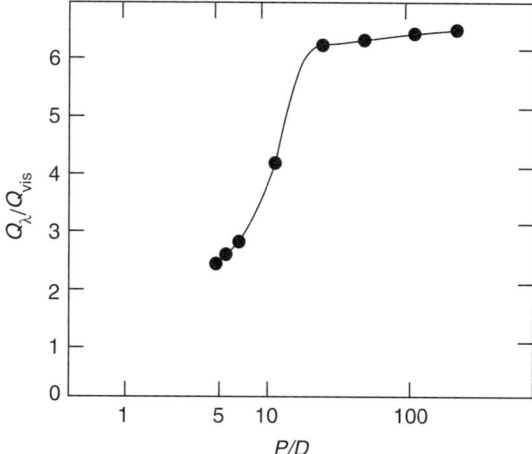

FIGURE 2.29. Variation of Q_λ/Q_{vis} ratio as a function of P/D [62].

2.2.6. Raman Spectra

Ethidium intercalation into DNA causes pronounced changes in the Raman spectrum of each component [64]. In Figure 2.30, the Raman spectra (450–1750 cm^{-1} of nucleosomal calf thymus DNA (160 bp), ethidium bromide, and a 1 : 10 complex (1 ethidium molecule to 10 bp of DNA) are shown, using near-infrared excitation (725 nm). The Raman spectra of DNA and ethidium are both rich in Raman-active vibrations. Their combination in the spectrum of the complex produces an elaborate Raman signature with many overlapping features.

By subtracting the spectrum of the complex from the sum (composite) of the spectra of the individual components, one obtains a difference spectrum that reveals the changes on binding. In this case, the difference spectrum reveals substantial changes in Raman marker bands for both DNA and ethidium bromide arising from complex formation (Fig. 2.31). However, because band overlap is so extensive, many of the difference features are difficult to assign to one or the other constituent. To assist band

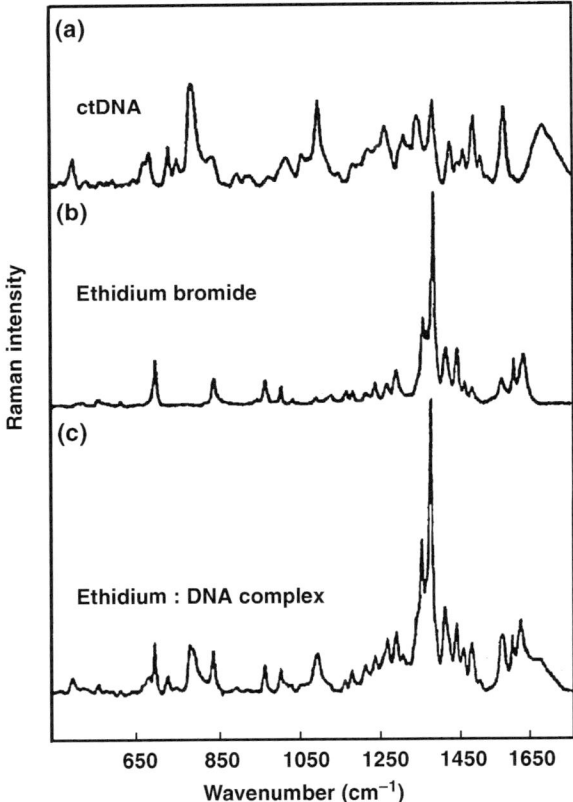

FIGURE 2.30. Raman spectra in the 450–1750 cm^{-1} of nucleosomal calf thymus DNA (a), ethidium bromide (b), and an ethidium : DNA (1 : 10) complex (c). For comparison, the ordinate of the DNA spectrum has been amplified twofold relative to those of ethidium bromide and the ethidium : DNA complex [64].

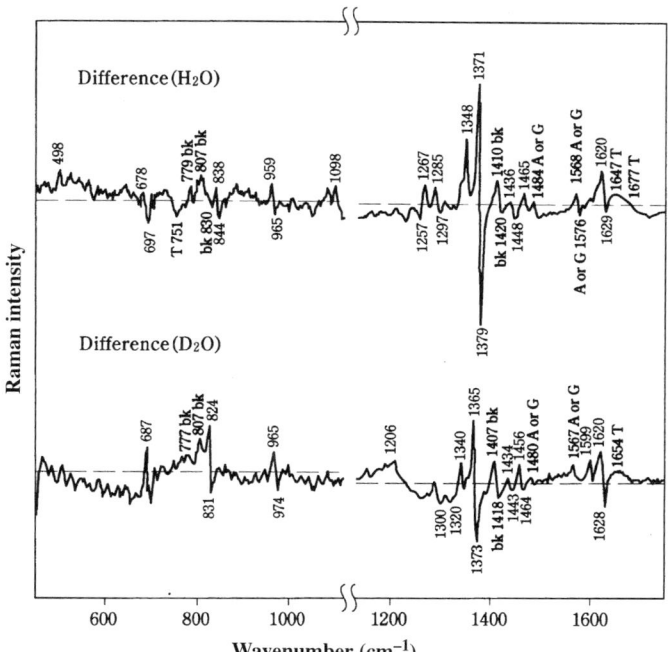

FIGURE 2.31. Raman difference spectra for H$_2$O (top) and D$_2$O (bottom) solutions of ethidium–DNA complexes, generated by subtracting in each case the spectrum sum of the constituents from the spectrum of the complex, specifically, complex−(DNA + ethidium). The ordinate scale for the 450–1100 cm^{-1} region is expanded three-fold relative to that of the 1200–1700 cm^{-1} region to enhance detail. Labels indicate bands (cm^{-1}) that can be assigned to either the DNA backbone (bk) or bases (A, C, G, T) [64].

assignments, complementary data were collected from corresponding D$_2$O solutions. The replacement of H with D (^{2}H) changes the vibrational frequencies of many modes, especially those that involve (directly or indirectly) hydrogens that are exchanged with the solvent. The additional Raman band shifts resulting from deuteration provided an additional frame of reference for band assignments in the difference spectra.

Comparison of the H$_2$O and D$_2$O Raman difference spectra of ethidium–DNA (Fig. 2.31) shows that deuteration causes frequency shifts for the majority of bands of the phenanthridinium ring. These shifts result from strong coupling of the phenanthridinium ring vibrations with those of the amino substituents at both ends of the ring. The phenyl marker at 1000 cm^{-1} was employed as an internal intensity standard. Additional bands that can be assigned to the phenyl ring at 1209 and 1602 cm^{-1} do not appear in either difference spectrum, indicating that bands of the phenyl ring are not perturbed on formation of the ethidium–DNA complex. These bands are also invariant to deuteration, which suggests limited vibrational coupling between the phenanthridinium ring and phenyl substituent.

The bands that can be assigned to DNA base vibrations are seen in the difference spectra of Fig. 2.31. The bands that are sensitive to changes in base hydrogen-bonding arrangements (bands near 1482, 1575, and 1650 cm^{-1}) are observed. This suggests that hydrogen bond formation between phenanthridinium amino substituents and the DNA bases may help to stabilize the intercalated ethidium molecule. Interestingly, although intercalation has been shown to dramatically increase the vertical stacking between base pairs, no significant recovery of Raman hypochromism is apparent. This suggests that reduced stacking between neighboring DNA bases at the intercalation site is effectively compensated by stacking between the bases and the intercalated phenanthridinium ring.

Changes in Raman bands of the DNA backbone (labeled "bk" in Fig. 2.31) are consistent with a major rearrangement of the DNA backbone conformation. The observed shifts (788 → 779, 830 → 807, and 1420 → 1410 cm^{-1}) are reminiscent of those reported for the B → A transformation of a poly(dA-dT) · poly(dA-dT) fiber as a function of relative humidity [65]. These bands are insensitive to deuteration and thus appear at nearly the same wavenumber values in both H$_2$O and D$_2$O difference spectra. Although the DNA backbone Raman markers in the ethidium–DNA complex more closely resemble those of A-DNA, rather than B-DNA, there is no Raman evidence of C3'-endo/anti deoxynucleoside conformations. This suggests that any changes in the DNA backbone conformation that may be required to accommodate the intercalated phenanthridinium ring, are highly localized in the phosphdiester moieties, and do not require changes in the sugar conformations. This also suggests fundamental differences between the molecular mechanisms governing ethidium intercalations in dinucleotides (mentioned above) and genomic long DNA.

This suggestion was further supported by the Raman experiments of Yuzaki and Hamaguchi [66]. By the use of 1064 nm excitation, and sample solutions with high concentrations of ethidium (up to ethidium/base pair ratio = 1/3), they found changes in the Raman spectra indicating a structural transition of DNA from the B form to an A-like form with increasing the ethidium concentration. Two Raman bands that are assignable to the deoxyribose-linked phosphodiester network (5'C-O-P-O-C3') appear around 800 cm^{-1}. In the B form, these bands are located at 835 and 790 cm^{-1}. In the A form these bands are at 808 and 785 cm^{-1}. The observed spectra were consistent with these known features.

It is interesting to note that, by polarized Raman spectroscopy [67], the phenanthridinium plane of the ethidium molecule intercalated in a long B-DNA is not found exactly perpendicular to the axis of the B-DNA. The Raman spectrum of ethidium DNA fiber (Fig. 2.32) shows that the in-plane bond stretching vibration of the phenanthridinium ring (1377 cm^{-1}) is much stronger for the orientation (bb) than for the orientation (cc). Because the polarizability oscillation of this vibration occurs mostly in the plane of the phenanthridinium rings, the fact that $I_{cc} < I_{bb}$ is considered to indicate that the phenanthridinium plane is oriented roughly perpendicular to the fiber axis (c axis), as expected for an intercalated ethidium molecule. That the (bb) orientation for this band is still fairly strong, however, suggests some tilting of the phenanthridinium plane within the fiber. For a quantitative analysis of this orientation, an examination of the Raman scattering tensor for the 1377 cm^{-1} band is required.

In an aqueous solution of ethidium bromide (1.6 mg/mL in 100 mM NaCl, pH 7, 20°C), the depolarizing ratio of the Raman band at 1377 cm^{-1} was found to be

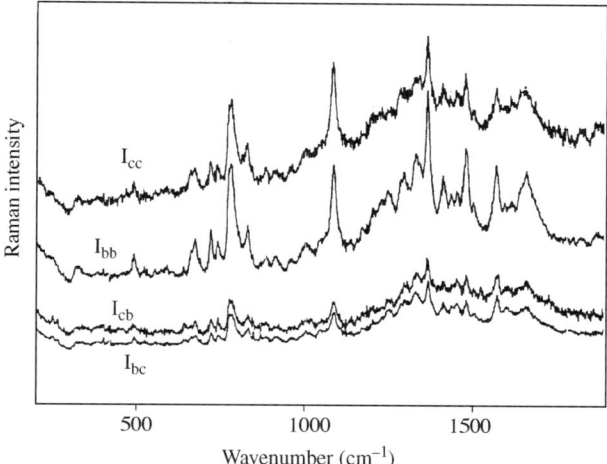

FIGURE 2.32. Polarized Raman spectra (785 nm) in the region 200–1900 cm^{-1} of an oriented fiber of the $R = 0.05$ EtBr:DNA complex. The spectral traces are labeled from top to bottom (I_{cc}, I_{bb}, I_{cb}, I_{bc}) in accordance with the polarized Raman intensity components defined in the text [67].

$\rho = I_{\parallel}/I_{\perp} = 0.15 \pm 0.01$; thus, the band is fairly polarized. Next, a single crystal of ethidium bromide was prepared, and it was confirmed to be monoclinic and in space group $P2_1/c$ with unit cell dimensions (a = 9.58 Å, b = 10.70 Å, c = 20.24 Å, β- = 106.3°) as reported previously [68]. The 785 nm excited Raman spectrum of this single crystal (Fig. 2.33) exhibits a dominant group of bands centered near 1377 cm^{-1}. Because a similar Raman signature has been observed for the planar phenanthrene molecule, the 1377 cm^{-1} band of ethidium is assigned to the planar phenanthridinium moiety. Specifically, the 1377 cm^{-1} band can be attributed a symmetric in-plane stretching vibration of conjugated C=C and C=N bonds of the phenanthridinium ring. On the basis of this assignment, the principal axes (xyz) of the 1377 cm^{-1} Raman tensor are selected as shown in the upper left corner of Fig. 2.33, such that y is along the line connecting atoms N23 and N24, x is perpendicular to y in the phenanthridinium plane, and z is perpendicular to the phenanthridinium plane.

In lieu of the single sharp band at 1377 cm^{-1} in the solution Raman spectrum of ethidium bromide, the crystal spectrum exhibits a closely spaced doublet with the center of gravity at 1377 cm^{-1} (Fig. 2.33). The twin peaks, which are resolved at 1373 and 1382 cm^{-1}, are attributed to crystal field splitting of the 1377 cm^{-1} vibration. The Raman spectrum of the ethidium bromide crystal also exhibits a sharp band at 1349 cm^{-1} as the crystal counterpart to the 1351 cm^{-1} band observed in the solution spectrum. This Raman marker is assigned to an ethidium vibrational mode that is distinct from the 1377 cm^{-1} mode. On the basis of these observations Tsuboi et al. [67] used the band intensity ratios $I_{aa}/I_{dd} = 0.26 \pm 0.01$ and $I_{aa}/I_{ad} = 2.5 \pm 0.5$ from Fig. 2.33 and $r = \rho = I_{\parallel}/I_{\perp} = 0.15$, as well as atomic coordinates given by Subramanian [68], to calculate the Raman tensor. The result was $r_1 = \alpha_{xx}/\alpha_{zz} = 23.5 \pm 2.5$ and $r_2 = \alpha_{yy}/\alpha_{zz} = 10.0 \pm 1.0$, as is shown in the upper left corner of Fig. 2.33.

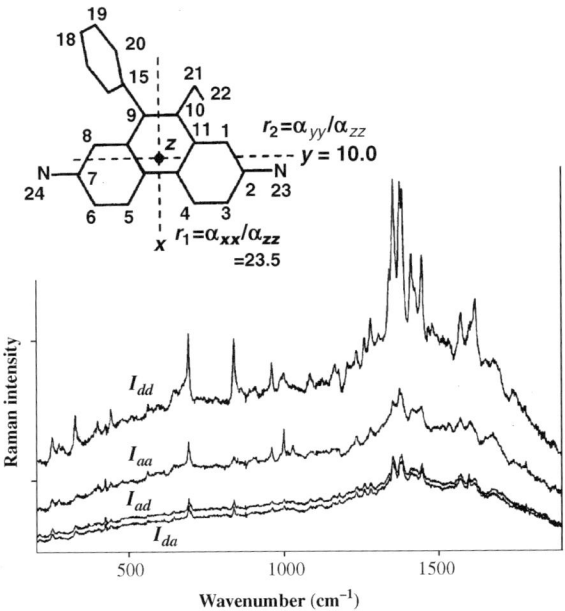

FIGURE 2.33. Polarized Raman spectra (785 nm excitation) of an oriented single crystal of ethidium bromide. The spectral traces are labeled from top to bottom (I_{dd}, I_{aa}, I_{ad}, I_{da}) in accordance with the polarized Raman intensity components defined in the text. Crystal field splitting is evident for the most intense bands of the I_{dd} spectrum. Structural formula of the ethidium ion of ethidium bromide (EtBr) and the Raman tensor axis system (xyz) selected for the 1377 cm^{-1} mode. are given in the upper left corner [67].

An analysis of the polarized Raman spectrum of ethidium bromide (Fig. 2.33) [67], along with the atomic coordinates [68], enabled the calculation of its Raman tensor. Using the polarized Raman data of the intercalated complex (Fig. 2.32), which indicates that $I_{cc}/I_{bb} = 0.3 \pm 0.1$ and $I_{cc}/I_{bc} = 1.5 \pm 0.5$ for the 1377 cm^{-1} band, it was shown that the allowed values of θ and χ are $θ = 35 \pm 5°$ and $χ = 30 \pm 10°$ (for the definitions of the orientation angles θ and χ; see Section 1.3.9). The experimental uncertainties in the polarized Raman intensities of Fig. 2.32 are relatively large because of the overlapping thymine band of DNA at 1375 cm^{-1}. However, the data clearly indicate that I_{cc}/I_{bb} is neither as high as 0.5 nor as low as 0.1, allowing the error limits to be set to ±5°, or less, for θ and ±10° for χ.

On the basis of this set of orientation angles (θ,χ), another model was proposed for the ethidium + DNA structure [67]. The new model involves the X-ray structure of ethidium + UpA · CpG) discussed in Section 2.2.1 [53,54], with no structural changes, except in the orientation of the drug in the binding site. As is shown in Fig. 2.34, the earlier model (Fig. 2.19) is rotated by 35° around the N−O line (which makes 30° with the y-axis in the phenanthridinium plane) as axis. This rotation makes the two base pairs located on the immediate upper and lower portions of the phenanthridinium plane tilted also by ∼35° from the perpendicular plane of the fiber axis. Such a base-pair orientation is just what is found in A-DNA. Thus, the new model

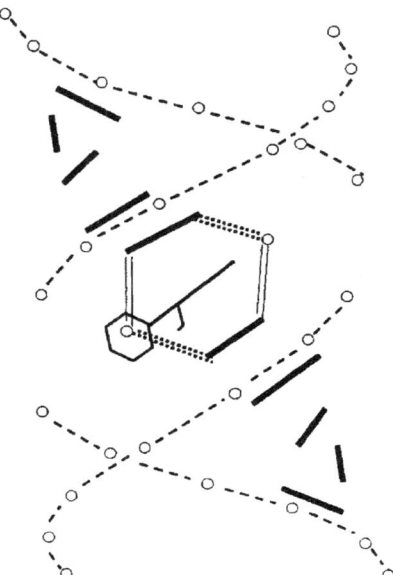

FIGURE 2.34. An ethidium–DNA binding model [67] derived from polarized Raman spectral data. and the Raman tensor ($r_1 = 23.5$ and $r_2 = 10.0$.) of the 1377 cm^{-1} band.

is essentially an A-DNA double helix with the ethidium molecule whose phenanthridinium plane is oriented with $\theta = 35°$ and $\chi = 30°$ with respect to the A-DNA axis. In other words, this model claims that ethidium intercalation into B-DNA causes an induction of a local A-DNA-like structure. However, the DNA chain located far from the intercalation site is considered to remain in (or restore) the original B-DNA structure. It is probable that the ethidium intercalation causes a reduction of the amount of H_2O molecules in the vicinity of the local DNA chain, and that this environment should make the A form more stable than the B form. However, as long as the environmental H_2O amount becomes higher around the DNA chain, apart from the ethidium location, the chain restores the B form.

2.2.7. Electrophoresis Studies of DNA Unwinding

As noted earlier, the intercalation of a drug into a DNA duplex will cause it to unwind. This can be observed by using electrophoresis [58] to examine the binding of ethidium bromide to a relaxed structure of pBR322—which is a closed, circular, plasmid DNA duplex, having 4362 bp. As the ethidium concentration increases, the total unwinding angle also increases. At an unwinding angle of 360°, the number (β) of helical turns is reduced by one, so that β changes from 436 to 435. Because the linking number (α) of this plasmid remains unchanged, the writhing number τ must increase by one in order to maintain the same local conformation. This relationship is defined by $\alpha = \beta + \tau$ (Section 1.1.10). The value of τ can be estimated by using topoisomerase I relaxation followed by gel electrophoresis. It can be determined

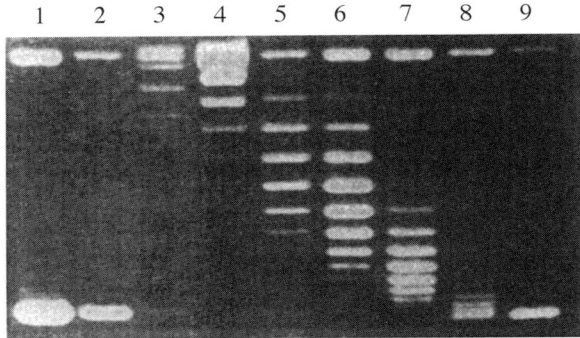

FIGURE 2.35. Electrophoresis analysis of topoisomers of pBR322 DNA, produced by ethidium bromide binding and the topoisomerase I relaxation, followed by the removal of the drub and enzyme. (1). Native intact pBR322 DNA; (2) native pBR322 DNA treated with incubation, phenol extract, and ethanol precipitation in buffer T (35 mM Tris-HCl, pH = 8.0, 73 mM KCl, 5 mM $MgCl_2$, 5 mM dithiothreitol, 5 mM spermidine, and 0.01% bovine albumine) containing no topoisomerase I and no ethidium bromide. (3–9). Ethidium bromide was added prior to topoisomerase. DNA concentration = 10 μg/mL. Ethidium concentrations (μg/mL): (3) 0; (4) 0.25; (5) 0.50; (6) 0.75; (7) 1.00; (8) 1.25; (9) 1.50 [71].

from Fig. 2.35 that an ethidium bromide concentration of 0.25 μg/mL causes an average linking number change of $\Delta\alpha = 2$ for the plasmid, while 0.50 μg/mL causes a change of $\Delta\alpha = 5$, 0.75 μg/mL causes a change of $\Delta\alpha = 7$, and 1.0 μg/mL causes a change of $\Delta\alpha = 10$. These observed $\Delta\alpha$ values are considered to be entirely attributable to the amount of unwinding ($\Delta\beta$) caused by the drug binding, where β is given by

$$\beta = \frac{N}{h} \qquad (2.4)$$

Here, N is the total number of base pairs involved in the closed circular duplex and h is the number of base pairs involved in one pitch of the DNA helix. It is possible to define an angle θ that measures the clockwise rotation (viewed along the helical duplex axis) needed to reach a base pair from the adjacent base pair located closer to the viewer:

$$\theta = \frac{360°}{h} = \frac{360°\beta}{N} \qquad (2.5)$$

The change in θ on drug binding is the difference of the angles between the unbound (θ_0) and bound (θ):

$$\Delta\theta = \theta - \theta_0 = \frac{360°\Delta\beta}{N} \qquad (2.6)$$

If the number drug molecules that bind to one closed circular duplex molecule is given by m, then the value of the unwinding angle (ϕ) due to one drug molecule should be

$$\phi = \frac{\Delta\theta}{m} = \frac{360°\Delta\beta}{m} \qquad (2.7)$$

TABLE 2.5. Angle of Unwinding of DNA Double Helix Caused by Ethidium Binding as Derived from Writhing Number (τ) and Equilibrium Constant

Sample Number	Total Drug Concentration		Bound Drug Concentration ($\times 10^{-5}$ M)	m	$\Delta\beta$	ϕ
	µg/mL	$\times 10^{-5}$ M				
4	0.25	=0.0635	0.0252	66°	2°	11°
5	0.50	=0.1269	0.0498	130°	5°	14°
6	0.75	=0.1904	0.0741	194°	7°	137°
7	1.00	=0.2538	0.0979	256°	10°	14°
8	1.25	=0.3173	0.1212	317°	15°	17°
9	1.50	=0.3807	0.1441	377°	20°	19°

Since the equilibrium constant K (in buffer-T, at 37°C) is known to be 40,000 M^{-1} (Section 2.2.3), the concentration of the bound drug can be calculated for each drug concentration, as shown for in Table 2.5. Since the number of base pairs of every plasmid pBR322 molecule is known to be 4362, the average number (m) of drug molecules bound to one closed circular duplex molecule can be calculated (fifth column of Table 2.5). The unwinding angle (ϕ) can be calculated by Eq. 2.7 from the $\Delta\beta$ (sixth column of Table 2.5) and m values, and the results are shown in the last column of Table 2.5. Thus, it is now concluded that ethidium binding causes unwinding of pBR322 DNA duplex in buffer-T at 37°C by ~13°, as long as the [bound drug]/[base pair] mole ratio is less than 1/10. As the amount of bound ethidium increases, however, the average unwinding angle caused by one drug molecule seems to increase.

Since the unwinding angle per intercalated ethidium molecule is used in a wide variety of calculations, this quantity was examined more extensively by Wang [69], who prepared a series of covalently closed bacteriophage PM2 DNA samples with varying degrees of superhelicity. The amount of bound ethidium per DNA nucleotide, needed for the removal of all superhelical turns, was determined for each sample by a number of methods. He concluded that ethidium unwinds the DNA helix by 26° per molecule bound.

2.2.8. Atomic Force Microscopy

Atomic force microscopy (AFM) is a powerful technique for direct observation of biological macromolecules and their assemblies. One of the advantages of AFM over other high-resolution microscopies (e.g., electron microscopy) is that sample preparation is relatively simple—it does not involve negative staining or shadow casting with a metal coating, In addition, the sample does not need to be kept under vacuum. Pope et al. [70] and Utsuno et al. [71] published studies of the changes in the pBR322 DNA tertiary structure that occur as a result of ethidium bromide binding. There were some of the differences between their sample morphologies. In the experiments by

Pope's group, the plasmid DNA solution was spotted on mica, and the mica was rinsed with H_2O, and blown dry with compressed N_2; In the experiments by Utsuno's group, the DNA solution was placed in a liquid cell, which was placed on mica, and the DNA sample, adsorbed on mica, was kept in the solution during the measurement without drying. It is another advantage of AFM that the imaging is permitted in buffer solution without drying the sample, allowing biological samples to remain intact, also allowing the results of the imaging to be correlated with other experimental results in solution. On the interaction of ethidium bromide and pBR322 DNA in solution, various pieces of information had been established; one molecule of ethidium bromide, when it is intercalated into a double-helix DNA, is known to unwind the DNA duplex so that the base pair–base pair angle (viewed along the helix axis) is reduced. Therefore, a relaxed closed circular pBR322 DNA molecule is predicted to change into a positively supercoiled form on ethidium bromide binding. One can also predict an approximate number of supercoils (which is nearly equal to writhing number (τ) in the plasmid molecules maintained at a given concentration in an aqueous solution that contains a given amount of ethidium bromide. The shape and appearance of the individual molecules in solution can be directly observed by AFM (Fig. 2.36).

For a given value of τ (writhing number), two kinds of supercoiled conformations are predictable: (1) an interwound form (or a plectonemic supercoiled form) and (2) a solenoid form (or a toroidal supercoiled form), in which the axis of the DNA duplex lies on an imaginary torus. Some of these kinds of superhelical forms are illustrated in Fig. 2.37. Each of these eight configurations shows a possible projection of a plasmid molecule on a two-dimensional plane. The line here represents an axis of a double-helix DNA chain. The number (N_τ) of the intersection points of the line within each configuration is supposed to be the number of supercoilings of the plasmid molecule. This number N_τ can be correlated with the writhing number τ, but these two are not exactly equal to each other. In general, τ is not an integer, whereas N_τ is an integer. It would be reasonable, however, to assume that N_τ is nearly equal to τ with an ambiguity of 1.0.

Figure 2.36 is a collection of topographic images recorded for plasmid pBR322 DNA molecules kept in solutions of 0.05 mg/mL. Here, the plasmid was completely relaxed by the topoisomerase I reaction, and this was confirmed by electrophoresis. Without ethidium bromide, some of the DNA molecules appear to be circular with no supercoil (Fig. 2.36a). Many of the chains, however, have one, two, or three intersection points. This is partly attributable to the solvent (10 mM HEPES + 1 mM $NiCl_2$) and temperature (25°C), in which the AFM experiment was made. These are different from the conditions (buffer-T at 37°C) in which the plasmid was relaxed. When ethidium bromide is added to the to the solution, apparently interwound forms are found to appear (Fig. 2.36b). In the range of ethidium bromide concentration from 0.25 to 0.75 μg/mL, however, the observed N_τ values are found to be mostly 2 or 3 only. When the concentration of ethidium bromide reaches 0.80 μg/mL, the plectonemic supercoiled forms with $N_\tau = 6$ or 7 are observed. However, many of the loops in the supercoiled forms could not be found resolved; they were observed as thick rods. Such a rod is interpreted as a condensed plectonemic supercoil (see e.g., Fig. 2.37g). In the solution with 1.0 μg/mL ethidium bromide, more of such thick rods appear

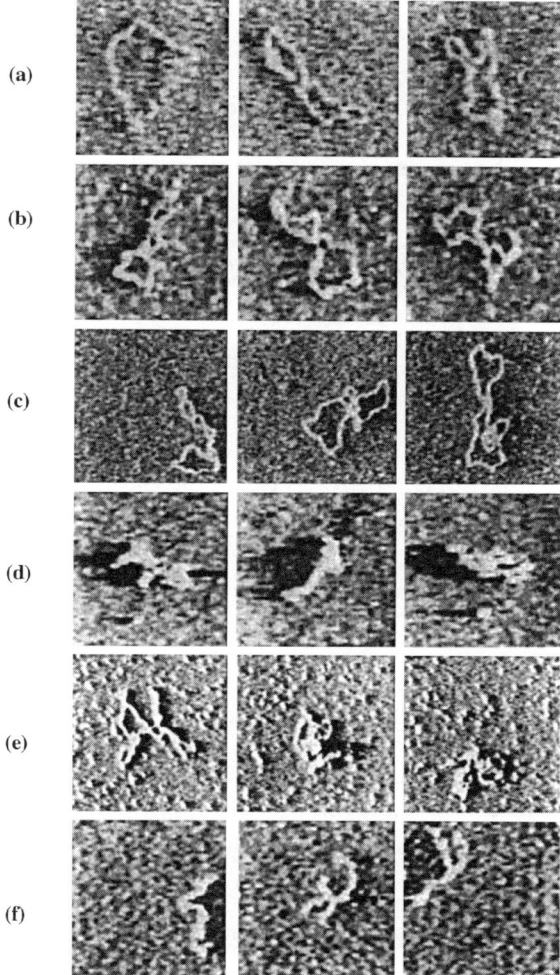

FIGURE 2.36. Atomic Force Microscopic images of plasmid pBR322 DNA in aqueous solutions, containing ethidium bromide. DNA concentration: 0.05 µg/mL. Ethidium bromide concentrations (µg/mL): (a) 0, (b) 0.75, (c) 0.80, (d) 1.0, (e) 2.0, (f) 10. Magnification—every square here has an area dimension of 1×1 µm [71].

(Fig. 2.36d,e). Still, one or two small loops are found in each molecule. In the solution where the ethidium bromide concentration is as high as 10 µg/mL, every plasmid molecule appears as a thick rod. Some molecules are found to be branched rods (Fig. 2.36f).

In surveying the observed AFM images, a general trend of increasing in the τ value with the ethidium bromide concentration is certainly confirmed. The upward gradient of τ, however, is much smaller than what is expected from the equilibrium constant ($K = 2.5 \times 10^5 \, M^{-1}$) and unwinding angle ($\phi = 15°$). At an ethidium bromide concentration of 0.80 µg/mL, for example, N_τ is found to be only 6 or 7 in AFM, whereas

164 INTERCALATING DRUGS

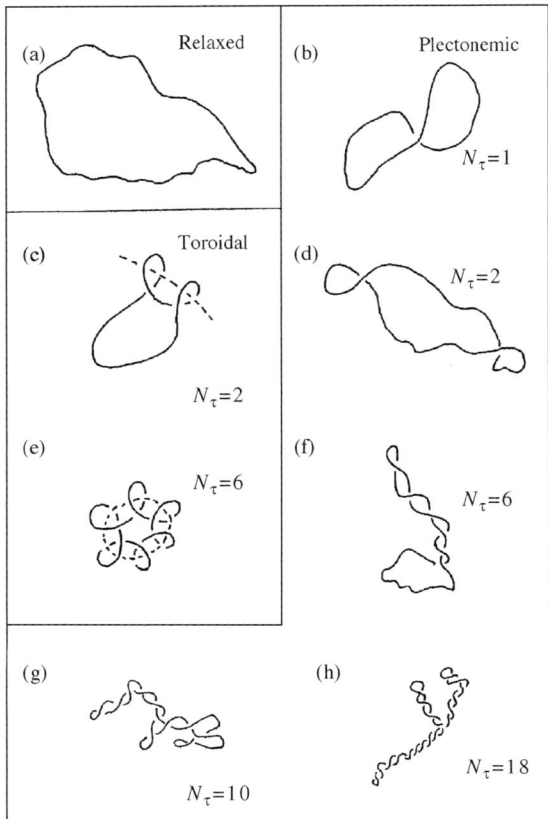

FIGURE 2.37. Diagramatic illustrations of possible tertiary structures of plasmid pBR322 molecules [71].

$\tau = 31$ is expected from the assumption of $K = 2.5 \times 10^5 \, \text{M}^{-1}$ and $\phi = 15°$. For such an inconsistency a detailed discussion was made by Utsuno et al. [71].

In a microscopic study, it is important not only to deduce a general trend from a survey of many images but also to observe individual images in detail. A special shape of a particular molecule could possibly be significant. The image in the leftmost frame of Figure 2.36c shows a plasmid pBR322 DNA molecule observed at an ethidium bromide concentration of 0.80 μg/mL. This is a plectonemically and positively supercoiled molecule with the number of supercoilings at 6 (compare with the configuration in Fig. 2.37f). The image in the central frame of Fig. 2.36c probably has the same number of ethidium bromide molecules and the same number of coilings. Its shape, however, is quite different; its two interwound small loops are crowded in the central portion, and its two large loops are placed outside. What is shown in the leftmost frame of Fig. 2.36d is an image of a plasmid molecule observed at an ethidium bromide concentration of 1.0 mg/mL. This is a plectonemically supercoiled molecule with a supercoil value of ∼10. Its interpretation is given by the configuration in Fig. 2.37g. The rightmost frame of Fig. 2.36f is an image for a branched plectoneme.

Here, the number of supercoils must be greater than 15. Its duplex chain axis is considered to be folded as shown in Fig. 2.37h.

It is interesting that, contrary to the observation of Pope et al. [70], Utsuno et al. [71] could not find any image that was assignable to a toroidal supercoiled form.

2.3. ACLACINOMYCIN

Aclacinmycin A (Fig. 2.38) was first found in a culture of *Streptomyces galilaeous* in 1975 [72], and has been used worldwide for more than 20 years as an antitumor drug with low cardiac toxicity. Its activity is believed to be related with its binding to DNA, and extensive studies have been made on its interaction with DNA.

2.3.1. Absorption Spectroscopy

An absorption spectrum of aclacinomycin A dissolved in buffer-T (a special Tris buffer, slightly alkaline at pH = 8.0) shows a peak at 514 nm (Fig. 2.39a). When 2.5 µL of 2.04×10^{-2} M pBR322 DNA solution was added to 3500 µL of aclacinomycin A solution, the peak height at 514 nm was lowered. Addition of a 2.5 µL DNA solution

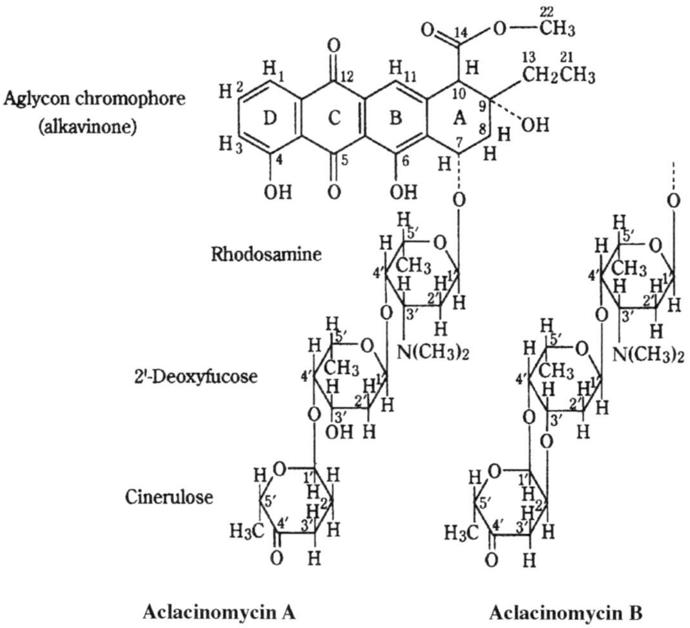

FIGURE 2.38. Molecular formulas of aclacinomycins A and B with the numbering system used in this book. The molecules contain an aglycon chromophore (alkavinone or K) with four fused rings (A–D). A trisaccharide is attached to the aglycon at the C7 position. The sugars of the trisaccharide are L-rhodosamine, 2'-deoxy-L-fucose, and cinerulose. In aclacinomycin B, deoxyfucose and cinerulose are fused through C3'- - O - - C2' bridge.

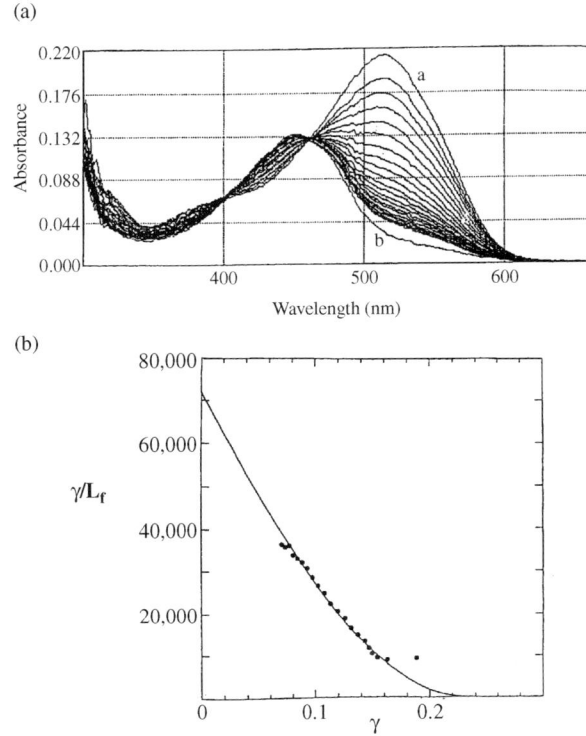

FIGURE 2.39. (a) Absorption spectra of aclacinomycin A at 37°C in buffer-T (which contains 35 mM Tris-HCl (pH = 8.0), 72 mM KCl, 5 mM $MgCl_2$, 5 mM dithiotreitol (DTT), 5 mM spermidine, and 0.01% bovine serum albumin). Curve a shows aclacinomycin A 2.3×10^{-5} M only. The unlabeled curves show the titration of pBR322 DNA into the drug solution. Each solution consists of 3500 µL of 2.3×10^{-5} M aclacinomycin A plus the stepwise addition of 2.5 µL × j of 2.04×10^{-2} M (in base pairs) of pBR322 DNA, where j = 1–21. Curve b shows aclacinomycin A 2.3×10^{-5} M (3500 µL) + 2.5 mg of pBR322 DNA. (b) A *Scatchard plot* (•) of aclacinomycin A bound to plasmid pBR322 DNA. Solid line shows a theoretical curve on the basis of Eq. (2.8), in which it was assumed that $K = 72,000$ M^{-1} and n = 4.0 [9].

was repeated 21 times until the total addition reached 52.5 µL. Each time the absorption spectrum of the mixture solution was recorded. As seen in Fig. 2.39, the 514 nm peak lowered gradually, and absorbance at 450 nm increased as more DNA was added. During these spectral changes, an isosbestic point appeared at 460 nm. This fact indicates that the spectral changes are caused by only two components: bands of free drug and one type of bound drug. The absorption spectra of the titration of pBR322 DNA into a solution of aclacinomycin are shown in Fig. 2.39a.

By examining the 514 nm peak height, the concentrations of free drug and bound drug (L_f and L_b, respectively) were evaluated for each mixture solution. Then, the binding ratio γ was calculated, and $γ/L_f$ was plotted against γ, giving the *Scatchard plot* (Section 2.2.3) shown in Fig. 2.39b. To search for the most probable set of K (equilibrium constant) and n (the number of base pairs occupied by one drug molecule) values, the following McGhee–von Hippel equation [59] was used:

$$\frac{\gamma}{L_f} = K(1-n\gamma)\frac{[1-n\gamma]^{n-1}}{[1-(n-1)\gamma]^{n-1}} \quad (2.8)$$

As can be seen in Fig. 2.39b, a McGhee–von Hippel curve fitted the experimental plot when it was assumed that $K = 72,000\ M^{-1}$ and $n = 4.0$ [58].

Similar examinations of the absorption spectra and equilibrium constants (K) have been made for aclacinomycin + linear double-helix DNA systems [72]. Some of the results are shown in Table 2.6. The hypochromicity of aclacinomycin A bound to DNA is also given in Table 2.6. This table indicates that the binding constant of aclacinomycin A with poly(dA-dT)$_2$ is the largest of those studied, and that it has the same hypochromicity as d(CCTAGG)$_2$. Thus, it is suggested that aclacinomycin A prefers the TpA sequence among the three possible intercalation sites in d(CCTAGG)$_2$.

2.3.2. ^{31}P NMR

The effects of intercalation on DNA structure can be seen by following the titration of aclacinomycin A into a solution of d(CCTAGG)$_2$, using ^{31}P NMR (Fig. 2.40) [73]. The assignments of the ^{31}P resonances of d(CCTAGG)$_2$, in the absence of aclacinomycin A, were made on the basis of its 2D-^{31}P-^{1}H through-bond correlation spectrum [5], and are given in the top spectrum of Fig. 2.40. The upfield peak at -4 ppm arises from the central TpA phosphorus atom in free d(CCTAGG)$_2$. As aclacinomycin A is added, this peak decreases in intensity, while two new ^{31}P resonances appear downfield (at higher ppm values) at -2.5 and -3.2 ppm, as indicated by the arrows; these are quite distinct and separated from those of d(CCTAGG)$_2$, and the other resonances from the DNA duplex remain almost unchanged. The two newly observed, downshifted resonances belong to the same two central TpA phosphodiester linkages, and suggest that the drug is intercalated between them. Although (CCTAGG)$_2$ is symmetric, these central P atoms (belonging to the two partner strands) become different in the complex because the aclacinomycin molecule is asymmetric. Thus, the overall symmetry of the aclacinomycin-d(CCTAGG)$_2$ complex is lost. It is concluded that the newly observed lower field resonances (-3.2 and -2.5 ppm) can be assigned to TpA phosphorus in the

TABLE 2.6. Binding Constants (K) of Aclacinomycin A (in Aqueous Solutions at Room Temperature, 0 M NaCl)

	d(CCTAGG)$_2$	Poly(dA-dT)·poly(dA-dT)	Poly(dG)·poly(dC)	Poly(dA-dG)·poly(dT-dC)
Binding constant (K)^a		$38 \times 10^5\ M^{-1}$	$19 \times 10^5\ M^{-1}$	$7 \times 10^5\ M^{-1}$
Hypochromicity^b	45%	45%	33%	33%

^{a}With polynucleotides.
^{b}Hypochromicity at 435 nm.
Source: Ref. 72

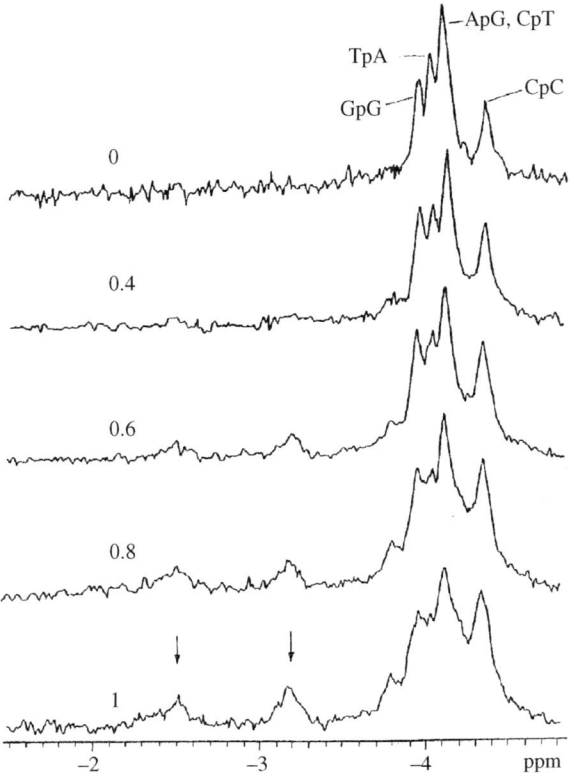

FIGURE 2.40. The 161.7 MHz ^{31}P NMR spectra of d(CCTAGG)$_2$ (top) and d(CCTAGG)$_2$ plus various amounts of aclacinomycin A. The figures at the left side indicate the molar ratio of added aclacinomycin A versus d(CCTAGG) double strand (the concentration of the latter is 700 OD/mL) [73].

complex state. These results strongly suggest that aclacinomycin A is intercalated at the 5'-TpA-3' site.

2.3.3. Exchangeable ^{1}H NMR Spectra

Figure 2.41a shows the ^{1}H NMR spectra of the exchangeable imino protons of d(CCTAGG)$_2$ in H$_2$O as a function of temperature [74]. The assignments are given on the bottom spectrum. The "melting" temperature of d(CCTAGG)$_2$ at this concentration is about 35°C. The two OH proton resonances of aclacinomycin A (positions 4 and 6, Fig. 2.38) were observed at 12.0 and 12.7 ppm in chloroform at 25°C, but cannot be observed in H$_2$O at the same temperature because of a rapid exchange with water.

Figure 2.41b shows the temperature dependence of the exchangeable proton NMR spectrum of aclacinomycin A with d(CCTAGG)$_2$ (molar ratio 1 : 1). Eight resonances were observed at 27°C (designated a–h). Six of these (peaks a–f) can be assigned to the imino protons from the complex of d(CCTAGG)$_2$ with aclacinomycin A, while the other two (peaks *g* and *h*) can be assigned to the chromophore OH protons of

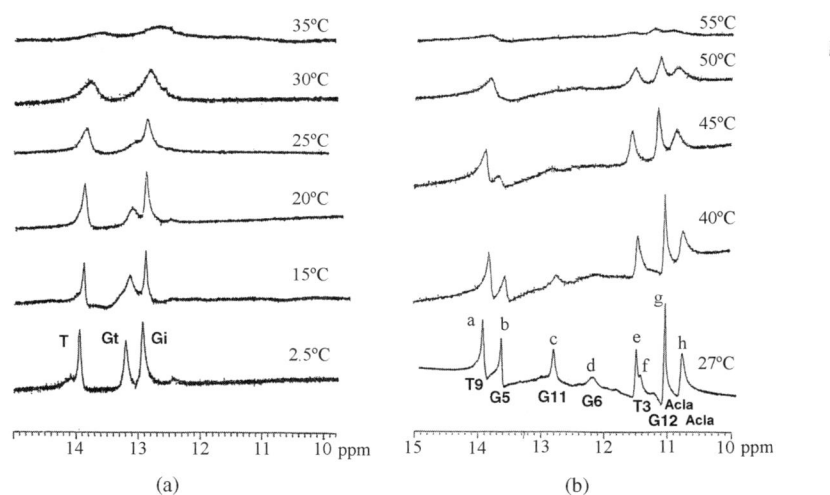

FIGURE 2.41. (a) The 399.95 MHz ^{1}H-NMR exchangeable imino proton spectra of d(CCTAGG)$_2$ (700 OD/mL) in H$_2$O as a function of temperature. (b) The 399.95 MHz exchangeable proton resonances in a mixture of aclacinomycin A with d(CCTAGG)$_2$ (molar ratio 1:1) in H$_2$O as a function of the temperature. The assignments are given along the bottom traces, with "Acla" denoting aclacinomycin. These were achieved on the basis of the fact that G-N1 imino protons occur in the 13.0–13.5 ppm region, while T-N3 imino protons generally resonate at values greater than 13.5 ppm, but are altered by the presence of the drug. In addition, the imino proton resonances from the terminal base pairs are broadened out at temperatures below the melting temperature due to "fraying" [74].

complexed aclacinomycin A, which exchange slowly with water [75]. On increasing the temperature, the resonance (d) and (f) are broadened, and disappear at 35°C; these can be assigned to the imino protons of the terminal base pairs (i.e., of G$_6$ and G$_{12}$). Resonance (c) disappears at 45°C, and the resonance (b) at 50°C. These resonances are from the base pairs that are one step in from the terminal (G$_5$ and G$_{11}$). The other four resonances are observed even at 50°C, indicating that complete dissociation has not yet occurred.

On the basis of both the ^{31}P and exchangeable NMR data, a schematic model of the aclacinmycin A–d(CCTAGG)$_2$ complex was constructed as shown in Fig. 2.42. In the absence of aclacinomycin A, the internal G-N1 and T-N3 imino proton signals are broadened at the same temperature (Fig. 2.41a). Since aclacinomycin A intercalates at the central base step, however, it stabilizes the helix in that part of the complex, resulting in a strengthening of the T-N3 imino hydrogen bonds. These proton signals broaden at higher temperatures in the complex, enabling the assignment of peaks (c) and (b) as the internal G-N1 protons (G$_5$ and G$_{11}$ in Fig. 2.42) The other four resonances are therefore assigned to the two T-N3 imino protons and two chromophore OH protons of aclacinomycin A. The sugar portion of aclacinomycin A seems to reach one of the G-C base pairs adjacent to a central A-T base pair, so that the duplex becomes extremely asymmetric. The broadening of the inner guanine N1-H signal takes place at 35°C in the free duplex (Fig. 2.41a), but only at 45 or 50°C in the

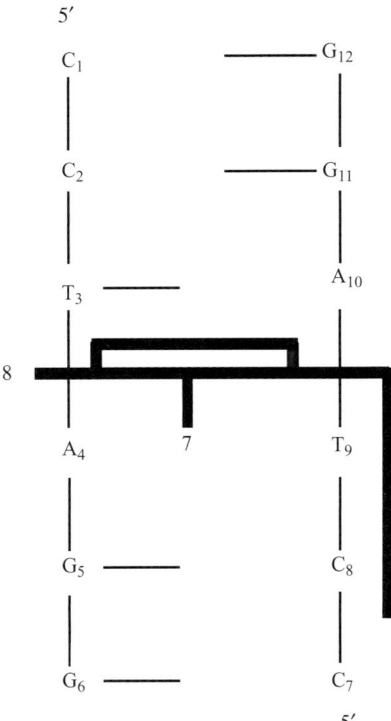

FIGURE 2.42. A schematic model of the aclacinomycin–d(CCTAGG)$_2$ complex. The aclacinomycin moiety is shown with thick lines. The base pairs are designated by the lowest numbered bases in the pair (i.e., 1–6), which constitute strand 1 of the duplex d(CCTAGG)$_2$, while 7 and 8 designate the chromophore OH protons at positions 6 and 4 of aclacinomycin A, respectively.

complex. One of the G-NH (G$_5$ in Fig. 2.42) must be located close to the sugar moiety of aclacinomycin and can interact with it, whereas the other (G$_{11}$ in Fig. 2.42) would not. Therefore, signal (b), which broadens at higher temperature (50°C, Fig. 2.41b), is assigned as G$_5$ (Fig. 2.42) and signal (c) to G$_{11}$. The broadening of the thymine-3-NH signals take place at 35°C in the free duplex, but only at 55°C in the complex.

2.3.4. Nonexchangeable ^{1}H NMR Spectra

It is interesting that the intercalation site of aclacinomycin A is not highly sequence-specific, as the preferred binding site changes depending on sequence. As decribed below aclacinomycin A prefers the terminal CpG step of d(CGTACG)$_2$ [76], whereas it is bound at the central TpA step of d(CCTAGG)$_2$ [75].

The one-dimensional ^{1}H NMR spectra [76] in Fig. 2.43 show that both the free aclacinomycin A and the free d(CGTACG)$_2$ have sharp *non*exchangeable resonances. When these are mixed in ratios of 0.5 : 1 and 1 : 1 (drug : DNA), the spectra display complex patterns. When the ratio reaches 2 : 1, however, the spectra become

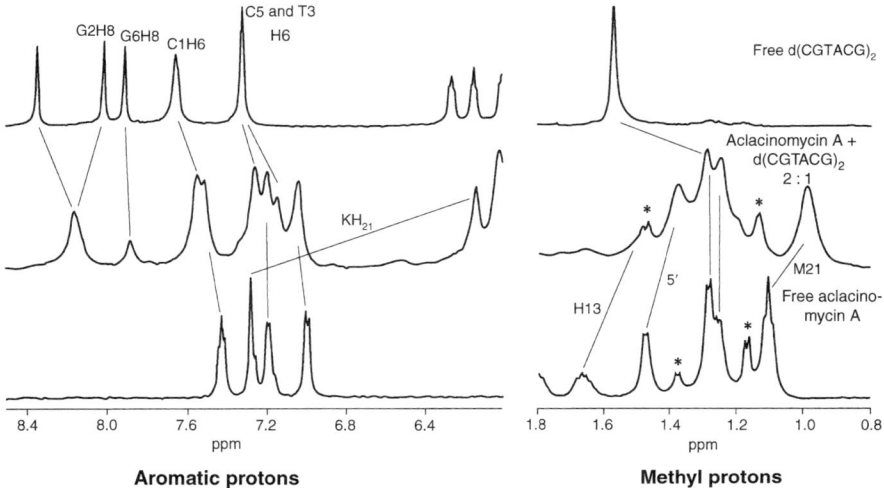

FIGURE 2.43. One-dimensional ^{1}H NMR spectra of aromatic and methyl regions of the free DNA, 2:1 aclacinomycin A-d(CGTACG)$_2$ complex, and free aclacinomycin A (a partial reproduction of Fig. 2 in Ref. 77). The spectra were recorded at 25°C. Resonances marked with asterisks are from the impurity of the drug sample.

considerably simplified, indicating that a stable and symmetric 2:1 complex is formed. The nonexchangeable resonances of both the free DNA and free drugs (aclacinomycin A and aclacinomycin B) were assigned by two-dimensional NOESY–walk analyses. Similar analyses were performed on the 2:1 drug–DNA complexes, aclacinomycin A–d(CGTACG)$_2$ (at 25°C and 40°C), and the aclacinomycin B–d(CGTACG)$_2$ (at 30°C and 40°C). The spectra of the complexes displayed numerous NOE cross-peaks, indicating that each has a well-defined structure. These spectra indicated that both aclacinomycin A and aclacinomycin B form very similar complexes [77]. Therefore, only the results obtained from the latter complex are given below.

The NOESY–walk strategy (Section 1.4.4) was used to assign and interpret the spectra, especially the aromatic H1' fingerprint region, and the aromatic H2'/H2" region (Fig. 2.44). The sequential connectivity in the aromatic H1' pathway can be traced without interruption. The interresidue cross-peak intensities between the C_1H1'-G_2H8 and C_5H1'-G_6H8 are significantly weaker than those of the other internucleotide cross-peaks (i.e., C_2H1'-T_3H6, T_3H1'-A_4H8, and A_4H1'-C_5H6). This suggests that the terminal base steps, the C_1pG_2 and C_5pG_6 steps, are distorted and no longer have the normal B-DNA backbone conformation. The aromatic H2'/H2" cross-peaks possess the same intensity pattern. Thus, the G_2H8-$C_1H2'/H2"$ and G_6H8-$C_5H2'/H2"$ cross-peaks are significantly weaker than the remaining corresponding cross-peaks. Normally, terminal base pairs are expected to fray, resulting in peaks that are often broad, weak, or unobservable. However, in this case, the terminal base steps were deemed to have too much structure for a frayed base pair. On the basis of this observation, as well as additional drug–DNA NOE cross-peaks (see below), it was

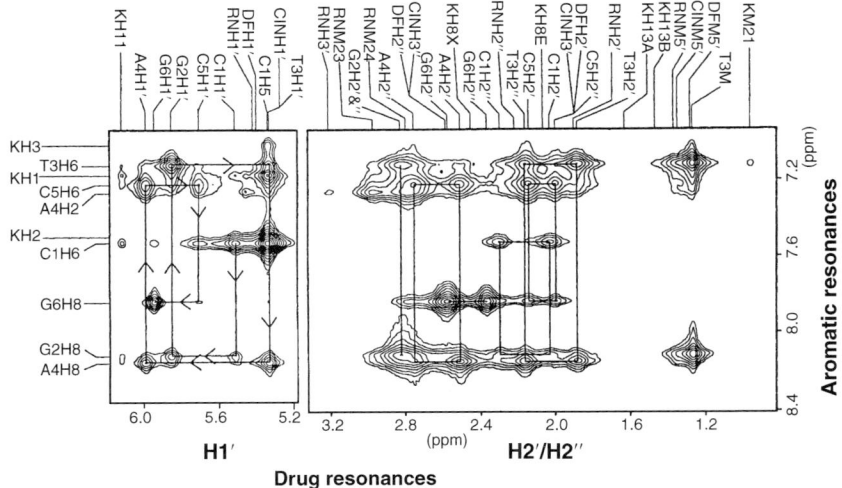

FIGURE 2.44. Expanded regions of the two-dimensional NOESY spectra, which provide key information in the assignments of the resonances. Left panels show cross-peaks between the aromatic protons (7.0–8.4 ppm) and the H1′ protons. The sequencial assignment pathway is indicated. Right panels show cross-peaks between the aromatic protons and H2′, H2″ protons. Each aromatic proton has cross-peaks to its own H2′/H2″ as well as to those from preceding (5′-side) nucleotide that are connected by a rectangular box. (Partial reproduction of Fig. 3 in Ref. 77.)

concluded that the 2:1 complex has its intercalation sites at the C_1pG_2 and C_5pG_6 steps.

Additional information on the placement of the drug in the intercalation site was also found in the NOESY spectra. Figure 2.45 shows the one-dimensional slices through the resonances of the alkavinone(K) proton (KH1, KH2, KH3, and KH11) (Fig. 2.38) of the two-dimensional NOESY spectrum of the 2:1 aclacinomycin B-d (CGTACG)$_2$ complex. The NOE intensities show that KH1 is close to both C_1H5 and C_5H5, KH2 is closer to C_1H5, and KH3 is closer to C_5H5.

The two-dimensional NOESY spectra (Figs. 2.44 and 2.45) reveal that the KH1 and KH3 protons, respectively, are spatially very close to C_1H5 and C_5H5, whereas KH2 is spatially close to both C_1H5 and C_5H5. But the last two proton pairs are further apart than those of KH1 to C_1H5 and KH3 to C_5H5. This suggests that ring D of alkavinone is sandwiched between C_1 and C_5. The position of the alkavinone aglycon ring is further defined by the NOE cross-peaks between the KH11 protons and many DNA protons, in particular those from the C_1 and G_2 residues. This clearly indicates that the alkavinone chromophore is oriented such that the ring edge containing H11 is facing toward the backbone of C_1pG_2.

NMR solution structures were determined from the integrated NOE cross-peak intensities using a procedure that did not directly calculate interatomic distances (r_{ij}). Instead, the procedure used [77] calculates an energy-based error function for each proton pair, based on a comparison of the simulated and experimental intensities [77,78]. The entire molecule was also required to have twofold symmetry, since

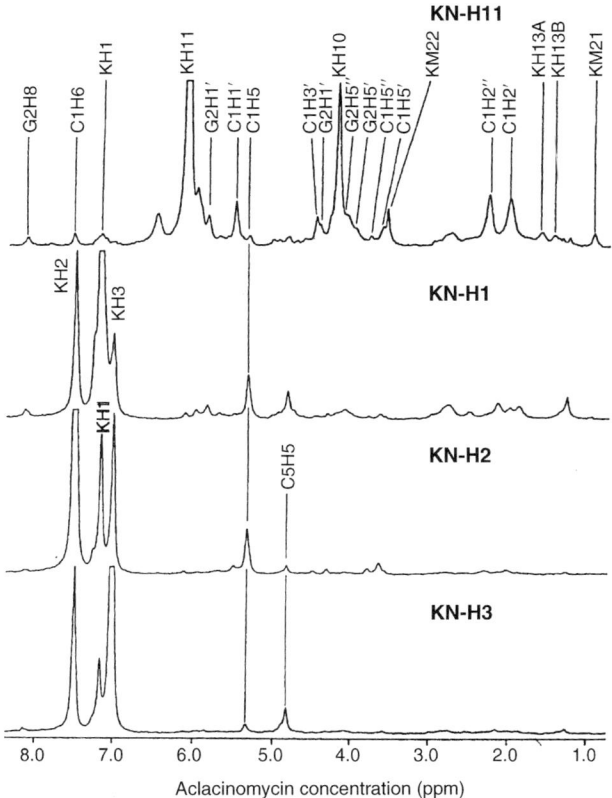

FIGURE 2.45. One-dimensional slices through the two-dimensional NOESY spectra of the 2:1 complex associated with four aclacinomycin B protons, alkavinone(K)-H1, -H2, -H3, and -H11 [77].

the NMR spectra indicated a single resonance for every proton, even though there are two d(CGTACG) strands and two aclacinomycins in the complex. An ensemble of the 10 refined structures of the 2:1 aclacinomycin B-d(CGTACG)$_2$ complex were obtained. The largest root mean square deviation (RMSD; see Section 1.7.1) among these 10 structures was ~0.7 Å. Some parts of the complex have higher deviations, which may be the result of either fast internal relative motions or an insufficient amount of reliable NOE distance information in those regions. For example, the G$_6$ residue has the largest RMS deviations, which is probably associated with fraying of the terminal base pairs. On the other hand, the aclacinomycin aglycon portion of the molecule (Fig. 2.38), which was shown to be intercalated in the CpG base step, has small RMS deviations, indicating that it has a well-defined binding conformation. The cinerulose B portion of aclacinomycin (Fig. 2.38) has large RMS deviations, which is the result of the sparse number of NOE cross-peaks between it and the rest of the complex. This lack of NOE data suggests that the cinerulose B does not bind tightly to d(CGTACG)$_2$.

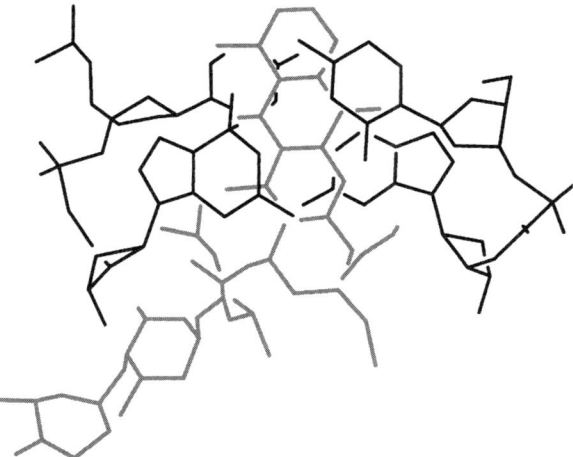

FIGURE 2.46. A view perpendicular to the alkavinione of the aclacinomycin B and d(CGTACG)$_2$ complex, showing the local surroundings of the intercalated drug (red) plus two base pairs (C$_1$pG$_2$ and C$_1$pG$_6$) of the hexamer helix. Hydrogen bonds between the drug and DNA are shown with dotted lines [77]. (See color insert.)

A conspicuous feature of the refined model is the large buckle of the two G:C base pairs that wrap around the aglycon. There are several hydrogen bonds between aclacinomycin and DNA. The O9 and O7 atoms of aclacinomycin, both of which are in the axial positions, are 3.44 and 2.99 Å, respectively, from the N2 of G$_2$. The O9 atom is 3.18 Å from the N3 of G$_2$. The trisaccharide lies in the minor groove. The first sugar coming off the alkavinone, the rhodosamine(R) ring, covers the G$_2$-C$_{11}$ base pair in the minor groove. The N-methyl groups (N3′ Me1 and N3′Me2) are proximal to the A$_{10}$-H2, H1′, and H4′ and the C$_{11}$-H1′, H4′, and H5′,5″ positions, consistent with the observed NOEs. Similarly the R-H3′ proton has a significant NOE to the A$_{10}$-H2 proton.

There are only a few definitive NOE cross-peaks between the 2′-deoxyfucose and cinerulose sugar rings and DNA. Nevertheless, the rigidity of these sugars, especially the fused 2′-deoxyfucose-cinerulose disaccharide, makes it possible to define the conformation of the trisaccharide tail.

The resulting model structure is of a DNA duplex that is clearly kinked (by 20°) at the T$_3$pA$_4$ step. This is due to the close contacts between the deoxyfucose sugar of aclacinomycin B and the deoxyribose of A$_4$, which forces the base pair to open up in the minor groove side. Since there are two aclacinomycin B molecules in the symmetric complex, the interactions resulting from both aclacinomycin B molecules may reinforce each other's role in the deformation of the helix.

Figure 2.46 shows in detail the intercalation geometry of aclacinomycin in the CpG step. The overall position of alkavinone ring is different from that of the DNA ring, with the D ring of alkavinone moving toward the minor groove direction by about 1.2 Å. This causes crowding between the drug sugar (deoxyfucose ring) and the A$_4$ deoxyribose, with the result that the DNA is kinked at the TpA step.

It is important to note that the structural perturbation of DNA by aclacinomycin is different from that caused by daunorubicin in several respects. The entire aclacinomycin covers four base pairs, and its trisaccharide sugar moiety projects farther into the surrounding solvent than does the daunosamine of daunorubicin. The complex is more hydrophobic. Also, the kink caused by aclacinomycin in the DNA double helix near the intercalation site is not seen in the daunorubicin–DNA complex. Such structural distortion may be recognized in a unique way by relevant enzymes such as topoisomerase II. Hence, the biological consequences in cells may be different for aclacinomycin versus daunorubicin.

2.3.5. Raman Spectroscopy

The Raman spectra of aclacinomycin A in H_2O and in D_2O are shown in Fig. 2.47 traces (a) and (b), respectively [80]. In D_2O solution, the three OH groups on the alkavinone ring of the drug exchange with the solvent and become OD groups, causing the observed shifts in the vibrational bands. The Raman spectrum in H_2O of aclacinomycin A plus a large excess ($\times 20$ moles of base pair) of calf thymus DNA is shown in Fig. 2.47 trace (c). As these spectra are acquired with an excitation source

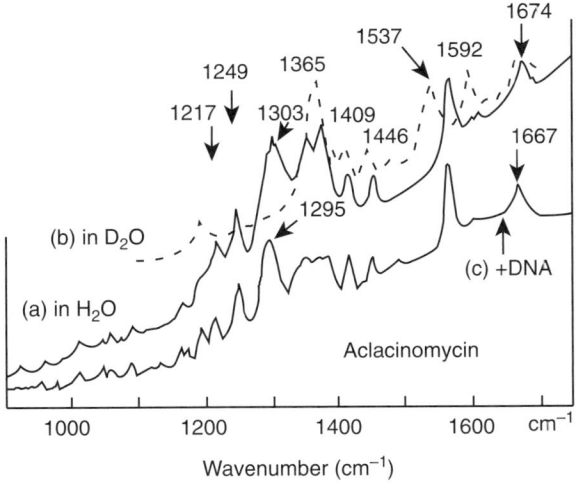

FIGURE 2.47. (a) Raman spectra of aclacinomycin (5×10^{-4} M) in H_2O (—) and (b) in D_2O (.....) excited at 406.7 nm; (c) Raman spectrum of aclacinomycin (5×10^{-4} M) + calf thymus DNA (1×10^{-2} M in base pairs) excited at 406.7 nm. These can be considered as resonance Raman spectra (sometimes called "near resonance") with respect to the drug, as the excitation wavelength (406.7 nm) is close the absorption envelope of the drug (maximum at 514 nm). However, it is off-resonance with respect to the DNA, which has an absorption maximum at ~260 nm. The following shifts are observed on drug binding to DNA—in H_2O, the free C=O stretching band at 1674 cm^{-1} shifts to 1667 cm^{-1} when bound, while in D_2O, the C=O band shifts from 1671 to 1666 cm^{-1}. In addition, the following shifts are detected: 1568 → 1565, 1454 → 1451, 1413 → 1415, 1372 → 1374, and 1451 → 1349 cm^{-1} in H_2O, and 1592 → 1590, 1537 → 1532, 1446 → 1444, 1409 → 1408, and 1365 → 1368 cm^{-1} in D_2O.

that was off-resonance with respect to the absorption spectrum of DNA, the DNA does not contribute significantly to this spectrum. On the other hand, these spectra are nearly on-resonance with respect to the drug, hence any observed spectral changes in these Raman bands must arise from the drug that has been distorted by the DNA to which it is bound.

Many of the Raman bands of aclacinomycin shift significantly on DNA binding. In H_2O, the free C=O stetching band at $1674\,cm^{-1}$ shifts to $1667\,cm^{-1}$ when bound. While in D_2O, the C=O band shifts from 1671 to $1666\,cm^{-1}$. This indicates that the C=O bond is appreciably perturbed by DNA. Similarly, there are many shifts in the vibrational bands between 1349 and $1592\,cm^{-1}$, which suggest that the bond-stretching force constants are generally lowered as a result of a slight amount of electron transfer from a DNA-base π orbital into the antibonding π^* orbital of the aclacinomycin chromophore. Three Raman bands at 1303, 1249, and $1217\,cm^{-1}$ of aclacinomycin disappear on deuteration [Fig. 2.47 trace (b)]; consequently, these are attributed to vibrations involving the exchangeable proton of C−O−H, especially those in which the in-plane bending motions are prominent. It is interesting that $1303\,cm^{-1}$ band shifts to $1295\,cm^{-1}$ on the DNA binding, as this indicates that one of its intramolecular hydrogen bonds (at position 4 or 6) weakens when it intercalates. A low-frequency Raman band at $492\,cm^{-1}$ of aclacinomycin (a C=O in-plane bending vibration) also shifts slightly to a lower frequency ($491\,cm^{-1}$), which is compatible with an idea that the C=O at position 5 weakens when it is stacked between the DNA bases.

2.3.6. Unwinding

One of the important parameters defining a geometric deformation of a DNA molecule, caused by a drug binding, is the angle of unwinding of DNA duplex. In an unbound and unperturbed B-DNA duplex (Section 1.1.8), base pairs may be superimposed on its adjacent neighbor by rotating it around the helical axis by an angle of $\sim 36°$ (on average), and simultaneously translating it by $3.4\,\text{Å}$ along the helical axis (see Table 1.1). If B-DNA deforms into A-DNA, this angle reduces from $36°$ to $32.7°$; if it deforms into Z-DNA, the angle is reduced from $36°$ to $-60°$.

The unwinding angle in B-DNA produced by the aclacinomycin binding can be determined by the methods discussed above. The plasmid pBR322, is a natural, closed circular DNA duplex having 4361 bp, and is purified from *Escherichia coli*. It is known to have a highly negative supercoiling, $\tau < -20$ (Section 1.1.8). The action of topoisomerase I, however, can bring it into a relaxed form ($\alpha = 428$, $\beta = 428$, and $\tau = 0$). So, by adding a proper amount of aclacinomycin (e.g., $2 \times 10^{-6}\,M$ for DNA $1.5 \times 10^{-5}\,M$), before introducing topoisomerase I, it is possible to prepare closed circle pBR322, which is moderately supercoiled (e.g., $\alpha = 420$, $\beta = 428$, and $\tau = -8$). The writhing number τ can be determined by electrophoresis experiment. Figure 2.48 shows the results of the electrophoresis experiments [60]. The purified, native pBR322 DNA was found to be highly supercoiled, containing a small amount of relaxed circular duplex (lane 1). The supercoiled pBR322 DNA was completely relaxed by the action of topoisomerase I (lane 2). This relaxed pBR322 DNA, however, showed a distribution of τ values from $+5$ to -1, instead of showing the single value $\tau = 0$.

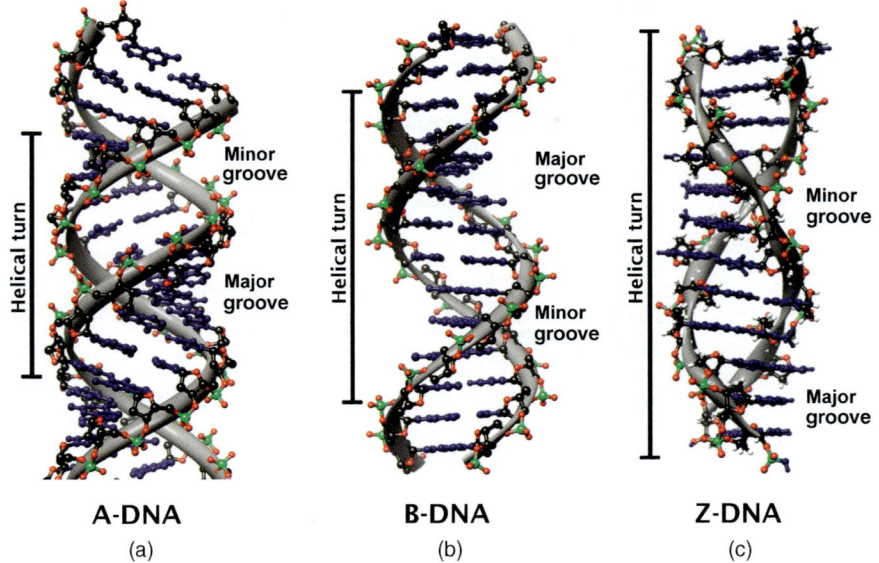

FIGURE 1.9. The side views of idealized fragments of (a) A-DNA, (b) B-DNA, and (c) Z-DNA, along with their approximate helical turns [276]. The A form has the tightest helical turn and hence is the shortest. Its base pairs also have the greatest amount of tilt with respect to the helical axis. The helical turn for the Z form is the least compact and stretches the longest distance. Note that the alternation of *syn* and *anti* glycosidic dihedral angles within the Z form is emphasized here as a twisting of the backbone ribbon. The B form has a moderate helical turn and length, with regular glycosidic dihedral torsion angles.

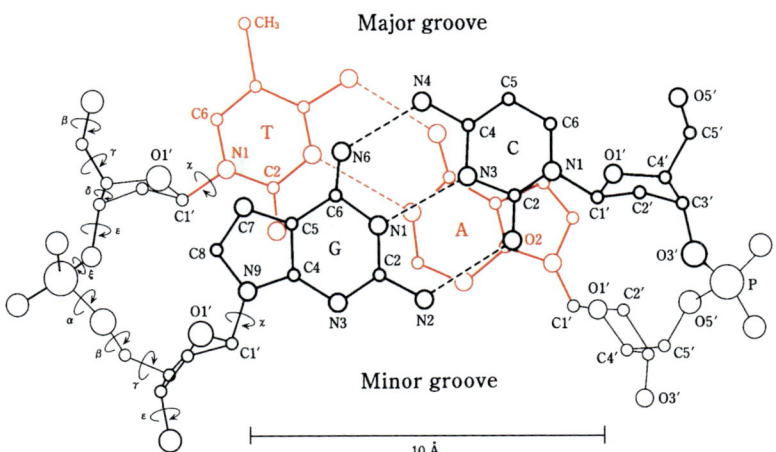

FIGURE 1.10. B-DNA structure, viewed along the helical axis. A G-C base pair is "above" an A-T base pair.

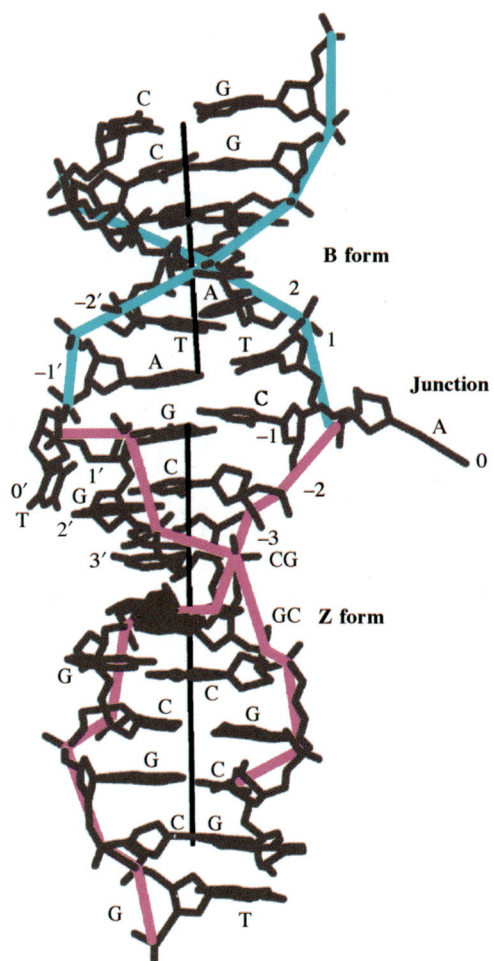

FIGURE 1.18. A 15-base-pair (15-bp) DNA structure containing a junction (black) between left-handed Z-DNA (magenta) and right-handed B-DNA (cyan) [58].

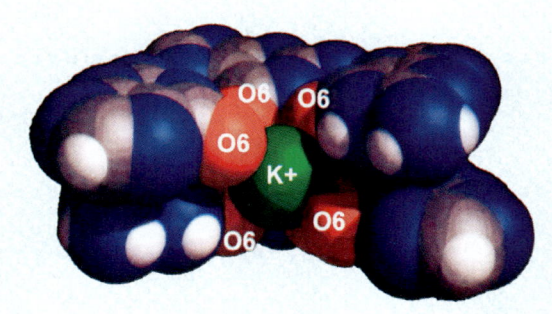

FIGURE 1.25. A side view of two G quartets stacked on top of each other, with an internal, coordinated cation in the center. The view of the stack is from one corner of the square, with the two closest guanosine bases removed to allow visualization of the core. The atom in the center is a K^+ ion, which can be seen coordinating with the O6 atoms in each of the surrounding bases. This figure is derived from experimental NMR structures determined by a combined approach using molecular modeling and experimental NMR data [108].

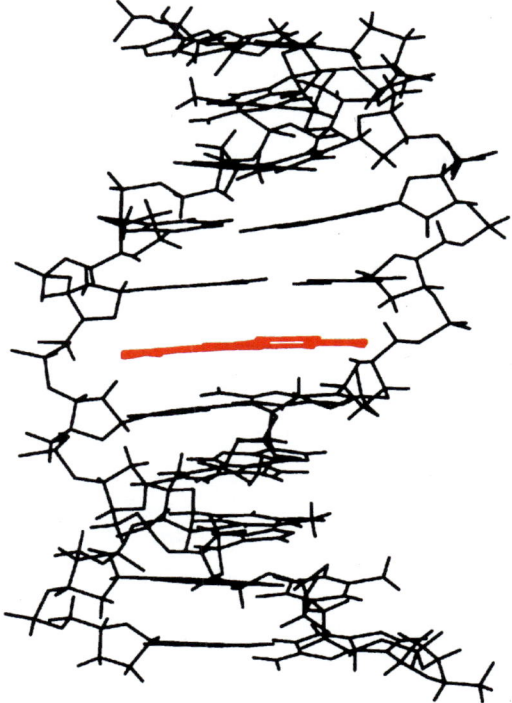

FIGURE 2.8. A molecular model of the complex of proflavine with the decamer d(GATAC-GATAC)$_2$ [17], viewed from the major groove of decamer. The proflavin molecule (red) is shown with its long axis (ordinate) aligned to the horizontal axis(abscissa). The structure was assembled from previously determined experimental pieces and energy minimized. The resulting structure deviated significantly from B-DNA. (*See text for full caption.*)

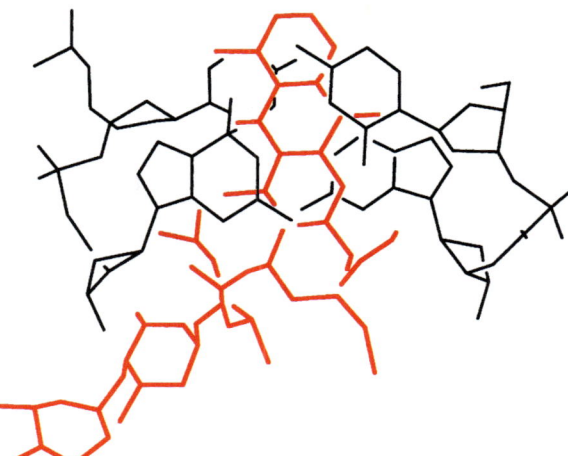

FIGURE 2.46. A view perpendicular to the alkavinione of the aclacinomycin B and d(CGTACG)$_2$ complex, showing the local surroundings of the intercalated drug (red) plus two base pairs (C$_1$pG$_2$ and C$_1$pG$_6$) of the hexamer helix. Hydrogen bonds between the drug and DNA are shown with dotted lines [77].

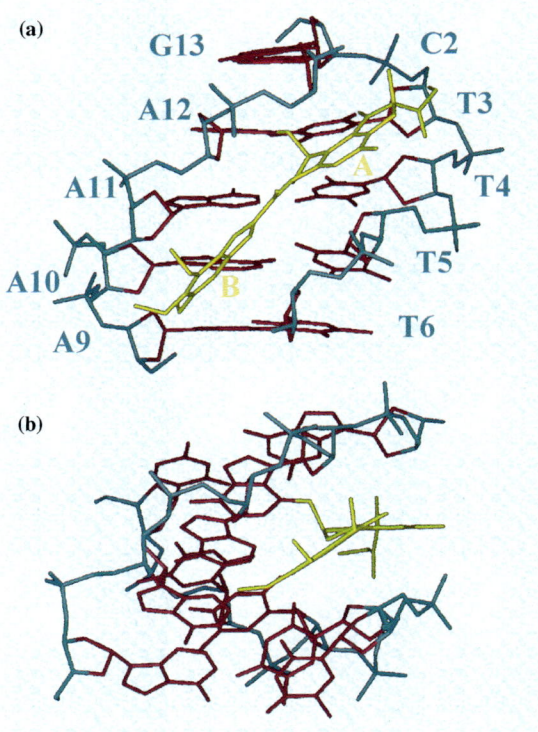

FIGURE 4.6. Two views of a relaxation matrix refined structure of duocamycin–hairpin duplex complex. The A and B rings of the drug are shown in yellow, and the d(C_2-T_2-T_4-T_5-T_6) d(A_9-A_{10}-A_{11}-A_{12}^*-G_{13}) segment is shown in magenta with the backbone in blue. The drug forms a covalent bond to the N3 atom of A_{12}. (a) View looking into the minor groove and normal to the helix axis; (b) view looking down the minor groove, indicating that the entire drug molecule is within the wall of the minor groove [11].

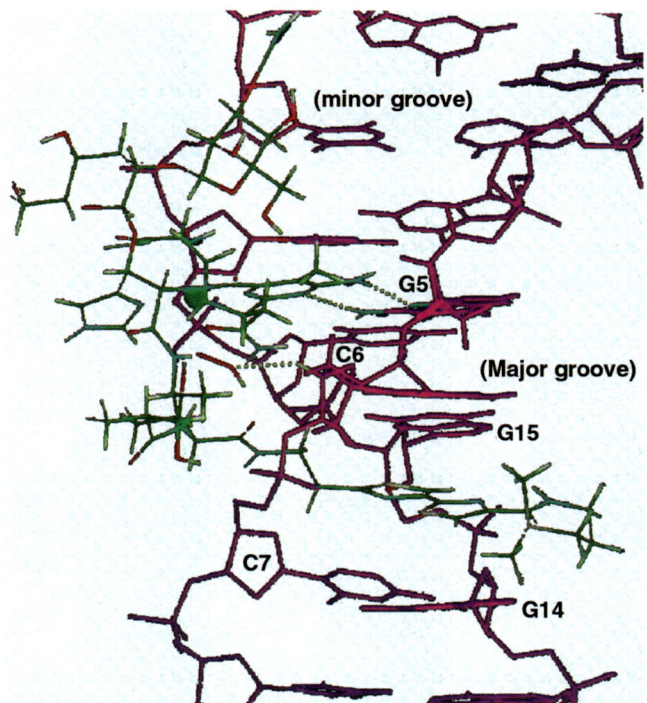

FIGURE 5.9. Structure of BLM A_2-Co(III)-OOH (shown in green) complexed to the 10-mer (in purple). The damaged strand is in the foreground running $5' \rightarrow 3'$ from upper right to lower left corner. The dotted lines indicate hydrogen-bonding interactions between the pyrimidine moiety of BLM and G_5. The proximity of the distal oxygen of the OOH group to the H4′ of C_6 (2.5 Å) is also indicated by the dotted line [22].

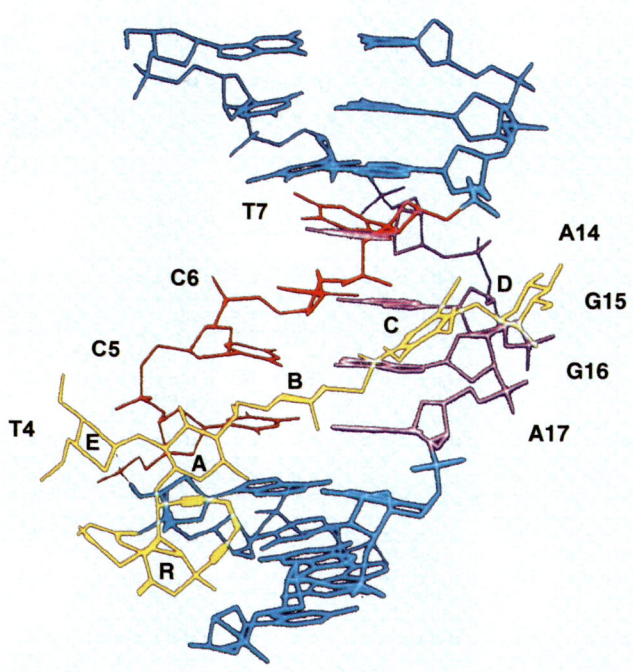

FIGURE 5.15. Molecular model of calicheamicin θ_1^I bound to the decamer in the central region. The drug (shown in yellow) and the central region of the duplex (in red and pink) are shown by different colors [38].

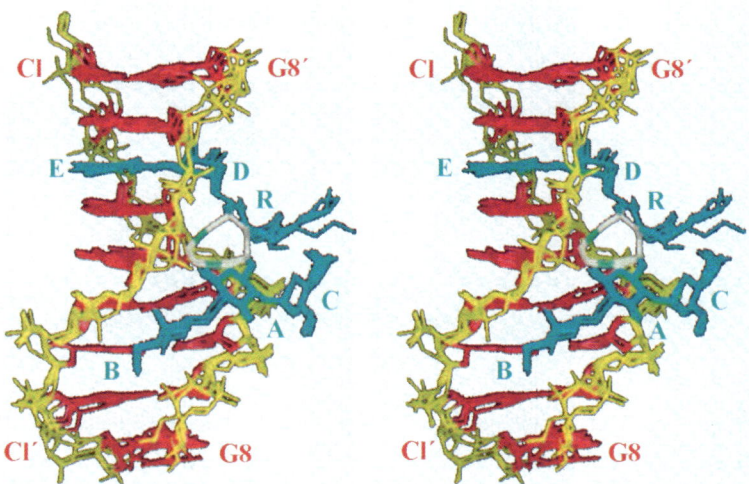

FIGURE 5.17. Stereo view of four superpositioned intensity refined structures of esperamicin A_1–d(CGGATCCG)$_2$ complex. The DNA backbone and base pairs are shown in yellow and magenta, respectively. The drug is shown in blue except for the enediyne ring, which is shown in white with its C3 and C6 carbon atoms in green [40].

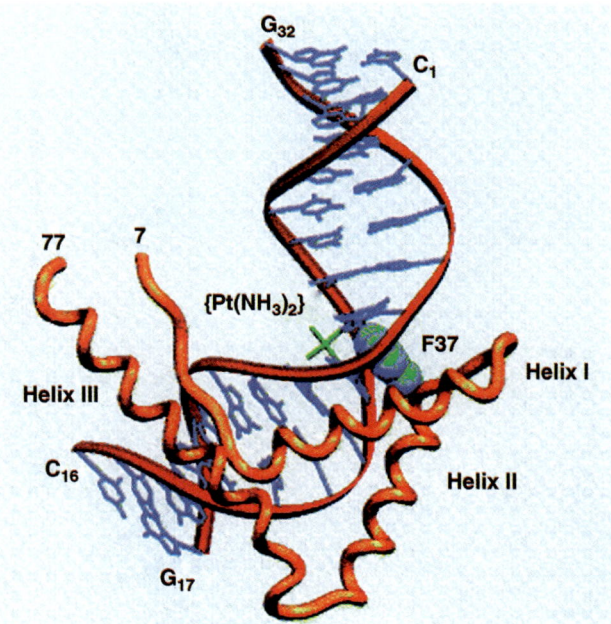

FIGURE 6.5. Overall structure of cisplatinated 16-mer complexed to domain A of the HMG1 protein. The protein backbone is shown in brown. The intercalatinphenylalanine (F37) residue is shown as van der Waals spheres, the 16-mer is shown in red and blue, and the cis-[Pt(NH$_3$)$_2$]$^{2+}$ moiety is shown in green [13,14].

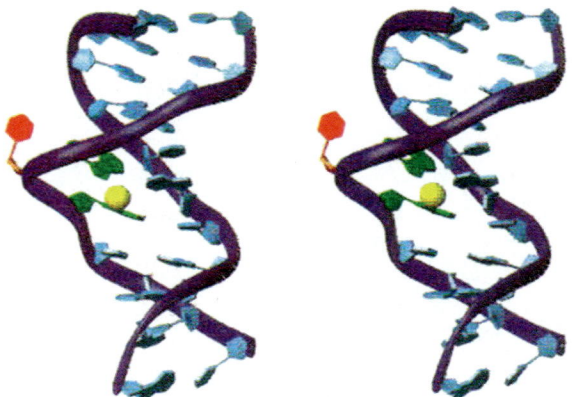

FIGURE 6.16. Stereo view into the major groove of one final structure created by molecular modeling. (*See text for full caption.*) [47].

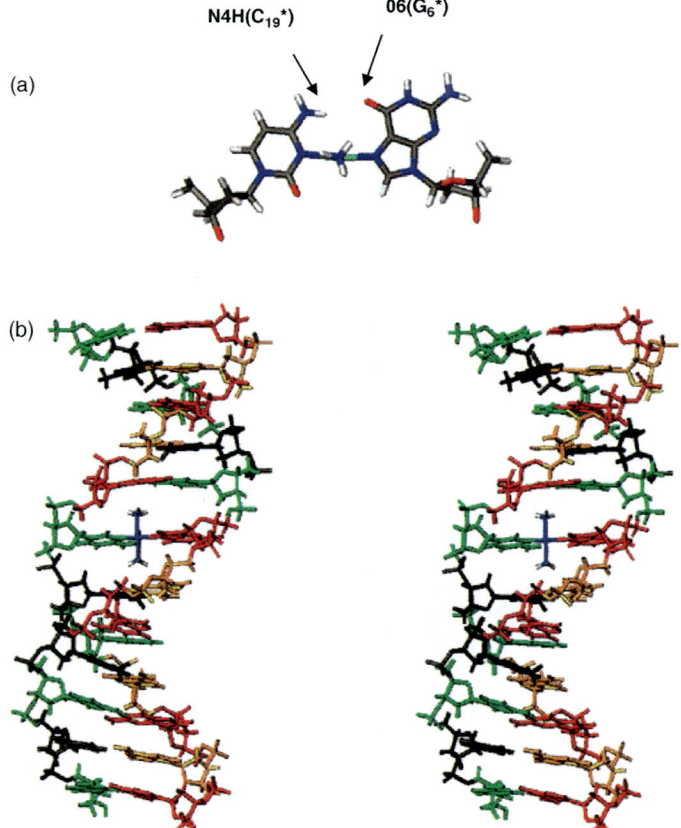

FIGURE 6.27. (a) View (from the top) of the platinated G_6^*-C_{19}^* base pair. (b) Stereo view of the crosslinked complex (*See text for full caption.*) [77].

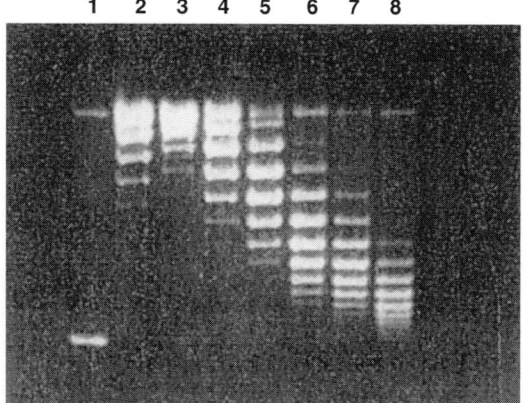

FIGURE 2.48. Electrophoresis analysis of topoisomers of pBR322 DNA, produced by an aclacinomycin A binding and the topoisomerase I relaxation, followed by removal of the drug and enzyme. Samples: (1) purified intact pBR322; (2) completely relaxed pBR322. (no drug was added); (3) aclacinomycin A had been added before topoisomerase was introduced (DNA concentration $= 1.52 \times 10^{-5}$ M, drug concentration 5.89×10^{-7} M); (4) the same (drug concentration $= 1.18 \times 10^{-6}$ M); (5) 1.77×10^{-6} M; (6) 2.36×10^{-6} M; (7) 2.95×10^{-6} M; (8) 3.54×10^{-6} M.

This range of τ values (± 3) is, in fact, predicted from the known values of elastic constants of DNA duplexes (bending force constant $k_b = 6.2 \times 10^{-19}$ N · m/rad and torsional force constant $k_t = 4.2 \times 10^{-19}$ N · m/rad [9]. That the central peak value of τ is $+2$, instead of 0, is likely due to the differences in temperature and solvent in the topoisomerase incubation, as compared to those in the electrophoresis experiment. The temperature was 37°C in the former and 20°C in the latter; solvent was buffer-T (35 mM Tris-HCl (pH = 8.0), 72 mM KCl, 5 mM MgCl$_2$, 5 mM dithiothreitol, 5 mM spermine, and 0.01% bovine serum albumin) in the former and TBE buffer (90 mM tris-borate and 2 mM EDTA) in the latter.

Sample 3 (lane 3) in the electrophoresis experiment [9] came from a topoisomerase solution containing 5.89×10^{-7} M aclacinomycin A as well as 1.52×10^{-5} M DNA (buffer-T, at 37°C). Because the equilibrium constant K is known to be 72,000 M^{-1} (see Section 2.3.1), the concentration of the bound drug can be calculated as 3.05×10^{-7} M. This means that the number of drug molecules bound per one pBR322 DNA molecule is $m = 87$ (Table 2.7). From the electrophoresis analysis, $\Delta\tau = |\tau_0 - \tau| = 2$. Then, the unwinding angle (ϕ) due to one drug molecule should be $\phi = 360\Delta\tau/m = 8°$. A similar experiment was repeated by increasing the concentration of aclacinomycin A stepwise. As seen in Table 2.7, the $\Delta\tau$ value also increased stepwise. The calculated m value also rose, but the calculated unwinding angle, ϕ, still remained at $\sim 8°$. Thus, it can be concluded that one molecule of aclacinomycin A bound to DNA duplex causes an unwinding of the helix by $8 \pm 2°$.

TABLE 2.7. Angle of Unwinding of DNA Double Helix Caused by Aclacinomycin A Binding as Derived from Writhing Number and Equilibrium Constant

Sample Number	C_T^a	C_B^b	m^c	τ^d	$\Delta\tau^e$	ϕ^f
3	5.89×10^{-7}	3.05×10^{-7}	87	+3 to −4	2±0.5	8.2±2.1
4	1.18×10^{-6}	6.05×10^{-7}	174	+1 to −5	4±0.5	8.3±1.0
5	1.77×10^{-6}	8.98×10^{-7}	258	0 to −7	5.5±0.5	7.7±0.7
6	2.36×10^{-6}	1.19×10^{-6}	340	−2 to −10	8±1	8.5±1.1
7	2.95×10^{-6}	1.47×10^{-6}	421	−3 to −11	9±1	7.7±0.9
8	3.54×10^{-6}	1.74×10^{-6}	500	−6 to −13	11.5±0.5	8.3±0.4

$^a C_T$ is the total concentration of aclacinomycin A (M).
$^b C_B$ is the concentration of aclacinomycin A bound to DNA (M).
$^c m$ is the number of aclacinomycin A molecules bound to one pBR322 (closed circular DNA duplex) molecule.
$^d \tau$ is the writhing number of pBR322 DNA, which comes from the topoisomerase + pBR322 DNA solution containing a given amount of aclacinomycin A (see Fig. 2.36).
$^e \Delta\tau$ is the change in writhing number of pBR322 DNA caused by the aclacinomycin A binding.
$^f \phi$ is the unwinding angle of pBR322 DNA due to one aclacinomycin molecule, calculated by the equation $\phi = \Delta\tau \times 360/m$.

The same method has been used for some other intercalators, and the unwinding angles ϕ have been determined: Daunomycin (daunorubicin) $\phi = 12 \pm 2°$ [9], ethidium bromide $\phi = 15 \pm 3°$[9], and chromomycin A$_3$, $\phi = 11.8 \pm 1.1°$ [81].

2.4. SEQUENCE PREFERENCE

Skorobogaty et al. [82] presented the first experimental evidince of the DNA-sequence specificity of daunomycin using DNasa I footprinting at reduced temperatures. This was performed using Hind III/Pvu II restriction fragment from pSP65, labeled at the 3′ terminus of the Hind III restriction site using [α-^{32}P]dATP and AMV reverse transcriptase. The nucleotide sequence of this restriction fragment is known [83] (Fig. 2.49). It is also known that the inhibition of enzymatic hydrolysis of DNA is dependent, in general, on both the DNA sequence and the drug/DNA ratio (v). This effect is much more pronounced at 5°C than at 25°C, and only the data taken at 5°C

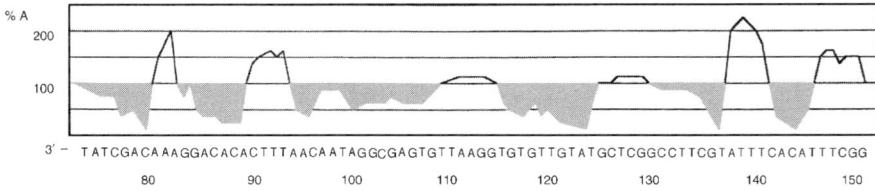

FIGURE 2.49. Densitometric analysis of the DNase I footprinting data. The ordinate represents the fractional (%) enzymatic cleavage ($n = 0.025$ compared to $n = 0$) plotted as a function of the nucleotide sequence visualized on the autoradiogram. The darkened areas correspond to the regions of diminished enzymatic cleavage and indicate drug-binding sites [82].

were presented. Below $v = 0.03$, the apparent sequence dependent inhibition of DNase I became inependent of v. Thus, to ensure that the highest affinity binding sites of daunomycin were being probed, densitometric analysis of the autoradiograms was confined to very low drug/DNA ratios. In Fig. 2.49, the ordinate represents the percentage ratio of the areas under complementary bands in a control lane ($v = 0$) to that of a suitable digest ($v = 0.025$). Domains of cleavage inhibition (ordinate values less than 100%) correspond to the drug-binding sites. There are 22 dinucleotide sequences (5′ or 3′)-CpA, in the visualized DNA fragment ($N = 22$), and 20 of them are found to be the binding sites ($n = 20$). Therefore, CA has a preference of $n/N = 91\%$ (Fig. 2.49). Likewise, it was found that for (5′ or 3′)-CG, $n = 5$ and $n/N = 55\%$; for CT, $n = 7$, $n/N = 50\%$; for GG $n = 2$, $n/N = 33\%$; for AT, $n = 2$, $n/N = 20\%$; and for AA, $n = 3$, $n/N = 21\%$.

DNase I footprinting patterns were also examined in the presence and absence of 2.5 M daunomycin at 4°C [84]. No simple consensus recognition sequence was apparent from the data, although many of the bound sites contain adjacent GC base pairs. They also examined DNase footprinting of distrisarubicin (Fig. 2.50), which was different from that produced by daunomycin. The best binding sites contained the dinucleotide step GpT (ApC) and were located in regions of alternating purines and pyrimidines.

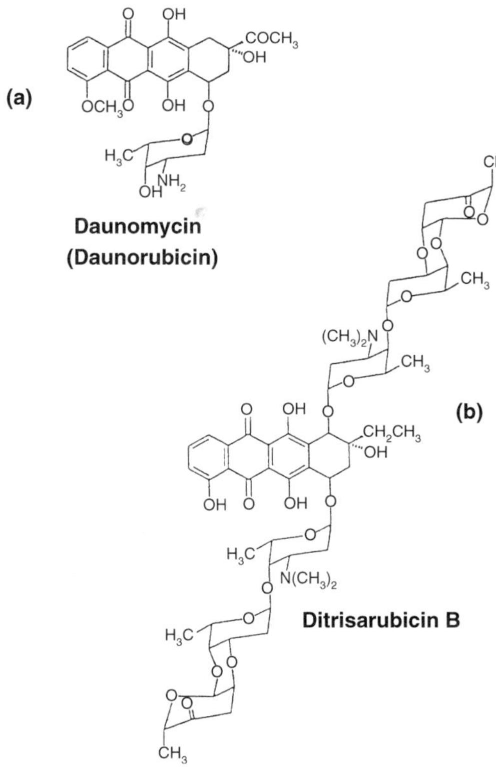

FIGURE 2.50. Structures of daunomycin (a) and ditrisarubicin.

180 INTERCALATING DRUGS

A more refined view of preferred Daunomycin binding sequences was obtained by a high-resolution DNase I footprinting titration [85]. An autoradiogram of the footprinting titration experiment for the interaction of daunomycin with the try T-DNA fragment is shown in Fig. 2.51. Sixty-five positions displayed single-band resolution in the 160 phosphodiester backbone bonds within the strand of the try T-DNA fragment. Four types of behavior were observed: (1) protection from DNase I cleavage; (2) protection, but only after reaching a critical total Daunomycin concentration; (3) enhanced cleavage; and (4) no effect of added drug. Ten sites were identified as the most strongly protected on the basis of the magnitude of the reduction of their digestion product band areas in the presence of daunomycin. These were

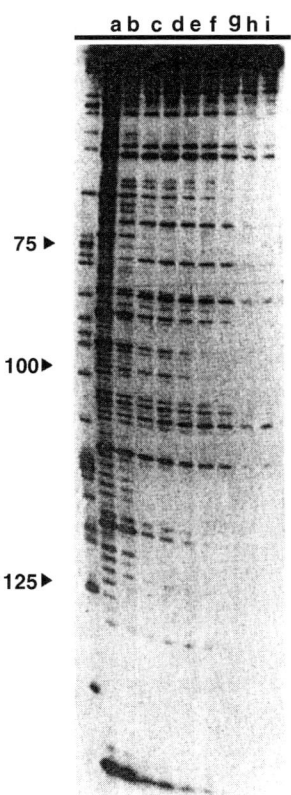

FIGURE 2.51. Autoradiogram of a footprinting titration experiment for the binding of daunomycin to the try T-DNA fragment [85]. The leftmost lane shows the products resulting from the Maxam–Gilbert dimethylsulfate–piperidine reaction and indicates the location of guanine residues within the sequence. The remaining lanes show the products resulting from limited DNase I digestion of the try T-DNA fragment in the presence of increasing concentration (expressed in micromolar units) of added daunomycin, as follows: (a) 0; (b) 0.25; (c) 0.5; (d) 0.75; (e) 1.0; (f) 1.5; (g) 2.0; (h) 2.5; (i) 3.0. The numbers on the left indicate the bond position within the sequence.

identified as the preferred Daunomycin binding sites. Seven of these 10 sites are found at the end of the triplet sequences 5'AGC, 5'TGC, 5'ACG, and 5'TCG.

Another way to determine the sequence preference of an intercalator is to review the experimentally determined molecular structures of intercalator–oligoDNA complexes. These experimental results are presented in Table 2.8, which lists the chemical structures of the intercalators and their binding sites in different DNA sequences, the along with related references [26,29,35,73,77,86–138]. This table makes it clear that drug intercalation is most likely to occur at pyrimidine-(5'-p-3')purine [or (Pyr)p(Pur)] base steps.

The primary reason for this specificity of drug–DNA interaction lies in the sequence-dependent conformation of the DNA double helix. This sequence-dependent conformation can be established by comparing the structures of many oligonucleotides with known sequences. It has been found that an anomalous conformation arises in a DNA duplex at the junction of a pyrimidine stack followed by a purine stack. Calladine [139] and Dickerson [140] proposed a series of generalized rules to describe the standard structure of duplex DNA [139,140]. On the basis of these findings, Takahashi et al. [74] suggested that at the (Py)p(Pu) junction, a purine–purine clash occurs in the minor groove of the DNA double helix. The (Py)p(Py) and (Pu)p(Pu) base steps form stable stacks without such clashes. The (Py)p(Py) clash can be alleviated if some of the local helical parameters adopt nonstandard values. One such parameter is the twist angle (t_g). This is the local residual rotation around the helical axis, and it is considered to be unusually low at the (Py)p(Pu) junction [90]. Several findings support this:

1. According to the Calladine–Dickerson rule [139,140], the t_g value in the (Py)p(Pu) junction is expected to be the lowest among the 12 phosphorus atoms of the four self-complementary tetramers [141] (Fig. 2.52).
2. The twist angle (t_g) in the CpG portion of d(CCGG)$_2$ was actually found to be extremely low in the crystalline state by X-ray analysis [142].
3. Ott and Eckstein [143] reported that the NMR signals of the third and ninth CpG phosphorus atoms in d(CGCGAATTCGCG)$_2$ were found at lower field strength than they had expected on the basis of their locations in sequence, and for these two portions Dickerson and Drew [144] found unusually low t_g values (27.4° and 30.3°, respectively) in the crystal of this dodecamer duplex.
4. Chen et al. [145], from their crystallographic analysis, found that the central CpG portion of d(GGCCGGCC)$_2$ has an unusually low t_g with unusually short interbase distance. The t_g values were found to be 35°, 28°, 45°, and 16°, respectively for the first, second, third, and fourth (central) phophodiester bridges from the 5' terminus.
5. Usually the chemical shift of phosphorus atom in an oligonucleotide is moved toward higher field on decrease in temperature [145]. However, one of the three phosphorus signals in d(CCGG)$_2$ shifts toward lower field on temperature reduction. Takahashi et al. [74] located the anomalous ^{13}P at CpG protion of d(CCGG)$_2$. An anomalous ^{13}P was also found in d(CCTAGG)$_2$ at the TpA portion [74].

182 INTERCALATING DRUGS

TABLE 2.8. Intercalators, Binding Sites, and Methodologies

Intercalator	Sequence and Binding Site	Method [Ref.]
Daunomycin (Daunorubicin)	C·G G·C T·A A·T C·G G·C	X-ray [86]
	C·G G·C T·A A·T C·G G·C	NMR [87]
	T·A T·A G·C G·C T·A A·T A·T T·A C·G C·G A·T A·T	X-ray [88]
	C·G G·C G·C C·G C·G G·C	X-ray [89]
	Arabinosyl C·G cytosine G·C aC·G G·Ca C·G G·C	X-ray [90]
Daunomycin	C·G G·C g5T·A A·Tg5 C·G G·C	X-ray [91]
β-D-Glucosylated		

TABLE 2.8. (*Continued*)

Intercalator	Sequence and Binding Site	Method [Ref.]
	A·T A·T T·A C·G G·C G·C G·C C·G C·G T·A T·A A·T	X-ray [92]
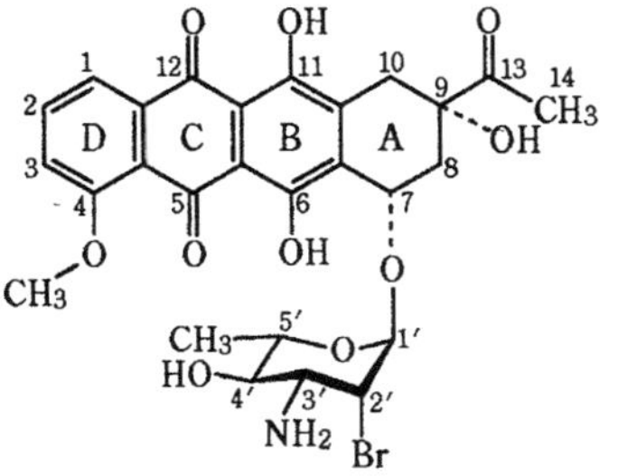	C·G G·C T·A A·T A·T G·C C·G G·C C·G G·C C·G G·C C·G G·C C·G G·C C·G A·T T·A T·A	NMR [93]
	T·A A·T C·G G·C T·A A·T	NMR [94]
MAR70 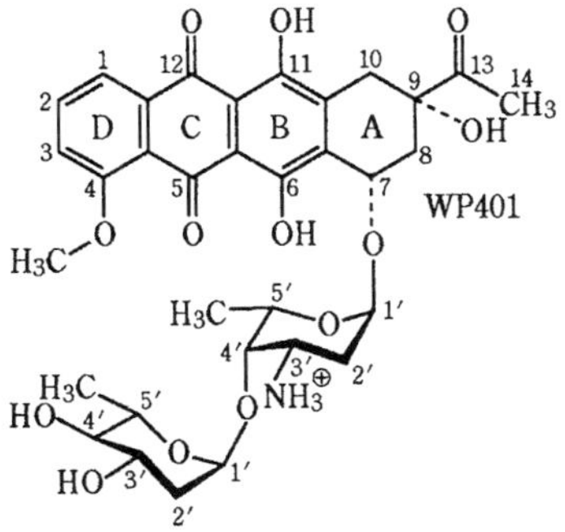	C·G G·C / 2-aminoadenine T·An / nA·T \ C·G G·C	X-ray [95]
WP401	T G C·G G·C G·C G·C G·Cbr5 C·G br5C·G C·G C·G G T G·C	X-ray [96]
3′-(3-Cyano-4-morpholinyl)- 3′-desaminodoxorubicin 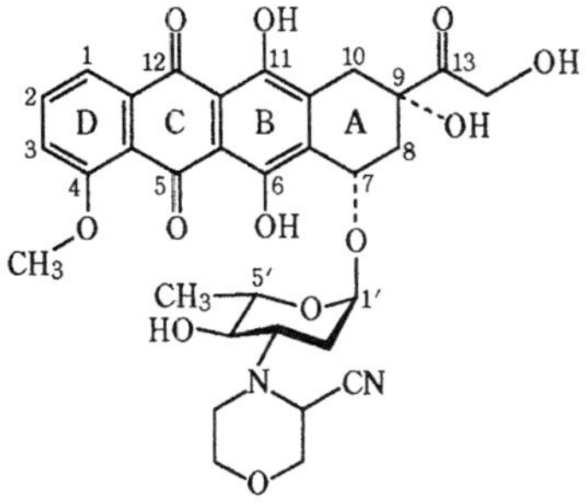	C·G G·C A·T T·A C·G G·C	X-ray [97]

(*continued*)

TABLE 2.8. (*Continued*)

Intercalator	Sequence and Binding Site		Method [Ref.]
N-Cyanomethyl-N-(2-methoxyethyl)daunomycin	C·G G·C A·T T·A C·G G·C		X-ray [98]
Iododoxorubicin	T·A G·C T·A A·T C·G A·T	C·G G·C A·T T·A C·G G·C	X-ray [99]
Idarubicin	T·A G·C A·T T·A C·G A·T		X-ray [100]
Adriamycin (Doxorubicin)	C·G G·C T·A A·T C·G G·C	C·G G·C A·T T·A C·G G·C	X-ray [101]
	C·G G·C T·A A·T C·G G·C		NMR [102]
	T·A G·C G·C C·G C·G A·T		X-ray [89]

TABLE 2.8. (*Continued*)

Intercalator	Sequence and Binding Site		Method [Ref.]
	C·G G·C T·A A·T C·G G·C		NMR [77]

Adriamycin
Daunomycin
  (=rabinomycin-ribomycin C)
4-Dimethoxydaunomycin
Morpholinodoxirubicin
  (=morpholine)
Methoxymorpholinodoxorubicin
9-Deoxydaunorubicin
9-Deoxydaunorubicin
	C·G
	G·C
	T·A
	A·T
	C·G
	G·C
			NMR [103]

4'-Epiadriamycin 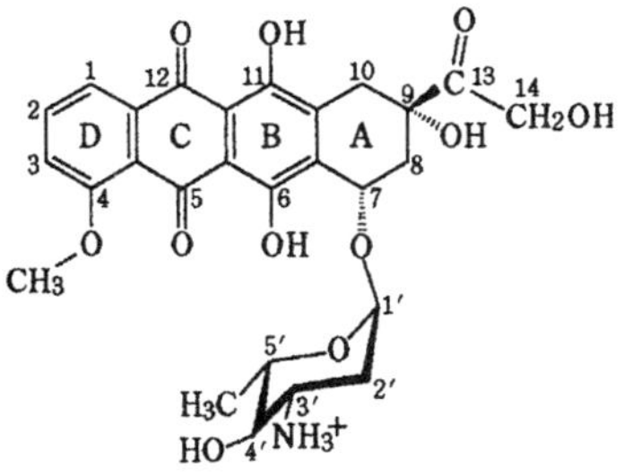	T·A G·C A·T T·A C·G A·T		X-ray [104]
	T·A G·C T·A A·T C·G A·T		X-ray [105]
Morpholinoadriamycin	C·G G·C T·A A·T C·G G·C		X-ray [106]
Morpholinodoxorubicin 3''-Cyanomorpholinodoxorubicin 2''-Methoxymorpholonodoxorubicin	C·G G·C T·A A·T C·G G·C	C·G G·C A·T T·A C·G G·C	X-ray [107]

(*continued*)

TABLE 2.8. (*Continued*)

Intercalator	Sequence and Binding Site	Method [Ref.]
Adriamycin	$\underline{\text{C·G}}$ $\underline{\text{C·G}}$ $\overline{\text{G·C}}$ $\overline{\text{G·C}}$ A·T T·A T·A A·T $\underline{\text{C·G}}$ $\underline{\text{C·G}}$ $\overline{\text{G·C}}$ $\overline{\text{G·C}}$	NMR [108]
Aclacinomycin A	C·G C·G $\underline{\text{T·A}}$ $\overline{\text{A·T}}$ G·C G·C	NMR [73]
Aclacinomycin B	$\underline{\text{C·G}}$ $\overline{\text{G·C}}$ T·A A·T $\underline{\text{C·G}}$ $\overline{\text{G·C}}$	NMR [77]
Actinomycin D	A·T T·A $\underline{\text{G·C}}$ $\overline{\text{C·G}}$ A·T T·A	NMR [109]
	A·T A·T A·T A·T A·T T·A T·A T·A T·A T·A A·T A·T A·T $\underline{\text{G·C}}$ $\underline{\text{G·C}}$ T·A C·G T·A $\overline{\text{C·G}}$ $\overline{\text{C·G}}$ $\underline{\text{G·C}}$ $\underline{\text{G·C}}$ A·T A·T $\underline{\text{G·C}}$ $\overline{\text{C·G}}$ $\overline{\text{C·G}}$ C·G T·A $\overline{\text{C·G}}$ A·T G·C $\underline{\text{G·C}}$ $\underline{\text{G·C}}$ A·T T·A T·A $\overline{\text{C·G}}$ $\overline{\text{C·G}}$ T·A A·T A·T G·C A·T T·A T·A T·A T·A A·T T·A A·T T·A	NMR [110]

TABLE 2.8. (*Continued*)

Intercalator	Sequence and Binding Site	Method [Ref.]
	A·T G·C ――― C·G T·A	NMR [111]
	A·T C·G G·C G·C ――― ――― C·G C·G T·A G·C	NMR [112]
Actinomycin D	A·T A·T A·T G·C ――― C·G T·A T·A T·A	NMR [113]
	G·C A·T A·T G·C ――― C·G T·A T·A C·G	X-ray [114]
	T·A C·G G·C ――― C·G G·C T T T T	NMR [115]
	G·C A·T A·T G·C ――― C·G T·A T·A C·G	X-ray [116]

(*continued*)

TABLE 2.8. (*Continued*)

Intercalator	Sequence and Binding Site	Method [Ref.]
	A·T A·T A·T G·C ――― C·G T·A T·A T·A	NMR [117]
Actinomycin D	G·C G·C G·C A·T A·T A A A·T T T A·T G·C G·C A·T ――― ――― C·G C·G G·C ――― T·A T T C·G no intercalation T·A T·A A A C·G C·G T·A C·G	NMR [118]
	A·T G·C T·A G·C ――― C·G T·A C·G G·C A·T G·C ――― C·G A·T C·G T·A	Raman [119]
Elliptinium (2-Methoxyl-9-hydroxyellipticinium acetate)	T·A A·T G·C C·G A·T T·A C·G ――― ――― G·C G·C C·G T·A A·T C·G G·C A·T T·A	NMR [120]

TABLE 2.8. (*Continued*)

Intercalator	Sequence and Binding Site	Method [Ref.]
Argomycin	G·C C·G --- A·T T·A --- G·C C·G	NMR [121]
Aminoacridine	C·G G·C --- A·T T·A	X-ray [122]
9-Amino-{N-(2-dimethylamino)ethyl}acridine-4-carboxamide 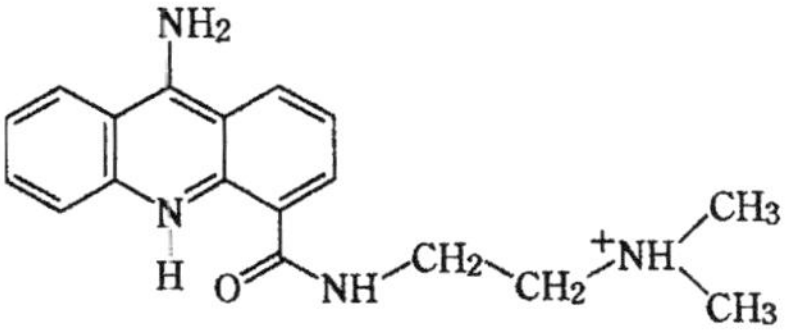	C·G --- G·C --- T·A A·T C·G --- G·C	X-ray [26]
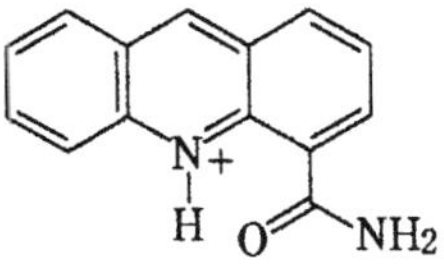	C·G --- G·C T·A A·T C·G --- G·C	X-ray [123]
N-(2-(Dimethylamino)ethyl)acridine-4-carboxamide 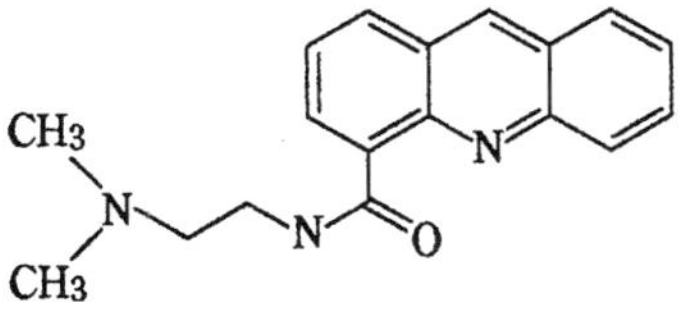	C·G --- G·C 5BrU·A A·U5Br C·G --- G·C	X-ray [29]
Amino acridine	C·G G·C --- A·T T·A	X-ray [35]

(*continued*)

190 INTERCALATING DRUGS

TABLE 2.8. (*Continued*)

Intercalator	Sequence and Binding Site	Method [Ref.]
1H-2,3-Dihydroindolizino [7,6,5-k]acridine chloride	A·T C·G G·C C·G G·C T·A	NMR [124]
Nogalamycin 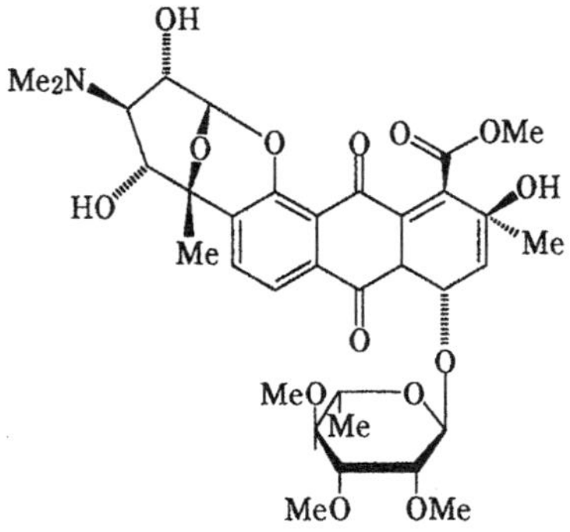	C·G G·C T·A A·T C·G G·C	NMR [125]
Nogalamycin	G·C C·G A·T T·A G·C C·G	NMR [126]
Nogalamycin Phosphothioate linkage ⟶	C·G G·Cm⁵ ₛT·Aₛ A·T C·G G·Cm⁵	X-ray [127]
Nogalamycin Phosphothioate linkage ⟶	m⁵C·G G·Cm⁵ ₛT·Aₛ A·T m⁵C·G G·Cm⁵	X-ray [128]

TABLE 2.8. (*Continued*)

Intercalator	Sequence and Binding Site	Method [Ref.]
Nogalamycin	$\overline{\text{C·G}}$ $\overline{\text{G·C}}$ $_s\text{T·A}_s$ $^s\text{A·T}^s$ $\overline{\text{C·G}}$ $\overline{\text{G·C}}$	X-ray [129]
Nogalamycin	$\overline{\text{m}^5\text{C·G}}$ $\overline{\text{G·Cm}^5}$ $_s\text{T·A}_s$ $^s\text{A·T}^s$ $\overline{\text{m}^5\text{C·G}}$ $\overline{\text{G·Cm}^5}$	X-ray [128,129]
Nogalamycin	A·T G·C $\overline{\text{C·G}}$ $\overline{\text{A·T}}$ $\overline{\text{T·A}}$ $\overline{\text{G·C}}$ C·G T·A	NMR [130]
Nogalamycin	G·C $\overline{\text{C·G}}$ $\overline{\text{A·T}}$ $\overline{\text{T·A}}$ $\overline{\text{G·C}}$ C·G	NMR [131]
Nogalamycin	G·C $\overline{\text{C·G}}$ $\overline{\text{G·C}}$ T·A	NMR [132]
	$\overline{\text{C·G}}$ $\overline{\text{G·C}}$ T·A A·T $\overline{\text{C·G}}$ $\overline{\text{G·C}}$	NMR [133]

(*continued*)

192 INTERCALATING DRUGS

TABLE 2.8. (*Continued*)

Intercalator	Sequence and Binding Site	Method [Ref.]
Manogaril (7-con-*O*-methylnogarol) 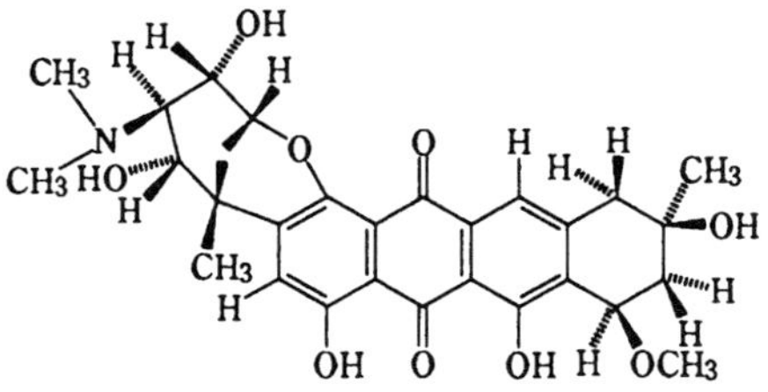	G·C A·T <u>C·G</u> A·T T·A <u>G·C</u> T·A C·G	NMR [134]
Nogalamycin	<u>T·A</u> G·C A·T T·A <u>C·G</u> A·T	X-ray [135]
Nogalamycin	<u>T·A</u> G·C T·A A·T <u>C·G</u> A·T	X-ray [136]
Nogalamycin	A·T T·A G·C C·G A·T T·A	NMR [137]
Nogalamycin	A G A C·G <u>G·C</u> T·A G·C A G A	NMR [138]

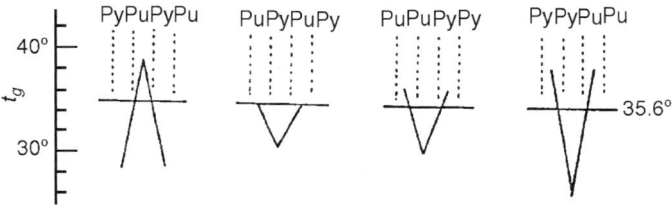

FIGURE 2.52. The local helix twist angle (t_g) for each portion of the B-form DNA fragments, calculated on the basis of the Calladine–Dickerson method [139–141]. t_g is the change in orientation of the C1′–C1′ vectors in two successive base pairs viewed down the helical axis. The standard B structure is assumed to have a value of $t_g = 34.6°$.

A review of Table 2.8 also suggests that the elements of a drug intercalation site are not only the two successive base pairs into which the chromophore is inserted, but also the third base pair located nextdoor to the two base pairs. The third base pair seems to take care of the other portion of the drug molecule in question, better than the intercalating chromophore. A few examples showing that such a situation is the case will be given below.

Among the most widely studied of anthracyclines, there are daunomycin, adriamycin, (doxorubicin), and 4′-epiadriamycin (Fig. 2.53). Various methods, including theoretical studies [145,146], DNA footprinting studies [88,147,148], and ultraviolet melting [105], were used to study the intercalations of anthracyclines with DNA and resulted in evidence that suggests preferential binding to d(CGA), d(CGT), d(TGA), and d(TGT) base-pair triplets, and d(YGG) and (YGC) base-pair triplets are particularly unfavorable sites (Y = any Py base). It is interesting that, opposing such a trend, Leonard et al. [89] constructed two single hexanucleotide–anthracycline complexes, d(CGGCCG)/daunomycin and d(TGGCCA)/adriamycin, and determined the structures using single-crystal X-ray diffraction techniques. In both cases the anthracycline

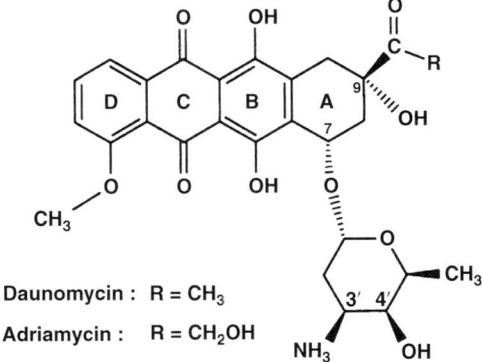

FIGURE 2.53. The molecular formula of daunomycin and adriamycin. The difference between adriamycin and 4′-epiadriamycin is an inversion of the stereochemistry at the 4′ position of the aminosugar. The aglycone ring consists of three unsaturated rings (B, C, D) and one semi-saturated ring (A).

194 INTERCALATING DRUGS

molecule is bound to nonpreferred d(YGG) base-pair triplet sites. Both complexes are essentially isomorphous with related structures but differ in terms of the number of favorable van der Waals interactions of the aminosugars of the drug molecules with the DNA duplexes and the formation in the minor groove of heterodromic pentagonal arrangements of hydrogen bonds involving water molecules that link the amino sugar to DNA. These differences in structure are used to rationalize the lack of affinity of Daunomycin-type anthracycline for d(YGG) and d(YGC) sites.

2.5. BIS- AND TRIS- INTERCALATORS

As shown in Table 2.8, a number of mono- intercalators were prepared to improve the efficacy of the drugs. However, these attempts were not successful in making marked improvements. More recently, several types of multiply-intercalating drugs have been synthesized to attain stronger binding and better selectivity of binding sites. These drugs contain two or more intercalating moieties connected by non intercalative linkers, and their modes of DNA binding vary depending on the length, rigidity, and functionality of the linker and experimental conditions employed.

For example, Atwell et al. [149] prepared three acridine derivatives shown in Fig. 2.54**a**. Compounds (a) and (b) are bis-acridines while (c) is the first tris-acridine

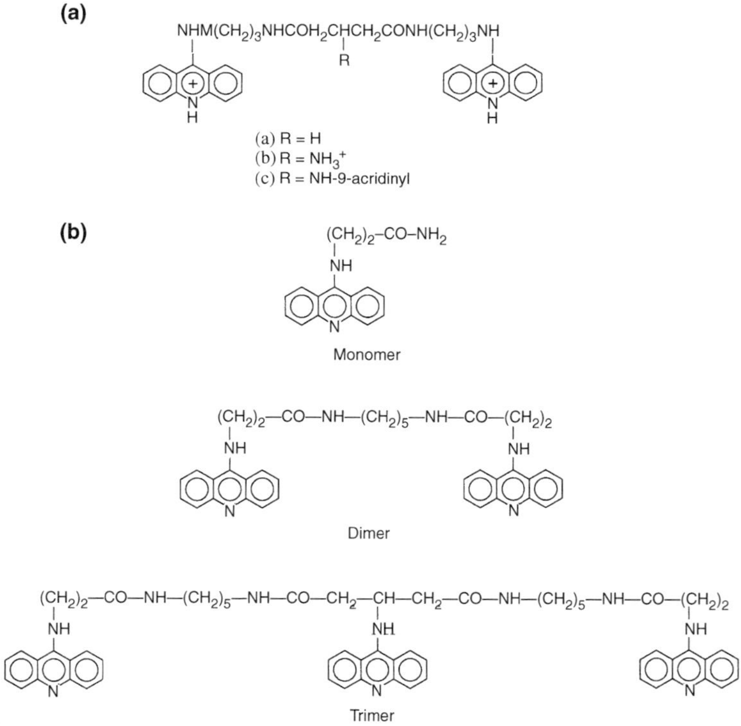

FIGURE 2.54. (**a**) Structures of bis- and tris- intercalating acridines; (**b**) stuctures of mono-, bis-, and tris- intercalating acridines.

reported. The complexes of all three acridines with DNA exhibit the UV–visible absorption maxima at 440 and 417 nm with absorption coefficients increasing in the order of (a) < (b) < (c), and the unwinding angles (degrees) increasing in the order of (a) (29°) < (b)(33°) < (c)(45°). These results seem to suggest tris- intercalation of (c) to DNA. Intermolecular multiple-intercalation between contiguous sites of DNA duplexes does not seem to occur in all cases because the unwinding angles (degrees) of mono- intercalation is much smaller (~15°) [150].

Gaugain et al. [151] prepared a series of the monomer, dimer, and trimer of acridines with the amide linker longer than that of Atwell et al. [149] (Fig. 2.54**b**). More molecular flexibility is expected for these compounds since the spacing between the acridine (~10.2 Å) is much longer than that reported by Atwell et al. (~7 Å). In this case, the observed unwinding angles for the monomer, dimer, and trimer were 17°, 32°, and 55°, respectively. Although their results again suggest tris- intercalation for the trimer, the DNA binding constants of the dimer and trimer were almost the same.

Hansen et al. [152] also compared the DNA binding of the three acridine derivatives connected by polymethylene linkages shown in Fig. 2.55. The results of viscosity, UV–visible absorption, fluorescence, and flow linear dichroism (FLD) studies suggest tris-intercalation of the trimer. However, biological activities of these three compounds did not differ appreciably. X-ray, crystallographic, and/or NMR studies are desired to determine the exact modes of binding of these bis- and tris (acridines) [153].

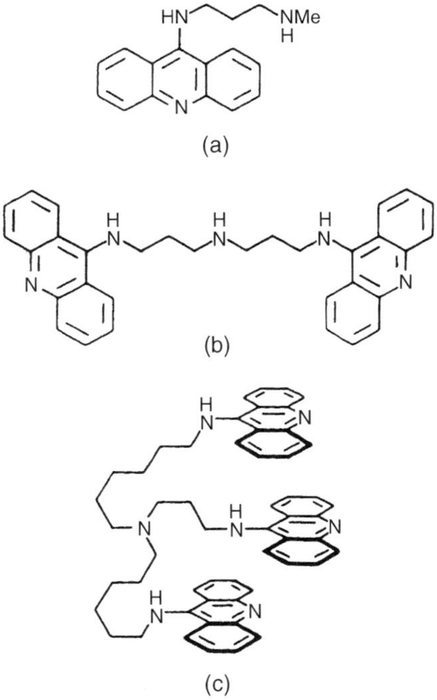

FIGURE 2.55. Structures of mono- (a), bis- (b), and tris- (c) acridines connected by polymethylene linkages.

The antitumor effectiveness for several families of intercalating drugs is correlated with their DNA binding affinity [38,154,155]. In view of this, many laboratories [156–163] synthesized drugs containing two chromophores joined by a molecular linker chain. The most widely studied class of potential bis- intercalating agents was bis (acridines). Compounds of this class were prepared by several groups as potential antitumor agents [164–166] and as probes of bifunctional ligand–DNA interaction [167–170]. A primary goal of the latter studies was to determine the importance of both the length and the nature (flexibility, polarity, hydrogen bond donor/acceptor ability, etc.) of the linker chain on the binding parameter of a potential bisintercalator.

Assa-Munt et al. [40] used high-field ^{1}H NMR spectroscopy to study the binding of the bis(acridines) to the synthetic deoxyribonucleotide d(AT)$_5$ · d(AT)$_5$, as mentioned in Section 2.1.4. A large perturbations (>1 ppm) of the low-field imino proton resonances that occur on ligand addition were monitored, and details of the mode of binding and kinetics of the bound chromophores could be determined. In particular, the method was able to effectively distinguish between mono- and bis- intercalation of the ligands and, in appropriate cases, distinguish between possible modes of bis-intercalation, as described below.

Figure 2.13 (trace d) shows the low-field spectrum of the decadeoxyribonucleotide nonaphosphate d(AT)$_5$ · d(AT)$_5$ at 6°C. The presence of resonances due to the hydrogen-bonded imino protons of the thymine residues in aqueous solution demonstrates that the molecule is in the duplex form, where the protons are protected from exchange with the solvent. At this temperature, it is evident that resonances from all 10 base pairs are observed. From the temperature dependence of the resonances, it can be concluded that the highest-field resonance at 13.1 ppm belongs to the imino protons of the terminal base pairs, which have been upfield shifted by exchange due to "fraying" even at this low temperature. Figure 2.13 shows the assignments of other resonances, as well as the low-field spectrum of the complexes of d(AT)$_5$ · d(AT)$_5$ with 9-aminoacridine. In a similar manner Assa-Munt et al. [40] showed the low-field spectra of complex with bis (acridine) having −(CH$_2$)$_8$− as the linker at different ligand : duplex ratios. As drug is added to d(AT)$_5$ · d(AT)$_5$, the new intensity appears at about 12.3 ppm, and there is a decrease in the intensity of the resonances at 13.2 ppm. The generation of an upfield-shifted peak is indicative of intercalative binding, in which the moderate shielding effect of an adjacent base pair is replaced by the somewhat greater shielding of the ligand chromophore. The intensity of the shifted peak relative to the unshifted peak is a diagnostic monitor of the drug-binding properties. With one drug per duplex, 2 resonances out of a possible 10 will be shifted for each chromophore bound (mono-intercalation). Thus 20% will be shifted. When two chromophores intercalate, 40% of the resonance intensity will be shifted upfield. The results with 9-aminoacridine provide a clear example of how the intercalation of two chromophores affects the spectrum of d(AT)$_5$ · d(AT)$_5$. Therefore, with a series of molecules containing this same chromophore, the NMR measurements provide a method for distinguishing between mono- and bis- intercalation. Using this technique, Assa-Munt et al. [40] measured the relative intensity of the upfield resonance observed in the spectrum of the 9AAcmplex with d(AT)$_5$ · d(AT)$_5$, and the result is summarized in Table 2.2.

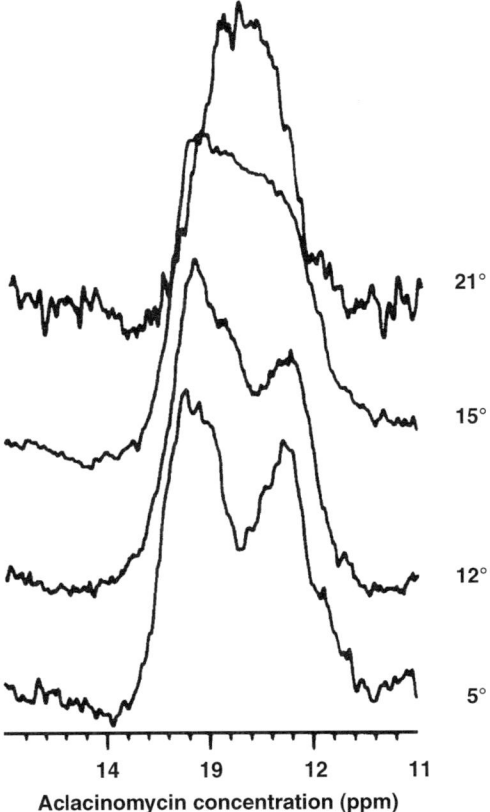

FIGURE 2.56. Effect of temperature on the low-field ^{1}H NMR spectrum of 1 : 1 complex of C8 [bis(acridine) with $(CH_2)_8$ in the linker] with $d(AT)_5 \cdot d(AT)_5$ [42].

Assa-Munt et al. [42] also examined dynamic properties of bis(acridine) complexes with $d(AT)_5 \cdot d(AT)_5$ by the use of ^{1}H NMR spectra and relaxation measurements at different temperatures. The effect of temperature on the 300 MHz spectra of a 1 : 1 complex of $d(AT)_5 \cdot d(AT)_5$ with a bis- intercalator containing a $(CH_2)_8$ or C8 linker is shown in Fig. 2.56. This compound is obviously in slow exchange at 5°C, but by 15°C the upfield-shifted resonances (corresponding to about 40% of the total intensity) broaden and coalesce, and by 21°C some loss of intensity, because of exchange with water, was evident. Next, they examined [42] the spin–lattice relaxation of the C8-$d(AT)_5 \cdot d(AT)_5$ complex at 6°C (Section 1.4.1). The original, unperturbed resonance at 13.2 ppm relaxed at a rate of 15 s^{-1}, whereas the separate broad upfield-shifted peak at 12.3 ppm exhibit both fast ($\sim$100 s^{-1}) and slow ($\sim$8 s^{-1}) decay components. They concluded, therefore, that two types of drug-shifted imino resonance are present at 12.3 ppm. At 21°C, the drug-shifted imino resonances and the unperturbed imino resonances coalesce (Fig. 2.56). The relaxation rates of resonances at $\sim$12.4 ($\sim$60 s^{-1}) and 12.9 ppm ($\sim$70 s^{-1}) were

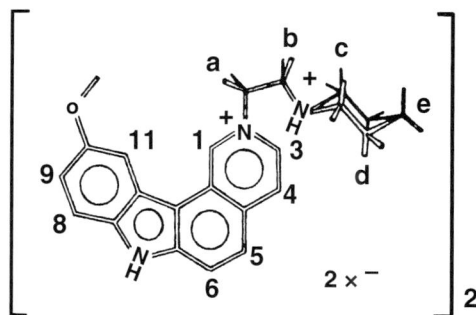

FIGURE 2.57. Ditercalinium. 2,2'-[4,4'-bipiperidine-1,1'-bis(ethane-1,2-dyl)]bis(10-methoxy-7H-pyrido[4,3-c]carbazolium)tetramethanesulfonate.

comparable. These proton exchange rates are much too slow to account for the substantial broadening of the resonances that was observed. The drug is considered to be in rapid exchange among the different binding sites as evidenced by a coalescence of resonances in the low-field spectrum. From similar measurements [171] on other bis(acridine) complexes with d(AT)$_5$ · d(AT)$_5$, an important conclusion was reached: namely, that certain bisintercalators rapidly migrate along DNA, despite having large binding constants ($K > 106$ M^{-1}).

Ditercalinium(2,2'-{[4,4'-biperidine]-1,1'-diyldi-2,1-ethanediyl}bis{10-methoxy-7H-pyrido [4,3-c]carbazolium} tetramethane sulfonate) (Fig. 2.57) is a rigid dimer created by connecting two 7H-pyido[4,3-c]carbazole moieties through quaternization

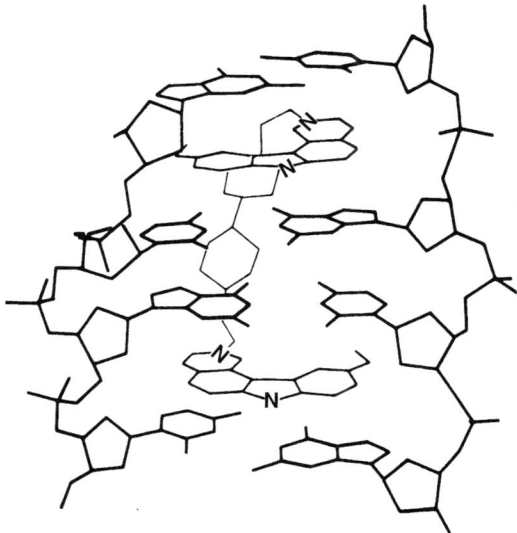

FIGURE 2.58. A model of a d(CpGpCpG)$_2$ minihelix bis-intercalated with a ditercalinium molecule, viewed from the minor groove side [172].

with a bis(ethyl bispiperidine) linker [172,173]. This is a DNA bis- intercalator that displays high DNA affinity and strong antitumor properties. The structures of the complexes formed by the self-complementary tetranucleotide d(CpGpCpG) with ditercalinium and with a related monomer were investigated in 0.1 M [^{2}H]acetate buffer (pH 5.5) by using 400 MHz ^{1}H NMR [172,173]. In both cases, d(CGCG) detained a right-handed duplex structure as shown by exchangeable-proton analysis and intramolecular nuclear Overhauser effect (NOE) measurements. According to the large upfield shifts measured on the base protons (including the imino proton) and on the aromatic protons of the pyridocarbazole rings, the monomer appears to mono-intercalate and the dimer, to bis-intercalate, into the teranucleotide duplex (Fig. 2.58).

REFERENCES

1. Lerman, L. S., "Structural considerations in the interaction of DNA and acridines," *J. Mol. Biol.* **3**:18–30 (1961).
2. Pritchard, N. J., Blake, A., and Peacocke, A. R., "Modified intercalation model for the interaction of amino acridines and DNA," *Nature* **212**:1360–1361 (1966).
3. Andreoni, A., Sacchi, L. A., and Svelto, O., "Structure studies on biological molecules via laser-induced fluorescence: Acridine-DNA complexes," in Moor, C. D., ed., *Chemical and Biochemical Applications of Lasers*, Academic Press, 1979, Vol. IV, p. 7.
4. Neidle, S., Achari, A., Taylor, G. L., Berman, H. M., Carrell H. L., Glusker, J. P., and Stallings, W. C., "Structure of dinucleoside phosphate-drug complex as model for nucleic acid-drug interaction," *Nature* **269**:304–307 (1977).
5. Waring, M. J., "DNA modification and cancer," *Ann. Rev. Biochem.* **50**:159 (1981).
6. Drug-nucleic acid interactions, in Chairs, J. B. and Waring, M. J., eds., *Methods in Enzymology*, Academic Press, San Diego, 2001, Vol. 340.
7. Cantor, C. R. and Schimmel, P. R. *Biophysical Chemistry, Techniques for the Study of Biological Structure and Function*, Freeman, San Francisco, 1980.
8. Biochemical spectroscopy, in Sauer, K., ed., *Methods in Enzymology*, Academic Press, San Diego, 1995, Vol. 246.
9. Utsuno, K. and Tsuboi, M., "Degree of DNA unwinding caused by binding of aclacinomycin A," *Chem. Pharm. Bull.* **45**:1551–1557 (1997).
10. Seeman, N. C., Day, R. O., and Rich, A., "Nucleic acid-mutagen interactions: Crystal structure of adenylyl-3', 5'-uridine plus 9-aminoacridine," *Nature* **253**:324–326 (1975).
11. Sakore, T. D., Reddy, R. S., and Sobell, H. M., "Visualization of drug-nucleic acids interactions at atomic resolution IV. Structure of an aminoacridine-dinucleotide monophosphate crystalline complex, 9-aminoacridine-5-iodocytidylyl(3',5')guanosine," *J. Mol. Biol.* **135**:763–785 (1979).
12. Wang, A. H.-J., Nathans, J., Van der Marel, G., van Boom, J. H., and Rich, A., "Molecular structure of a double helical DNA fragment intercalator complex between deoxy CpG and a terpyridine platinum compound," *Nature* **276**:471–474 (1978).
13. Jain, S. C., Bhandary, K. K., and Sobell, H. M., "Visualization of drug-nucleic acids interactions at atomic resolution. VI. Structure of two drug-dinucleoside monophosphate crystalline complexes, elliptosine-5-Iodocytidylyl(3',5')guanosine

and 3,5,6,8-tetramethyl-*N*-methyl phenanthrolinium-5-iodocytidylyl(3′,5′)guanosine," *J. Mol. Biol.* **135**:813–840 (1979).

14. Shieh, H.-S., Berman, H. M., Dabrow, M., and Neidle, S., "The structure of drug-deoxydinucleodide phosphate complex: Generalized conformational behavior of intercalation complexes with RNA and DNA fragments," *Nucl Acids Res.* **8**:85–97 (1980).

15. Aggarwal, A., Islam, S. A., Kuroda, R., and Neidle, S., "X-ray crystallographic analysis of a ternary intercalation complex between proflavine and the dinucleoside monophosphates CpA and UpG," *Biopolymers* **23**:1025–1041 (1984).

16. Westhof, E., Rao, S. T., and Sundaralingam, M., "Crystallographic studies of drug-nucleic acid interaction: Proflavine intercalation between the non-complementary base-pairs of cytidyl-3′,5′-adenine," *J. Mol. Biol.* **142**:331–361 (1980).

17. Neidle, S., Pearl, L. H., Herzyk, P., and Berman, H. M., "A molecular model for proflavine-DNA intercalation," *Nucl. Acids Res.* **16**:8999–9016 (1988).

18. Wang, A. H.-J., Quigley, G. J., and Rich, A., "Atomic resolution analysis of a 2:1 complex of CpG and acridine orange," *Nucl. Acids Res.* **6**:3879–3890 (1979).

19. Reddy, B. S., Seshadri, T. P., Sakore, T. D., and Sobell, H. M., "Visualization of drug-nucleic acid intercalations at atomic resolution," *J. Mol. Biol.* **135**:787–812 (1979).

20. Berman, H. M., Neidle, S., and Stodola, R. K., "Drug-nucleic acid interactions: Conformational flexibility at the intercalation site," *Proc. Natl. Acad. Sci. USA* **75**:828–832 (1978).

21. Berman, H. M., Stallings, W., Carrell, H. L., Glusker, J. P., Neidle, S., Taylor, G., and Achari, A., "Molecular and crystal structure of an intercalation complex: Proflavin-cytidyl(3′,5′)guanosine," *Biopolymers* **18**:2405–2429 (1979).

22. Sthof, E., Rao, S. T., and Sundaralingam, M., "Crystallographic studies of drug-nucleic acid interactions: Proflavin intercalation between the non-complementary base-pairs of cytidyl-3′,5′-adenosine," *J. Mol. Biol.* **142**:331–361 (1980).

23. Bhandary, K. K., Sakore, T. D., and Sobell, H. M., "Visualization of drug-nucleic acid interactions at atomic resolution. IX. Structures of two *N,N*-dimethylproflavine: 5-Iodocytidylyl(3′-5′)guanosine crystalline complexes," *J. Biomol. Struct. Dyn.* **1**:1195–1217 (1984).

24. Sakore, T. D., Bhandary, K. K., and Sobell, H. M., "Visualization of drug-nucleic acid interactions at atomic resolution X. Structure of a *N,N*-dimethylproflavin: Deoxycytidylyl (3′-5′)deoxyguanosine crystalline complex," *J. Biomol. Struct. Dyn.* **1**:1219–1227 (1984).

25. Adams, A., "Crystal structures of acridines complexed with nucleic acids," *Curr. Med. Chem.* **9**:1667–1675 (2002).

26. Adams, A., Guss, J. M., Collyer, C. A., Denny, W. A., and Wakelin, L. P. G., "Crystal structure of toisomerase II poison 9-amino-[*N*-(2-dimethylamino)ethyl]acridine-4-carboxamide bound to the DNA hexanucleotide d(CGTACG)$_2$," *Biochemistry* **38**:9221–9233 (1999).

27. Denny, W. A., Atwell, G. J., Rewcastle, G. W., and Baguley, B. C., "Potential antitumor agents. 49, 5-Substituted derivatives of *N*-[2-(dimethylamino)ethyl]-9-aminoacridine-4 carboxiamide with *in vivo* solid-tumor activity," *J. Med. Chem.* **30**:658–663 (1987).

28. Ren, B., Ye, S., Liang, X., and Komiyama, M., "Acridine-bearing PNA for efficient protection of designated site of DNA from nuclease S1," *Chem. Lett.* **32**:714–715 (2003).

REFERENCES

29. Todd, A. K., Adams, A., Thorpe, J. H., Denny, W. A., Wakelin, P. G., and Cardin, C. J., "Major groove binding and 'DNA-Induced' fit in the intercalation of a derivative of the mixed topoisomerase I/II poison N-(2-(dimethylamino)ethyl)acridine-4-carboxamide (DACA) into DNA: X-ray structure complexed to d(CG(5-BrU)ACG)2 at 1.3 resolution," *J. Med. Chem.* **42**:536–540 (1999).
30. Gupta, G., Dhingra, M. M., and Sarma, R. H., "Left-handed intercalated DNA double helix: Rendezvous of ethidium and actinomycin D in the Z-helical conformatio space," *J. Biomol. Struct. Dyn.* **1**:97–113 (1983).
31. Westhof, E., Hosur, M. V., and Sundaralingam, M., "Nonintercalative binding of proflavin to Z-DNA: Structure of a complex between d(5BrC-G-5BrC-G) and proflavin," *Biochemistry* **27**:5742–5747 (1988).
32. Perry, P. J., Gowan, S. M., Reszka, A. P., Polucci, P., Jenkins, T. C., Kelland, L. R., and Neidle, S., "1,4- and 2,6-disubstituted amidoanthracene-9,10-dione derivatives as inhibitors of human telomerase," *J. Med. Chem.* **41**:3253–3260 (1998).
33. Perry, P. J., Peszka, A. P., Wood, A. A., Read, M. A., Gowan, S. M., Dosanjh, H. S., Trent, J. O., Jenkins, T. C., Kelland, L. R., and Neidle, S., "Human telomerase inhibition by regioisomeric disubstituted aminoanthracene-9,10-diones," *J. Med. Chem.* **41**:4873–4884 (1998).
34. Read, M. A. and Neidle, S., Structural characterization of a guanine-quadruplex ligand complex," *Biochemistry* **39**:13422–13432 (2000).
35. Neville, D. M., Jr. and Davies, D. R., "The interaction of acridine dyes with DNA: An X-ray diffraction and optical investigation," *J. Mol. Biol.* **17**:57–74 (1966).
36. Waring, M., "Variation of the supercoils in closed circular DNA by binding of antibiotics and drugs: Evidence for molecular models involving intercalation," *J. Mol. Biol.* **54**:247–279 (1970).
37. Feigon, J., Denny, W. A., Leupin, W., and Kearns, D. R., "Interaction of antitumor drugs with natural DNA: ^{1}H NMR study of binding mode and kinetics," *J. Med. Chem.* **27**:450–465 (1984).
38. Denny, W. A., Cain, B. F., Atwell, G. J., Hansch, C., Panthananickal, A., and Leo, A., "Potential antitumor agents. 36. Quantitative relationships between experimental antitumor activity, toxicity, and structure for the general class of 9-anilinoacridine antitumor agents," *J. Med. Chem.* **25**:276–315 (1982).
39. Baguley, B. C., Denny, W. A., Aiwell, G. J., and Cain, B. F., "Potential antitumor agents. 35. Quantitative relationships between antitumor (L1210) potency and DNA binding for 4′-(9-acRidylamino)methane sulfon-m-anisidide analoogues," *J. Med. Chem.* **24**:520–525 (1981).
40. Assa-Munt, N., Denny, W. A., Leupin, W., and Kearns, D. R., "1H NMR study of the binding of bis(acridines) to d(AT)$_5$-d(AT)$_5$. 1. Mode of binding," *Biochemistry* **24**:1441–1449 (1985).
41. Kearns, D. R., "High-resolution nuclear magnetic resonance studies of double helical polynucleotides," *Annu. Rev. Biophys. Bioeng.* **6**:477–523 (1977).
42. Assa-Munt, N., Leupin, W., Denny, W. A., and Kearns, D. R., "^{1}H NMR study of the binding of bis(acridine) to d(AT)$_5$ · d(AT)$_5$. 2. Dynamic aspects.," *Biochemistry* **24**:1449–1460 (1985).
43. Hogan, M., Dattagupta, N., and Crothers, D. M., "Transient electric dichroism studies of the structure of the DNA complex with intercalated drugs," *Biochemistry* **18**:280–288 (1979).

44. Lerman, L. S., "The structure of the DNA-acridine complex," *Proc. Natl. Acad. Sci. USA* **49**:94–102 (1963).
45. Zhou, B. W., Puga, E., Sun, J. S., Garestler, T., and Helene, C., "Stable triple helices formed by aciridine-containing oligonucleptides with oligopurine tracts of DNA interrupted by one or 2 pyrimidines," *J. Am. Chem. Soc.* **117**:10425–10428 (1995).
46. Dalgleish, D. G., Peacock, A. R., Fey, G., and Harvey, C., "The circular dichroism in the ultraviolet of aminoacridines and ethidium bromide bound to DNA," *Biopolymers* **10**:1853–1863 (1971).
47. Luedtke, N. W., Liu, Q., and Tor, Y., "RNA-ligand interactions: Affinity and specificity of aminoglycoside dimers and acridine conjugates to the HIV-1 Rev response element," *Biochemistry* **42**:11391–11403 (2003).
48. Newton, B. A., "The mode of action of phenanthridines: The effect of ethidium bromide on cell division and nucleic acid synthesis," *J. Gen. Microbiol.* **17**:718–730 (1957).
49. Elliott, W. H., "The effects of antimicrobial agents on deoxyribonucleic acid polymerase," *Biochem. J.* **86**:562–567 (1963).
50. Waring, M. J., "Complex formation with DNA and inhibition of Escherichia coli RNA polymerase by ethidium bromide," *Biochim Biophys. Acta* **87**:358–361 (1964).
51. Tsai, C., Jain, S. C., and Sobell, H. M., "Visualization of drug-nucleic acid interactions at atomic resolution. I. Structure of ethidium/dinucleoside monophosphate crystalline complex, ethidium: 5-Iodo uridylyl(3′-5′)adenosine," *J. Mol. Biol.* **114**:301–315 (1977).
52. Jain, S. C., Tsai, C., and Sobell, H. M., "Visualization of drug-nucleic acid interactions at atomic resolution. II. Structure of ethidium/dinucleoside monophosphate crystalline complex, ethidium: 5-Iodocytidylyl(3′-5′)guanosine," *J. Mol. Biol.* **114**:317–331 (1977).
53. Jain, S. C. and Sobell, H. M., "Visualization of drug-nucleic acid interactions at atomic resolution. VII. Structure of an ethidium/dinucleoside monphosphate crystalline complex, ethidium: Uridylyl(3′-5′)adenosine," *J. Biomol. Struct. Dyn.* **1**:1161–1177 (1984).
54. Sobell, H. M., Tsai, C., Jain, S. C., and Gilbert, S. G., "Visualization of drug-nucleic acid interactions at atomic resolution. III. Unifying structural concepts in understanding drug-DNA interactions and their broader implications in understanding protein-DNA interactions," *J. Mol. Biol.* **114**:333–365 (1977).
55. Fuller, W. and Waring, M. J., "A molecular model for the interaction of ethidium bromide with deoxyribonucleic acid," *Ber. Bunseng. Phys. Chem.* **68**:805–808 (1964).
56. Bond, P. J., Langridge, R., Jannette, K. W., and Lippard, S. J., "X-ray fiber diffraction evidence for neighbor exclusion binding of a platinum metallointercalation regent to DNA," *Pro. Natl. Acad. Sci. USA* **72**:4825–4829 (1975).
57. Arnott, S., "Polynucleotide secondary structures: an historical perspective," in Neidle, S., ed., *Oxford Handbook of Nucleic Acid Structure*, Oxford Univ. Press, 1999, p. 7.
58. Waring, M. J., "Complex formation between ethidium bromide and nucleic acids, *J. Mol. Biol.* **13**:269–282 (1965).
59. McGhee, J. D. and von Hippel, P. H., "Theoretical aspect of DNA-protein interactions: Co-operative and non-co-operative binding of large ligands to a one-dimensional homogeneous lattice," *J. Mol. Biol.* **86**:469–489 (1974).
60. Guo, Q., Lu, M., Marky, L. A., and Kellenback, N. R., "Interaction of dye ethidium bromide with ANA containing guanine repeats," *Biochemistry* **31**:2451–2455 (1992).
61. Williamson, J. R., Raghuraman, M. K., and Thomas, R. C., "Monovalent cation-induced structure of telomeric DNA: G-Quartet model," *Cell* **59**:871–880 (1989).

62. Lepecq, J.-B. and Paoletti, C., "A fluorescent complex between ethidium bromide and nucleic acids. Physical-chemical characterization," *J. Mol. Biol.* **27**:87–106 (1967).
63. Weill, G. and Calvin, M., "Optical properties of chromophore-macromolecule complexes: Absorption and fluorescence of acridine dyes bound to polyphosphates and DNA," *Biopolymers* **1**:401–417 (1963).
64. Benevides, J. M. and Thomas, G. J., Jr., "Local conformational changes induced in B DNA by ethidium intercalation," *Biochemistry* **44**:2993–2999 (2005).
65. Thomas, G. J., Jr. and Benevides, J. M., "An A-helix structure for poly(dA-dT) · poly(dA-dT)," *Biopolymers* **24**:1101–1105 (1985).
66. Yuzaki, K. and Hamaguchi, H., "Intercalation-induced structural change of DNA as studied by 1064 nm near-infrared multichannel Raman spectroscopy," *J. Raman Spectrosc.* **35**:1013–1015 (2004).
67. Tsuboi, M., Benevides, J. M., and Thomas, G. J., Jr., "Orientation of the ethidium molecule intercalated into DNA as revealed by a polarized Raman spectroscopy," paper presented at Annual Meeting of the Biophysical Society, Baltimore 2003.
68. Subramanian, E., Trotter, J., and Bugg, C. E., "Crystal structure of ethidium bromide," *J. Cryst. Mol. Struct.* **1**:3–15 (1971).
69. Wang, J. C., "The degree of unwinding of the DNA helix by ethidium I. Titration of twisted PM2 molecules in alkaline cesium chloride density gradients," *J. Mol. Biol.* **89**:783–801 (1974).
70. Pope, L. H., Davies, M. C., Laughton, C. A., Roberts, C. J., Tendler, S. J. B., and Williams, P. M., "Atomic force microscopy studies of intercalation-induced changes in plasmid DNA tertiary structure," *J. Microsc.* **199**:68–78 (2000).
71. Utsuno, K., Tsuboi, M., Katsumata, S., and Iwamoto, T., "Viewing of complex molecules of ethidium bromide and plasmid DNA in solution by atomic force microscopy," *Chem. Pharm. Bull.* **49**:413–417 (2001).
72. Oki, T., Matsuzawa, Y., Yoshimoto, A., Numata, K., Kitamura, I., Hori, S., Takamatsu, A., Umezawa, H., Ishizuka, M., Naganawa, H., Suda, H., Hamada, M., and Takeuchi, T., "New antitumor antibiotics, aclacinomycins A and B," *J. Antibiot.* **28**:830–834 (1975).
73. Takahashi, S., Nagashima, N., Nishimura, Y., and Tsuboi, M., "Nuclear magnetic resonance study on the interaction of aclacinomycin-A with a deoxyribo-hexanucleotide d(CCTAGG)$_2$ in aqueous solution," *Chem. Pharm. Bull.* **34**:4494–4499 (1986).
74. Takahashi, S., Nagashima, N., Nishimura, Y., and Tsuboi, M., "Anomalous temperature dependence of the phosphorus-31 nuclear magnetic resonance chemical shift in d(CCGG) and d(CCTAGG) at the junction of the pyrimidine stack followed by the purine stack," *Chem. Pharm. Bull.* **34**:3987–3993 (1986).
75. Patel, D. J., "Peptide antibiotic-oligonucleotide interactions. nuclear magnetic resonance investigations of complex formation between actinomycin D and d-ApTpGpCpApT in aqueous solution," *Biochemistry* **13**:2396–2402 (1974).
76. Patel, D. J., Kozlowski, S. A., and Rice, J. A., "Hydrogen bonding, overlap geometry, and sequence specificity in anthracycline antitumor antibiotic. DNA complexes in solution," *Proc. Natl. Acad. Sci. USA* **78**:3333–3337 (1981).
77. Yang, D., and Wang, A. H.-J., "Structure by NMR of antitumor drugs aclacinomycin A and B complexed to d(CGTACG)," *Biochemistry* **33**:6595–6604 (1994).
78. Robinson, H. and Wang, A. H.-J., "A simple spectral-driven procedure for the refinement of DNA structures by NMR spectroscopy," *Biochemistry* **31**:3624–3633 (1992).

79. Niglies, M., Habazetti, J., Brüger, A. T., and Holak, T. A., "Relaxation matrix refinement of the solution structure of squash trypsin inhibitor," *J. Mol. Biol.* **219**:499–510 (1991).
80. Nonaka, Y., Tsuboi, M., and Nakamoto, K., "Comparative study of aclacinomycin versus adriamycin by means of resonance Raman spectroscopy," *J. Raman Spectrosc.* **21**:133–141 (1990).
81. Utsuno, K., Kojima, K., Maeda, Y., and Tsuboi, M., "The average unwinding angle of DNA duplex produced by the binding of chromomycin A_3," *Chem. Pharm. Bull.* **46**:1667–1671 (1998).
82. Skorobogaty, A., White, R. J., Phillips, D. R., and Reiss, J. A., "The 5′-CA DNA-sequence preference of daunomycin," *FEBS Lett.* **227**:103–106 (1988).
83. Melton, D. A., Krieg, P. A., Rebagliatti, M. R., Mniatis, T., Zinn, K., and Green, M. R., "Efficient *in vitro* synthesis of biologically active RNA and RNA hybridization probes from plasmids containing a bacteriophage SP6 promoter," *Nucl. Acids Res.* **12**:7035–7056 (1986).
84. Fox, K. R. and Kunimoto, S., "Sequence selective binding of ditrisarubicin B to DNA: Comparison with daunomycin," *FEBS Lett.* **250**:323–327 (1989).
85. Chaires, J. B., Herrera, J. E., and Waring, M., "Preferential binding of daunomycin to 5′(A or T)CG and 5′(A or T)GC sequences revealed by footprinting titration experiments," *Biochemistry* **29**:6145–6153 (1990).
86. Wang, A. H.-J., Ughetto, G., Quigley, G. J., and Rich, A., *Biochemistry* **26**:1152–1163 (1987).
87. Ragg, E., Mondelli, R., Battistini, C., Garbesi, A., and Colonna, *FEBS Lett.* and **36**:231–234 (1988).
88. Nunn, C. M., Van Meervelt, L., Zhang, S., and Kennard, O., *J. Mol. Biol.* **222**:167–177 (1991).
89. Leonard, G. A., Hambley, T. W., McAuley-Hecht, K., Brown, T., and Hunter, W. N. *Acta Cryst.* **D49:** 458–467 (1993).
90. Zhang, H., Gao, Y. G., Van der Marel, G. A., van Boom, J. H., and Wang, A. H.-J., *J. Biol. Chem.* **268**:10095–10101 (1993).
91. Gao, Y., Robinson, H., Wijsman, E. R., Van der Marel, G. A., van Boom, J. H., and Wang, A. H.-J., *J. Am. Chem. Soc.* **119**:1496–1497 (1997).
92. Eaton, R. J., Veselkov, D. A., Pakhomov, V. I., Baranovskii, S. F., Bolotin, P. A., Osetrov, S. G., Djimant, L. N., Davies, D. B., and Veselkov, A. N., *Mol. Biol.* **33**:709–718 (1999).
93. Davies, D. B., Eaton, R. J., Baranovskii, S. F., and Veselkov, A. N., *J. Biomol. Struct. Dyn.* **17**:887–901 (2000).
94. Veselkov, A. N., Eaton, R. J., Pakhomov, V. I., Semanin, A. V., Baranovskii, S. F., Djimant, L. N., and Davies, D. B., *Mol. Biol.* **35**:740–749 (2001).
95. Gao, Y. G., Liaw, Y. C., Li, Y. K., Van der Marel, G. A., van Boom, J. H., and Wang, A. H.-J., *Proc. Natl. Acad. Sci USA* **88**:4845–4849 (1991).
96. Dutta, R., Gao, Y. G., Priebe, W., and Wang, A. H.-J., *Nucl. Acids Res.* **26**:3001–3005 (1998).
97. Ettorre, A., Cirilli, M., and Ughetto, G., *Eur. J. Biochem.* **258**:350–354 (1998).
98. Saminadin, P., Dautant, A., Mondon, M., d'Estaintot, B. L., Courseille, C., and Precigoux, G., *Eur. J. Biochem.* **267**:457–464 (2000).

99. Berger, T., Su, L., Spitzner, J. R., Kang, C., Burke, T. G., and Rich, A., *Nucl. Acids Res.* **23**:4488–4494 (1995).
100. Gallois, B., Langlois, d'E. B., Brow, T., and Hunter, W. N., *Acta Cryst.* **D49**:311–317 (1993).
101. Frederick, C. F., Williams, L. D., Ugetto, G., Van der Marel, G. A., van Boom, J. H., and Rich, A., *Biochemistry* **29**:2538–2549 (1990).
102. Odefey, C., Westendorf, J., Dieckman, T., and Oschkiat, H., *Chem. –Biol. Interact.* **85**:117–126 (1992).
103. Bortolini, R., Mazzini, S., Mondelli, R., Bagg, E., Uibricht, C., Vioglio, S., and Penco, S., *Appl. Magn. Reson.* **7**:71–87 (1994).
104. d'Estaintot, B. L., Gallois, B., Brown, T., and Hunter, W. N., *Nucl. Acids Res.* **20**:3561–3566 (1992).
105. Leonard, G. A., Brown, T., and Hunter, W. N., *Eur. J. Biochem.* **204**:69–74 (1992).
106. Cirilli, M., Bachechi, F., Ughetto, G., Collonna, F. P., and Capobino, M. L., *J. Mol. Biol.* **230**:878–889 (1993).
107. Gao, Y. and Wang, A. H.-J., *J. Biomol. Struct. Dyn.* **13**:103–118 (1995).
108. Mazzini, S., Mondelli, R., and Bagg, E., *J. Chem. Soc. Perkin Trans.* **2**:1983–1992 (1998).
109. Brown, S. C., Mullis, K., Levenson, C., and Shafer, R. H., *Biochemistry* **23**:403–408 (1984).
110. Jones, R. L., Scott, E. V., Zone, G., Marzilli, L. G., and Wilson, W. D., *Biochemistry* **27**:6021–6026 (1988).
111. Gorenstein, G. B., Lai, K., and Shah, D. O., *Biochemistry* **23**:6717–6723 (1984).
112. Delepierre, M., Van Heijenoort, C., Igolen, J., Pothier, J., LeBret, M., and Roques, B. P., *J. Biomol. Struct. Dyn.* **7**:557–589 (1989).
113. Liu, X., Chen, H., and Patel, D. J., *J. Biomol. NMR* **1**:323–347 (1991).
114. Kamitori, S., and Kakusagawa, F., *J. Mol. Biol.* **225**:445–456 (1992).
115. Brown, D. R., Kurz, X., Hsu, V., and Kearns, D. R., *Biochemistry* **33**:651–664 (1994).
116. Kamitori, S., and Takasagawa, F., *J. Am. Chem. Soc.* **116**:4154–4165 (1994).
117. Chen, H., Liu, X., and Patel, D. J., *J. Mol. Biol.* **258**:457–479 (1996).
118. Lian, C., Robinson, H., and Wang, A. H.-J., *J. Am. Chem. Soc.* **118**:8791–8801 (1996).
119. Toyama, A., Miyagawa, Y., Yoshimura, A., Fujimoto, N., and Takeuchi, H., *J. Mol. Struct.* **598**: 85–91 (2001).
120. Mauffret, O., Rene, B., Convert, O., Monnot, M., Lescot, F., and Fermandjiam, S., *Biopolymers* **31**:1325–1341 (1991).
121. Searle, M. S., Bicknell, W., Wakelin, L. P. G., and Denny, W. A., *Nucl. Acids Res.* **19**:2897–2906 (1991).
122. Luzatti, V., Masson, F., and Lerman, L. S., *J. Mol. Biol.* **3**:631 (1961).
123. Adams, A., Guss, J. M., Collyer, C. A., Denny, W. A., and Wakelin, L. P. G., *Nucl. Acids Res.* **23**:4244–4253 (2000).
124. Bostock-Smith, C. E., Gimenetz-Amau, E., Missailidic, S., Stevens, M. G., Malcolm, F. G., and Searle, M. S., *Biochemistry* **38**:6723–6731 (1999).
125. Robinson, H., Liaw, Y.-C., Van der Marel, G. A., van Boom, J. H., and Wang, A. H.-J., *Nucl. Acids Res.* **18**:4851–4858 (1990).

126. Searle, M. S., Hall, J. G., Denny, W. A., and Wakelin, L. P. G., *Biochemistry* **27**:4340–4349 (1988).
127. William, J. D., Egli, M., Gao, Q., Bash, P., Van der Marel, G. A., Gijs, A., Van Boom, J. H., Rich, A., and Frederick, C. A., *Proc. Natl. Acad. Sci. USA* **87**:2225–2229 (1990).
128. Egli, M., Williams, L. D., Frederick, C. A., and Rich, A., *Biochemistry* **30**:1364–1372 (1991).
129. Liaw, Y., Gao, Y., Robinson, H., Van der Marel, G. A., van Boom J. H., and Wang, A. H., J. *Biochemistry* **28**:9913–9918 (1989).
130. Zang, X., and Patel, D. T., *Biochemistry* **29**:9451–9466 (1990).
131. Searle, M. S., and Lane, A. N. *FEBS Lett.* **297**:292–296 (1992).
132. Van Houte, L.P.A., Van Garderen, C. J., and Patel, D. P., *Biochemistry* **32**:1667–1674 (1993).
133. Robinson, H., Yang D., and Wang, A. H., J., *Gene* **149**:179–188 (1994).
134. Chen H., and Patel, D. J., *J. Am. Chem. Soc.* **117**:5901–5913 (1995).
135. Smith, C. K., Davies, G. J., Dodson, E. J., and Moore, M. H., *Biochemistry* **34**:415–425 (1995).
136. Smith, C. K., Brannigan, J. A., and Moore, M. H., *J. Mol. Biol.* **263**:237–258 (1996).
137. Williams H. E.L., and Searle, M. S., *J. Mol. Biol.* **290**:699–716 (1999).
138. Colgave, M. L., Williams, H. E. L., and Searle, M. S., *Chem. Commun.* 315–316 (2001).
139. Calladine, C. R., "Mechanics of sequence-dependent stacking of bases in B-DNA," *J. Mol. Biol.* **161**:343–352 (1982).
140. Dickerson, R. E., "Base sequence and helix structure variation in B and A DNA," *J. Mol. Biol.* **166**:419–441 (1983).
141. Nishimura, Y. and Tsuboi, M., "A possible correlation between DNA conformation and the mode of action of restriction enzymes," *J. Biochem.* (Tokyo) **96**:1807–1811 (1984).
142. Conner, B. N., Takano, T., Tanaka, S., Itakura, K., and Dickerson, R. E.," The molecular structure of d(^{I}CpCpGpG)," *Nature* **295**:294–299 (1982).
143. Ott, J. and Eckstein, F., "31P NMR spectral analysis of the dodecamer d(CGCGAATT CGCG)," *Biochemistry* **24**:2530–2535 (1985).
144. Dickerson, R. E. and Drew, H. R., "Structure of B-DNA dodecamer. II. Influence of base sequence on helixc structure," *J. Mol. Biol.* **149**:761–786 (1981).
145. Chen, K. Y., Gresh, N., and Pullman, B., "A theoretical investigation on the sequence selective binding of daunomycin to double-stranded polynucleotides," *J. Biomol. Struct. Dyn.* **3**:445–466 (1985).
146. Pullman, B., in Sarma, R. H., and Sarma, M. H., eds., *Structure and Expression*, Academic Press, New York, 1987, Vol. 2. pp. 237–249.
147. Pullman, B., *Anticancer Drug Design* **7**:97–105 (1987).
148. Skavobogaty, A., White, R. J., Phillips, D. R., and Reiss, J. A., "Elucidation of the DNA sequence preferences of daunomycin," *Drug. Design Deliv.* **3**:125–152 (1988).
149. Atwell, G. J., Leupin, W., Twigden, S. J., and Denny, W. A., "Triacridine derivative: First DNA tris-intercalating ligand," *J. Am. Chem. Soc.* **105**:2913–2914 (1983).

150. Atwell, G. J., Stewart, G. M., Leupin, W., and Denny, W. A., "A diacridine that binds by bisintercalation at two contiguous sites on DNA," *J. Am. Chem. Soc.* **107**:4335–4337 (1985).

151. Gaugain, B., Marcovits, J., LePecq, J. B., and Reques, B. P., "DNA polyintercalation: Comparison of DNA binding properties of an acridine dimmer and trimer," *FEBS Lett.* **169**:123–126 (1984).

152. Hansen, J. B., Koch, T., Buchardt, O., Nielsen, P. E., Norden, B., and Wirth, M., "Trisintercalation in DNA by N-[3-(9-Acridylamino)propyl]-N-N-bis{6-(4-acridynyl amino)hexyl}amine," *J. Chem. Soc. Chem. Commun.* **1984**:509–511 (1984).

153. Denny, W. A., "Acridine-based antitumour agents," in Wilman, ed., *The Chemistry of Antituour Agents*, Blankie, London, 1990, pp. 1–29.

154. LePecq, J. B., Xuong, N. D., Gosse, C., and Paoletti, C., "A new antitumoral agent: 9-Hydroxyellipticine. Possibility of a rational design of anticancerous drugs in the series of DNA intercalating drugs," *Proc. Natl. Acad. Sci. USA* **71**:5078–5082 (1974).

155. Baguley, B. C., Denny, W. A., Atwell, G. J., and Cain, B. F., "Potential antitumor agents. 34. Quantitative relationships between DNA binding and molecular structure for 9-anilinoacridines substituents in the anilino ring," *J. Med. Chem.* **24**:170–177 (1981).

156. LePecq, J. B., LeBret, M., Barbet, J., and Roques, B., "DNA polyintercalting drugs: DNA binding of diacridine derivatives," *Proc. Natl. Acad. Sci. USA* **72**:2915–2919 (1975).

157. Canellakis, E. S., Bono, V., Bellantone, R. A., Krakow, J. S., Fico, R. A., and Schulz, R. M., "Diacridines: Bifunctional intercalators III. Definition of the general site of action," *Biochim. Biophys. Acta* **418**:300–314 (1976).

158. Canellakis, E. S., Shaw, Y. H., Hanners, W. E., and Schwartz, R. A., "Diacridines: Bifunctional intercalators. I. Chemistry, physical chemistry and growth inhibitory properties," *Biochim. Biophys. Acta* **418**:277–289 (1976).

159. Cain, B. F., Baguley, B. C., and Denny, W. A., "Potential antitumor agents. 28. Deoxyribonucleic acid polyintercalating agents," *J. Med. Chem.* **21**:658–668 (1978).

160. Gaugain, B., Markovits, J., and Roques, B. P., "Hydrogen bonding in deoxyribonucleic acid base recognition. 1. Proton nuclear magnetic resonance studies of dinucleotide-acridine alkylamide complexes," *Biochemistry* **20**:3035–3042 (1981).

161. Wakelin, L. P. G., Romanos, M., Chen, T. K., Glaubiger, D., Canellakis, E. S., and Waring, M. J., "Structural limitations on the bifunctional intercalation of diacridines into DNA," *Biochemistry* **17**:5057–5063 (1978).

162. Becker, M. M. and Dervan, P. B., "Molecular recognition of nucleic acid by small molecules. Binding affinity and structural specificity of bis(methidium)spermine," *J. Am. Chem. Soc.* **101**:3664–3666 (1979).

163. Kuhlman, K. F. and Mosher, C. W., "New dimeric analogues of ethidium: Synthesis, interaction with DNA, and antitumor activity," *J. Med. Chem.* **24**:1333–1337 (1981).

164. Fico, R. M., Chen, T. K., and Canellakis, E. S., "Bifunctional intercalators: Relationship of antitumor activity of diacridines to the cell membrane," *Science* **198**:53–56 (1977).

165. Chen, T. K., Fico, R. M., and Canellakis, E. S., "Diacridines, bifunctional intercalators, chemistry and antitumor activity," *J. Med. Chem.* **21**:868–874 (1978).

166. Denny, W. A., Atwell, G. J., and Baguley, B. C., "Potential antitumor agents. 40. Orally active 4,5-disubstituted derivatives of ammsacridine," *J. Med. Chem.* **27**:363–367 (1984).

167. Lown, J. W., Gunn, B. C., Chang, R. Y., Majumdar, K. C., and Lee, J. S., *Can. J. Biochem.* **56**:1006–1015 (1978).

168. Capelle, N., Barbet, J., Dessen, P., Blanquet, Roques, B. P., and LePecq, J. B., "Deoxyribonucleic acid bifunctional intercalators: Kinetic investigation of binding of several acridine dimers to deoxyribonucleic acid," *Biochemistry* **18**:3354–3362 (1979).
169. Wright, R. G. McR., Wakelin, L. P., Fields, A., Aceson, R. M., and Waring, J., "Effects of ring substituents and linker chains on bifunctional intercalation of diacridines into deoxyribonucleic acid," *Biochemistry* **19**:5825–5836 (1980).
170. King, H. D., Wilson, W. D., and Gabbay, E. J., "Intercalations of some novel amide-linked bis(acridine) with deoxyribonucleic acid," *Biochemistry* **21**:4982–4989 (1977).
171. Roques, B. P., Peleprat, D., Le Guen, I., Porcher, G., Gosse, C., and LePecq, J. B., "DNA bifunctional intercalators: antileukemic activity of new pyridocarbazole dimers," *Biochem. Pharmacol.* **28**:1811–1815 (1979).
172. Delbarre, A., Delepierre, M., Garbay, C., Igolen, J., LePecq, J. B., and Roques, B. P., "Geometry of the antitumor drug ditercalinium bisintercalated into d(CpGpCpG)$_2$ by ^{1}H NMR," *Pro. Natl. Acad. Sci. USA* **84**:2155–2159 (1087).
173. Delbarre, A., Delepierre, M., D'Estaintot, B. L., Igoren, J., and Roques, B. P., "Bisintercalation of ditercalinium into a d(CpGpCpG)2 minihelix: Structure and dynamics aspects—a 400 MHz ^{1}H-NMR study," *Biopolymers* **26**:1001–1033 (1987).

CHAPTER 3

Groove-Binding Drugs

The double-helix structure of DNA displays the major and minor grooves having different width and depth, and one end of the base pair is exposed to the major groove while the other end is exposed to the minor groove (Fig. 1.10). As shown in Fig. 3.1, natural antibiotics such as netropsin and distamycin are long, flexible, and crescent-shaped molecules containing two and three N-methylpyrrole rings, respectively, which are connected via peptide linkage to the positively charged amidinium $(C(NH_2)(NH_2)^+)$ group at the terminal position(s). As shown in the following sections, they can fit tightly into the minor groove in the AT-rich region of B-DNA, and their adducts are stabilized by hydrogen-bonding, electrostatic, and van der Waals interactions. These drugs cannot bind to the GpC sequence, however, because the NH_2 group of guanine protrudes into the minor groove, thus preventing their access.

Netropsin binds to the AT-rich region in the following order of sequence preference:

$$d(AAAA) > d(AATT) > d(ATAT) > d(ACAC)$$

Thus far, the major interest of research has focused on the origin of such sequence specificity and design of synthetic drugs that exhibit preferences for other and longer sequences. For this purpose, a number of analogs of distamycin and netropsin have been synthesized and their sequence specificities studied using a variety of techniques. In general, they are classified into three types: (1) lengthening of the peptide chain via end-to-end linkage and crosslinking, (2) replacement of N-methylpyrrole by imidazole and other rings, and (3) shortening of the carboxamide (NH—C=O) chain by replacing it with the keto or amino linkage. More recent developments in this area have been reviewed extensively [1,2].

Drug–DNA Interactions: Structures and Spectra by Kazuo Nakamoto, Masamichi Tsuboi, and Gary D. Strahan
Copyright © 2008 John Wiley & Sons, Inc.

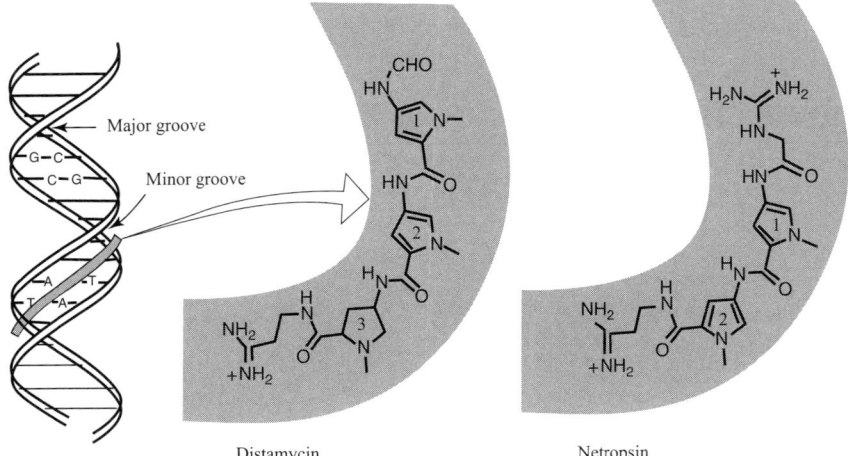

FIGURE 3.1. Conceptual drawing of groove binding, and structures of distamycin and netropsin.

3.1. NETROPSIN AND DISTAMYCIN

Dickerson and coworkers [3,4] determined the crystal structure of the 1:1 complex of netropsin complexed to the dodecamer duplex:

```
         1   2   3   4   5   6   7   8   9  10  11  12
5'-d(C   G   C   G   A   A   T   T'  C   G   C   G)-3'      T' : 5-bromothymine
3'-d(G   C   G   C   T'  T   A   A   G   C   G   C)-5'
        24  23  22  21  20  19  18  17  16  15  14  13
```

The drug molecule binds to the central AATT region in the minor groove by replacing the water molecules of the spine of hydration (Section 1.1.6). As seen in Fig. 3.2 the amide NH groups (N4, N6, and N8) of the drug molecule form three *bifurcated* hydrogen bonds to the adenine N3 or thymine O2 atoms along the floor of the groove. Although these hydrogen-bonding and electrostatic interactions are responsible for the strong binding, the sequence specificity is determined by close van der Waals contacts between adenine C2H and the pyrrole ring CH(C5 and C11) or the CH_2(C2 and C16) protons of the drug. The NH_2 (N1 and N10) atoms of the cationic terminal groups are hydrogen-bonded to the N3 atoms of adenine (A_5 and A_{17}) and not to the O atoms of the phosphate backbone. As will be shown later, this structure is in agreement with that obtained previously by NMR spectroscopy. Similar "bifurcated" hydrogen bonds were found by X-ray analysis of distamycin bound to d(CGCAAATTTGCG)$_2$ [5]. The structure described above is called "class I" type [4].

On the other hand, somewhat different structures were found by X-ray analysis of netropsin bonded to d(CGCGATATCGCG)$_2$ [6] and d(CGCG'AATTCGCG)$_2$ (G'=O6-ethylguanine) [7]. In these cases ("class II" type), the structures are disordered, suggesting that the drug molecules bind in two different orientations. In addition, the position of the drug molecule in the minor groove is shifted by about

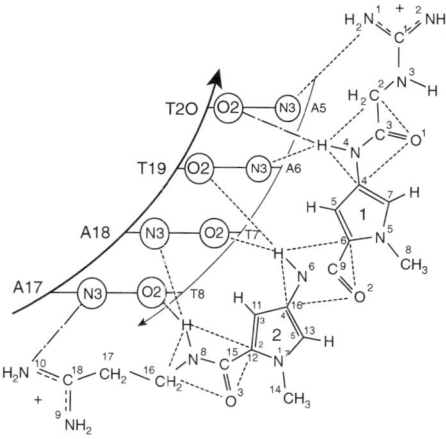

FIGURE 3.2. Diagram showing netropsin binding to the dodecamer. Only adenine N3 and thymine O2 atoms in the minor groove ladder are indicated. Dotted–dashed lines indicate that N-to-N and N-to-O distances are short enough to be standard hydrogen bonds, whereas dotted lines indicate distances of 3.2 Å or more [3].

a half a base-pair step and there are no bifurcated hydrogen bonds found in class I adducts. Thus, the AT-sequence specificity of netropsin cannot be explained with class II structure.

These differences were attributed to the low resolution of X-ray analysis of class II relative to those of Class I [4]. Molecular dynamics simulation of netropsin complexed to d(CGCGAATTCGCG)$_2$ supports class I structure [8]. The solution structure of netropsin bound to d(GGAATTCC)$_2$ was also found to be of class I type by 2D NMR spectroscopy [9].

Pelton and Wemmer [10] determined the structure of distamycin complexed to d(CGCGAATTCGCG)$_2$ by 2D NMR spectroscopy and molecular dynamics calculations. Figure 3.3a illustrates the hydrogen-bonding scheme in the central region. The drug molecule fits into the AATT region of the minor groove snugly, and forms bifurcated hydrogen bonds seen previously for the netropsin adduct (Fig. 3.2). Pyrrole rings 1 and 2 are almost parallel to each other, but ring 3 is rotated significantly to fit into the groove. The complex is characterized by van der Waals contacts between adenine C2H and pyrrole H3 protons. It was suggested that, in addition to hydrogen bonding and electrostatic interactions at the charged end, stacking interactions between DNA sugar O1′ atoms and the three pyrrole rings contribute to the stability of the adduct.

In addition to the ^{1}H NMR studies cited above [9], the interaction of netropsin with d(GGTATACC)$_2$ was investigated by ^{13}C NMR spectroscopy [11,12]. The results are consistent with the hydrogen-bonding scheme mentioned above, and provide information about the effect of drug binding on sugar puckering and glycosyl torsion angles in the central TATA region.

Distamycin forms a 2:1 complex with the 11-mer duplex, d(CGCAAATTGGC)·d(GCCAATTTGCG), at high drug/DNA molar ratios (still below 1/1). Pelton and Wemmer [13] determined the solution structure of this complex by 2D NMR spectroscopy together with model-building-energy refinement techniques. As shown in

GROOVE-BINDING DRUGS

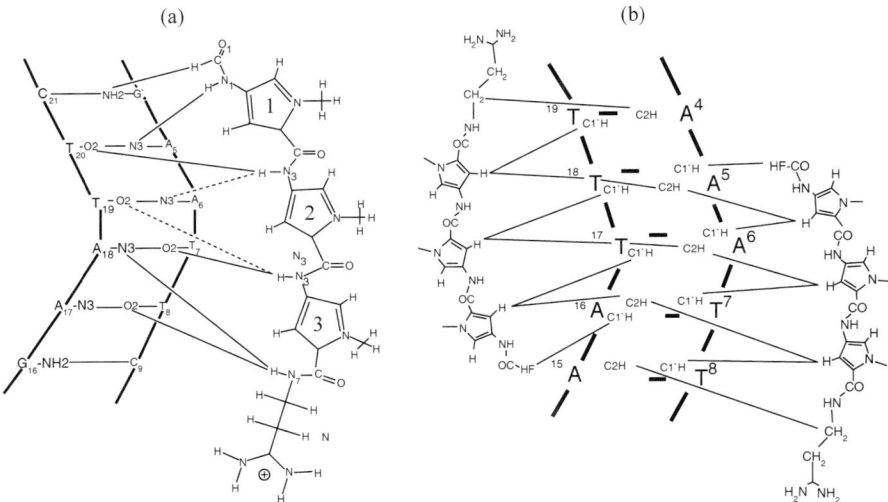

FIGURE 3.3. (a) Hydrogen-bonding scheme of distamycin-d(CGCGAATTCGCG)$_2$ complex. Solid lines indicate hydrogen bonds in the highly constrained structure, while dotted lines denote the additional hydrogen bonds that formed when the NMR constraints were relaxed [10]. (b) Schematic drawing of distamycin binding site on d(CGCAAATTGGC)$_2$·d(GCCAATTTGCG). Solid lines indicate intermolecular contacts [13].

Fig. 3.3b, two drug molecules bind to the 5'-AAATT-3' region and take the side-by-side antiparallel orientation so that electrostatic repulsion between the two positively charged ends is mimimal. Molecular modeling studies show that the minor groove must be expanded to accommodate two drug molecules in this orientation. Later, X-ray analysis of the 2:1 complex of distamycin bonded to d(ICATATIC)$_2$ (I : inosine) confirmed the presence of such an antiparallel orientation. The minor groove was expanded by the thickness of one drug (~3.4 Å) to accommodate the second drug molecule [14].

Burckhardt et al. [15] studied the interactions of netropsin and related compounds with poly(dA-dT)$_2$ and other alternating duplex polymers by CD spectroscopy. Figure 3.4 shows the CD spectra of netropsin mixed with poly(dA-dT)$_2$ in 0.01 M Tris-buffer (pH = 7.0). The r (molar ratio of drug/base pair) value is indicated for each spectrum. It is seen that two distinctly different sets of spectra are obtained for low (0–0.08) (Fig. 3.4a) and high (0.11–0.57) (Fig. 3.4b) r values, and that they are characterized by the two isodichroic points at 288–290 and 312–314 nm. It was suggested that the first binding step (low r) causes some subtle conformational changes of the duplex, and that elctrostatic interaction and hydrogen bonding do not play dominant roles. This step is not affected by ionic strength and temperature. On the other hand, the second step (high r) is seen for alternating duplexes such as poly(dA-dT)$_2$ but not for poly(dA)·poly(dT) Such sequence specificity may arise from hydrogen bonding as shown by X-ray studies [3,4] or involve the formation of a symmetric antiparallel arrangement of the 2 : 1 complexes mentioned previously [14].

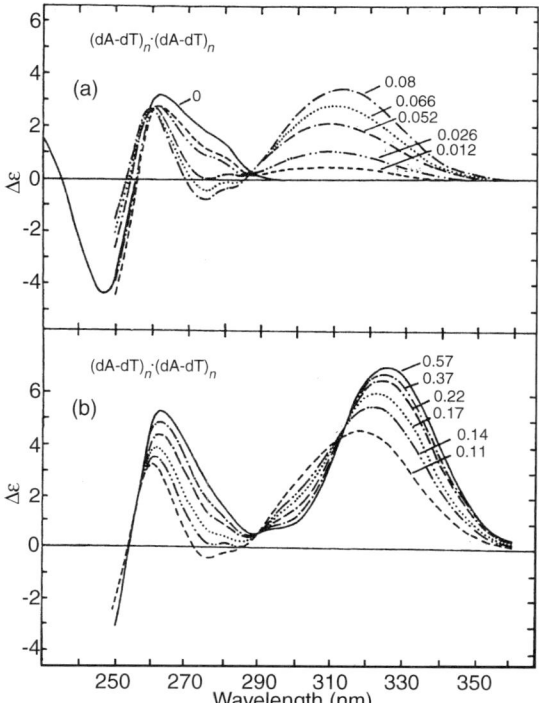

FIGURE 3.4. CD spectra of netropsin bonded to poly(dA-dT)$_2$ in 0.01 M Tris buffer, pH 7.0: (a) at low drug concentrations; and (b) at high drug concentrations. The r values (molar ratio of drug/base pair) are given for each spectrum [15].

Liquier et al. [16] carried out an FTIR study on netropsin bonded to poly(dA-dT)$_2$ and poly(dA)·poly(dT) in aqueous solutions. On addition of netropsin, the IR spectrum of poly(dA-dT)$_2$ in the 1750–1550 cm^{-1} region (in D$_2$O) changes as seen in Fig. 3.5a. The 1696 cm^{-1} band in trace a is the C2=O stretching of thymine of poly(dA-dT)$_2$. When netropsin is added at $r = 0.25$, a new band appears at $\sim$1680 cm^{-1} (trace b), and this trend becomes more prominent at $r = 0.33$ (trace c). However, similar experiments with poly(dA)·poly(dT) (Fig. 3.5b) shows no appearance of a new band, although the C2=O stretching band of thymine at 1696 cm^{-1} becomes weaker by adding more drug. In panels (**a**) and (**b**), the 1582 cm^{-1} band becomes stronger as the drug concentration increases since it originates in the drug. Panels (**c**) and (**d**) show the IR spectra of the same samples in the 1400–1270 cm^{-1} region (in H$_2$O). Vibrations in this region are due mainly to deoxyriboses coupled with those of the bases through the glycosidic linkage. In the case of poly(dA-dT)$_2$ (Fig. 3.5c), a doublet at 1301 and 1298 cm^{-1} is strongly enhanced by adding the drug; the former is possibly downshifted. Again, no marked changes were observed for poly(dA)·poly(dT) (Fig. 3.5**d**) except that the band at 1328 cm^{-1} becomes weaker as the drug is added. The observed spectral changes in panel (**c**) may be attributed to the effect of hydrogen bonding between the drug and poly(dA-dT)$_2$ as found in the netropsin–dodecamer complex by

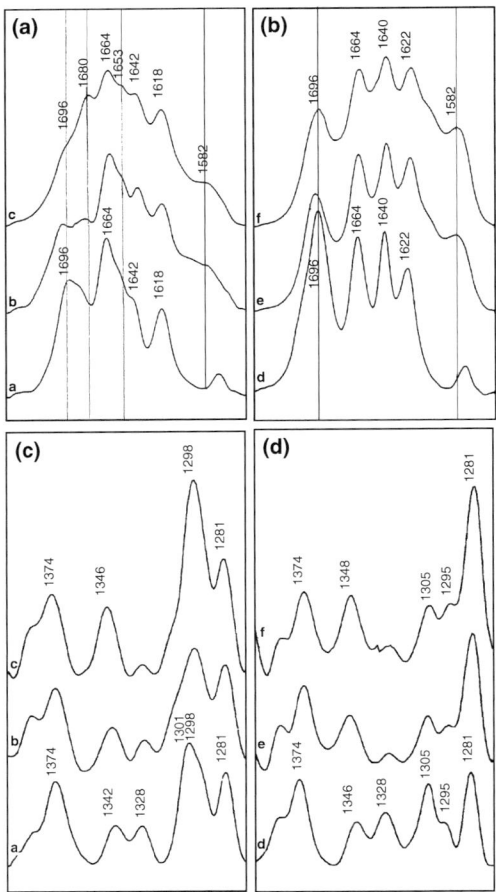

FIGURE 3.5. FTIR spectra in the (**a, b**) 1750–1550 cm^{-1} region (D$_2$O solution) and (**c, d**) 1400–1270 cm^{-1} region (H$_2$O solution). Poly(dA-dT)$_2$ (a) and its mixtures with netropsin at the mixing ratios (r) of $r = 0.25$ (b) and $r = 0.33$ (c). Poly(dA) • poly(dT) (d) and its mixtures with netropsin at $r = 0.25$ (e) and $r = 0.33$) (f). Here, r denotes the molar ratio of netropsin/base pair [16].

X-ray analysis [3,4]. Such hydrogen bonding is expected to be much weaker in the case of poly(dA) • poly(dT) because of its twisted structure (Fig. 1.13). The IR spectra in the 1270–800 cm^{-1} region (not shown) exhibit two strong bands at around 1227 and 1089 cm^{-1} that are due to the antisymmetric and symmetric PO$_2$ stretching of the phosphate backbone. However, no significant changes are observed on addition of netropsin since it does not alter their structures.

Lu et al. [17] studied the interaction of distamycin with polynucleotides by Raman spectroscopy. The Raman spectra were obtained by 488 nm excitation so that only the vibrations that originated in the drug molecule are enhanced. Vibrational assignments were based on those of model compounds of the three functional groups constituting distamycin (*N*-methylpyrrole, peptide, and acetamidinium ion; see Fig. 3.1). Figure 3.6, panels (a) and (b) show the Raman spectra of distamycin and its mixture

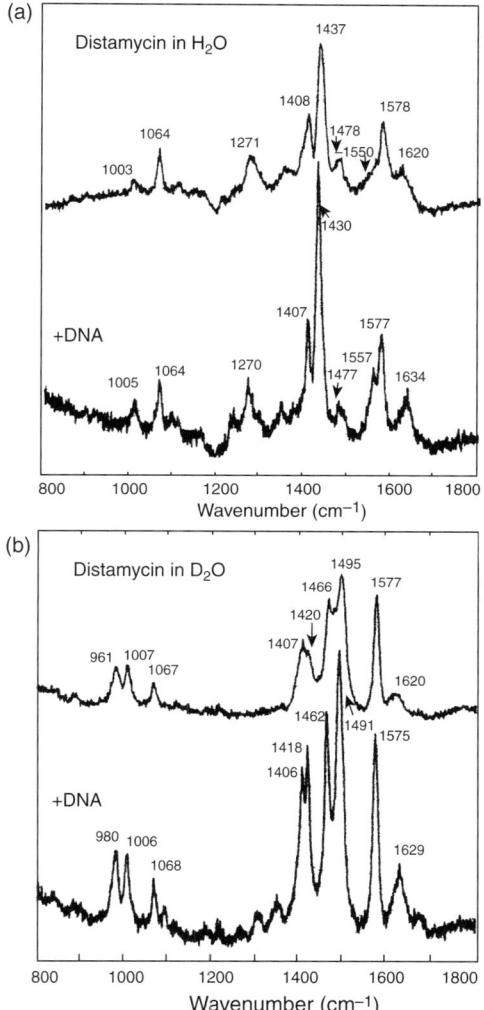

FIGURE 3.6. Raman spectra (488 nm excitation) of distamycin and distamycin mixed with CT-DNA in H_2O (a) and D_2O (b) at room temperature. All the spectra were obtained by subtracting H_2O (or D_2O) bands and the background spectra using a divided rotating cell. The molar ratio of distamycin to base pair is in the range of 0.1–0.2 [17].

with calf thymus (CT)-DNA ($r = 0.1$–0.2) in H_2O and D_2O, respectively. On addition of DNA, two distinct changes were noted in these spectra:

1. In H_2O (Fig. 3.6a), the amide I band (80% C=O stretching) at 1620 cm^{-1} was upshifted to 1634 cm^{-1} while the pyrrole ring vibration coupled with amide II at 1437 cm^{-1} was downshifted to 1430 cm^{-1}. Similar changes were observed in D_2O (Fig. 3.6b); the band at 1620 cm^{-1} was upshifted to 1629 cm^{-1} and the 1495 cm^{-1} band was downshifted to 1491 cm^{-1}.

2. The Raman spectrum of distamycin becomes sharper as a result of DNA binding. This is particularly apparent in the 1620–1550 cm^{-1} region in H$_2$O.

These spectral changes suggest that the pyrrole ring and the peptide group of distamycin are nearly coplanar in the free state, and that this coplanarity is lost when distamycin is accommodated inside the minor groove of DNA. If these two groups were coplanar, electron migration would be expected to occur from the pyrrole ring to the peptide group through the C–C(O) linkage, and this would lower the amide I frequency and raise the pyrrole ring mode frequency via the resonance:

[chemical structure showing resonance between two forms of the pyrrole-peptide group]

When this resonance is disrupted on DNA binding, the amide I band would shift to a higher and the pyrrole ring mode to a lower frequency. This resonance disruption is also responsible for the redshift of the UV absorption band of distamycin at 304–319 nm with a marked intensity increase.

Furthermore, this conformational change accounts for the observed sharpness of distamycin bands in the bound state. In the free state, all the internal rotation angles around the C–C and C–N bonds connecting the pyrrole ring and the peptide group fluctuate in a narrow range centered around 0°. However, this dynamic fluctuation causes broadening of the vibrational bands. When distamycin is bound to DNA, the conformation of the drug is fixed by steric requirements in the minor groove. As a result, the bands become sharper and the internal rotation angles show considerable deviations from 0° that destroy the coplanarity of the pyrrole ring and the peptide group.

As stated above, the degree of conformational changes of distamycin resulting from the interaction with a polynucleotide can be measured by the shifts of amide I and pyrrole ring vibrations, sharpening and intensification of these bands, and the redshift of the UV absorption band. On the basis of these criteria, the degree of conformational changes of distamycin was found to decrease in the following order:

$$\text{poly}(dA) \cdot \text{poly}(dT) > \text{DNA} > \text{poly}(dA\text{-}dT)_2 \gg \text{poly}(dG\text{-}dC)_2$$

Thus, molecular distortion of distamycin is more severe when complexed with poly(dA)·poly(dT) than with poly(dA-dT)$_2$. This is expected from the severely distorted propeller-like structure of the former. On the other hand, conformational changes of polynucleotides resulting from interaction with the drug is less in the former than in the latter. This is also supported by previous IR studies [16] showing that the C2=O stretching band of thymine of the latter at 1696 cm^{-1} is shifted to 1680 cm^{-1} whereas no band shifts are observed when the former is mixed with distamycin.

Interactions of distamycin with poly(dA-dT)$_2$ and poly(dA)·poly(dT) were also studied by UV resonance Raman spectroscopy (320 and 200 nm excitation) [18], and the role of hydrogen bonding mentioned above was discussed with respect to the observed shift of the C=O stretching (amide I) vibration.

3.2. DERIVATIVES OF NETROPSIN AND DISTAMYCIN

A number of derivatives of netropsin and distamycin, both natural and synthetic, are known, and their interactions with oligonucleotides and DNA have been studied using a variety of physico-chemical techniques. Most of their derivatives preferentially bind to the AT-rich region of DNA, whereas their binding to GpC sequences is unfavorable because of steric interferences between the H3 protons of pyrrole rings and the exocyclic NH_2 group of guanine. However, some derivatives can bind to 5–7 bp regions containing both A-T and G-C base pairs.

As stated in the preceding section, netropsin containing three amide (NH–C=O) units can recognize four base pairs in successive sequences. Dervan and coworkers [19,20] have shown that an oligo(N-methylpyrrolecarboxamide) having n amide units ($n - 1$ pyrrole rings) can recognize ($n + 1$) base sequences., and this "$n + 1$ rule" holds up to $n = 7$ since polypeptides longer than $n = 7$ can no longer fit into the natural twist of B-form DNA. This rule was confirmed by DNA cleavage experiments using a series of N-methylpyrrolecarboxamides of $n = 4$–7 to which a cleavage agent, Fe(II)-EDTA, (Section 5.1.4) is attached at the end.

Dervan, Wemmer, and coworkers [21] carried out an extensive investigation of oligopeptides containing both pyrrole and imidazole rings. They first synthesized 1-methylimidazole-2-carboxamide-netropsin (2-ImN) (Fig. 3.7b) and determined the

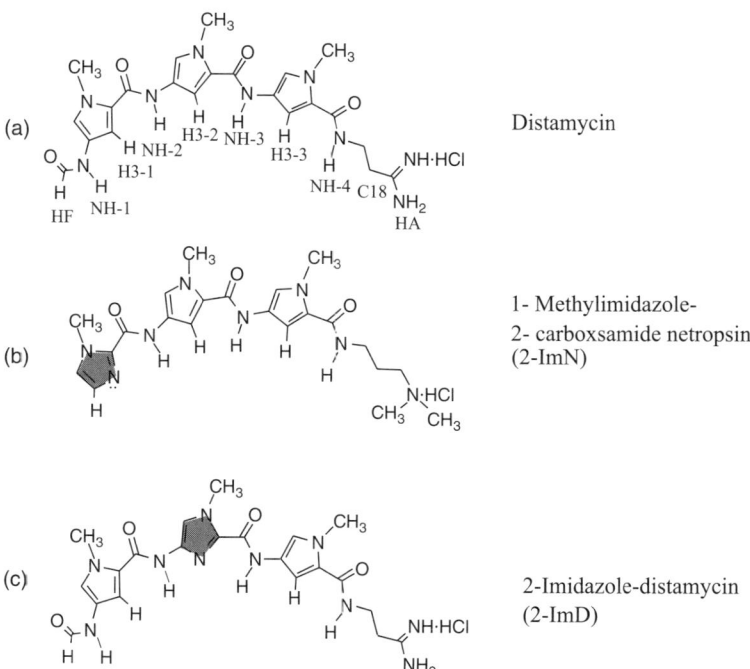

FIGURE 3.7. Structures of distamycin (a), 2-ImN (b), and 2-ImD (c). Imidazole (Im) rings are shaded.

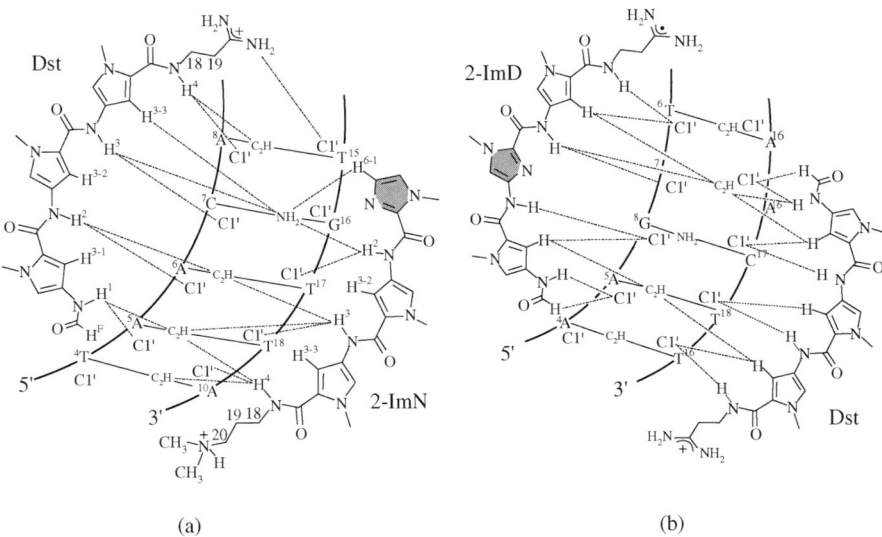

FIGURE 3.8. Schematic drawings of selected intermolecular NOEs (a) between distamycin, 2-ImN, and d(GCCTAACAAGG)·d(CCTTGTTAGGC) in 1 : 1 : 1 ratio: (b) between distamycin, 2-ImD, and d(CGCAAGTTGGC)·d(GCCAACTTGCG) in 1 : 1 : 1 ratio. In the latter, NOE contacts between amide protons of the drug and 11-mer protons are indicated by dashed lines, while those between other protons of the drug and 11-mer protons are shown by stippled lines. Drug-11-mer contacts involving the amidinium and methylene protons as well as drug–drug contacts are not shown. In both (a) and (b), imidazole (Im) rings are shaded [22,23].

solution structure of the 2 : 1 adduct of 2-ImN with d(GCATGACTCGG)·d(CCGAGTCATGC) by 2D NMR spectroscopy. This work was extended to the 1 : 1 : 1 adduct (2-ImN/distamycin/DNA) with d(GCCTAACAAGG)·d(CCTTGTTAGGC) [22], and to the 1 : 1 : 1 adduct (2-ImD/distamycin/DNA) with d(CGCAAGTTGGC)·d(GCCAACTTGCG), where 2-ImD is 2-imidazole-distamycin (Fig. 3.7c) [23]. Figure 3.8 shows observed intermolecular NOEs for the latter two cases. In part (a), 2-ImN spans over the 5′-TGTTA-3′ sequence with the N3 atom of the imidazole ring close enough to interact with the NH_2 group of guanine, while distamycin spans along the 5′-TAACA-3′ sequence in an antiparallel side-by-side configuration. In this case, NOEs between the imidazole H4 proton and the nearby NH_2 group of guanine were observed. In part (b), 2-ImD spans over the 5′-AAGTT-3′ sequence, although NOEs between any protons of the drugs and the amino protons of G_6 were not observed. The bonding modes of the 2 : 1 adduct of 2-ImN [21] and the above mentioned two 1 : 1 : 1 adducts [22,23] are illustrated in Fig. 3.9 [configurations (a)–(c)], where light and shaded circles denote pyrrole (P) and imidazole (Im) rings, respectively [24]. It is seen that P and Im preferentially interact with A/T and G bases, respectively. On the basis of this concept, Geierstanger et al. [24] synthesized the oligopeptide consisting of two P and two Im rings in the order shown in Fig. 3.9 (ImPImP). As expected, it forms the 2 : 1 adduct with each ligand bonded to the 5′-(A,T)GCGC(A,T)-3′ sequence of $d(CGTAGCGCTACG)_2$ (Fig. 3.9d).

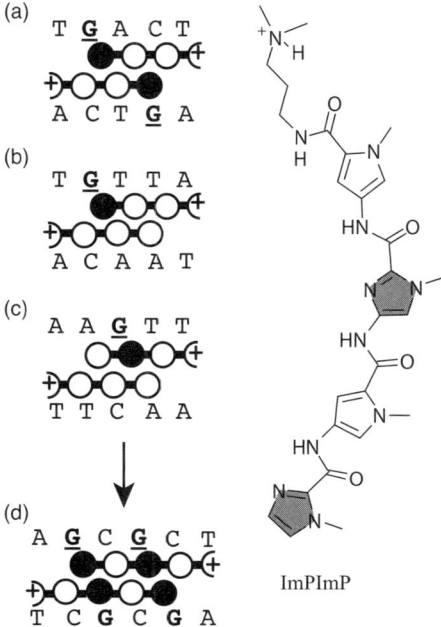

FIGURE 3.9. (Left) Specific 2:1 complexes with mixed A/T and G/C sequences; (a) 2-ImN homodimer; (b) 2-ImN-distamycin heterodimer; (c) 2-ImD-distamycin heterodimer; (d) the designed four-ring ImPImP homodimer. The black and white circles denote imidazole (Im) and pyrrole (P) rings, respectively. The upper strands of the sites read 5′ to 3′, with specifically targeted guanine underlined. (Right) the structure of four-ring crescent-shaped peptide, ImPImP with imidazole rings shaded [24].

Lee et al. [25] synthesized a series of imidazole-containing and C-terminus-modified analogs shown in Fig. 3.10 (**4, 5, 6,** and **7** for $n = 1, 2, 3,$ and 4, respectively). Although all these polypeptides bind to CT-DNA and poly(dG-dC)$_2$, the binding constants of **4** and **5** are smaller than that of Distamycin, whereas those of **6** and **7** are larger than or comparable to that of distamycin. However, relative to distamycin, **6** and **7** bind significantly more weakly to poly(dA-dT)$_2$ but slightly more strongly to poly(dG-dC)$_2$. These results suggest that the acceptance of the GpC sequence is facilitated by the introduction of imidazole rings. The binding constants of **7** are slightly smaller than those of **6** for all the nucleotides tested, indicating that the optimum number of imidazole rings in the series is $n = 3$. Figure 3.10 also shows CD spectra that indicate the effects of adding **6** to poly(dA-dT)$_2$ (a), poly(dG-dC)$_2$ (b), and CT-DNA (c). The positive bands near 335 nm and the negative bands near 305 nm are presumably due to the π–π* transitions of the ligand bound to the oligonucleotides, and their appearance provides a clear evidence for the adduct formation since the ligand itself has no CD bands in the 340–300 nm region. Although similar results are obtained for **5** and **7**, their ligand-induced CD bands are much stronger for CT-DNA (42% GC) and poly(dG-dC)$_2$ than for poly(dA-dT)$_2$. Thus, these oligoimidazole ligands have

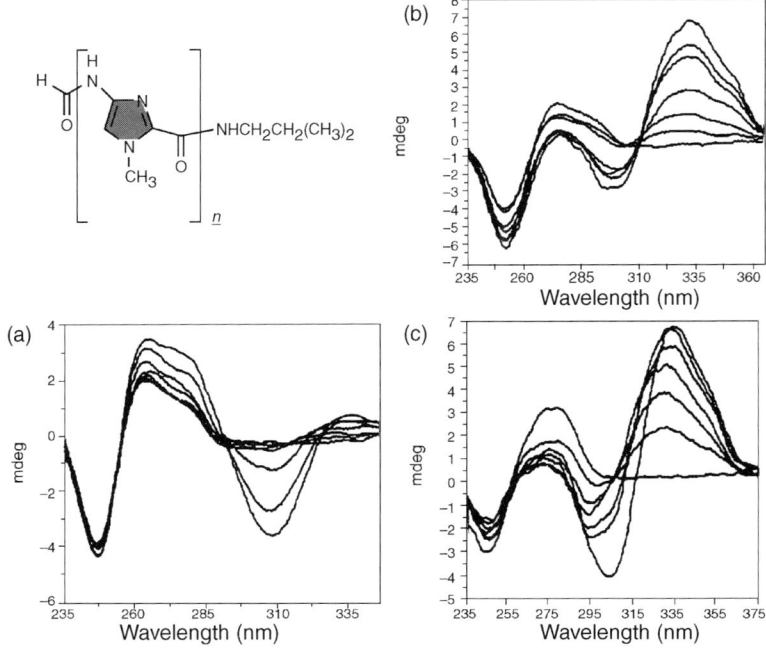

FIGURE 3.10. Structures of oligo(imidazolecarboxamide) of $n = 1, 2, 3, 4$ (**4, 5, 6,** and **7**, respectively), and CD titration of **6** to poly(dA-dT)$_2$ (a), poly(dG-dC)$_2$ (b), and CT-DNA (c) The traces correspond to r values (molar ratio of the drug/DNA base pair) of 0, 0.05, 0.10, 0.20, 0.40, 0.60, 0.80, and 1.0 for poly(dA-dT)$_2$ and CT-DNA. For poly(dG-dC)$_2$, r values are 0, 0.12, 0.25, 0.35, 0.50, and 0.60 [25].

more affinity with GC-rich sequences than distamycin. Mixing of **4** ($n = 1$, monoimidazole) with these nucleotides did not cause any changes in CD spectra, indicating that there is minimal interaction between them. In all cases, the CD spectra of the adducts exhibit positive and negative bands at ∼270 and ∼245 nm, respectively, which are characteristic of B-form DNA.

As stated previously, the design of sequence-specific drugs based on the motif of the 2 : 1 adduct of side-by-side antiparallel orientation [13] is limited by the length of the oligopeptide (up to seven amide units). Another approach is to connect two peptide chains at the ends via a proper linkage. A series of such *heterodimers* that connect the terminal carboxylic acid of 2-ImN and the terminal amine of tris(N-methylpyrrolecarboxamide) (P3) via amino acid linkers (Fig. 3.11a) were prepared by Mrksich et al. [26]. The binding affinity of these heterodimers to the 5'-TGTTA-3' sequence was estimated to be at least two orders of magnitude larger than that of the individual components. White et al. [27] extended this "hairpin" concept to the two oligopeptides containing pyrrole, 3-hydroxypyrrole, and imidazole shown in Fig. 3.12 (**1, 2,** and **3**). Columns **4** and **5** illustrate their binding schemes with 5'-d(TGGTCA)-3'(**4**) and 5'-d(TGGACA)-3' (**5**), respectively. The dissociation constant of the complex formed between **2** and **4** (abbreviated as **2–4**) was 18 times

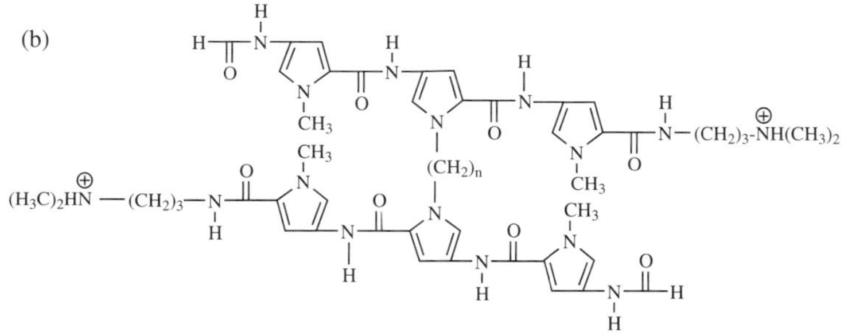

FIGURE 3.11. Structures of (a) hairpin peptides [26] and (b) crosslinked peptides [28].

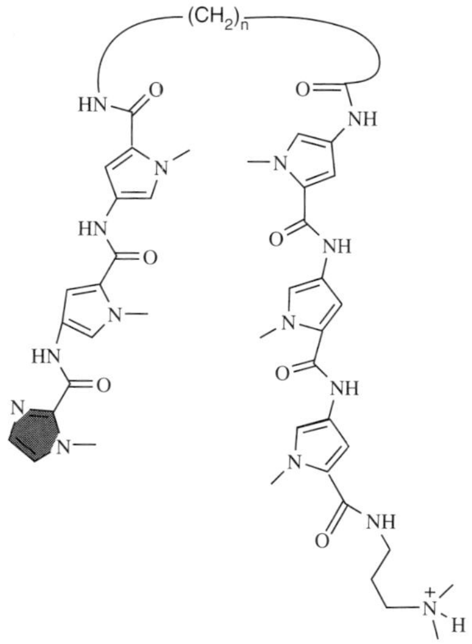

larger than that of the complex **2–5**, while the dissociation constant of **3–5** was 77 times larger than that of **3–4**. Thus, it is possible to distinguish between the two hexamers **4** and **5** (GpTpC versus GpApC sequence) using the oligopeptides such as **2** and **3**.

Crosslinked dimers shown in Fig. 3.11b are another modification of the side-by-side bonding scheme. It is expected that such central crosslinking restrains two peptide chains into a particular conformation and prevents the slippage or wobbling in the minor groove. Their interactions with poly(dA)·poly(dT) and poly(dA-dT)$_2$ were studied by CD spectroscopy, focusing on the effect of linker length on the binding mode [28].

In dansylated distamycins shown in Fig. 3.13 [structures (a) and (b)], the dansyl group (N,N'-dimethylaminophthalene sulfonamide) is attached directly to the conjugated N-methylpyrrole carboxamide chain, and their bindings with poly(dA)·poly (dT), poly(dA-dT)$_2$, CT-DNA, and poly(dG-dC)$_2$ were studied by fluorescence spectroscopy [29]. The intensity of their fluorescence emission near 530 nm increased 10–20 times with the former three oligonucleotides but did not increase with

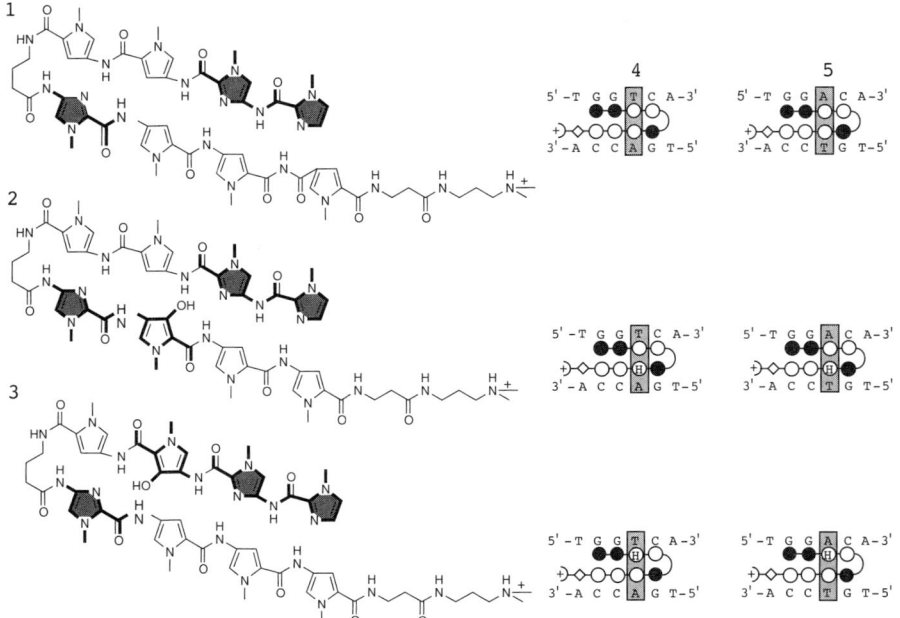

FIGURE 3.12. Structures of three hairpin peptides containing pyrrole (○), imidazole (●), and 3-hydroxypyrrole (⊕) (**1**, **2**, and **3** respectively), and their modes of binding with 5'-TGGTCA-3'(column **4**) and 5'-TGGACA-3'(column **5**) [27].

poly(dG-dC)$_2$. These fluorescent distamycin-like drugs have been used to monitor DNA melting and salt-induced adduct dissociation, although their binding affinities were low relative to the undansylated analogs.

Flow linear dichroism (FLD) and CD spectroscopy were used to study the binding of a furan analog of distamycin (FDst, shown in Fig. 3.13c) with CT-DNA [30]. The results reveal the formation of three types of adducts having 1:5, 4:5, and 3:1 (drug:base pair) ratios. The 1:5 adduct is *groove-binding* similar to distamycin, but is much less stable against ionic strength. The average angle between the long axis of FDst and the helical axis of DNA was estimated by FLD analysis. However, the structures of other adducts having different binding ratios were not clear.

Lown et al. [31] synthesized a series of oligopeptides in which two netropsin or distamycin moieties are connected by a flexible linker (polymethylene bridge), and tested their antitumor and antiviral activities. For example, the compound shown in Fig. 3.14a was found to be about 30 times more potent than distamycin, and its ID$_{50}$ values (50% inhibition dose) against murine leukaemia (L1210) and two other tumor cell lines ranged from 2.1 to 3.3 µg/mL, which is much smaller than those of distamycin (24–31 µg/mL). Ermishov et al. [32] studied the interaction of the dimeric netropsins (Fig. 3.14b) with three AT-rich 20-mers by Raman and surface-enhanced Raman (SER) spectroscopy. The results show that the binding causes

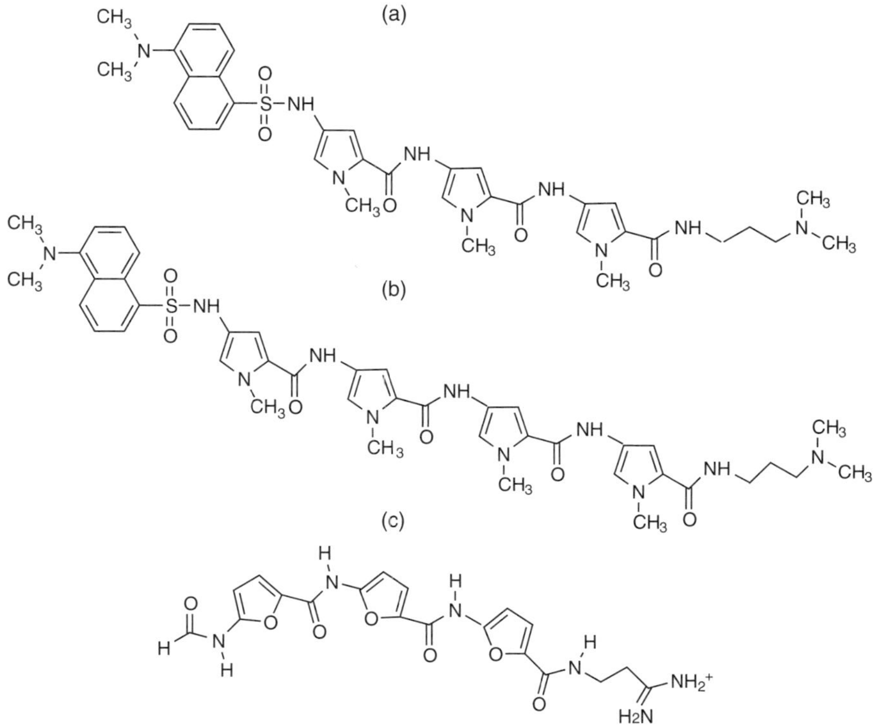

FIGURE 3.13. Structures of dansylated distamycins (a, b) and furan analoge of distamycin (c).

conformational distortions in both components. Although the oligonucleotides maintain overall B form, local distorstions are noted at the binding sites and G bases, and the torsional angles of the peptide bonds of the dimeric netropsin are varied to fit into the minor grooves of these 20-mers. In general, the Raman and SER spectra are similar, but the pyrrole ring vibrations are more strongly enhanced in the latter spectra, indicating that the pyrrole rings of the dimeric drug are orientated parallel to the SER-active surface.

The effect of chiral centers on drug–DNA binding has been studied by using ^{1}H NMR spectroscopy, DNA footprinting, and thermodynamic methods. For example, both natural (4S)-(+)-anthelvencin A and its synthetic (4R)-(−) enantiomer (Fig. 3.14c) form 1:1 complexes via binding to the 5'-AATT-3'sequence of d(CGCAATTGCG)$_2$. Although the drug configuration and location in the minor groove differed between the two enantiomers, these differences did not alter binding affinities significantly [33]. This work was extended to (4S)-(+)- and (4R)-(−)-dihydrokikumycin B, which has one less pyrrole moiety than does anthelvencin A. Thermodynamic data show that natural (4S)-(+) isomer binds to poly(dA-dT)$_2$ more efficiently than does its (4R)-(−) isomer [34]. Debart et al. [35] synthesized a number of netropsin analogs, and measured their binding constants with polynucleotides and

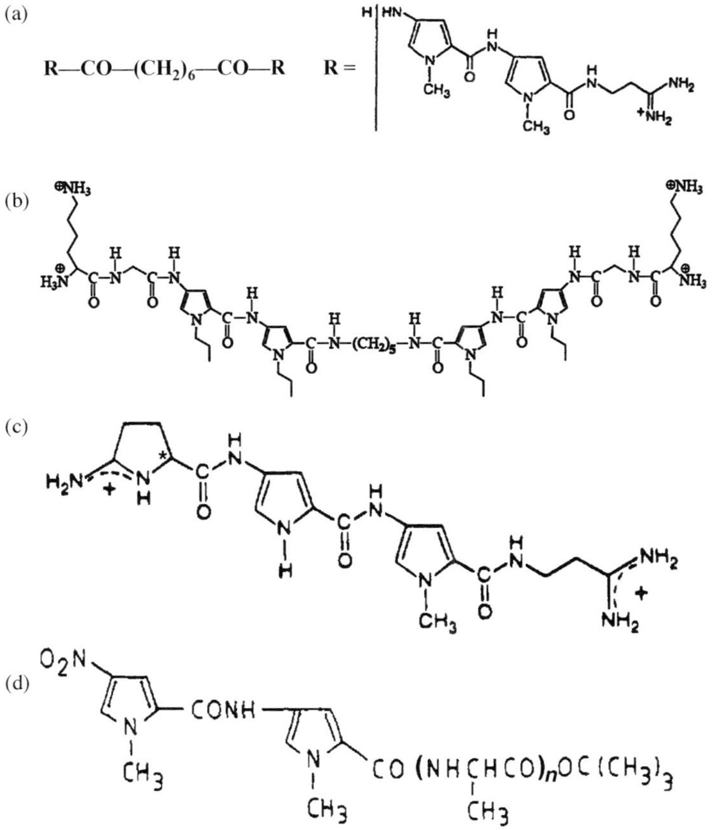

FIGURE 3.14. Structures of netropsin derivatives (a–d). The chiral center in (c) is marked by an asterisk (*).

cytotoxicities against murine and human tumor cell lines. The ID_{50} values of the compounds shown in Fig. 3.14d (where $n = 1, 2$) were substantially smaller than that of netropsin.

3.3. HOECHST 33258, SN6999, AND THEIR DERIVATIVES

Many synthetic drugs contain a variety of aromatic and/or heterocyclic rings instead of N-methylpyrrole rings of distamycin and netropsin. These compounds also assume a crescent-shaped conformation, and bind to the AT-rich region in the minor groove of DNA via hydrogen bonding, van der Waals contacts, and Coulombic interaction. Hoechst 33258, shown in Fig. 3.15a, is a typical example. It contains two benzimidazole rings at the center with a phenolic and a positively charged N-methylpiperazine group at the ends. Crystal structures of Hoechst 33258 and its derivatives bound to oligonucleotides have been determined by several workers. For example, the crystal structure of Hoechst 33258 complexed to d(CGCGAATTCGCG)$_2$

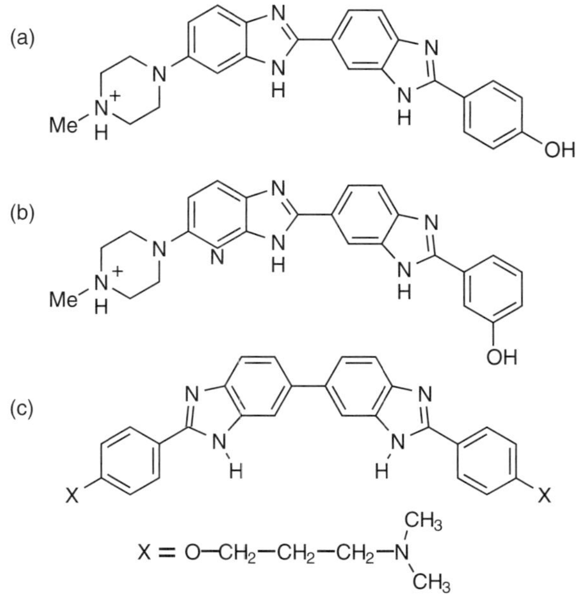

FIGURE 3.15. Structures of Hoechst 33258 (a), its *meta*-hydroxyl analog, (b) and symmetric bisbenzimidazole binder (c).

```
              1    2    3    4    5    6    7    8    9   10   11   12
    5'-d(C    G    C    G    A    A    T    T    C    G    C    G)-3'
    3'-d(G    C    G    C    T    T    A    A    G    C    G    C)-5'
             24   23   22   21   20   19   18   17   16   15   14   13
```

was determined, and its binding to the central AATT region in the minor groove was confirmed [36]. Sriram et al. [37] compared the crystal structures of Hoechst 33258 and 33342 (the OH group of Hoechst 33258 is replaced by the OC_2H_5 group) complexed to $d(CGCG'AATTCGCG)_2$ (G' = O6-ethylguanine) and its nonethylated 12-mer. In all cases, Hoechst drugs are bound to the central 5'-AATT-3' region. However, the G'_4-C_{21} and C_9-G'_{16} base pairs have different pairing schemes; the former pairing is close to a normal Watson–Crick type except with bifurcated hydrogen bonds between G'_4 and C_{21} and the ethyl group in the proximal orientation, whereas the latter pairing takes a wobble configuration with the ethyl group in the distal orientation. The occurrence of a dynamic equilibrium between these two G'-C base pair configurations may explain why thymine is preferentially incorporated across the G' lesion site during the cell replication process, because thymine can pair with G' in only one way, which is similar to normal Watson–Crick G-C pairing.

Figure 3.16 shows the crystal structure of a *meta*-hydroxy analog of Hoechst 33258 (Fig. 3.15b) complexed to $d(CGCGAATTCGCG)_2$, determined by Clark et al. [38].

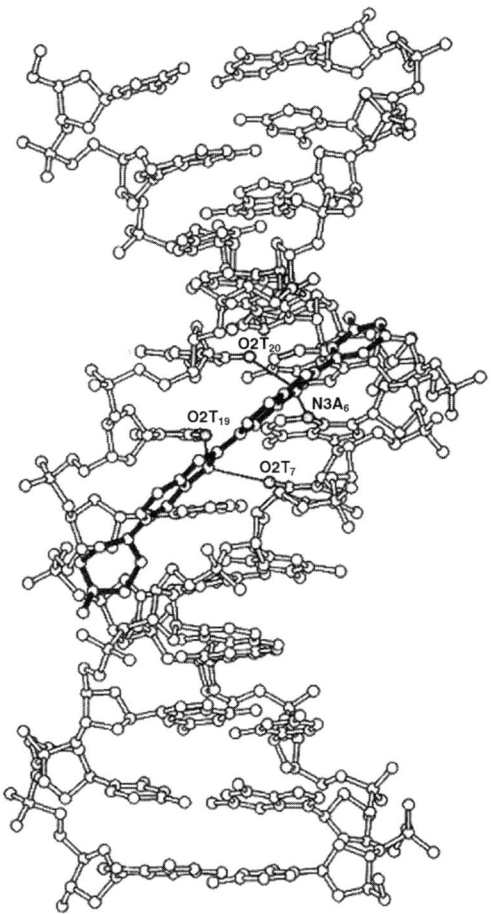

FIGURE 3.16. Crystal structure of the *meta*-hydroxyl analog of Hoechst 33258 complexed to d(CGCGAATTCGCG)$_2$. The drug molecule shown by filled-in bonds is located in the 5′-AATTC site, with the phenol ring lying toward the 5′ end of the sequence and the bulky piperazine ring close to the C$_9$-G$_{16}$ base pair (the "pip-down" orientation). Drug–DNA hydrogen bonds involving imidazole groups are shown as thin lines [38].

The overall structure of this complex is similar to that of Hoechst 33258 bonded to the same 12-mer [36]. The major differences are in the conformation of the drug and the resulting increase in minor groove width over the binding site, particularly at the phenol binding site because of the high ring twist and shallow binding depth. However, the *meta*-hydroxyl group is not involved in hydrogen bonding to the exocyclic NH$_2$ group of a guanine base. Mann et al. [39] synthesized a symmetric bisbenzimidazole derivative (see Fig. 3.15c) and determined the crystal structure of its complex with the same 12-mer duplex. This drug gave IC$_{50}$ values (concentration in μM at 50% inhibition) in the range of 0.12–0.38 against ovarian cancer cell lines, which are much smaller than those of Hoechst 33258 (9.5 – >100). Furthermore, it was active against two cisplatin-resistant cell lines and against a P-glycoprotein overexpressing doxorubicin-resistant cell line.

Utsuno et al. [40] examined the effect of Hoecht 33258 binding on the geometry of a DNA duplex (plasmid pBR322) using topoisomerase II relaxation followed by gel electrophoresis. Fluorescence, optical absorption, and calorimetric measurements were performed on this drug–DNA system. The concentrations of both drug and DNA were varied using the same buffer as that for topoisomerase reaction. These studies reveal that there are two modes of drug–DNA interaction. When the drug concentration is much lower than the DNA base-pair concentration, Hoechst 33258 binds in the minor groove of the DNA duplex and occupies a site of a five-base sequence containing no G-C pair. In this case, the equilibrium constant K_1 is $1.8 \times 10^7 \, M^{-1}$ (at 37°C), and the enthalpy of binding ΔH_1 is -865 cal/mol. When the drug concentration is much higher, Hoechst 33258 shows another binding mode that is much weaker with $K_2 = 2.25 \times 10^4 \, M^{-1}$ and $\Delta H_2 = -464$ cal/mol. It gives fluorescence quenching, has no base-pair preference, and causes an unwinding by 1°. These results are consistent with a self-association, or coaggregation, process in which the drug molecule is indiscriminately attracted to itself and the DNA (regardless of sequence), probably via intercalation. Some electrostatic attraction may be involved as well.

Utsuno et al. [41] also examined the changes in the tertiary structure of DNA duplex induced by Hoechst 33258 binding, using atomic force microscopy (Section 2.2.5). When the drug concentration is as high as 0.5 µg/mL, Hoechst 33258 seems to function as a clamp for two DNA chains and forms a condensate (aggregate) that has been found to have a toroidal shape. By surveying more than 100 microscopic images of such condensates formed in a 1 µg/mL drug solution, a mechanism of toroidal condensate formation was discussed.

Interactions of four types of minor groove binders (Fig. 3.17) with polynucleotides in D_2O were studied by Adnet et al. [42] using FTIR spectroscopy. As seen in Figure 3.18a (trace 2), free poly(dA-dT)$_2$ exhibits two bands at 1696 and 1662 cm^{-1}, which were assigned to the C2=O and C4=O stretching vibrations of thymidine, respectively. On addition of **A2** (see Fig. 3.17), a new band appeared at 1683 cm^{-1} and its intensity increased as more ligand was added (traces 3–5). Since the ligand alone does not exhibit such a band (trace 1), it was assigned to one of the C=O stretching vibrations of thymidine bonded to the drug. Similar results were obtained for **A1** and **B1** (Fig. 3.17), and together they indicated that the thymidine C=O groups are hydrogen-bonded to the NH groups of these drugs in the minor groove, as shown by many other studies. As seen in Fig. 3.18b, these new bands do not appear when **B2** (Fig. 3.17) is added to poly(dA-dT)$_2$ or poly(dG)•poly(dC) (traces 2 and 4) because no binding interaction occurs between them. However, the addition of **B2** (Fig. 3.17) to poly(dG)•poly(dC) shifts the bands at 1688 and 1650 cm^{-1} of the polymer to 1696 and 1654 cm^{-1}, respectively (traces 3 and 4 in Fig. 3.18b). These two bands were assigned to coupled vibrations of the C6=O stretching of guanosine and C2=O stretching of cytosine. Thus, the presence of adjacent cytosine residues in the same strand of DNA is favored by **B2** (Fig. 3.17). In all cases, no spectral changes were noted in other regions,

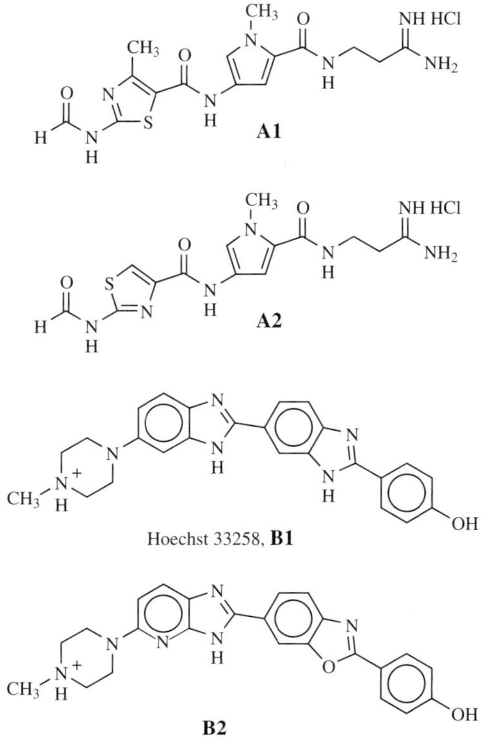

FIGURE 3.17. Structures of four minor groove binders used for FTIR studies [42].

indicating that B-DNA conformations of the oligonucleotides are not altered by these groove binders.

Figure 3.19 shows the structures of other types of minor groove binding antitumor drugs. The solution structure of SN6999 (Fig. 3.19a) complexed to d(GGTTAATGCGGT)·d(ACCGCATTAACC) was determined by 2D NMR spectroscopy [43]. The observed intermolecular NOEs can be rationalized by assuming that the drug molecule bound to the central d(TTAAT)·d(ATTAA) region takes two orientations relative to the long helical axis and that these two orientations are in rapid exchange. Intermolecular NOEs are extended to one residue neighboring this region (in the 3′ direction), indicating the contribution of the quaternary ammonium groups at both ends of SN6999 to minor groove binding.

Interactions of SN6999 and SN18071 (Fig. 3.19b) with polynucleotides were studied by CD and UV spectroscopy and other techniques [44]. It is well established that hydrogen bonding plays the major role in groove binding. Since SN18071 has no N–H groups, its DNA binding is expected to be much weaker than that of SN6999. Figure 3.20 compares the CD spectra of the three polynucleotides complexed with these two drugs. The intensities of the drug-induced positive bands that appear near 385 nm for SN6999 and near 405 nm for SN18071 serve as a measure of their bound states. Although both drugs exhibit strong affinity to poly(dA-dT)$_2$, the former shows

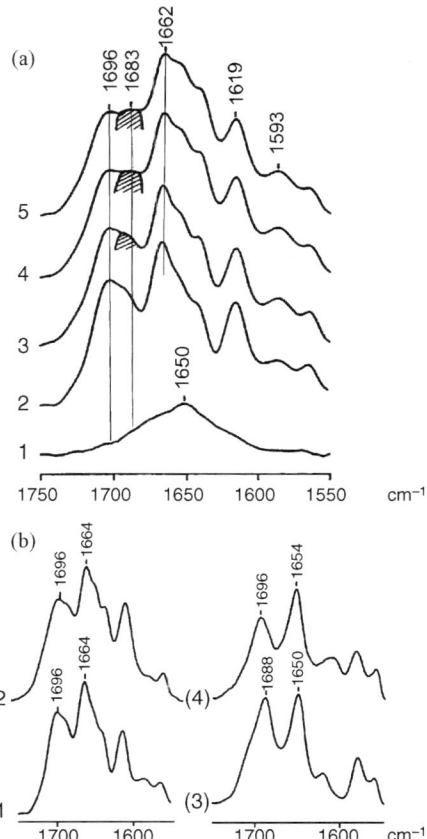

FIGURE 3.18. (a) FTIR spectra (in D_2O) of **A2** (see Fig. 3.17) + poly(dA-dT)$_2$ mixtures obtained by increasing r values(molar ratio of ligand/base pairs). Traces: (1) free ligand; (2) free (dA-dT)$_2$; (3) $r = 0.125$; (4) $r = 0.25$; (5) $r = 0.40$. (b) FTIR spectra (in D_2O). Traces: (1) poly(dA-dT)$_2$; (2) poly(dA-dT)$_2$ + **B2** (Fig. 3.17) ($r = 0.25$); (3) poly(dG)·poly(dC); (4) poly(dG)·poly(dC) + **B2** ($r = 0.25$) [42].

more affinity than the latter as seen in their intensities compared at similar r (molar ratio of drug/base pair) values. The same trend is seen for CT-DNA. However, SN6999 shows no binding to poly(dG-dC)$_2$, whereas SN18071 seems to bind to some extent. Interactions of these drugs with a DNA triple helix, poly(dA)·2 poly(dT), have also been studied by CD spectroscopy [45].

SN7167 is another analog of SN6999 in which the hydrogen atoms at the two positions marked by asterisks in Fig. 3.19a are substituted by the NH$_2$ groups. The crystal structure of its complex with d(CGCGAATTCGCG)$_2$ [46] shows that there are very few close contacts between the drug and the floor of the minor groove, and that the exocyclic NH$_2$ groups of the drug are not involved in hydrogen bonding to the 12-mer. Yet, this drug exhibits some biological activity, although the reason for it is not clear.

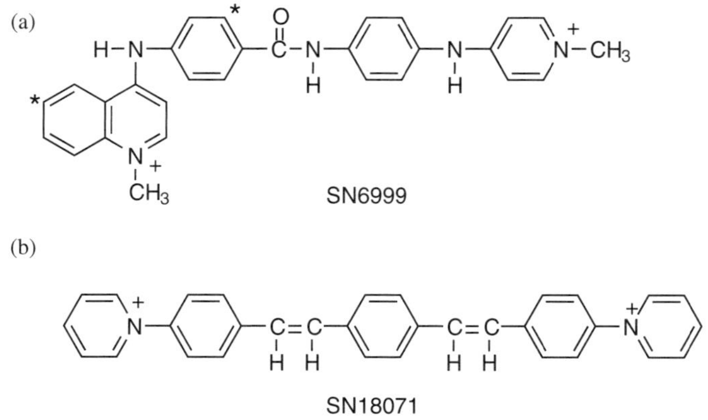

230 GROOVE-BINDING DRUGS

FIGURE 3.19. Structures of (a) SN6999 and (b) SN18071. The two asterisks in (a) indicate that the hydrogen atoms at these positions are substituted by the NH$_2$ groups in SN7167.

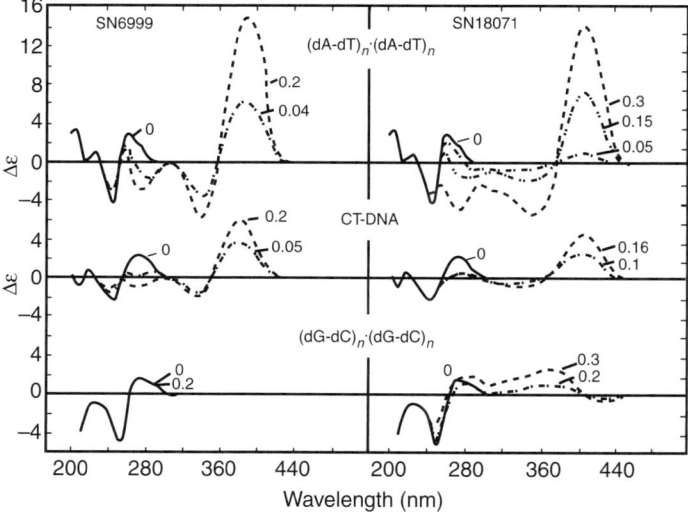

FIGURE 3.20. CD spectra of poly(dA-dT)$_2$, CT-DNA, and poly(dG-dC)$_2$ titrated with SN6999 (left) and SN18071 (right) in 0.1 M NaCl solution. The number given to each spectrum indicates the r value (molar ratio of drug/base pair) [44].

3.4. CHROMOMYCIN, MITHRAMYCIN, AND OTHER GC BINDERS

Chromomycin A$_3$ and mithramycin, shown in Fig. 3.21, are antitumor antibiotics extracted from *Streptomyces grisius* and *Streptomyces plicatus*, respectively. These drugs consist of an aglycon ring (chromophore) to which the disaccharide (A-B), the trisaccharide (C-D-E), and a hydrophilic sidechain are attached. However, they differ

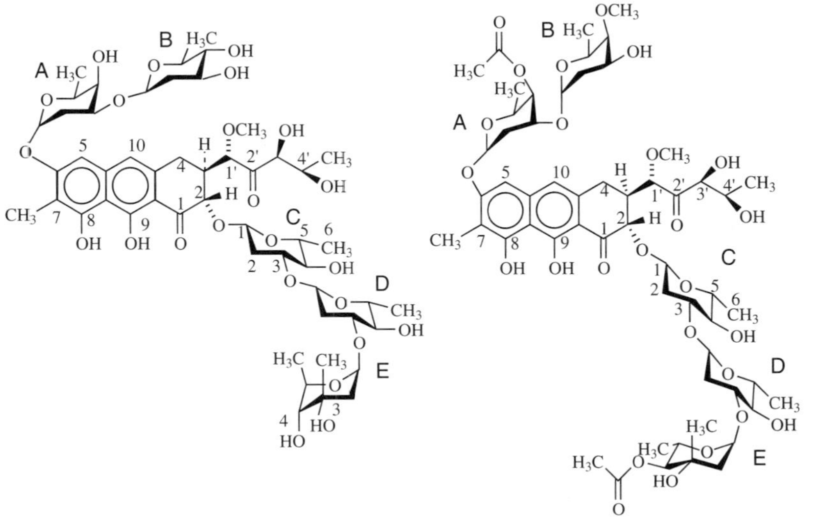

FIGURE 3.21. Structures of mithramycin and chromomycin A_3 [47].

in the structures of the pyranoses that are connected to each end of the aglycon ring. Unlike the AT-groove-binding drugs such as distamycin and netropsin discussed in the preceding sections, chromomycin (A_3) and mithramycin bind to the GC-rich region of an oligonucleotide in the minor groove with a stoichiometry of 2 : 1 (drug/duplex) molar ratio in the presence of a divalent metal ion such as Mg^{2+}.

The solution structure of the chromomycin–d(TTGGCCAA)$_2$ complex was first determined by 2D NMR spectroscopy [47]. The results show that the drug binds to the central GGCC region as a symmetric dimer and induces a conformational change in B-DNA double helix, resulting in a wider and shallower minor groove at the binding site. This is in contrast to netropsin and distamycin, which bind to the AT-rich region in the minor groove with minimal changes in groove width. It was suggested that the GC specificity is associated with hydrogen bonding between the C8 OH group of the aglycon ring and the N3 atom (acceptor) or NH_2 group (donor) of guanosine.

NMR studies were also carried out on mithramycin and chromomycin bound to d(ATGCAT)$_2$ [48]. Nuclear Overhauser effect (NOE) contacts from the A, B, C, and D pyranoses of mithramycin to several deoxyribose protons indicate that the A and B rings are along the sugar–phosphate backbone of G_3pG_4, whereas the C and D rings are along that of A_5pT_6 sequence. The overall structure of the corresponding chromomycin complex is similar. However, NOESY experiments at 1 : 1 molar ratio (drug/duplex) indicate that mithramycin is in slightly faster chemical exchange between the free and bound statres than is chromomycin, suggesting that the former has slightly lower affinity with the hexamer than the latter.

Figure 3.22 is a stereo view of the structure of the Mg(II)-coordinated chromomycin bonded to d(AAGGCCTT)$_2$ in a 1 : 2 : 1(Mg(II)/drug/duplex) ratio that was

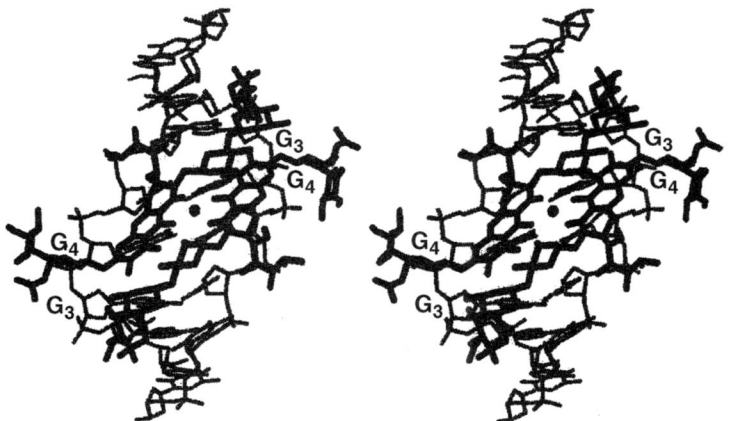

FIGURE 3.22. Stereo view of the 2/1/1 (drug/Mg^{2+}/octamer) complex. The dot at the center indicates the Mg^{2+} ion [G6,49].

derived by Gao et al. from 2D NMR studies [49,G6]. Their results reveal the following:

1. The two drug molecules (ligands) are aligned in a head-to-tail orientation with a slight tilt relative to each other, and form a 75° angle between their planes. The C-D-E trisaccharide chain of each ligand assumes an extended conformation, projecting in opposite directions in the complex.
2. The Mg^{2+} ion is coordinated to the O1 carbonyl and O9 enolate atoms of the two slightly tilted aglycon chromophores with the O9−Mg−O9 (intermolecular) angle of 170° while all other O−Mg−O angles are near 95°.
3. The minor groove at the central GGCC segment of the binding site is wider and shallower than B-DNA and is close to A-DNA conformation.
4. The sequence specificity results from intermolecular hydrogen bonding of the OH group at C8 of the chromophore with the NH_2 and N3 atom of G_4, and of the O1 atom of the E-ring sugar with the NH_2 group of G_3.

NMR investigations were also carried out on the Mg^{2+}-coordinated mithramycin dimer bonded to d(TGGCCA)$_2$ and d(TCGCGA)$_2$ [50]. The global structures of these complexes are similar to that of the corresponding chromomycin complex described above [49]. Both drugs bind to the minor groove in the central GC region, with the OH group at C8 of the aglycon hydrogen-bonded to the NH_2 group of guanosine. However, they differ in two respects: (1) In the chromomycin complex, all four residues in the central GGCC segment adopt the sugar puckers and glycoside torsion angles of A helix, whereas in the mithramycin complex, only the central cytosine residue in the GpC sequence adopts such a conformation; and (2) the interacting E-ring sugar is more hydrophilic in the mithramycin complex than in the corresponding chromomycin complex because the former has an additional OH group and lacks a CH_3 group.

As stated above, these drugs tend to convert B-DNA into A-DNA conformation, which has a wider and shallower minor groove, and their sequence specificity is attributed to hydrogen bonding between the OH group of C8 of the aglycon ring and NH_2 group or N3 atom of guanosine. Poly(dG)·poly(dC) takes the A form, in which the closely spaced NH_2 groups are exposed in the minor groove. Thus its binding to the drug does not require any conformational change of DNA. On the other hand, poly $(dG-dC)_2$ requires this B → A conformational change to bind the drug. More detailed studies on the interactions of chromomycin and mithramycin with these polynucleotides were undertaken using UV absorption, fluorescence, and CD spectroscopy [51].

The Mg-coordinated mithramycin dimer also forms a 4 : 2 : 1 complex (drug/Mg^{2+}/octamer) with d$(ACCCGGGT)_2$, and its solution structure was determined by 2D NMR spectroscopy [52]. This structure has two dimers of Mg-coordinated mithramycins, which are in two identical off-center binding sites, and retains the overall C_2 symmetry of the complex. As a result, only one set of peaks is observed for both the drugs and the octamer. However, chromomycin can form only a 2 : 1 : 1 complex with the same oligonucleotide. It was suggested that hydrogen bonding between the 4OH group and the 4O atom of two E-ring saccharides in the dimers might be the driving force in forming the 4 : 2 : 1 complex of mithramycin; chromomycin cannot form such a complex because its 4OH group of the E-ring sugar is acetylated and the stereochemistry of the 3OH group in the E ring is different between these two drug molecules.

Utsuno et al. [53] examined the effect of chromomycin binding on the geometry of the DNA duplex (plasmid pBR332) using topoisomerase I relaxation followed by gel electrophoresis (Section 2.2.7) To determine the equilibrium constant of this drug–DNA binding reaction under the same conditions as those employed for the topoisomerase reaction (10^{-5} M at 37°C), fluorescence measurements were performed, and analyzed by a *Scatchard plot* with the McGhee–von Hippel exclusion site model. The binding constant was found to be $3.8 \times 10^5 \, M^{-1}$ under these conditions, and the number of base pairs involved in the site of one chromomycin molecule on the duplex was found to be 5. It was concluded that one chromomycin molecule, bound to the duplex, unwinds it by $11.8 \pm 1.1°$. In addition, the enthalpy of binding was determined to be -31.81 kJ/mol using a titration calorimeter with a more concentrated solution (6.2 mM).

Topotecan (Tpt), shown in Fig. 3.23a, is another minor-groove binder at the GpC site. Streltsov et al. [54] confirmed it by FLD, CD, and Raman spectroscopy. Figure 3.23b shows the Raman spectra with band assignments of free topotecan (trace 1) and its CT-DNA complex (trace 2). The following changes are noted on DNA binding: (1) the band at $1309 \, cm^{-1}$ is downshifted to $1304 \, cm^{-1}$, (2) the band at $1563 \, cm^{-1}$ is downshifted to $1559 \, cm^{-1}$, (3) the intensity ratio of the Fermi resonance doublet at 1650 and $1658 \, cm^{-1}$ (I_{1658}/I_{1650}) increases from 0.6 to 0.9, and (4) the intensity ratio of the two bands at 1619 and $1563 \, cm^{-1}$ (I_{1619}/I_{1563}) decreases from 1.0 to 0.7. Similar spectral changes are observed when Tpt is mixed with guanosine (dG) (trace 3). However, the Raman spectrum of Tpt mixed with inosine (dI) (trace 4) is different from that of traces 2 and 3 in the following respects: (1) the $1309 \, cm^{-1}$ band is not shifted by Tpt., (2) the I_{1619}/I_{1563} ratio increases, (3) the

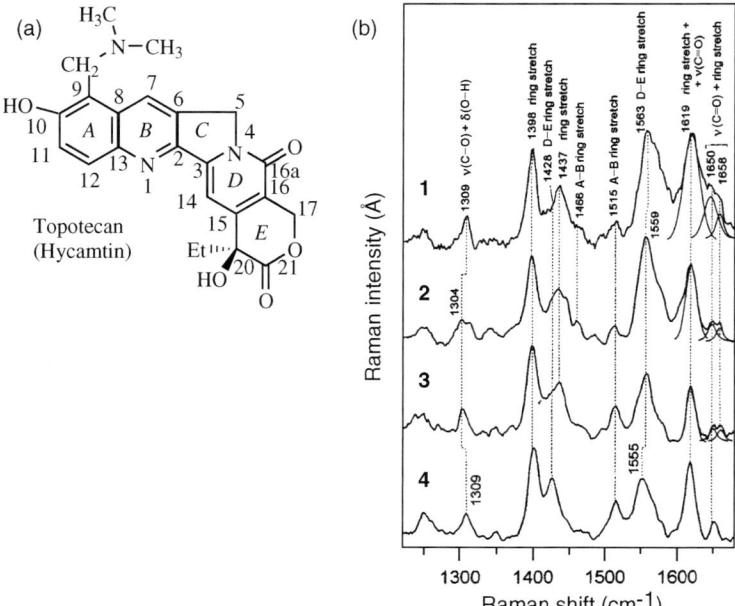

FIGURE 3.23. (a) Structure of topotecan (Tpt); (b) Raman spectra (647.1 nm excitation, pH = 6.8) of free Tpt (1), Tpt with CT-DNA (2), Tpt with dG (3), and Tpt with dI (4). The spectra were obtained by subtracting those of free nucleosides or CT-DNA from the corresponding mixtures (Tpt = 10^{-4} M). The molar ratios of Tpt/DNA base and Tpt/nucleoside were 1/25 and 1/23, respectively [54].

1658 cm^{-1} band disappears or becomes very weak, and (4) the 1437 cm^{-1} band disappears and a new band emerges at 1428 cm^{-1}. Thus, the Tpt-dI complex is very different from the Tpt-dG and Tpt-DNA complexes because of the absence of the exocyclic NH$_2$ group in dI. Other nucleosides, dA, dT, and dC, also cause spectral changes similar to that observed for Tpt-dI. These results suggest that Tpt preferentially binds to dG sites of DNA in the minor groove by forming hydrogen bonds between its CO group of E-ring and the NH$_2$ group of dG, and between its OH group of E ring at C20 and the N atom of dG. Their FLD and CD studies also support this conclusion.

3.5. GROOVE BINDING AND INTERCALATION

As stated in Chapter 2, actinomycin D is a typical *intercalator* at the GpC site whereas netropsin is a typical *groove binder* at the ApT site. Using NMR spectroscopy, Patel and coworkers [55,56] demonstrated that these two drugs can bind simultaneously to the self-complementary 12-mer:

```
5'-d(C  G  C  G  A  A  T  T  C  G  C  G)-3'
      1  2  3  4  5  6  6  5  4  3  2  1
3'-d(G  C  G  C  T  T  A  A  G  C  G  C)-5'
```

Figure 3.24a, trace **A** shows the imino proton spectrum of this duplex measured at 27°C. Although the imino proton of the terminal GC base pair (1) is broadened as a result of fraying at this temperature, the remaining imino proton signals can be assigned as indicated by the numbers in the figure. The signals at 13.79 (5) and 13.67 (6) ppm are assigned to the AT base pairs while those at 13.10 (2), 12.945 (3), and 12.74 (4) ppm are assigned to the GC base pairs. When two equivalents of actinomycin D are intercalated between base pairs 2 and 3 near the end of the duplex, two GC base pair signals (2 and 3) show marked upfield shifts and the AT base-pair signals (5 and 6) are upshifted slightly by a small perturbation caused by the intercalation at the adjacent GpC sites (Fig. 3.24a, trace **B**). In the mixture of one equivalent of netropsin per the dodecamer, the H3 imino proton signal of thymidine at 13.79 ppm (5) is markedly downshifted to 14.25 ppm, while that at 13.67 ppm (6) shows a slight upfield shift (Fig. 3.24a, trace **C**). Addition of another equivalent of

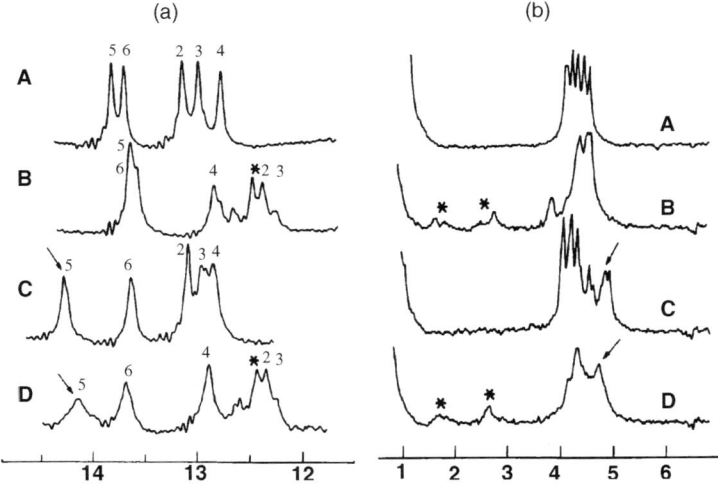

FIGURE 3.24. (a) Imino proton NMR spectra of (**A**) the 12-mer duplex (pH = 6.8), (**B**) the complex of two actinomycins per duplex (pH = 6.75), (**C**) the complex of one netropsin per duplex (pH = 6.89), and (**D**) the complex of two actinomycins plus one netropsin per duplex (pH = 6.75). The upshifts of two gaunosine H1 imino protons on actinomycin binding are marked by asterisks. The downshifts of one thymine H3 imino proton on netropsin binding are marked by arrows. (b) ^{31}P NMR spectra of (**A**) the 12-mer duplex (pH = 7.5), (**B**) the complex of two actinomycins per duplex (pH = 7.85), (**C**) complex of one netropsin per duplex (pH = 7.70), and (**D**) complex of two actinomycins plus one netropsin per duplex (pH = 7.8). The downshifts at the actinomycin *intercalation* sites are marked by asterisks, and the upshifts at the netropsin *groove-binding* site are marked by arrows. Chemical shift values on the abscissa are given by upfield shifts from trimethylphosphate [55].

netropsin causes no spectral changes, indicating that this dodecamer duplex has only one binding site for netropsin. Figure 3.24a, trace **D** shows the spectrum obtained for the complex of two equivalents of actinomycin D and one equivalent of netropsin with one equivalent of the duplex. It exhibits the upfield shifts of the two GC imino protons (base pairs 2 and 3) and marked downfield shift of the AT imino proton of thymidine (5) as observed for the dodecamer mixed with each drug separately (Fig. 3.24a, traces **B** and **C**). These results confirm that two actinomycin D and one netroposin can bind simultaneously to the dodecamer, although there is an interaction between them.

These results were further confirmed by ^{31}P NMR spectroscopy. As seen in Fig. 3.24b, trace **A**, the dodecamer exhibits partially resolved signals of phosphodiesters at 4.05–4.51 ppm. When two equivalents (2 eq) of actinimycin D are added, some of these signals are markedly downshifted to 1.64 and 2.54 ppm (marked by asterisks in Fig. 3.24b, trace **B**), which are assigned to the phosphodiesters facing the benzenoid and quinonoid edges of the phenoxazone ring. On the other hand, addition of 1 eq of netropsin causes upshifts of two signals to 4.8 and 4.9 ppm (Fig. 3.24b, trace **C**). These are assigned to the phosphodiesters of AT base pairs at the netropsin binding site. On mixing of 2 eq of actinomycin D and 1 eq of netropsin with the dodecamer, two broad signals are observed at 1.72 and 2.64 ppm (Fig. 3.24b, trace **D**). This is similar to the case shown in Fig. 3.24b, trace **B**, in which actinomycin D alone is added. The phosphodiester signals of AT base pairs previously observed for netropsin alone at 4.8 ppm is slightly shifted to 4.72 ppm as a result of mutual interaction between actinomycin D and netropsin at their binding sites.

A variety of hybrid drugs (combilexins) in which *groove binders* are connected *covalently* to *intercalators* via linkers have been synthesized by Bailly and co-workers [57–60]. These drugs are expected to be more stable and sequence-specific than those of nonhybrid components. As an example, Fig. 3.25a shows the structure of ThiaNetGA in which a thiazole-containing analog of netropsin (ThiaNet) is linked to the glycylanilino-9-aminoacridine chromophore (GA). Using UV–visible absorption, fluorescence, and electric linear dichroism (ELD) spectroscopy, Plouvier et al. [57] have shown that the ThiaNet moiety is bonded to the *minor groove* while the GA moiety is partially *intercalated*. Here, ELD denotes a LD spectrum measured under the influence of a short electric field pulse by which DNA molecules are orientated in solution to make it optically anisotropic. Linear dichroism (ΔA; see Section 1.2.3) is the difference between the absorbance (A) for light polarized parallel ($A_{\|}$) and perpendicular ($A_{\perp}$) to the applied field at a selected wavelength. The reduced dichroism ($\Delta A/A$) is defined as $(A_{\|} - A_{\perp})/A$, where A is the absorbance in the absence of electric field at the same wavelength. Figure 3.25b shows the ELD spectrum of the ThiaNetGA(hybrid)–DNA complex. The negative band in the 380–490 nm region is due to the GA moiety, while the positive band in the 300–330 nm region originates in the ThiaNet moiety of the hybrid, with a minor contribution from the GA moiety. Figure 3.25c compares the electric field dependence of $\Delta A/A$ values for DNA alone (□ at 260 nm), ThiaNet–DNA (■ at 310 nm), hybrid–DNA (○ at 310 nm), GA-DNA (▲ at

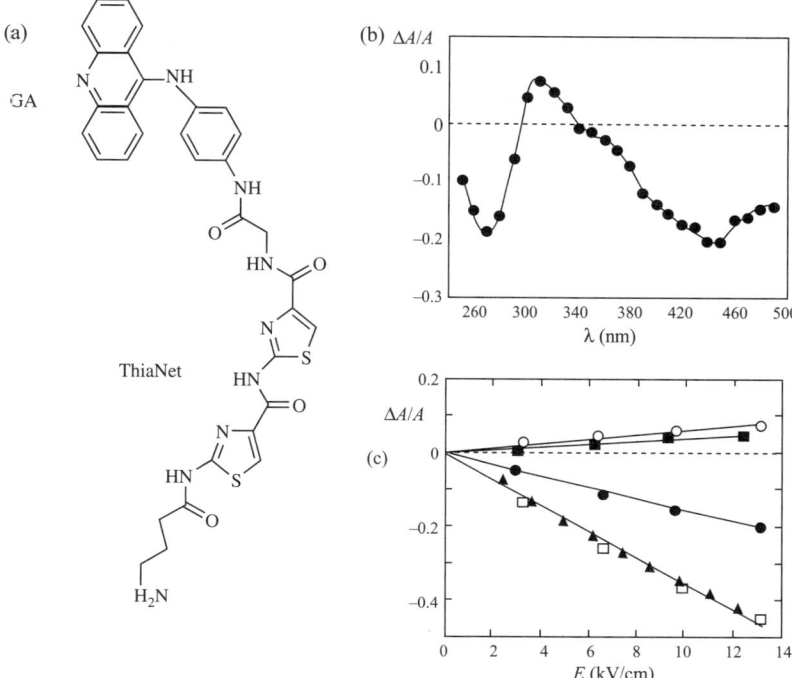

FIGURE 3.25. (a) Structure of the hybrid ThiaNetGA; (b) ELD spectrum of the ThiaNetGA-DNA complex at 13 kV/cm; (c) electric field dependence of the reduced dichroism ($\Delta A/A$) measured for CT-DNA alone at 260 nm (□), its complex with ThiaNetGA at 440 nm (●) and 310 nm (○), and for its complex with ThiaNet (310 nm) (■) and with GA at 440 nm (▲). The drug : DNA ratio is 0.1 : 1 mM, Na cacodylate buffer, pH = 6.5 [57].

440 nm) and hybrid–DNA (● at 440 nm). It is noted that the field dependences of ThiaNet-DNA (■) and hybrid-DNA (○) at 310 nm are almost identical, indicating that the GA moiety of the hybrid does not interfere with *minor-groove binding* of the ThiaNet moiety. On the other hand, $\Delta A/A$ values measured at 440 nm for GA-DNA (▲) are much lower than that of hybrid-DNA (●) (e.g., −0.43 vs. −0.2 at 13 eV/cm), and the former is almost identical to that of DNA alone (□ at 260 nm). These results indicate that the transition moment of GA in the GA-DNA complex lies parallel to the base pairs of DNA, but the transition moment of GA in the hybrid–DNA complex is inclined to the helical axis and its angle is estimated to be about 65° [57]. This inclination is attributed to the short glycyl linker between the ThiaNet and GA moieties, which results in only *partial intercalation* of the latter in the hybrid-DNA complex. A spin-labeled ThiaNetGA was also prepared to study its cellular distribution by ESR spectroscopy. This hybrid drug is nontoxic and exhibits only moderate antitumor activity against P388 leukemia cells in mice.

According to Wilson et al. [61], 4,6-bis[4′-(4″-methylpiperazino)phenyl]pyrimidine (structure **a**) and its 5-methyl derivative(**b**) shown in Fig. 3.26a have

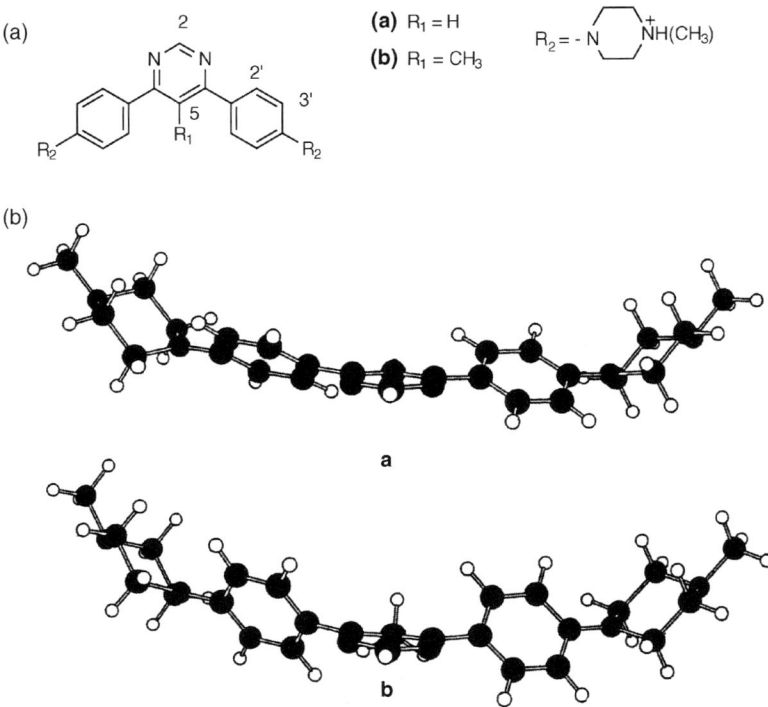

FIGURE 3.26. (a) Structures of unfused aromatic systems containing terminal piperazino substituents (structures **a** and **b**); (b) molecular mechanics minimized conformations of structures **a** (top) and **b** (bottom). The view into the central pyrimidine ring is the same for both compounds. The larger twist of the phenyl–pyrimidine–phenyl aromatic system can be seen for structure **b** [54].

markedly different interactions with DNA. Figure 3.26b compares the molecular models of structures **a** and **b** obtained by molecular mechanics calculations. The torsional angles of the phenyl–pyrimidine ring plane are 25–30° for structure **a** but much larger for structure **b** (50–60°) because of steric hindrance due to its central methyl group, and the calculated barrier to a planar three-ring aromatic system is approximately 4 kJ/mol for structure **a** but 33 kJ/mol for structure **b**. As a result, an acceptable *intercalated model* in which both terminal piperazino groups are in the *major groove* can be built for structure **a**, but not for structure **b** because it has a significant twist of the phenyl–pyrimidine ring plane. Furthermore, structure **b** does not fit well in the minor groove because of its size and conformation, although it can be fit into the major groove. A typical intercalator consists of an aromatic fused ring system with protonated nitrogens atom or protonated sidechains attached to it (Chapter 2). Thus, it is unusual that a molecule such as structure **a** can form an intercalated complex with DNA. Their conclusions were supported by viscosity measurements as well as spectroscopic studies (UV–visible, CD, and NMR).

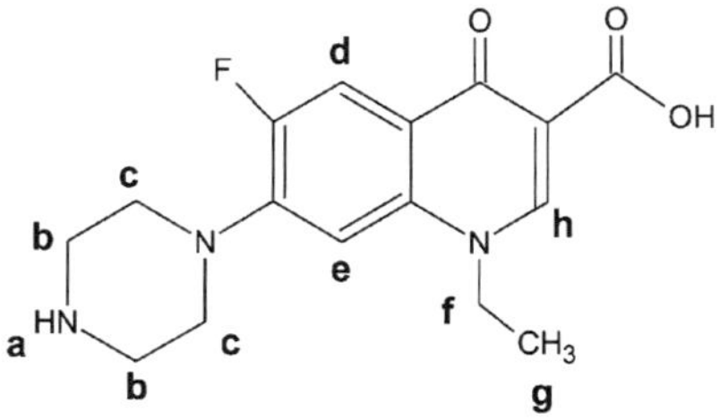

```
            1 2 3 4 5 5 4 3 2 1
DNA-1   5'- C G C G A T C G C G -3'
        3'- G C G C T A G C G C -5'

DNA-2   5'- C G C A C G T G C G -3'
        3'- G C G T G C A C G C -5'

DNA-3   5'- C G C A G C T G C G -3'
        3'- G C G T C G A C G C -5'

DNA-4   5'- G C G C A T G C G C -3'
        3'- C G C G T A C G C G -5'

DNA-5   5'- G C G A C G T C G C -3'
        3'- C G C T G C A G C G -5'

DNA-6   5'- G C G A G C T C G C -3'
        3'- C G C T C G A G C G -5'

DNA-7   5'- C A G G A T C C T G -3'
        3'- G T C C T A G G A C -5'
```

FIGURE 3.27. Structures of norfloxacin and seven decamers containing 5'-CG-3', 5'-GC-3,' and 5'-GG-3' sequences [62].

Interactions of norfloxacin with seven different decamers containing 5'-CpG-3', 5'-GpC-3', and 5'-GpG-3' sequences shown in Fig. 3.27 were studied by Sandström et al. [62] using ^{1}H NMR spectroscopy. On mixing with norfloxacin, the guanine imino protons of the central 5'-CpG-3' steps of **DNA-1, 2, 5** (marked by boldface letters) were broadened and shifted to high field, indicating that the drug interacts with these imino protons preferentially. However, this interaction cannot be attributed to "classical intercalation" because their binding constants are much smaller compared with those of typical intercalators such as acridine [63]. Thus, norfloxacin probably interacts via "partial intercalation" with the 5'CpG-3' sequence at or near the center of these oligomers. The imino protons of the terminal CpG steps were not observed because of the end-fraying effect. Such selective broadening did not occur for **DNA-3, 4, 6, 7** since they do not contain central or near-central 5'-CpG-3' sites. Furthermore, broadening of the aromatic proton signal of adenine (H2) and an upfield shift of the guanine amino proton signal (both protons in the minor groove) were observed regardless of the sequence of these decamers. 2D NOESY spectral studies showed that no significant structural changes of the oligomers occurred on drug binding. Thus, it was concluded that the planar two-ring system of norfloxacin is partially intercalated at the central 5'-CpG-3' steps and that the drug binds to the minor groove in all cases.

REFERENCES

1. Bailly, C. and Chaires, J. B., "Sequence-specific DNA minor groove binders. Design and synthesis of netropsin and distamycin analogues," *Bioconj. Chem.* **9**:513–538 (1998).
2. Gallmeier, H.-C. and König, B., "Heteroaromatic oligoamides with dDNA affinity," *Eur. J. Org. Chem.* 3473–3483 (2003).
3. Kopka, M. L., Yoon, C., Goodsell, D., Pjura, P., and Dickerson, R. E., "The molecular origin of DNA-drug specificity in netropsin and distamycin," *Proc. Natl. Acad. Sci. USA* **82**:1376–1380 (1985).
4. Goodsell, D. S., Kopka, M. L., and Dickerson, R. E., "Map interpretation," *Biochemistry* **34**:4983–4993 (1995).
5. Coll, M., Frederick, C. A., Wang, A. H.-J., and Rich, A., "A bifurcated hydrogen-bonded conformation in the d(A·T) base pairs of the DNA dodecamer d(CGCAAATTTGCG) and its complex with distamycin," *Proc. Natl. Acad. Sci. USA* **84**:8385–8389 (1987).
6. Coll, M., Aymami, J., van der Marel, G. A., van Boom, J. H., Rich, A., and Wang, A. H.-J., "Molecular structure of the netropsin-d(CGCGATATCGCG) complex: DNA conformation in an alternating AT segment," *Biochemistry* **28**:310–320 ((1989).
7. Sriram, M., Van der Marcel, G. A., Roelen, H. L. P. F., van Boom, J. H., and Wang, A. H.-J., "Structural consequences of a carcinogenic alkylation lesion on DNA effect of O^6 ethylguanine on the molecular structure of the d(CGC[e^6G]AATTCGCG)-netropsin complex," *Biochemistry* **31**:11823–11834 (1992).
8. Wellenzohn, B., Winger, R. H., Hallbrucker, A., Mayer, E., and Liedl, K. R., "Simulation of EcoRI dodecamer netropsin complex confirms class I complexation mode," *J. Am. Chem. Soc.* **122**:3927–3931 (2000).
9. Patel, D. J. and Shapiro, L., "Sequence-dependent recognition of DNA duplexes," *J. Biol. Chem.* **261**:1230–1240 (1986).
10. Pelton, J. G. and Wemmer, D. E., "Structural modeling of the distamycin A-d(CGCGAATTCGCG)$_2$ complex using 2D NMR and molecular mechanics," *Biochemistry* **27**:8088–8096 (1988).
11. LaPlante, S. R. and Borer, P. N., "Changes in ^{13}C NMR chemical shifts of DNA as a tool for monitoring drug interactions," *Biophys. Chem.* **90**:219–232 (2001).
12. Boudreau, E. A., Pelczer, I., Borer, P. N., Heffron, G. J., and LaPlante, S. R., "Changes in drug ^{13}C NMR chemical shifts as a tool for monitoring interactions with DNA," *Biophys. Chem.* **109**:333–344 (2004).
13. Pelton, J. G. and Wemmer, D. E., "Structural characterization of a 2:1 distamycin A·d(CGCAAATTGGC) complex by two-dimensional NMR," *Proc. Natl. Acad. Sci. USA* **86**:5723–5727 (1989).
14. Chen, X., Ramakrishnan, B., and Sundaralingam, M., "Crystal structures of the Side-by-side binding of distamycin to AT-containing DNA octamers d(ICITACIC) and d(ICATATIC)," *J. Mol. Biol.* **267**:1157–1170 (1997).
15. Burckhardt, G., Votavova, H., Sponar, J., Luck, G., and Zimmer, C., "Two binding modes of netropsin are involved in the complex formation with poly(dA-dT)·poly(dA-dT) and other alternating DNA duplex polymers," *J. Biomol. Struct. Dyn.* **2**:721–736 (1985).

16. Liquier, J., Mchami, A., and Taillandier, E., "FTIR study of netropsin binding to poly d(A-T) and poly dA·poly dT," *J. Biomol. Struct. Dyn.* **7**:119–126 (1989).

17. Lu, D. S., Nonaka, Y., Tsuboi, M., and Nakamoto, K., "Molecular distortion of distamycin on binding to DNA as revealed by Raman spectroscopy," *J. Raman Spectrosc.* **21**:321–326 (1990).

18. Grygon, C. A. and Spiro, T. G., "Ultraviolet resonance Raman spectroscopy of distamycin complexes with poly(dA)-poly(dT) and poly(dA-dT): Role of H-bonding," *Biochemistry* **28**:4397–4402 (1989).

19. Youngquist, R. S. and Dervan, P. B., "Sequence-specific recognition of B-DNA by oligo (N-methylpyrrolecarboxamide)s," *Proc. Natl. Acad. Sci. USA* **82**:2565–2569 (1985).

20. Dervan, P. B., "Design of sequence-specific DNA-binding molecules," *Science* **232**:464–471 (1986).

21. Mrksich, M., Wade, W. S., Dwyer, T. J., Geierstanger, B. H., Wemmer, D. E., and Dervan, P. B., "Antiparallel side-by-side dimeric motif for sequence-specific recognition in the minor groove of DNA by the designed peptide 1-methylimidazole-2-carboxamide netropsin," *Proc. Natl. Acad. Sci. USA* **89**:7586–7590 (1992).

22. Geierstanger, B. H., Jacobsen, J. P., Mrksich, M., Dervan, P. B., and Wemmer, D. E., "Structural and dynamic characterization of the heterodimeric and homodimeric complexes of distamycin and 1-methylimidazole-2-carboxamide-netropsin bound to the minor groove of DNA," *Biochemistry* **33**:3055–3062 (1994).

23. Geierstanger, B. H., Dwyer, T. J., Bathini, Y., Lown, J. W., and Wemmer, D. E., "NMR characterization of a heterocomplex formed by distamycin and its analog 2-ImD with d(CGCAAGTTGGC):d(GCCAACTTGCG): Preference for the 1:1:1 2-ImD: Dst: DNA complex over the 2:1 2-ImD:DNA and the 2:1 Dist:DNA complexes," *J. Am. Chem. Soc.* **115**:4474–4482 (1993).

24. Geierstanger, B. H., Mrksich, M., Dervan, P. B., and Wemmer, D. E., "Design of a G·C-specific DNA minor groove-binding peptide," *Science* **266**:646–650 (1994).

25. Lee, M., Rhodes, A. L., Wyatt, M. D., Forrow, S., and Hartley, J. A., "GC base sequence recognition by oligo(imidazolecarboxamide) and C-terminus-modified analogues of distamycin deduced from circular dichroism, proton nuclear magnetic resonance, and methidiumpropylethylenediamine-tetraacetate-iron (II) footprinting studies," *Biochemistry* **32**:4237–4245 (1993).

26. Mrksich, M., Parks, M. E., and Dervan, P. B., "Hairpin peptide motif. A new class of oligopeptides for sequence-specific recognition in the minor groove of double-helical DNA," *J. Am. Chem. Soc.* **116**:7983–7988 (1994).

27. White, S., Szewczyk, J. W., Turner, J. M., Baird, E. E., and Dervan, P. B., "Recognition of the four Watson-Crick base pairs in the DNA minor groove by synthetic ligands," *Nature* **391**:468–471 (1998).

28. Chen, Y.-H., Yang, Y., and Lown, J. W., "Optimization of cross-linked lexitropsins," *J. Biomol. Struct. Dyn.* **14**:341–355 (1996).

29. Bhattacharya, S. and Thomas, M., "DNA binding properties of novel dansylated distamycin analogues in which the fluorophore is directly conjugated to the N-methyl-pyrrole carboxamide backbone," *J. Biomol. Struct. Dyn.* **19**:935–945 (2002).

30. Mikheikin, A. L., Nikitin, A. M., Strel'tsov, S. A., Leinsoo, T. A., Chekhov, V. O., Brusov, R. V., Zhuze, A. L., Gurskii, G. V., Shafer, R. and Zasedatelev, A. S., "Interaction of a furancarboxamide analog of antibiotic distamycin A with DNA," *Mol. Biol.* **31**:854–861 (1997).

31. Lown, J. W., Krowicki, K., Balzarini, J., Newman, R. A., and De Clercq, E., "Novel linked antiviral and antitumor agents related to netropsin and distamycin: Synthesis and biological exaluation," *J. Med. Chem.* **32**:2368–2375 (1989).
32. Ermishov, M., Sukhanova, A., Kryukov, E., Grokhovsky, S., Zhuze, A., Oleinikov, V., Jardillier, J. C., and Nabiev, I., "Raman and surface-enhanced Raman scattering spectroscopy of bis-netropsins and their DNA complexes," *Biopolym. (Biospectr.)* **57**:272–281 (2000).
33. Lee, M., Shea, R. G., Hartley, J. A., Kissinger, K., Pon, R. T., Vesnaver, G., Breslauer, K. J., Dabrowiak, J. C., and Lown, J. W., "Molecular recognition between oligopeptides and nucleic acids: sequence-specific binding of the naturally occurring antibiotic (4S)-(+)-anthelvencin A and its (4R)-(−) enantiomer to deoxyribonucleic acids deduced from ^{1}H NMR, footprinting, and thermodynamic data," *J. Am. Chem. Soc.* **111**:345–354 (1989).
34. Lee, M., Shea, R. G., Hartley, J. A., Lown, J. W., Kissinger, K., Dabrowiak, J. C., Vesnaver, G., Breslauer, K. J., and Pon, R. T., "Molecular recognition between oligopeptides and nucleic acids. Sequence specific binding of (4S)-(+)- and (4R)-(−)-dihydrokikumycin B to DNA deduced from ^{1}H NMR, footprinting studies and thermodynamic data," *J. Mol. Recogn.* **2**:6–17 (1989).
35. Debart, F., Perigaud, C., Gosselin, G., Mrani, D., Rayner, B., Le Ber, P., Auclair, C., Balzarini, J., De Clercq, E., Paoletti, C. and Imbach, J.-L., "Synthesis, DNA binding, and biological evaluation of synthetic precursors and novel analogues of netropsin," *J. Med. Chem.* **32**:1074–1083 (1989).
36. Teng, M.-K., Usman, N., Frederick, C. A., and Wang, A. H.-J., "The molecular structure of the complex of Hoechst 33258 and the DNA dodecamer d(CGCGAATTCGCG)," *Nucl. Acids Res.* **16**:2671–2690 (1988).
37. Sriram, M., van der Marel, G. A., Roelen, H. L. P. F., van Boom, J. H., and Wang, A. H.-J., "Conformation of B-DNA containing O^6-ethyl-G-C base pairs stabilized by minor groove binding drugs: Molecular structure of d(CGC[e^{6}G]AATTCGCG) complexed with Hoechst 33258 or Hoechst 33342," *EMBO J.* **11**:225–232 (1992).
38. Clark, G. R., Squire, C. J., Gray, E. J., Leupin, W., and Neidle, S., "Designer DNA-binding drugs: the crystal structure of a *meta*-hydroxy analogue of Hoechst 33258 bound to d(CGCGAATTCGCG)$_2$," *Nucl. Acids Res.* **24**:4882–4889 (1996).
39. Mann, J., Baron, A., Opoku-Boahen, Y., Johansson, E., Parkinson, G., Kelland, L. R., and Neidle, S., "A new class of symmetric bisbenzimidazole-based DNA minor groove-binding agents showing antitumor activity," *J. Med. Chem.* **44**:138–144. (2001).
40. Utsuno, K., Maeda, Y., and Tsuboi, M., "How and how much can Hoechst 33258 cause unwinding in a DNA duplex?" *Chem. Pharm. Bull.* **47**:1363–1368 (1999).
41. Utsuno, K., Tsuboi, M., Katsumata, S., and Iwamoto, T., "Visualization of complexes of Hoechst 33258 and DNA duplexes in solution by atomic force microscopy," *Chem. Pharm. Bull.* **50**:216–219 (2002).
42. Adnet, F., Liquier, J., Taillandier, E., Singh, M. P., Rao, K. E., and Lown, J. W., "FTIR study of specific binding interactions between DNA minor groove binding ligands and polynucleotides," *J. Biomol. Struct. Dyn.* **10**:565–575 (1992).
43. Chen, S.-M., Leupin, W., Rance, M., and Chazin, W. J., "Two-dimensional NMR studies of d(GGTTAAGCGT)·d(ACCGCATTAACC) complexed with the minor groove binding drug SN-6999," *Biochemistry* **31**:4406–4413 (1992).

44. Luck, G., Reinert, K.-E., Baguley, B., and Zimmer, Ch., "Interaction of the nonintercalative antitumour drugs SN-6999 and SN-18071 with DNA: Influence of ligand structure on the binding specificity," *J. Biomol. Struct. Dyn.* **4**:1079–1094 (1987).

45. Förtsch, I., Birch-Hirschfeld, E., Schütz, H., and Zimmer, Ch., "Different effects of nonintercalative antitumor drugs on DNA triple helix stability: SN-18071 promotes triple helix formation," *J. Biomol. Struct. Dyn.* **14**:317–329 (1996).

46. Squire, C. J., Clark, G. R., and Denny, W. A., "Minor groove binding of a bis-quaternary ammonium compound: The crystal structure of SN 7167 bound to d(CGCGAATTCGCG)$_2$," *Nucl. Acids Res.* **25**:4072–4078 (1997).

47. Gao, X. and Patel, D. J., "Solution Structure of the chromomycin-DNA complex," *Biochemistry* **28**:751–762 (1989).

48. Banville, D. L., Keniry, M. A. and Shafer, R. H., "NMR investigation of mithramycin A binding to d(ATGCAT)$_2$: A comparative study with chromomycin A$_3$," *Biochemistry* **29**: 9294–9304 (1990).

49. Gao, X., Mirau, P., and Patel, D. J., "Structure refinement of the chromomycin dimer-DNA oligomer complex in solution," *J. Mol. Biol.* **223**:259–279 (1992).

50. Sastry, M. and Patel, D. J., "Solution structure of the mithramycin dimer-DNA complex," *Biochemistry* **32**:6588–6604 (1993).

51. Majee, S., Sen, R., Guha, S., Bhattacharyya, D., and Dasgupta, D., "Differential interactions of the Mg^{2+} complexes of chromomycin A$_3$ and mithramycin with poly(dG-dC) · poly(dC-dG) and poly(dG) · poly(dC)," *Biochemistry* **36**:2291–2299 (1997).

52. Keniry, M. A., Banville, D. L., Simmonds, P. M., and Shafer, R., "Nuclear magnetic resonance comparison of the binding sites of mithramycin and chromomycin on the self-complementary oligonucleotide d(ACCCGGGT)$_2$; evidence that the saccharide chains have a role in sequence specificity," *J. Mol. Biol.* **231**:753–767 (1993).

53. Utsuno, K., Kojima, K., Maeda, Y., and Tsuboi, M., "The average unwinding angle of DNA duplex produced by the binding of chromomycin A$_3$," *Chem. Pharm. Bull.* **46**:1667–1671 (1998).

54. Streltsov, S., Sukhanova, A., Mikheikin, A., Grokhovsky, S., Zhuze, A., Kudelina, I., Mochalov, K., Oleinikov, V., Jardillier, J.-C., and Nabiev, I., "Structural basis of topotecan-DNA recognition probed by flow linear dichroism, circular dichroism and Raman spectroscopy," *J. Phys. Chem., B* **105**:9643–9652 (2001).

55. Patel, D. J., Kozlowski, S. A., Rice, J. A., Broka, C. and Itakura, K., "Mutual interaction between adjacent dG · dC actinomycin binding sites, and dA · dT netropsin binding sites on the self-complementary d(C-G-C-G-A-A-T-T-C-G-C-G) duplex in solution," *Proc. Natl. Acad. Sci. USA* **78**:7281–7284 (1981).

56. Patel, D. J., Pardi, A., and Itakura, K., "DNA conformation, dynamics, and interactions in solution," *Science* **216**:581–590 (1982).

57. Plouvier, B., Houssin, R., Hecquet, B., Colson, P., Houssier, C., Waring, M. J., Hénichart, J.-P., and Bailly, C., "Antitumor combilexin. A thiazole-containing analogue of netropsin linked to an acridine chromophore," *Bioconj. Chem.* **5**:475–481 (1994).

58. Bourdouxhe-Housiaux, C., Colson, P., Houssier, C., Waring, M. J., and Bailly, C., "Interaction of a DNA-threading netropsin-amsacrine combilexin with DNA and chromatin," *Biochemistry* **35**:4251–4264 (1996).

59. Helissey, P., Giorgi-Renault, S., Colson, P., Houssier, C., and Bailly, C., "Sequence-recognition and cleavage of DNA by a netropsin-phenazine-di-N-oxide conjugate," *Bioconj. Chem.* **11**:219–227 (2000).
60. Carrasco, C., Helissey, P., Haroun, M., Baldeyrou, B., Lansiaux, A., Colson, P., Houssier, C., Giorgi-Renault, S. and Bailly, C., "Design of a composite ethidium-netropsin-anilinoacridine molecule for DNA recognition," *ChemBiochem.* **4**:50–61 (2003).
61. Wilson, W. D., Barton, H. J., Tanious, F. A., Kong, S.-B., and Strekowski, L., "The interaction with DNA of unfused aromatic systems containing terminal piperazino substituents. Intercalation and groove-binding," *Biophys. Chem.* **35**:227–243 (1990).
62. Sandström, K., Wärmländer, S., Leijon, M., and Gräslund, A., "^{1}H NMR studies of selective interactions of norfloxacin with double-stranded DNA," *Biochem. Biophys. Res. Commun.* **304**:55–59 (2003).
63. Son, G. S., Yeo, J.-A., Kim, M.-S., Kim, S. K., Holmén, A., Åkerman, B. and Nordén, B., "Binding mode of norfloxacin to calf thymus DNA," *J. Am. Chem. Soc.* **120**:6451–6457 (1998).

CHAPTER 4

Covalent Bonding Drugs

A number of antitumor drugs inhibit DNA replication and transcription via *covalent bonding* to DNA bases. In this chapter, interactions of covalent bonding drugs (natural antibiotics as well as their synthetic analogs) with DNA are reviewed using selected examples. These drugs are also called *alkylating agents* because, on adduct formation, nucleotides are alkylated by drug molecules. Covalent bonding drugs containing metals such as cisplatin and its derivatives are discussed in Chapter 6.

4.1. (+)CC1065 AND RELATED DRUGS

(+)CC1065 is a chiral highly potent antitumor drug extracted from *Streptomyces zelensis*. As shown in Fig. 4.1, it consists of three subunits, two of which (**B** and **C**) are similar, and the remaining **A** subunit contains a pyrrolidine ring. When reacted with DNA, the C4 atom of the **A** subunit forms a *covalent bond* (C4–CH$_2$–N) with the N3 atom of adenine by opening the cyclopropane ring, as indicated by the arrows in the figure. According to X-ray analysis by Chidester et al. [1], the molecule is curved and somewhat twisted with hydrogen-bond donors and acceptors on the outer periphery of the curve and with a lipophilic surface on the inner periphery as seen in Fig. 4.1. These workers also studied its interaction with DNA by CD spectroscopy, and noted that the strongest interaction occurs with poly(dA-dT)$_2$.

Hurley and coworkers [2] compared the ^{13}C NMR sptectra of (+)CC1065 and its adenine adduct, and observed large downfield shifts of the ^{13}C signals of the drug atoms C4 (21.7 → 54.8 ppm) and C4a (21.8 → 40.2 ppm), as expected from the opening of the cyclopropane ring. This provided a definitive confirmation of the covalent bonding site mentioned above. Changes in chemical shifts of atoms C6a, C7, and C8 on adduct formation provided additional evidence. These workers obtained the adenine adduct by heating the (+)CC1065-DNA adduct at 100°C for 30 min.

According to Reynolds et al. [3], this adduct facilitates the thermally induced breakage of DNA between the deoxyribose at the covalently bound adenine site and the phosphate on the 3′ side. Sequence specificity studies on the (+)CC1065-DNA,

---

Drug–DNA Interactions: Structures and Spectra by Kazuo Nakamoto, Masamichi Tsuboi, and Gary D. Strahan
Copyright © 2008 John Wiley & Sons, Inc.

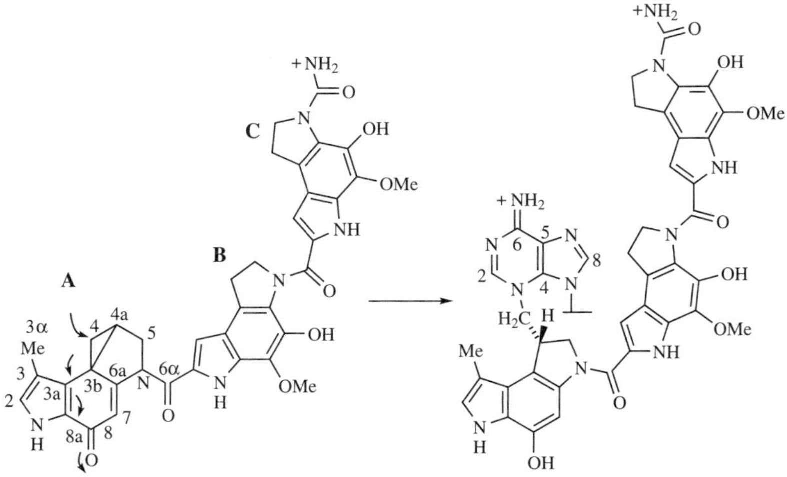

FIGURE 4.1. Structures of CC1065 and its adenine adduct. The natural drug assumes the (+) conformation at the C4a chiral center. The synthetic enantiomer, (−)-CC1065, has different biological activity. The cyclopropane ring in subunit **A** opens on adduct formation, according to the mechanism indicated by the arrows [2].

(SV40 fragments) show that the drug preferentially binds to the 5-bp sequences, 5′-AAAAA-3′ and 5′-PuNTTA-3′, where Pu is a purine base and N represents any base. Figure 4.2 illustrates a stereo view of a molecular model of (+)CC1065 docked to 5′-d(CGGAGTTAGG)-3′ sequence, which was constructed using the known atomic coordinates of both components. It was found that the drug molecule with the chiral configuration of C3b (R) and C4a (S) fits comfortably in the minor groove and spans over the 5-bp region mentioned above.

A more detailed 2D NMR study (^{13}C and ^{1}H) on the (+)CC1065-adenine adduct was carried out by Scahill et al. [4], who not only confirmed the previous results [2] but also extended their work to the adduct with the octamer:

```
              1   2   3   4   5    6   7   8
    5′-d(C   G   A   T   T   A*   G   C)-3′
    3′-d(G   C   T   A   A   T    C   G)-5′
             16  15  14  13  12   11  10  9
```

Here, A* indicates the alkylation site. Changes in chemical shift values on adduct formation and observed NOEs between the drug and the octamer indicate that the C4a atom of the drug is bonded to the N3 atom of A6 via the CH$_2$−N linkage, and that the drug molecule extends into the minor groove as illustrated in Fig. 4.3.

To determine the tautomeric form of the *covalently modified* adenine in the (+) CC1065-DNA adduct, Lin and Hurley [5] studied ^{1}H and ^{15}N NMR spectra of the adduct with d(GGCGGAGTTA*GG)·d(CCTAACTCCGCC). Here, A* denotes 6-^{15}N deoxyadenosine. The results show that the adenine in the adduct takes the amino form in which an extra positive charge on the N6 amino group is delocalized over the entire adenine and the NH$_2$−C6 bond takes a partial double-bond character as shown in Fig. 4.1.

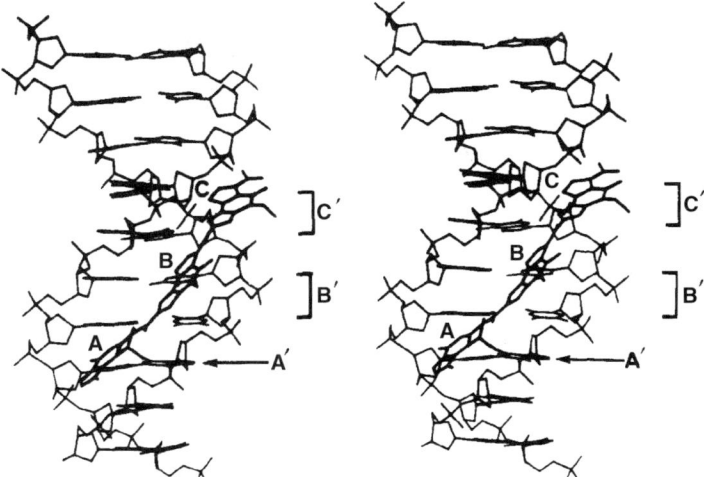

FIGURE 4.2. Stereo view of the CC1065-DNA adduct. The base-pair sequence from top to bottom is 5'-d(CGGAGTTAGG)-3'. Structures **A**, **B**, and **C** indicate the subunits of the drug shown in Fig. 4.1. Structure **A'** is the adenine covalent binding site, **B'** is a pair of highly conserved bases immediately to the 5' side of the adenine covalent binding site, and **C'** is a pair of less well conserved bases at the 5' end of the drug-binding site [3].

It was previously suggested [2] that the hydrophilic substituents on the outer periphery of the drug molecule might be hydrogen-bonded to the phosphate group of DNA via a layer of water molecules. On the basis of ^{1}H NMR studies in combination with ^{17}O-labeled 12-mer and ^{17}O-water, Lin et al. [6] proposed that a water molecule bridges between the 8-phenolic proton of the **A** subunit of the covalently bonded drug

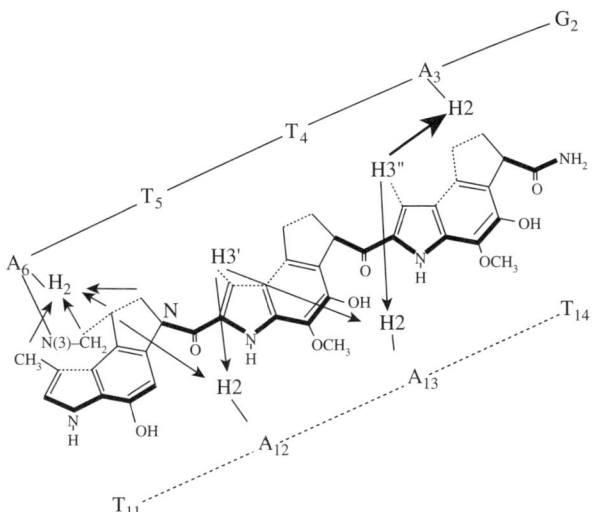

FIGURE 4.3. Disposition of the drug in the minor groove of the octamer duplex. Arrows indicate some observed NOEs between the drug and the duplex [4].

and an anionic oxygen of the phosphate of the 12-mer, and that another water molecule forms a bridge between one of the NH_2 protons of the covalently modified adenine and the C4=O group of the neighboring thymine. This is supported by the differences in T_1 relaxation time of these protons between normal water and 40.5% ^{17}O-enriched water.

Alkylation of guanine by (+)CC1065 is much slower than that of adenine. Park and Hurley [7,8] studied the ^{1}H NMR spectra of the adduct with the 12-mer in which G_9^*(N3) is alkylated:

```
        1   2   3   4   5   6   7   8   9   10  11  12
5'-d(G  C   G   C   A   A   T   T   G*  C   G   C)-3'
3'-d(C  G   C   G   T   T   A   A   C   G   C   G)-5'
        24  23  22  21  20  19  18  17  16  15  14  13
```

The results indicate the following: (1) the imino proton signal (N1) of the alkylated G_9^* is absent and the amino proton signals (N4) of base-paired C_{16} are downshifted by ~2.0 ppm; (2) the H5 and H6 protons of C_{16} also show large downfield shifts, suggesting that the C_{16} is in the protonated state; and (3) this protonated cytosine exhibits a single broad signal at 14.6 ppm at 5°C and 15°C that originates in the N3 imino proton. These workers proposed two possible mechanisms involving the G^- - C^+ ion pair [pathway (a)] and a concerted attack at the N3 of G with the transfer of the imino proton (N1) of G to N3 of C [pathway (b)] as illustrated in Fig. 4.4. Since protonated cytosine is deaminated to uracil (U) more easily than unprotonated cytosine, protonation of the cross-strand C_{16}, resulting from the alkylation of G_9, may cause a C → U transition mutation and contribute to the mutagenic and carcinogenic properties of these guanine-alkylating drugs.

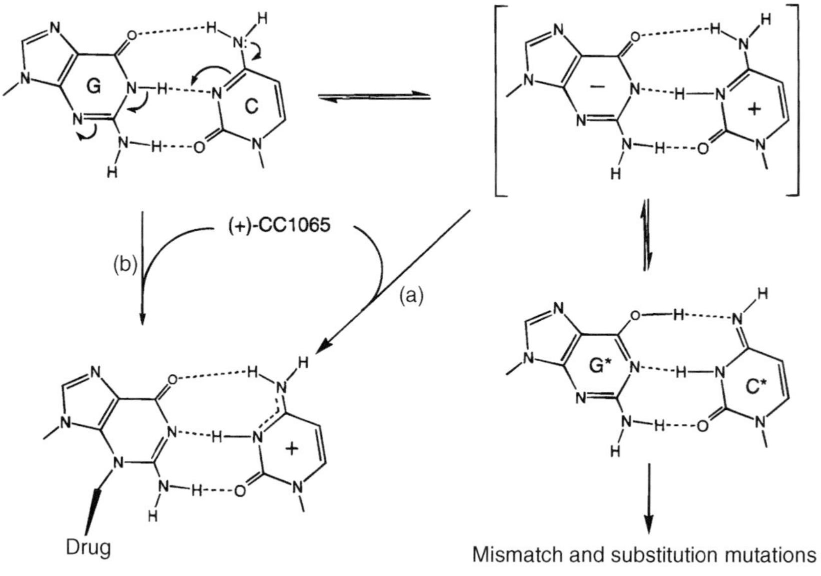

FIGURE 4.4. Proposed mechanisms for the alkylation of N3 guanine: (a) tautomerization of the G-C base pair to give the enol-G–imino-C tautomer; (b) a concerted mechanism [7].

Although (+)CC1065 is cytotoxic against several tumor cells, it also causes delayed death in mice and rabbits at less than therapeutic doses. For this reason, its clinical use was abandoned. To understand the origin of its cytotoxicity and to design better drugs, a number of derivatives have been synthesized and their cytotoxicities tested. For example, Li et al. [9] studied the structure–activity relationship in a series of (+)CC1065 derivatives in which the **A** subunit is unchanged and the **B-C** subunits are modified. In a series shown in Table 4.1, the size of the **B-C** subunit decreases in the order:

$$(+)CC1065 > U\text{-}68{,}415 > U\text{-}66{,}694 > U\text{-}68{,}819 > U\text{-}66{,}664$$

The same order was found in the strength of their interaction with CT-DNA measured by CD difference spectra (ΔCD) and the increase in their thermal melting points (ΔT_m). The ID_{50} values of these derivatives for L1210 leukemia cells also follow

TABLE 4.1. Structure–Activity Relationship of (+)CC1065 Derivatives

Structure (B-C subunit)	ΔCD^a	ΔT_m^b	ID^c
(+)CC1065	245	25	0.021~0.050
U-68,415	70	13.7	0.015 ± 0.005
U-66,694	36	10.2	0.41 ± 0.06
U-68,819	10	9	1.6
U-66,664	0	1.6	17.4 ± 1.6

[a] CD difference spectra (spectrum of drug–DNA minus that of drug) at molar ratio of DNA (as measured by phosphate)/drug = 13 in 0.01 M phosphate buffer (pH = 7.2). The drug concentration was 0.85×10^{-5} M.
[b] Difference in thermal melting point of CT-DNA with and without drug in 0.01 M phosphate buffer (pH = 7.2). Other conditions are the same as in footnote a.
[c] Drug concentration required for a 50% inhibition of cell growth. This value was determined by comparing the cell number of the control to that of drug-treated sample after incubation at 37°C for 3 days.
Source: Ref. 9.

the same order. Thus, the biological activities of these derivatives (racemic) are controled not only by the **A** subunit containing a cyclopropane ring but also by the structure of the **B-C** subunits.

CPI-CDPI$_2$ is a derivative of (+)CC1065 (Fig. 4.1) where the OH and OCH$_3$ groups of the **B-C** subunits are all replaced by hydrogen atoms. The solution structure of the adduct of CPI-CDPI$_2$ with the self-complementary decamer.

```
          1   2   3   4   5   6   7   8   9   10
  5'-d(C  G   C   T   T   A   A   G   C   G)-3'
  3'-d(G  C   G   A*  A   T   T   C   G   C)-5'
         20  19  18  17  16  15  14  13  12   11
```

was elucidated by combining 2D ^{1}H and ^{31}P NMR spectroscopy with model building [10]. Formation of covalent bonding between the drug and the N3 atom of A$_{17}$ was confirmed by the observation that the H8 (0.5 ppm), H3' (0.54 ppm) and H4' (0.85 ppm) signals of A$_{17}$ are all upshifted. In addition, aromatic proton signals of the neighboring nucleotide bases, A$_{16}$ and G$_{18}$, are upshifted by 0.2 ppm. Since this drug binds assymmetrically to the symmetric oligonucleotide, it causes a doubling of most of the DNA resonances in the NMR spectrum. Molecular modeling studies suggest that the duplex is kinked by ~60° at the alkylating site and that **A** and **B** subunits lie in the minor groove slightly rotated from an edge-on to a partial face-on orientation. This kinking probably inhibits the binding of the second drug molecule at the A$_7$ site.

Similar to (+)CC1065, (+)-duocarmycin A (DA) and (+)-duocarmycin SA (DSA), shown in Fig. 4.5, contain a cyclopropane ring that alkylates the N3 atom of adenine in the minor groove. Duocarmycins are known to bind preferentially to 5'-(A/T)-A-A-A* -3' and 5'-(A/T)-T-T-T-A*-Pu –3' sequences. With this in mind, Lin and Patel [11] studied the solution structure of DA bound to the hairpin nucleotide:

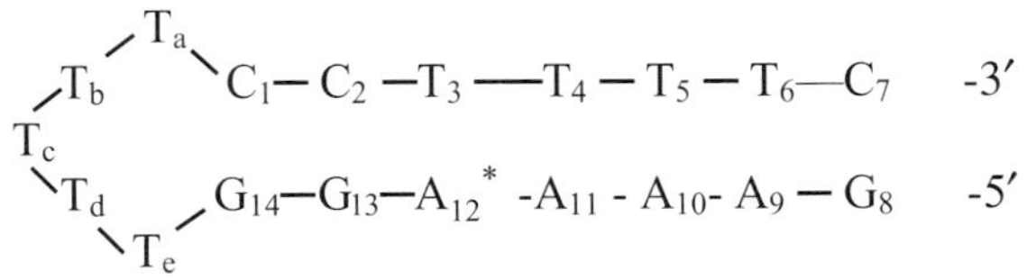

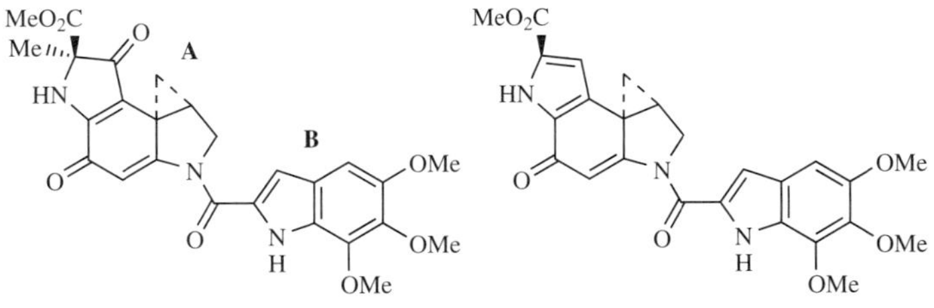

$$\begin{array}{l} \quad\quad T_a \\ \quad T_b \diagdown C_1-C_2-T_3-T_4-T_5-T_6-C_7 \quad -3' \\ T_c \\ \quad\diagdown T_d \diagup G_{14}-G_{13}-A_{12}{}^* -A_{11}-A_{10}-A_9-G_8 \quad -5' \\ \quad\quad\diagdown T_e \end{array}$$

FIGURE 4.5. Structures of (+)-duocarmycin A and (+)-Duocarmycin SA. Cyclopropane rings are indicated by dotted lines. The chiral center is at C4a atom [12].

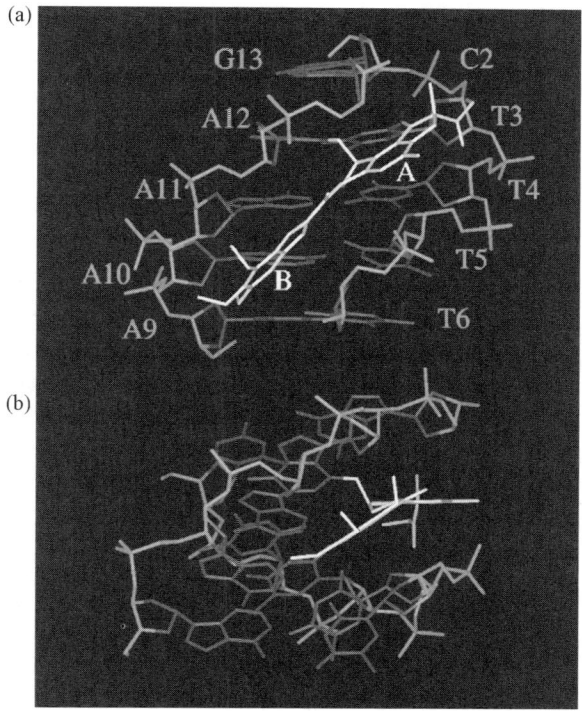

FIGURE 4.6. Two views of a relaxation matrix refined structure of duocarmycin–hairpin duplex complex. The A and B rings of the drug are shown in yellow, and the $d(C_2\text{-}T_2\text{-}T_4\text{-}T_5\text{-}T_6)$ $d(A_9\text{-}A_{10}\text{-}A_{11}\text{-}A_{12}^*\text{-}G_{13})$ segment is shown in magenta with the backbone in blue. The drug forms a covalent bond to the N3 atom of A_{12}. (a) View looking into the minor groove and normal to the helix axis; (b) view looking down the minor groove, indicating that the entire drug molecule is within the wall of the minor groove [11]. (See color insert.)

by 2D NMR spectroscopy. Covalent binding (Fig. 4.1) at A_{12} is supported by (1) the upfield shift (2.0 ppm) of the imino proton of T_3, (2) the downfield shift (0.87 ppm) of the imino proton of T_4, (3) weak NOEs between T_3 (imino proton) and A_{12} (amino proton), and (4) the upfield shifts of ^{31}P signals at $A_{11}\text{-}A_{12}$ (5.29 ppm), $A_{12}\text{-}G_{13}$ (5.05 ppm), $T_4\text{-}T_5$ (5.29 ppm), and $T_5\text{-}T_6$ steps (4.70 ppm). Molecular dynamics studies including NOE-based intensity refinements lead to a molecular model shown in Fig. 4.6; the **A** and **B** rings are located deep in the minor groove with the **B** ring directed toward the 5′ end of the modified strand. DA takes an extended conformation and its long axis aligned $\sim 45°$ relative to the helix axis with its nonpolar edges sandwiched within the walls of the minor groove. The **A** ring spans over the $C_2\text{-}T_3\text{-}T_4$ and $G_{13}\text{-}A_{12}\text{-}A_{11}$ segments, while the **B** ring spans over $T_4\text{-}T_5\text{-}T_6$ and $A_{11}\text{-}A_{10}\text{-}A_9$ segments. The whole duplex still maintains B-DNA conformation, although $A_{12}^*\text{-}T_3$ pairing is weak relative to other base pairings.

As seen in Fig. 4.5, the differences between DA and DSA are the lack of the methyl group at the C2 and the carbonyl group at the C3 position in the latter. Thus, similar structures are expected for their DNA adducts in the AT-rich region. The structure of the DSA adduct with the 11-mer

```
        1   2   3   4   5   6   7   8   9   10  11
5'-d(G  A   C   T   A   A   T   T   G   A   C)-3'
3'-d(C  T   G   A*  T   T   A   A   C   T   G)-5'
        22  21  20  19  18  17  16  15  14  13  12
```

was determined by 2D NMR spectroscopy with high resolution [12]. In this case, the drug is covalently bonded to the N3 atom of A_{19}. Although the overall structure is similar to that of the DA-hairpin nucleotide adduct [11], significant differences are noted in the minor groove width at the binding site, reflecting the difference in the hybridization at the C2 atom of the drug. The sequence selectivities of DA and DSA are similar, and the order of preference in the 3-bp sequence is 5'-AAA* > 5'-TTA* > 5'-TAA* > 5'-ATA*. In addition, A or T is prefered over G or C as the fourth 5' base. A similar order also holds for (+)CC1065. The sequence selectivities of these drugs, their synthetic enantiomers, and the relationships between structure, functional reactivity, and biological properties have been reviewed by Boger and Johnson [13].

DSI is an indole analog of DSA that lacks three methoxy groups on the **B** ring of DSA. The structure of the DSI adduct with the same 11-mer as that used for the DSA adduct [12] was determined by ^{1}H NMR spectroscopy [14]. In the free state, the **A** and **B** subunits of these drugs (Fig. 4.5) are coplanar. On alkylation, however, they are twisted, and the relative twist angle of the two subunits is 45° for DSA and 37° for DSI. The larger the twist, the more disruption of the conjugate system. As a result, DSA is more reactive, and its rate of alkylation reaction is faster than that of DSI.

Although DA alone cannot alkylate guanine, DA with distamycin A (Dist) (Section 3.1) can cooperatively cause alkylation of guanine residues in GC-rich sequences [15] A detailed NMR study on such "concerted DNA recognition" was carried out by Sugiyama et al. [16], who determined the structure of the ternary alkylation complex, DA/Dist/octamer, by 2D NMR spectroscopy. Here, the octamer is

```
        1   2   3   4   5   6   7   8
5'-d(C  A   G   G   T   G*  G   T)-3'
3'-d(G  T   C   C   A   C   C   A)-5'
        16  15  14  13  12  11  10  9
```

Marked shifts of the imino protons of T_5 (upfield to 13.2 ppm) and G_6 (downfield to 14.2 ppm) confirmed the formation of a *covalent bond* between the C4 atom of DA and the N3 atom of G_6, which fixes the position of DA in the minor groove. The position of Dist was determined by intermolecular NOEs between Dist protons and DNA sugar protons, and the relative positions of DA and Dist were derived from the

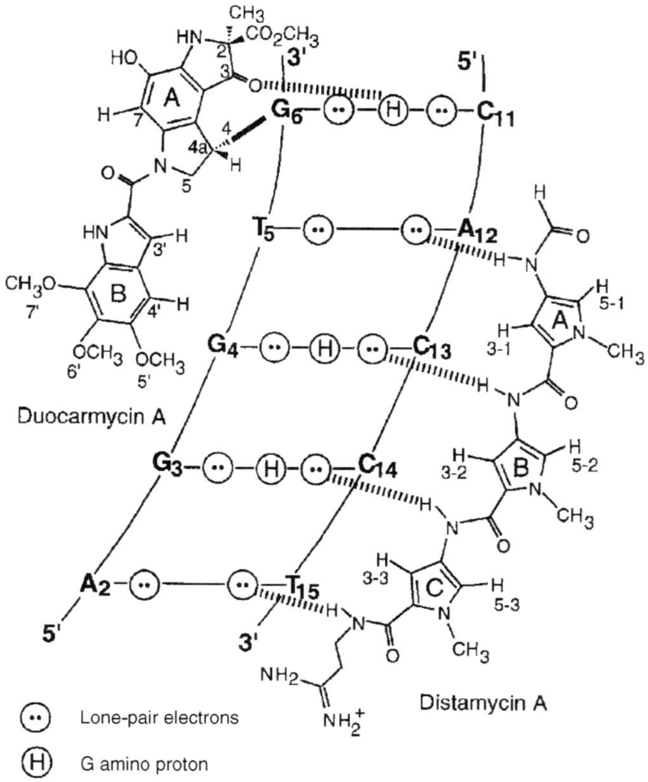

FIGURE 4.7. Schematic drawing of the heterodimeric binding of duocarmycin A and distamycin to the minor groove of d(CAGGTGGT)$_2$. Hydrogen bonds between the drugs and the octamer are indicated by dashed lines. Hydrogen bonding between the O3 of duocarmycin A and the N2 amino group of G$_6$ helps in anchoring the drug in its position (bold face line) [16].

NOEs between these two drug molecules. Figure 4.7 illustrates the side-by-side orientation of the two drugs, which is stabilized by five hydrogen bonds. This structure is responsible for the preferential binding of the DA-Dist pseudodimer to the 5′-AGGTG-3′ sequence.

Another derivative of (+)CC1065 is bizelesin. It consists of two **A** subunits of (+) CC1065 that are connected by a rigid linker, and binds to two adenine bases on opposite strands as shown in Fig. 4.8. The solution structure of its adduct with the decamer

	1	2	3	4	5	6	7	8	9	10	
5′-d(C	G	T	A	A	T	T	A*	C	G)-3′
3′-d(G	C	A*	T	T	A	A	T	G	C)-5′
	20	19	18	17	16	15	14	13	12	11	

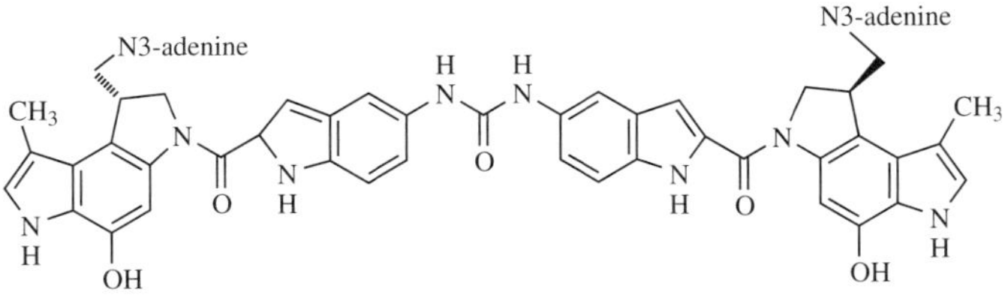

254 COVALENT BONDING DRUGS

FIGURE 4.8. Structure of bizelesin bonded to adenines [17].

was determined by ^{1}H NMR spectroscopy [17]. In this case, covalent crosslinking occurs between A_8 and A_{18}, and results in two major adducts that differ in the central 5′-AATT-3′ region; one (major) adduct contains an ApT step in which both adenines are *syn*-oriented and the A-T base-pairing is of Hoogsteen type, whereas the other contains *anti*-oriented ApT-step adenines that show no evidence of hydrogen bonding with pairing thymines.

4.2. ANTHRAMYCIN AND TOMAYMYCIN

Figure 4.9 shows the strucures of the well-known antitumor antibiotics, anthramycin and tomaymycin, which were isolated from various strains of *Streptomyces* and belong to the family of chemicals classified as pyrrolo[1,4]-benzodiazepines. These drugs form *covalent* bonds between their C11 atom and the exocyclic NH_2 group of guanine residue in the minor groove. The reaction is slow, and the suggested mechanism [18] involves the imine intermediate shown in Fig. 4.9. Kopka et al. [19] carried out an

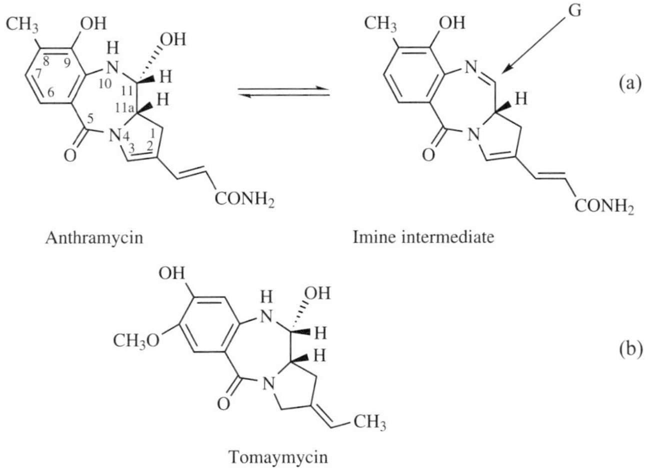

FIGURE 4.9. Structures of anthramycin and tomaymycin [19].

X-ray analysis of anthramycin bonded to the self-complementary decamer in a 2 : 1 (drug/decamer) ratio:

```
        1   2   3   4   5   6   7   8   9   10
5'-d(C  C   A   A   C   G   T   T   G*  G)-3'
3'-d(G  G*  T   T   G   C   A   A   C   C)-5'
        20  19  18  17  16  15  14  13  12  11
```

where the exocyclic NH$_2$ (N2 amine) group of G$_9$* is covalently bonded to the C11 atom of the drug. Two drug molecules are bonded to the penultimate guanines in the minor groove, and their acrylamide tails extend toward the center of the helix. The 3D structure of one end of the adduct is shown in Fig. 4.10. The drug molecule takes the C11(*S*), C11a (*S*) chiral conformation, and the CH(drug)–NH$_2$ (guanine) covalent bond is reinforced by hydrogen bonding (shown by dotted lines) from C9-OH, N10-NH, and the NH$_2$ group of the acrylamide tail of the drug to the base pairs of the decamer.

Graves et al. [20] studied the interaction of anthramycin methyl ether (C11-OCH$_3$) with d(ATGCAT)$_2$ by ^{13}C NMR spectroscopy. These workers substituted the C11 atom of the drug by ^{13}C, and observed the upfield shift of this ^{13}C signal by 16 ppm on addition of the hexamer duplex, as expected for the formation of a covalent bond involving the

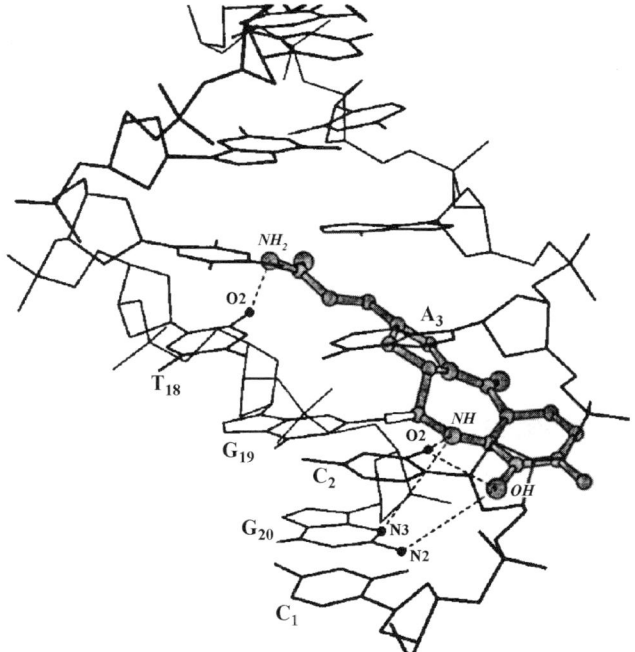

FIGURE 4.10. Three-dimensional structure of one end of the anthramycin–decamer adduct. Hydrogen bonds between the drug and the decamer are indicated by dotted lines [19].

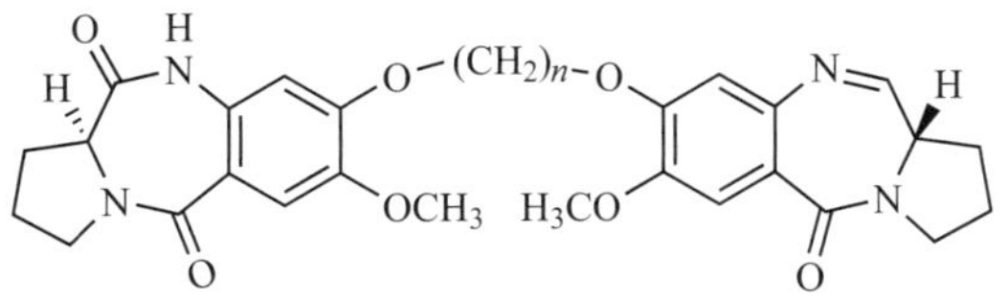

FIGURE 4.11. Structures of pyrrolobenzodiazepine dimers [24].

C11 atom. Since this bonding occurs only with G_3 of the hexamer and not with G_9 of its complementary strand, several nucleotide signals are doubled because of the loss of helical C_2 symmetry. The NH proton signal of the C11-N2 amine (G_3) linkage was observed between 7.2 and 7.0 ppm at 20–40°C. This work was followed by a 2D NOE study [21] that revealed the orientation of the drug molecule in the minor groove.

The structure of the 2:1 complex of tomaymycin with d(CICG*AATTCICG)$_2$ (I: inosine) was determined by 2D NMR, molecular modelling, and fluorescence spectroscopy [22]. The overall structure is similar to that of the 2:1 complex of anthramycin with d(CCAACGTTG*G)$_2$ described previously [19]. Thus the C11(S) atoms of two drug molecules are covalently bonded to the exocyclic NH$_2$ groups of guanines (G_4* and G_{16}*) of the 12-mer, and their aromatic rings are oriented to the 3′ side of G* in the minor groove. Structure–activity relationship studies on a number of derivatives of anthramycin and tomaymycin revealed that structural modification of the A ring does not improve the cytotoxicities of these drugs [23].

Kamal et al. [24] synthesized a series of pseudodimeric pyrrolobenzodiazepines (shown in Fig. 4.11) and compared the DNA binding ability and antitumor activity of these drugs ($n = 3,4,5,8$). They observed that, when $n = 5$, the melting point of CT-DNA shows the largest increase, suggesting that $n = 5$ stabilizes the DNA adduct the most. Also cytotoxicity studies against various human tumor cell lines show that $n = 5$ exhibits an interesting profile of activity and selectivity. Molecular modeling studies on the adducts of these drugs with the 15-mer DNA containing a central AGA as the preferred binding site indicate that further increase in the linker length decreases the stability. In general, their stabilities are governed by van der Waals and Coulombic interactions together with covalent bonding between the drug and the guanine base.

4.3. ECTEINASCIDINS

Ecteinascidins (Ets) are antitumor drugs isolated from marine tunicate *Ecteinascida turbinata*. As shown in Fig. 4.12a, these drugs consist of three subunits, with C subunit attached to the rigid bis(tetrahydroisoquinoline) A-B subunit "scaffold" via a flexible 10-membered lactone ring. Guan et al. [25] carried out X-ray analyses on the N12-formyl derivative of Et729 and the N12-oxide of Et743, and determined the absolute configurations of chiral centers; C1(R), N2(R), C3(R), C4 (R), C11(R), C13(S), C21(S), and C22(R). The asymmetric unit cell of Et contains two independent molecules, and Fig. 4.12b shows the 3D structure of one of them for Et729.

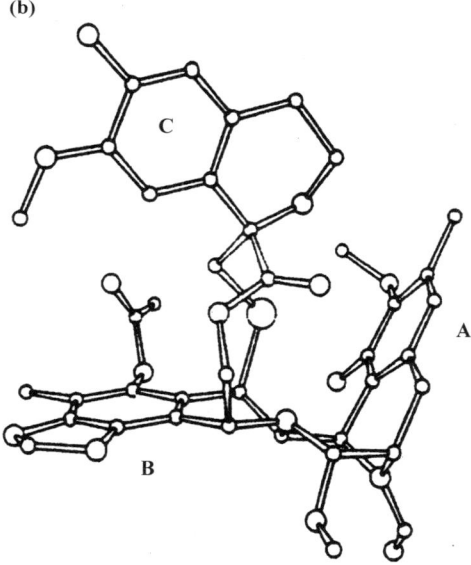

FIGURE 4.12. (a) Structures of Et729 ($R_1 = OCH_3$, $R_2 = CHO$) and Et743 ($R_1 = OH$, $R_2 = CH_3$); (b) configuration of one of the two independent molecules of Et729 [25].

Similar to anthramycin, Et's react with DNA slowly to form a covalent bond between the C21 atom of the drug and the N2 (amine) of guanine in the minor groove. Figure 4.13 shows a computer-generated stereo view of the adduct of Et729 with d(TTGG*GAA)·d(AACCCTT) obtained by these workers. The adduct is stabilized by extensive hydrogen bonding between the three rings of Et and base pairs in the minor groove.

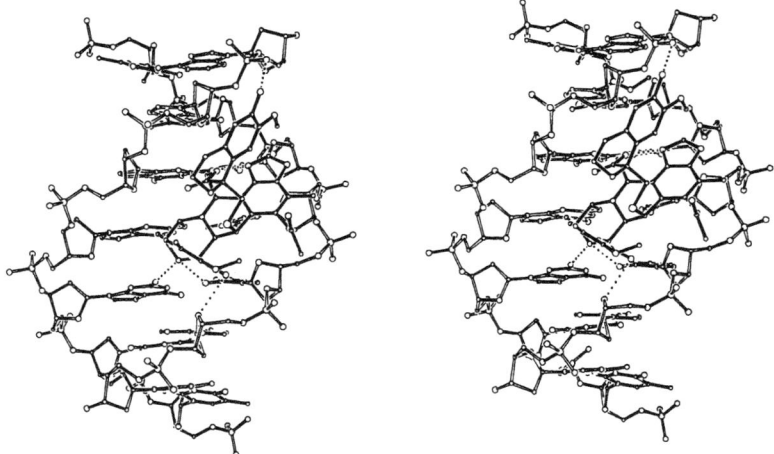

FIGURE 4.13. Stereo view of a molecular model of Et729- d(TTGG*GAA)·d(TTCCCAA) adduct. Hydrogen bonding between the drug and the heptamer is shown by dotted lines [25].

To investigate the DNA sequence selectivity of Et743 and Et736 (the C subunit of Et743 is replaced by tetrahydro-β-carboline), Hurley and coworkers [26,27] carried out ^{1}H NMR studies on the interactions of these drugs with the 12-mer

```
          1    2    3    4    5    6    7    8    9   10   11   12
5'-d(C    G    T    A    A    G*   C    T    T    A    C    G)-3'
3'-d(G    C    A    T    T    C    G    A    A    T    G    C)-5'
         24   23   22   21   20   19   18   17   16   15   14   13
```

and three derivatives in which the central 5'-AG*C-3' sequence was replaced by 5'-CG*G-3', 5'-GG*G-3' and 5'-AG*T-3' sequences. Formation of covalent bonding between the C21 atom of Et and the NH$_2$ group of G$_6$ is manifested by the splitting of proton signals of A$_5$, G$_6^*$, C$_7$, and T$_8$ into two sets corresponding to the complexed and uncomplexed strands. NOESY cross-peak studies on the Et736-12-mer adduct revealed that signals of B and C subunits are connected to those of G$_6^*$, C$_7$, and T$_8$, and signals of A and B subunits are connected to those of T$_{21}$, T$_{20}$, and C$_{19}$. Similar results were obtained for the Et743 adduct. Both drugs have the identical A and B subunits, which are the major sites of interactions with the 12-mer. To substantiate the possibility of hydrogen bonding between the drug molecule and the 12-mer, these workers assigned the signals of exchangeable protons attached to the hydrogen-bond donors and other protons near the proposed acceptors. Their molecular dynamics studies based on the NMR data revealed that the A-B scaffold of Et743 forms hydrogen bonds with the bases in the 5'-XGY-3' region as shown in Fig. 4.14. It was found that the sequences 5'-AGC-3' and 5'-CGG-3', which show strong reactivity, can take favorable positions for maximizing the possibility of hydrogen bonding. On the other hand, the sequences 5'-GGG-3' and 5'-AGT-3', which show moderate and poor reactivities, respectively, take less favorable positions for forming hydrogen bonding with these

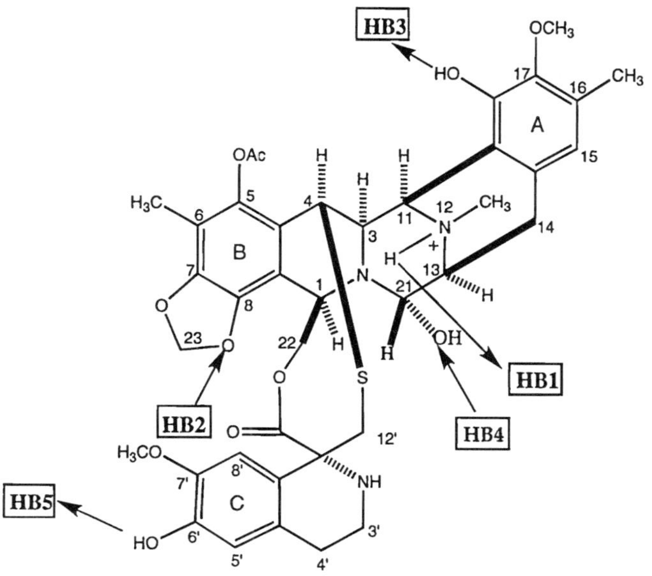

FIGURE 4.14. Hydrogen-bonding interaction (HB1 to HB5) between Et743 and the 12-mer. Direction of arrow indicates the path from donor to acceptor. HB1 to $T_{20}(O2)$, HB2 from $G_{18}(N2)$, HB3 to $T_{20}(O3')$, HB4 from $G_6(N2)$, and HB5 to phosphate oxygen (O1) of T_9 [26].

bases. A more extensive *sequence selectivity* study of Et743 involving 16 different 5'-XGY-3' sequences was made using DNA footprinting techniques [28]. The results show that, in the 5'-XGG-3' series, the efficiency of Et743-induced alkylation is in the order of X = C > T > G > A.

4.4. MITOMYCINS

Mitomycins extracted from *Streptomyces caespitosus* are antitumor antibiotics currently in clinical use. Several review articles are available on mitomycins [G4–G6]. The structure of a typical mitomycin, mitomycin C (MC), is shown in Fig. 4.15 (structure **a**). It contains an aziridine ring, an indoloquinone ring, and a carbamate group. Thus far, over 1000 analogs have been synthesized to improve its clinical properties. Unlike other drugs, mitomycins are reductive alkylating agents. Thus, a one-electron reduction produces a monofunctional adduct, whereas a two-electron reduction yields a bifunctional adduct with DNA (Fig. 4.15, structures **d** and **f**, respectively). In the former, a covalent bond is formed between the C1 atom of the drug and the N2 amine group of guanine, whereas in the latter, N2 atoms of two guanines on complementary strands are crosslinked via the C1 and C10 atoms of the drug. It is known that the first alkylation step prefers guanine at d(CpG) sequence over other G-containing sequences. The reduction is carried out chemically or enzymatically, and, in the case of $Na_2S_2O_4$ reduction, the distribution of mono- and bifunctional adducts depends on the initial conditions used. Figure 4.15 illustrates a probable mechanism of

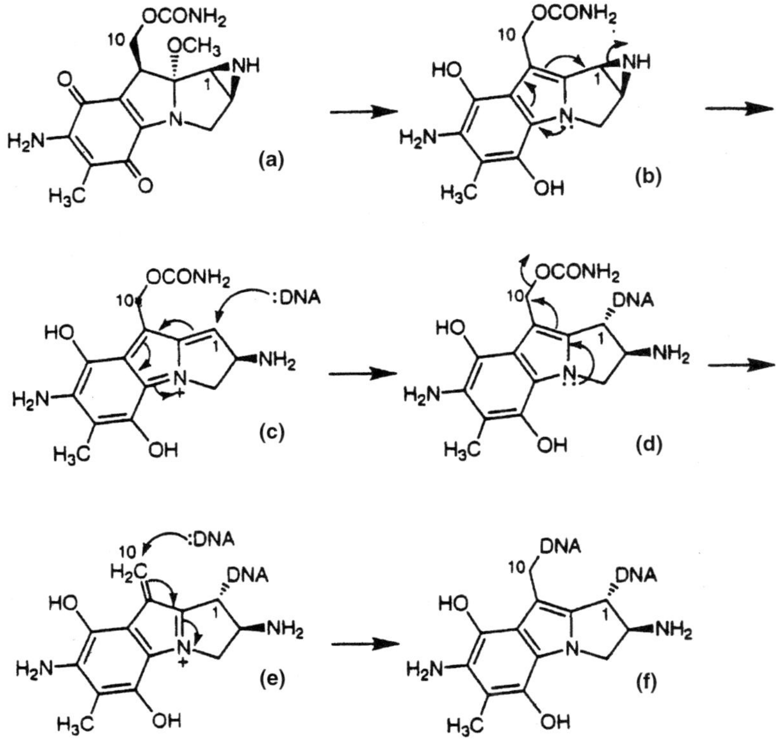

FIGURE 4.15. Probable mechanism of alkylation to form mono- and bifunctional adducts by reduced mitomycin C [G5].

alkylation that is triggered by the reduction of the quinone ring [G5,29]. A more detailed mechanism was proposed later [30].

Sastry et al. [31] determined the structure of the monoalkylated adduct of MC with the 9-mer duplex

```
         1   2   3   4   5   6   7   8   9
5'-d(I   C   A   C   G*  T   C   I   T)-3'
3'-d(C   G   T   G   C   A   G   C   A)-5'      (I : inosine)
         18  17  16  15  14  13  12  11  10
```

by 2D ^{1}H and ^{31}P NMR spectroscopy. Figure 4.16 illustrates the structure of the mono adduct of MC with guanine (G_5). Here, the C atom positions of MC are marked by double primes to distinguish them from those of the sugar (single primes).

The atomic positions of MC in the minor groove were determined by NOEs between MC and the 9-mer. Thus, the location of the five-membered ring bonded to guanine was determined by the NOEs between MC (H1″, H2″, and H3″$_{a,b}$) and the nucleotide (G_5, T_6, A_{13}, G_{15}), and the location of the quinone ring was determined by the NOEs between MC (CH_3-6″, NH_2-7″) and the nucleotide (G_{15}, T_{16}). Finally, the

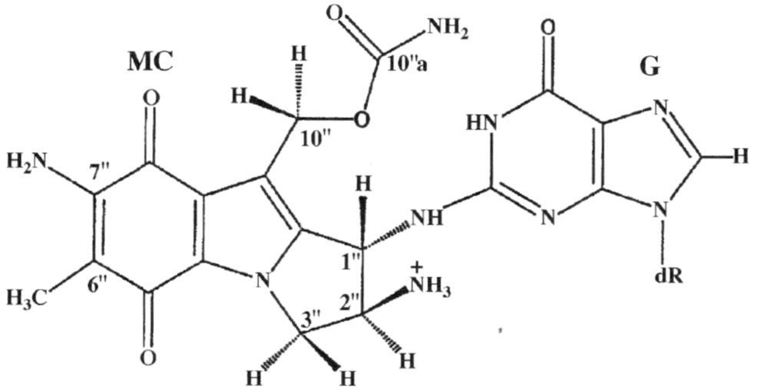

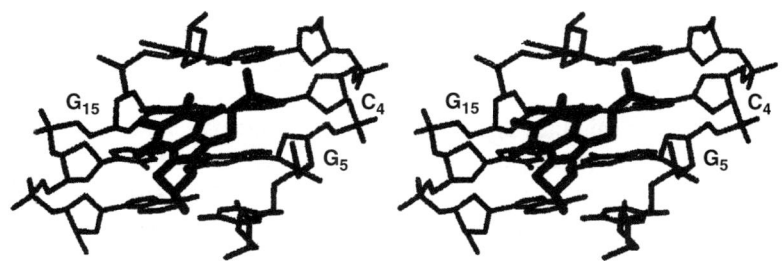

FIGURE 4.16. Structure of monoalkylation adduct of MC with guanine [31].

carbamate group was positioned by the NOEs between MC (H10″$_{a,b}$) and G$_5$ (H1′). The stereo view of a molecular model of the adduct in the central region shown in Fig. 4.17 was derived based on changes in chemical shifts due to adduct formation and NOEs between the drug and the 9-mer signals mentioned above. It is seen that the aromatic ring of the drug is parallel to the sugar–phosphate backbone, the puckered five-membered ring connected to the indoloquinone is centered near the G$_5$-C$_{14}$ base pair, and the indoloquinone ring is directed toward the C$_4$-G$_{15}$ base pair. The carbamate oxygen is close to the NH$_2$ group of G$_{15}$, suggesting the possibility of *interstrand crosslinking*.

Two-dimensional NMR (^{1}H and ^{31}P) spectroscopy combined with minimized potential energy calculations [32] was employed to elucidate the structure of the bifunctional MC adduct with the hexamer containing the d(CpG) sequence on both strands:

$$\begin{array}{cccccc}
1 & 2 & 3 & 4 & 5 & 6 \\
5'\text{-d(T} & A & C & G^* & T & A)\text{-}3' \\
3'\text{-d(A} & T & G^* & C & A & T)\text{-}5' \\
12 & 11 & 10 & 9 & 8 & 7
\end{array}$$

In this case, MC is crosslinked via the N2 amine groups of G$_4$ and G$_{10}$. The results of computational studies suggest that MC is crosslinked in a widened minor groove with its chromophore ring near the G$_{10}$*-T$_{11}$ sequence.

FIGURE 4.17. Stereo view of the MC-9-mer adduct in the central region. The drug is alkylated at the N2 of G$_5$ [31,G6].

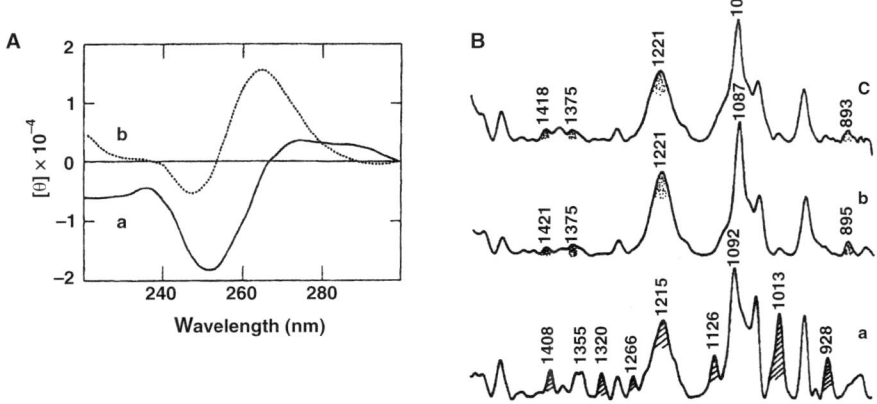

FIGURE 4.18. (A) CD spectra of (a) poly(dG-dC)$_2$ and (b) poly(dG-dC)$_2$ modified by MC; (B) FTIR spectra of (a) Z form, (b) B form of poly(dG-dC)$_2$, and (c) poly(dG-dC)$_2$ modified by MC. Bands characteristic of B and Z conformations are dotted and shaded, respectively [33].

Jolles et al. [33] studied the effect of MC on the conformation of poly(dG-dC)$_2$ by several spectroscopic methods. Figure 4.18A compares the CD spectrum of the polymer in its B form (trace **a**) with that modified by adding MC at a high binding ratio (trace **b**). The latter CD spectrum resembles that of Z-DNA (Fig. 6.38). Previously, Tomasz et al. [34] made several observations indicating that this is not the B → Z conversion. For example, the ethanol-induced B → Z transition is not facilitated but rather inhibited by MC, probably as a result of the interstrand crosslinking. These workers attributed the observed CD modification to a drug-induced left-handed but non-Z conformational change or to the superposition of an induced CD on that of B-DNA resulting from drug–base electronic interaction.

Additional evidence to support the Jolles group's view [33] was provided by FTIR spectra shown in Fig. 4.18B. It is seen that none of the bands characteristic of the Z form (trace **a**) appear in the spectrum of poly(dG-dC)$_2$ modified by MC (trace **c**). In fact, the latter is identical to the spectrum of the B form (trace **b**). Figure 4.19 shows the UV resonance Raman (RR) spectra obtained by 257 nm excitation. Since both the polymer and MC have strong electronic absorption bands in this region, the vibrations originating in these chromophores are resonace-enhanced with this excitation line. In Fig. 4.19, trace **a** is the RR spectrum of the polymer modified by MC, and trace **b** is the RR spectrum of MC metabolites obtained from the Na$_2$S$_2$O$_4$ reductive system. The bands due to the metabolites appear as shoulder bands near 1486 and 1584 cm^{-1}. Trace **c** is the difference spectrum obtained by subtracting trace **b** from trace **a**. Thus, trace **c** represents the spectrum of the polymer modified by MC. The main differences between trace **c** and the unmodified polymer (trace **d**) are the increase in intensity at 1642 cm^{-1} and the decrease in intensity at 1486 cm^{-1}. The intensity ratio of these two bands (I_{1486}/I_{1642}) decreases as a result of covalent bonding of MC to the N2 atom of guanine. The 1642 cm^{-1} band is the only vibration [C2=O stretching] of cytosine of

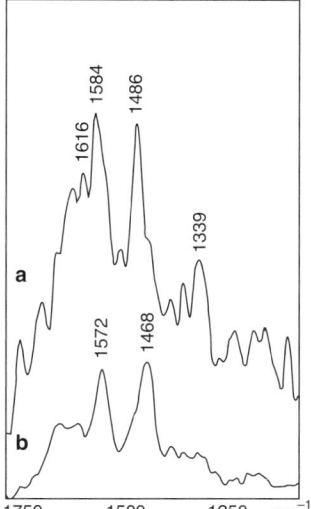 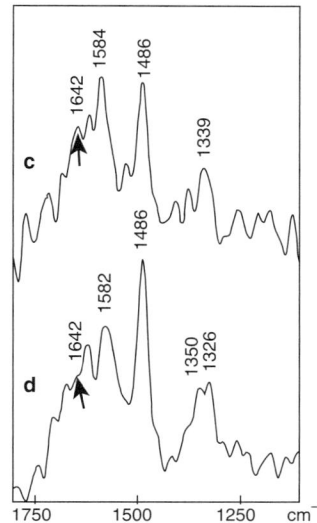

FIGURE 4.19. Resonance Raman spectra (257 nm excitation): (**a**) poly(dG-dC)$_2$ modified by MC, (**b**) MC metabolites, (**c**) subtraction of (**b**) from (**a**), and (**d**) poly(dG-dC)$_2$ [33].

poly(dG-dC)$_2$ observed by 257 nm excitation, and appears only in double-stranded polynucleotides. Its enhancement has been attributed to an interbase vibrational coupling of the carbonyl groups, which could be correlated to increased rigidity of the duplex caused by interstrand crosslinking [33].

Porfiromycin is a derivative of MC (Fig. 4.15, structure **a**) in which the NH group of the aziridine ring of MC is replaced by the N-CH$_3$ group, and its biological activity is known to be lower than that of MC. According to the X-ray analysis by Arora et al. [35], the overall conformation of porfiromycin is similar to those of other mitomycins. Thus, the five-membered ring attached to the benzoquinone ring is approximately planar and the other five-membered ring adopts an envelope conformation. However, they are different in the orientation of the aziridine ring and carbamate sidechain. Molecular modeling studies show that monofunctional binding prefers the d(CpG) over the d(GpC) sequence, whereas bifunctional binding has no clear-cut preference. The latter result may suggest that bifunctional binding is kinetic rather than thermodynamic.

2,7-Diaminomitosene (2,7-DAM) shown in Fig. 4.20a is the major metabolite of MC and the structure of its adduct with the self-complementary 12-mer:

```
         1   2   3   4   5   6   7   8   9  10  11  12
5'-d(G   T   G   G*  T   A   T   A   C   C   A   C)-3'
3'-d(C   A   C   C   A   T   A   T   G*  G   T   G)-5'
```

was determined by 2D NMR spectroscopy [36]. It is a 2 : 1 (2 drug/1 duplex) complex in which the C10″ atom of the drug is covalently bonded to the N7 atom

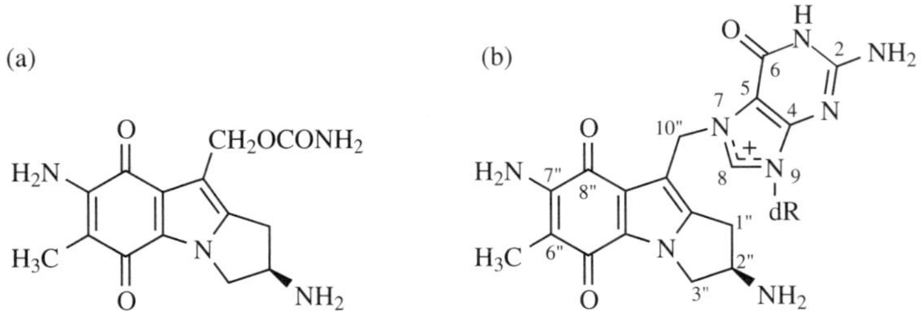

FIGURE 4.20. Structures of 2,7-diaminomitosene(2,7-DAM) (a) and its covalent adduct with guanine (b) [36].

of guanine (G^*) (Fig. 4.20b) so as to retain the overall C_2 symmetry. The unique feature of this adduct is that 2,7-DAM is anchored in the major groove, which may be responsible for its low cytotoxicity relative to other MC derivatives.

As stated earlier, mitomycins exhibit no appreciable binding properties to DNA unless they are reductively activated. In contrast, 2,7-DAM can bind to DNA without reductive activation. This was confirmed by Kumar et al. [37] using UV spectroscopy. On titration with CT-DNA, the UV spectra of MC show almost no changes (Fig. 4.21a) whereas the bands of 2,7-DAM at ∼313 and ∼350 nm exhibit large hypochromic and relatively small bathochromic effects (Fig. 4.21b). The binding parameters of 2,7-DAM with three natural DNA with varying GC contents as well as poly(dA-dT)$_2$ and poly(dG-dC)$_2$ were calculated from the *Scatchard plots* of these titration experiments under different conditions, and no apparent binding specificity was noted. Studies on pH dependence indicate that the binding affinity involves a strong electrostatic component. The results of viscosity measurements show conclusively that 2,7-DAM intercalates into DNA in a nonspecific manner.

The structure of mitomycin A (MA) is similar to that of MC (Fig. 4.15, structure **a**); the only difference is that the amino group attached to the benzoquinone ring of MC is replaced by the methoxy (OCH$_3$) group. MA is more cytotoxic by two or three orders of magnitude than MC, and this trend was correlated with the higher redox potential of MA (−0.19 V) relative to MC (−0.40 V) [38]. It was suggested that the higher cytotoxicity of MA is due to its propensity for reductive activation by cellular thiols, whereas MC is resistant to thiol activation.

4.5. INTERCALATING ALKYLATORS

Structural and spectroscopic studies on intercalating drugs such as adriamycin (doxorubicin) and Daunomycin (daunorubicin) have been discussed in Chapter 2. Incorporation of an alkylating agent into these drugs may enhance their antitumor properties. One of such intercalating alkylators is a natural antitumor antibiotic, hedamycin, shown in Fig. 4.22. It consists of a planar anthrapyrantrione chromophore (A-B-C-D ring) to which one bis-epoxide chain and two aminosugar rings (E, F) are

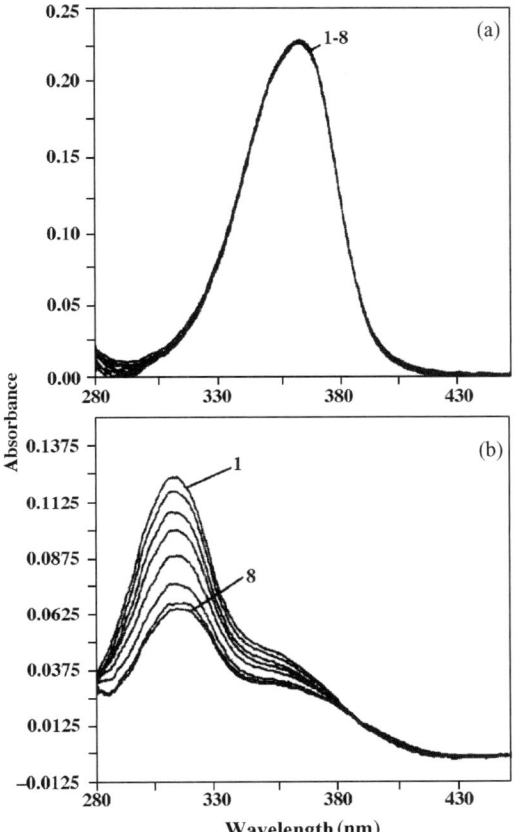

FIGURE 4.21. Changes in UV absorption spectra of mitomycin (a) and 2,7-DAM (b) on titration with CT-DNA in citrate–phosphate buffer, $[Na^+] = 10$ mM, pH $= 6.05 \pm 0.05$. In (a), curves 1–8 are the spectra of 10.11 µM of MC on titration with 0, 7.3, 30, 60, 120, 240, 479, and 720 µM of CT-DNA, respectively. In (b), curves 1–8 are the spectra of 11.9 µM of 2,7-DAM on titration with 0, 7.1, 30.0, 59.5, 119, 238, 476, and 714 µM of CT-DNA, respectively [37].

attached. Using 2D ^{1}H NMR spectroscopy. Pavlopoulos et al. [39] determined the structure of the adduct of hedamycin with the hexamer duplex:

```
         1    2    3    4    5    6
   5'-d(C    A    C    G    T    G)-3'
   3'-d(G    T    G*   C    A    C)-5'
        12   11   10   9    8    7
```

The absence of NOEs between C_3 and G_4 and between G_{10} and C_9 signals indicates that the drug molecule is intercalated between C_3-G_{10} and G_4-C_9 base pairs. Since hedamycin contains two epoxide rings, probable sites of covalent bonding are at C16, C17, and C18. On adduct formation, however, H18 showed a much larger change in

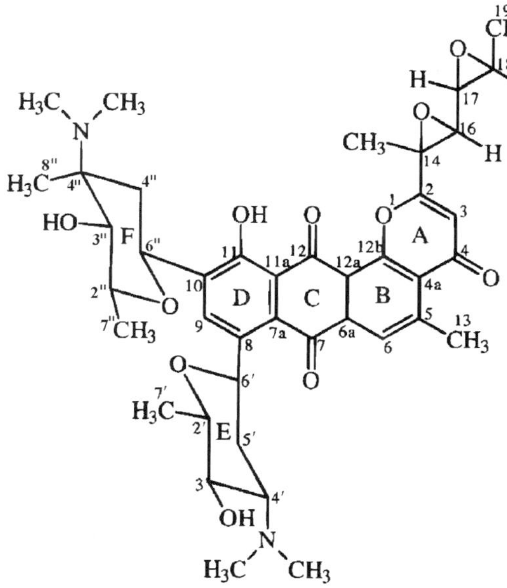

FIGURE 4.22. Structure of hedamycin [39].

chemical shift (~2 ppm) than did H16 and H17 (less than 0.5 ppm). It is known from previous studies that the N7 atom of a guanine residue is involved in covalent bonding. The disappearance of intraresidue NOEs of G_{10} between H8 and H1' (also H2', H2") on adduct formation suggested G_{10} to be the site of alkylation. Thus, the intercalated hedamycin molecule forms a covalent bond between its C18 atom and the N7 atom of G_{10}.

The orientation of the drug molecule in the duplex was elucidated largely on the basis of the following information. As illustrated in Fig. 4.23, the intermolecular NOEs are observed between B ring signals (H6 and $CH_3$13) and those of C_3, G_4, and G_{10}. Thus, B ring must be stacked between these bases. The NOEs between the B-ring H6 and other protons of the C_3 sugar moiety indicate that the edge of the chromophore is opposite to the bis-epoxide and the F ring is closer to the nonmodified strand than to the alkylated strand. The NOEs from $CH_3$13 of the drug to H5 of C_3 and the absence of any NOEs from $CH_3$13 to the minor groove protons indicate that one end of the chromophore (A-B) protrudes into the major groove as shown in the figure. The presence of NOEs between H9 (of the D ring) and G_{10}(H5', H5") and between H9 and T_{11} (H4') (not shown for clarity in Fig. 4.23) suggests that the D ring is in the minor groove and located closer to the modified strand than to the nonmodified strand. Since the B ring is close to the modified strand, the long axis of the drug molecule is not perpendicular to the long axis of the flanking base pairs but forms an angle with that of the C_3-G_{10} base pair plane. The locations of sugar rings (E and F) in the duplex are defined by NOEs between their signals and the hexamer duplex; the E ring is orientated in the 3'-direction while the F ring is orientated in the 5' direction.

A stereo view of a molecular model based on these NOEs and distance-restrained molecular dynamics is shown in Fig. 4.24. The main conclusions are that (1) the

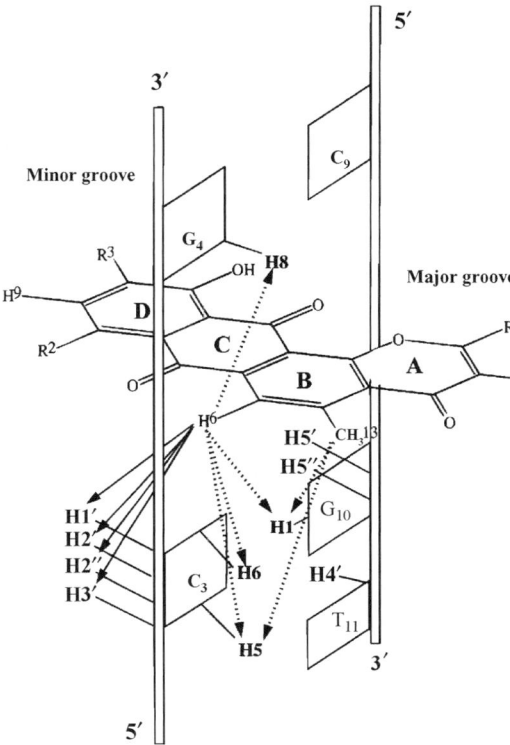

FIGURE 4.23. Schematic representation of intermolecular NOEs between hedamycin and the hexamer. The left and right bars represent nonmodified and alkylated strands, respectively. Contacts to major groove protons are shown by dotted arrows, and those to the minor groove protons are shown by solid-line arrows. For the purpose of clarity, contacts of the H9 proton of the D-ring with G_{10}(H5′,H5″) and T_{11}(H4′) are not shown [39].

intercalation site is slightly wedge-shaped, (2) the C18 atom of the drug assumes the R conformation, (3) the torsional angles around the C8-C6′ (E-ring sugar) and C10-C6″ (F-ring sugar) bonds are significantly large to bring the E ring toward the 3′ side and the F ring toward the 5′ side of the alkylation site, and (4) a higher degree of

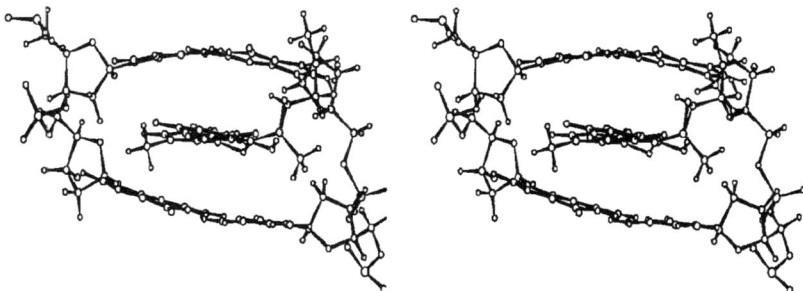

FIGURE 4.24. Stereo view of the intercalation site of hedamycin–hexamer adduct. The drug takes a wedge-shaped conformation at the binding site [39].

complementarity is found between the walls and floor of the minor groove and the orientation of the sugar group for 5′-CpG-3′ than for the 5′-TpG-3′ sequence. The latter result suggests that this complementarity is a factor to determine the sequence selectivity of hedamycin.

Altromycin B is another intercalating alkylator that belongs to the same pluramycin family of antibiotics as hedamycin. It has the same anthrapyrantrione chromophore as hedamycin, but its peripheral substituents at C2, C5, C8, and C10 are different from those of hedamycin. The altromycin B-guanine adduct was isolated from CT-DNA after thermal depurination of the alkylated DNA, and the covalent bonding site was elucidated from ^{1}H and ^{13}C NMR studies by Sun et al. [40]. These workers proposed a general reaction mechanism for the drugs of this type; the drug first intercalates into DNA, followed by insertion of the disaccharide into the minor groove and positioning the epoxide in the major groove near the N7 atom of guanine. Then, nucleophilic attack from the N7 atom leads to an acid-catalyzed opening of the expoxide ring, resulting in *covalent bonding* between the C16 and N7(G) atoms.

Anthracycline antiobiotics such as daunorubicin (DAU) and doxorubicin (DOX), shown in Fig. 4.25, become intercalating alkylators in the presence of formaldehyde (HCHO). Wang and coworkers [41,42] carried out X-ray analysis (1.5 Å resolution) to determine the crystal structures of the adducts of DAU and DOX with hexamers. In the 2 : 1 adduct of DAU with the hexamer duplex

$$
\begin{array}{cccccccc}
 & 1 & 2 & 3 & 4 & 5 & 6 & \\
5'\text{-d(C} & G & C' & G^* & C & G)\text{-}3' \\
3'\text{-d(G} & C & G^* & C' & G & C)\text{-}5' \\
 & 12 & 11 & 10 & 9 & 8 & 7 & \\
\end{array}
$$

(where C′ denotes arabinosylcytosine, shown in Fig. 4.25), the tetracycline chromophore of DAU is intercalated between C_5 and G_6 and a covalent bond is formed between the 3′-NH$_2$ group of the amino sugar of DAU and the 2-NH$_2$ of G_4, with the

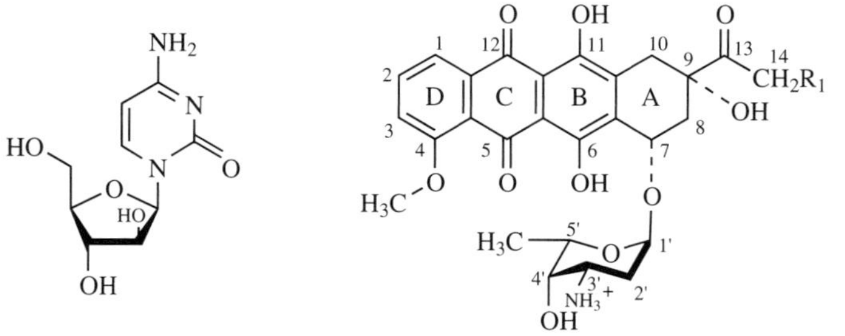

FIGURE 4.25. Structures of daunorubicin, doxorubicin, and arabinosylcytosine [41].

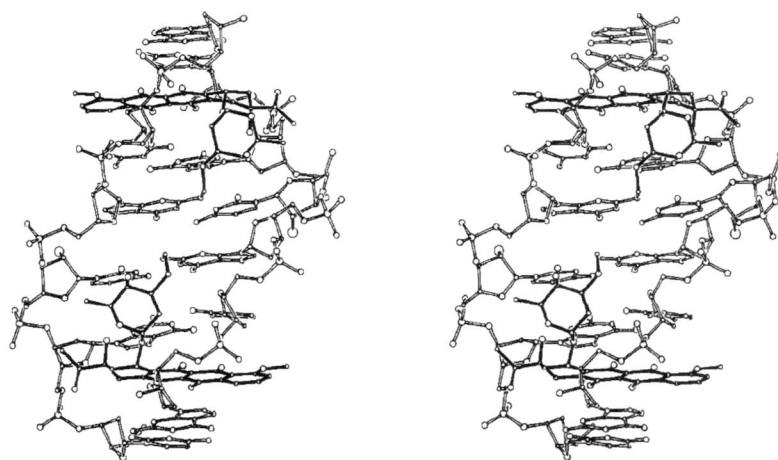

FIGURE 4.26. Stereo view of the 2:1 complex of daunorubicin–hexamer along the C_2 symmetry axis from the minor groove side [41].

CH_2 group from HCHO serving as a bridge ($NH-CH_2-NH$ bonding; see Fig. 4.27). Another DAU molecule is bonded in the same fashion, resulting in the overall structure of C_2 symmetry. As seen in the stereo view of the 3D structure shown in Fig. 4.26, the orientation of the tetracycline ring is skewed with respect to the neighboring base pair, with the D ring reaching the major groove and the sugar moiety lying in the minor groove. The 9-OH group (A ring) of DAU is hydrogen-bonded to the $N8(G_2)$ and $N2(G_8)$ atoms. All the sugars are $C2'$-*endo*, and the glucosidic bonds have *anti*-conformation as found in B-DNA. The structure of the 2:1 adduct of DOX with $d(CAC'GTG)_2$ is similar to that of the DAU adduct. In this case, one drug molecule is intercalated at the $5'$-T_5G_6-$3'$ sequence and covalently bonded to G_4, while another molecule is bonded to the hexamer so as to maintain the overall C_2 symmetry.

Zeman et al. [43] prepared the 1:1 adduct of doxorubicin with the hexamer

```
         1   2   3   4   5   6
5'-d(A   T   G*  C   A   T)-3'
3'-d(T   A   C   G   T   A)-5'
        12  11  10   9   8   7
```

via HCHO-mediated reaction and oxidoreductive (Fenton) reaction in Tris buffer, and confirmed that the adducts obtained by these two different reactions take an identical structure [44]. Detailed structural information on the adduct was obtained by 2D NMR spectroscopy [43]. The absence of NOEs between C_4 and A_5 and between G_9 and T_8 base pairs indicates the intercalation site of the drug in the duplex. As shown in Fig. 4.27, a covalent bond is formed between the $3'$-NH_2 (sugar) of the drug and 2-NH_2 of G_3 via the CH_2 bridge. This is definitively confirmed by the NMR spectra of isotopically labeled adducts. Thus, the adduct obtained by using H ^{13}CHO shows NOEs between ^{13}C and each of its geminal protons, which, in turn, show NOEs with

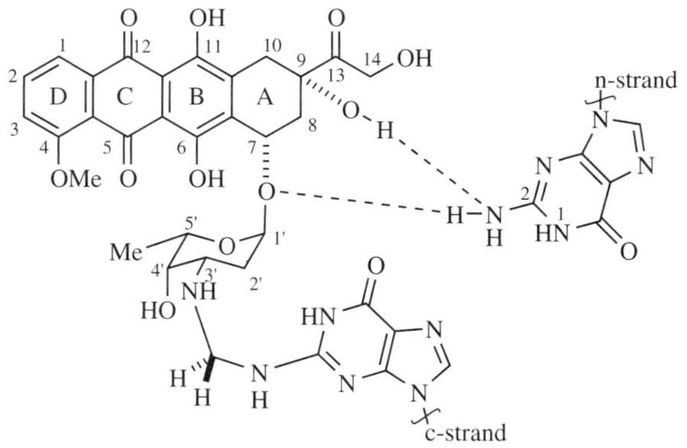

FIGURE 4.27. Modes of bonding between DOX(adriamycin) and duplex DNA. The drug is covalently bonded to the "c-strand" and noncovalently bonded to the "n-strand." The covalent methylene linkage between the drug and c-strand of guanine derives from formaldehyde. Rings B–D are intercalated into the duplex from the minor groove [43].

$5'$-CH_3 of the drug and protons of C_{10} (2'H, 6H) and A_{11} (2H, 8H). These NOEs disappear in the adducts obtained with HCHO and DCDO.

The structure of another natural anthracycline antibiotic, SN07, complexed to an unusual DNA triple helix at acidic pH, was determined by NMR spectroscopy [44]. Figure 4.28 shows the structures of SN07, the triple helix, and the mode of binding. The

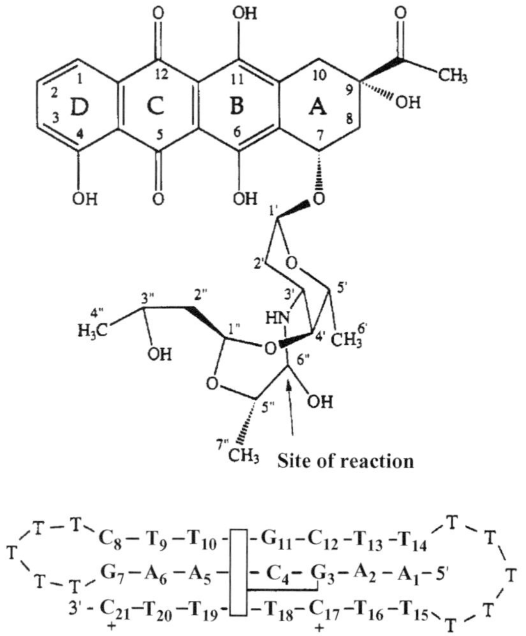

FIGURE 4.28. Structures of SN07 (anthracycline) and its bonding scheme to a DNA triple helix [44].

aglycone chromophore is intercalated between the T_{19}-A_5-T_{10} and T_{18}-C_4-G_{11} triples, and the covalent bond is formed between the NH_2 group of G_3 and the aminosugar residue of the drug. The covalent adduct was formed without disruption of the third strand since downfield-shifted amino protons from protonated C_{21} and C_{17} characteristic of an intact triplex were observed.

REFERENCES

1. Chidester, C. G., Krueger, W. C., Mizsak, S. A., Duchamp, D. J., and Martin, D. G., "The structure of CC-1065, a potent antitumor agent, and its binding to DNA," *J. Am. Chem. Soc.* **103**:7629–7635 (1981).
2. Hurley, L. H., Reynolds, V. L., Swenson, D. H., Petzold, G. L., and Scahill, T. A., "Reaction of the antitumor antibiotic CC-1065 with DNA: Structure of a DNA adduct with DNA sequence specificity," *Science* **226**:843–844 (1984).
3. Reynolds, V. L., Molineux, I. J., Kaplan, D. J., Swenson, D. H., and Hurley, L. H., "Reaction of the antitumor antibiotic CC-1065 with DNA. Location of the site of thermally induced strand breakage and analysis of DNA sequence specificity," *Biochemistry* **24**:6228–6237 (1985).
4. Scahill, T. A., Jensen, R. M., Swenson, D. H., Hatzenbuhler, N. T., Petzold, G., Wierenga, W., and Brahme, N. D., "An NMR study of the covalent and noncovalent interactions of CC-1065 and DNA," *Biochemistry* **29**:2852–2860 (1990).
5. Lin, C. H. and Hurley, L. H., "Determination of the major tautomeric form of the covalently modified adenine in the (+)-CC-1065-DNA adduct by ^{1}H and ^{15}N NMR studies," *Biochemistry* **29**:9503–9507 (1990).
6. Lin, C. H., Beale, J. M., and Hurley, L. H., "Structure of the (+)-CC-1065-DNA adduct: Critical role of ordered water molecules and implications for involvement of phosphate catalysis in the covalent reaction," *Biochemistry* **30**:3597–3602 (1991).
7. Park, H.-J. and Hurley, L. H., "Covalent modification of N3 of guanine by (+)-CC-1065 results in protonation of the cross-strand cytosine," *J. Am. Chem. Soc.* **119**:629–630 (1997).
8. Park, H.-J., "Evidence for a common molecular basis for sequence recognition of N3-guanine and N3-adenine DNA adducts involving the covalent bonding reaction of (+)-CC-1065," *Arch. Pharm. Res.* **25**:11–24 (2002).
9. Li, L. H., Wallace, T. L., DeKoning, T. F., Warpehoski, M. A., Kelly, R. C., Prairie, M. D., and Krueger, W. C., "Structure and activity relationship of several novel CC-1065 analogs," *Investig. New Drugs* **5**:329–337 (1987).
10. Powers, R. and Gorenstein, D. G., "Two-dimensional ^{1}H and ^{31}P NMR spectra and restrained molecular dynamics. Structure of a covalent CPI-CDPI$_2$-oligodeoxyribonucleotide decamer complex," *Biochemistry* **29**:9994–10008 (1990).
11. Lin, C. H. and Patel, D. J., "Solution structure of the covalent duocarmycin A-DNA duplex complex," *J. Mol. Biol.* **248**:162–179 (1995).
12. Eis, P. S., Smith, J. A., Rydzewski, J. M., Case, D. A., Boger, D. L., and Chazin, W. J., "High resolution solution structure of a DNA duplex alkylated by the antitumor agent duocarmycin SA," *J. Mol. Biol.* **272**:237–252 (1997).

13. Boger, D. L. and Johnson, D. S., "CC-1065 and the duocarmycins: Unraveling the keys to a new class of naturally derived DNA alkylating agents," *Proc. Natl. Acad. Sci. USA* **92**:3642–3649 (1995).
14. Schnell, J. R., Ketchem, R. R., Boger, D. L., and Chazin, W. J., "Binding-induced activation of DNA alkylation by duocarmycin SA; insights from the structure of an indole derivative-DNA adduct," *J. Am. Chem. Soc.* **121**:5645–5652 (1999).
15. Yamamoto, K., Sugiyama, H., and Kawanishi, S., "Concerted DNA recognition and novel site-specific alkylation by duocarmycin A with distamycin A," *Biochemistry* **32**:1059–1066 (1993).
16. Sugiyama, H., Lian, C., Isomura, M., Saito, I., and Wang, A. H.-J., "Distamycin A modulates the sequence specificity of DNA alkylation by duocarmycin A," *Proc. Natl. Acad. Sci. USA* **93**:14405–14410 (1996).
17. Seaman, F. C. and Hurley, L., "Interstrand cross-linking by bizelesin produces a Watson-Crick to Hoogsteen base-pairing transition region in d(CGTAATTACG)$_2$," *Biochemistry* **32**:12577–12585 (1993).
18. Lown, J. W. and Joshua, A. V., "Antitumor antibiotics. XVI. Molecular mechanism of binding of pyrrolo(1,4)benzodiazepine antitumor agents to deoxyribonucleic acid. Anthramycin and tomaymycin," *Biochem. Pharmacol.* **28**:2017–2026 (1979).
19. Kopka, M. L., Goodsell, D. S., Baikalov, I., Crzeskowiak, K., Cascio, D., and Dickerson, R. E., "Crystal structure of a covalent DNA-drug adduct: Anthramycin bound to C-C-A-A-C-G-T-T-G-G and a molecular explanation of specificity," *Biochemistry* **33**:13593–13610 (1994).
20. Graves, D. E., Pattaroni, C., Krishnan, B. S., Ostrander, J. M., Hurley, L. H., and Krugh, T. R., "The reaction of anthramycin with DNA. Proton and carbon nuclear magnetic resonance studies on the structure of the anthramycin-DNA adduct," *J. Biol. Chem.* **259**:8202–8209 (1984).
21. Graves, D. E., Stone, M. P., and Krugh, T. R., "Structure of the anthramycin-d(ATGCAT)$_2$ adduct from one- and two-dimensional proton NMR experiments in solution," *Biochemistry* **24**:7573–7581 (1985).
22. Boyd, F. L., Stewart, D., Remers, W. A., Barkley, M. D., and Hurley, L. H., "Characterization of a unique tomaymycin-d(CICGAATTCICG)$_2$ adduct containing two drug molecules per duplex by NMR, fluorescence, and molecular modeling studies," *Biochemistry* **29**:2387–2403 (1990).
23. Thurston, D. E., Bose, D. S., Howard, P. W., Jenkins, T. C., Leoni, A., Baraldi, P. G., Guiotto, A., Cacciari, B., Kelland, L. R., Foloppe, M.-P., and Rault, S., "Effect of A-ring modifications on the DNA-binding behavior and cytotoxicity of pyrrolo[2,1-c][1,4]benzodiazepines," *J. Med. Chem.* **42**:1951–1964 (1999).
24. Kamal, A., Ramesh, G., Laxman, N., Ramulu, P., Srinivas, O., Neelima, K., Kondapi, A. K., Sreenu, V. B., and Nagarajaram, H. A., "Design, synthesis, and evaluation of new noncrosslinking pyrrolobenzodiazepine dimers with efficient DNA binding ability and potent antitumor activity," *J. Med. Chem.* **45**:4679–4688 (2002).
25. Guan, Y., Sakai, R., Rinehart, K. L., and Wang, A. J.-H., "Molecular and crystal structures of ecteinascidins: Potent antitumor compounds from the Caribbean tunicate *Ecteinascidia turbinata*," *J. Biomol. Struct. Dyn.* **10**:793–818 (1993).
26. Seaman, F. C. and Hurley, L. H., "Molecular basis for the DNA sequence selectivity of ecteinascidin 736 and 743: Evidence for the dominant role of direct readout via hydrogen bonding," *J. Am. Chem. Soc.* **120**:13028–13041 (1998).

27. Moore, B. M., Seaman, F. C., and Hurley, L. H., "NMR –based model of an ecteinascidin 743-DNA adduct," *J. Am. Chem. Soc.* **119**:5475–5476 (1997).
28. Pommier, Y., Kohlhagen, G., Bailly, C., Waring, M., Mazumder, A., and Kohn, K. W., "DNA sequence- and structure-selective alkylation of guanine N2 in the DNA minor groove by ecteinascidin 743, a potent antitumor compound from the caribbean tunicate *Ecteinascidia turbinate*," *Biochemistry* **35**:13303–13309 (1996).
29. Iyer, V. N. and Szybalski, W., "Mitomycins and porfiromycin: Chemical mechanism of activation and cross-linking of DNA," *Science* **145**:55–58 (1964).
30. Tomasz, M., Chawia, A. K., and Lipman, R., "Mechanism of monofunctional and bifunctional alkylation of DNA by mitomycin C," *Biochemistry* **27**:3182–3187 (1988).
31. Sastry, M., Fiala, R., Lipman, R., Tomasz, M., and Patel, D. J., "Solution structure of the monoalkylated mitomycin C-DNA complex," *J. Mol. Biol.* **247**:338–359 (1995).
32. Norman, D., Live, D., Sastry, M., Lipman, R., Hingerty, B. E., Tomasz, M., Broyde, S., and Patel, D. J., "NMR and computational characterization of mitomycin cross-linked to adjacent deoxyguanosines in the minor groove of the d(T-A-C-G-T-A)·d(T-A-C-G-T-A) duplex," *Biochemistry* **29**:2861–2875 (1990).
33. Jolles, B., Laigle, A., Liquier, J., and Chinsky, L., "Evaluation of the structural modifications induced by mitomycin C on nucleic acids," *Biophys. Chem.* **46**:179–185 (1993).
34. Tomasz, M., Barton, J. K., Magliozzo, C. C., Tucker, D., Lafer, E. M., and Stollar, B. D., "Lack of Z-DNA conformation in mitomycin-modified polynucleotides having inverted circular dichroism," *Proc. Natl. Acad. Sci. USA* **80**:2874–2878 (1983).
35. Arora, S. K., Cox, M. B., and Arjunan, P., "Structural, conformational, and theoretical binding studies of antitumor antibiotic porfiromycin (N-methylmitomycin C), a covalent binder of DNA, by X-ray, NMR, and molecular mechanics," *J. Med. Chem.* **33**:3000–3008 (1990).
36. Subramaniam, G., Paz, M. M., Kumar, G. S., Das, A., Palom, Y., Clement, C. C., Patel, D. J., and Tomasz, M., "Solution structure of a guanine-N7-linked complex of the mitomycin C metabolite 2,7-diaminomitosene and DNA. Basis of sequence selectivity," *Biochemistry* **40**:10473–10484 (2001).
37. Kumar, G. S., He, Q.-Y., Behr-Ventura, D., and Tomasz, M., "Binding of 2.7-diaminomitosene to DNA; model for the precovalent recognition of DNA by activated mitomycin C," *Biochemistry* **34**:2662–2671 (1995).
38. Paz, M. M., Das, A., Palom, Y., He, Q.-Y., and Tomasz, M., "Selective activation of mitomycin A by thiols to form DNA cross-links and monoadducts: Biochemical basis for the modulation of mitomycin cytotoxicity by the quinone redox potential," *J. Med. Chem.* **44**:2834–2842 (2001).
39. Pavlopoulos, S., Bicknell, W., Craik, D. J., and Wickham, G., "Structural characterization of the 1:1 adduct formed between the antitumor antibiotic hedamycin and the oligonucleotide duplex d(CACGTG)$_2$ by 2D NMR spectroscopy," *Biochemistry* **35**:9314–9324 (1996).
40. Sun, D., Hansen, M., Clement, J. J., and Hurley, L. H., "Structure of the altromycin B (N7-guanine)-DNA adduct. A proposed prototypic DNA adduct structure for the pluramycin antitumor antibiotics," *Biochemistry* **32**:8068–8074 (1993).
41. Zhang, H., Gao, Y.-G., van der Marel, G. A., van Boom, J. H., and Wang, A. H.-J., "Simultaneous incorporations of two anticancer drugs into DNA. The structures of formaldehyde-cross-linked adducts of daunorubicin-d(CG(araC)GCG) and doxorubicin-d(CA(araC)GTG) complexes at high resolution," *J. Biol. Chem.* **268**:10095–10101 (1993).

42. Wang, A. H.-J., Gao, Y.-G., Liaw, Y.-C., and Li, Y.-K., "Formaldehyde cross-links daunorubicin and DNA efficiently: HPLC and X-ray diffraction studies," *Biochemistry* **30**:3812–3815 (1991).
43. Zeman, S. M., Phillips, D. R., and Crothers, D. M., "Characterization of covalent adriamycin-DNA adducts," *Proc. Natl. Acad. Sci. USA* **95**:11561–11565 (1998).
44. Ye, X., Kimura, K., and Patel, D. J., "Site-specific intercalation of an anthracycline antitumor antibiotic into a Y·RY DNA Triplex through covalent adduct formation," *J. Am. Chem. Soc.* **115**:9325–9326 (1993).

CHAPTER 5
Strand-Breaking Drugs

5.1. BLEOMYCINS

In 1966, Umezawa and coworkers [1] discovered a family of glycopeptide antibiotics, bleomycins (BLMs), which were isolated as copper complexes from *Streptomyces verticillis*. Metal-free BLMs can cleave the DNA strand preferentially at the 5'-GpC-3' and 5'-GpT-3' sequences in the presence of ferrous ion and molecular oxygen. Figure 5.1 shows the structure of a typical bleomycin, BLM-A_2. It consists of the DNA-binding domain and the metal-binding domain, which are connected by the linker. The former domain is the bithiazole (BTZ) ring to which the terminal substituent containing the dimethylsulfonium group (R group) is attached. BLM-A_2 analogs such as BLM-B_2 and pepleomycin differ only in the structure and/or the charge of the R group. The latter domain is the β-aminoalaninopyrimidine–β-hydroxyhistidyl moiety, which coordinates around the metal ion such as Fe(II) and Co(II) via the five nitrogen atoms (marked by a dot in Fig. 5.1) to form a distorted square–pyramidal structure. Coordination of O_2 to the vacant axial site leads to the formation of "activated BLM," and this shortlived intermediate produces a radical species that extracts the 4' hydrogen atom of the deoxyribose and results in DNA strand scission. Extensive investigations have been made to determine the mode of binding of BLMs to the DNA strand and to elucidate the mechanism of metal-ion-mediated DNA cleavage reactions. Although many review articles are available on these subjects, only those published after 1990 are cited here [2–6].

5.1.1. Metal-Free Bleomycins

Although isolated as Cu(II) complexes, bleomycins are administered to patients in the metal-free form. *In vivo*, however, metal-free BLMs react with the Fe(II) ion and O_2 in the cell to form "activated BLMs," which cause the DNA backbone cleavage. *In vitro*, the BLM-DNA interaction can be studied in the absence of these cofactors. In this case,

---

Drug–DNA Interactions: Structures and Spectra by Kazuo Nakamoto, Masamichi Tsuboi, and Gary D. Strahan
Copyright © 2008 John Wiley & Sons, Inc.

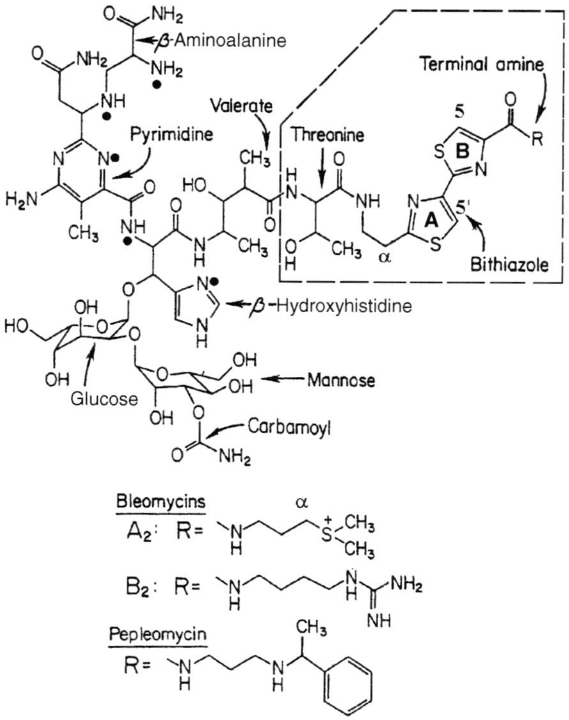

FIGURE 5.1. Structures of bleomycins and pepleomycin. The DNA-binding domain is delineated by dashed lines. Five nitrogen atoms marked by dots are used to coordinate around the metal atom (see Fig. 5.5).

the main interest of research has been to elucidate the mode of interaction of the BTZ (bithiazole) domain with DNA duplex.

The BTZ domain may interact with DNA via (partial) intercalation, groove binding, and their combinations. Gamcsik et al. [7] carried out NMR studies to determine the mode of interaction of BLM-A_2 with the hexamer duplex

$$5'\text{-d}(C_1 \quad G_2 \quad C_3 \quad G_4 \quad C_5 \quad G_6)\text{-}3'$$
$$3'\text{-d}(G_6 \quad C_5 \quad G_4 \quad C_3 \quad G_2 \quad C_1)\text{-}5'$$

Interaction of BLM with this hexamer caused broadening of all the proton signals of BLM, and in addition, the C5H (-0.294), C5'H (-0.226) and $CH_2(\alpha)$ (-0.150) of the BTZ ring and the $CH_2(\alpha)$ of the dimethylsulfonium group (-0.115) showed upfield shifts given in parentheses (in ppm). The magnitudes of these shifts are consistent with *partial* intercalation, but not with *full*, intercalation of the BTZ ring system between base pairs of the hexamer duplex [7]. Furthermore, these shifts are temperature-dependent; the values listed above (at 5°C) gradually decrease on raising temperature, and become zero at ~70°C, indicating that BLM dissociates from the duplex at this temperature. Also, an increase in ionic strength decreases the magnitude of the upfield shift, suggesting the decrease in ionic interaction between the BTZ domain and the duplex.

Among the proton signals of the hexamer duplex (B form), those of H1'(G_6) (−0.166 ppm) and H5(C_1)(−0.162 ppm) show the largest shifts on addition of BLM, indicating that the BTZ domain of BLM preferentially binds to the 5'-CpG-3' sequence at the flexible end of the duplex. In a short oligonucleotide such as the hexamer described above, the increased ease of access to the base pairs at the ends may be more important to the binding process than the base-pair sequence.

Interaction of BLM-A_2 with poly(dA-dT)$_2$ was studied as a function of temperature by NMR spectroscopy [8]. Among the perturbed BLM signals, the C5H and C5'H signals of BTZ showed the largest upfield shifts and line broadening, which reached maxima just below the melting temperature of the duplex (∼60°C). The C2H(A), C8H(A), and C6H(T) of the oligomer showed downfield shifts when mixed with BLM. However, all these shifts were much smaller than those observed for the BLM-d(CGCGCG)$_2$ adduct discussed above. It was not possible to draw definitive conclusions about the mode of bonding although two types of interaction (partial intercalation and groove binding) were suggested.

Rajani et al. [9] studied the interaction of metal-free BLM($A_2 + B_2$) with CT-DNA at 19°C and 30°C by Raman spectroscopy. Using curvefitting techniques, these workers resolved overlapping bands of DNA into those due to individual bases (G, C,A,T) and sugar moieties as shown in Fig. 5.2. At 30°C, significant hypochromism was observed for nearly all base modes, suggesting intercalation or at least partial intercalation. At 19°C, hypochromism of G and C bases was less than that at 30°C in the 1290–1160 cm^{-1} region. Their work was extended the low-frequency region (within 950–550 cm^{-1}) [10]. On interaction of CT-DNA with BLMs, the population of C3'-endo–anti conformation of the sugar (A-DNA) was found to increase at both temperatures. This may indicate local desolvation of the minor groove resulting from minor groove binding of the BTZ moiety.

Weselucha-Birczynska et al. [11] studied the binding modes of BLM-A_2 and a model compound of its BTZ moiety, 2-(2-amino-4-thiazolyl)-4-thiazolecarboxylic acid (ATTC) (Fig. 5.3b) with poly(dA-dT)$_2$, poly(dG-dC)$_2$, and CT-DNA by Raman (406.7 nm excitation) and ultraviolet resonance Raman (UVRR) (274 nm excitation) spectroscopy. As seen in Fig. 5.3a, b, BLM-A_2 and ATTC exhibit almost identical spectra by both excitation lines; the strongest band is at 1540 cm^{-1} with satellite bands at 1592, 1560, and 1497 cm^{-1}. These bands were assigned to the BTZ ring vibrations involving the stretching mode of the central C−C bond connecting the two thiazole rings.

DNA base vibrations are selectively resonance-enhanced by UV excitation at 274 nm since the polynucleotides (DNA) have absorptions at 254–262 nm. As seen in Fig. 5.3c, relative intensities of the two thymine bands of poly(dA-dT)$_2$ at 1655 and 1580 cm^{-1} change markedly when mixed with BLM or ATTC. Similar trends are seen for the 1657 and 1576 cm^{-1} bands of DNA (Fig. 5.3d). In contrast, no obvious changes in intensity are observed when poly(dG-dC)$_2$ is mixed with BLM or ATTC (not shown). These results suggest that the interaction occurs at the thymine bases. In the case of DNA, participation of adenine bases in the binding mechanism is indicated by the observation that the Raman intensity of the adenine band at 1336 cm^{-1} relative to that of the 1482 cm^{-1} band (an overlap of adenine and guanine bands) changes when

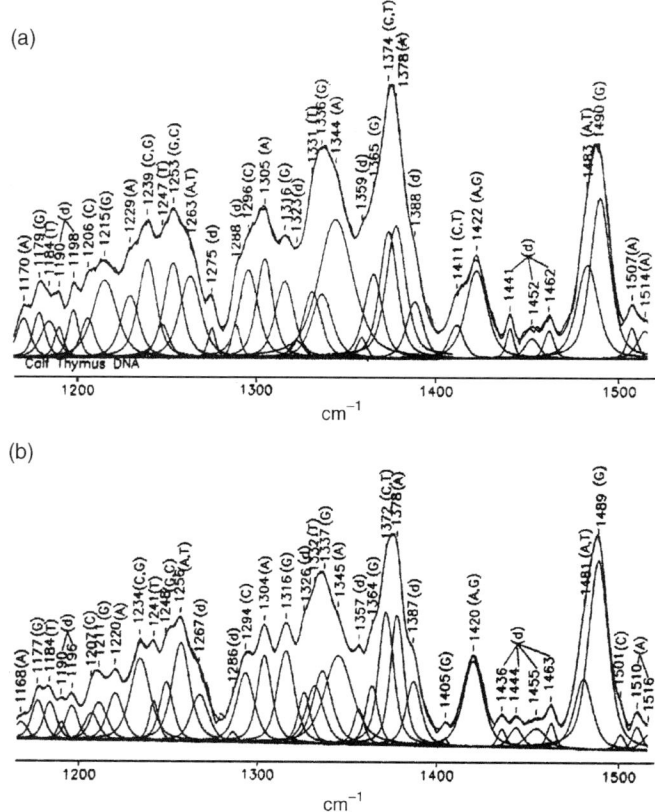

FIGURE 5.2. Raman spectra (514.5 nm excitation, 30°C) and curvefit results in the 1525–1160 cm^{-1} region: (a) CT-DNA and (b) its complex with metal-free bleomycin [9].

mixed with ATTC or BLM (Fig. 5.3d). However, the relative intensity of these adenine bands remains almost unchanged when poly(dA-dT)$_2$ is mixed with ATTC or BLM (Fig. 5.3c). Thus, these drugs must be interacting with some A/T bases of DNA in sequence other than alternating ATAT sequence. No band shifts were observed throughout these experiments. This may suggest that the BTZ domain of BLM and ATTC prefer to interact via groove binding at AT-rich region of DNA with minimal intercalation.

According to Chien et al. [12], the fluorescence spectrum of BLM-A$_2$ exhibits two maxima at 353 and 405 nm that originate in the BTZ and 4-aminopyrimidine rings, respectively. The former band is quenched on addition of DNA, and the degree of quenching was used to determine the binding constant of the BTZ portion of BLM-A$_2$ with DNA.

The modes of binding of metal complexes of BLM discussed in the following sections may be different from those of metal-free BLM because the latter can take a more flexible conformation than can the former.

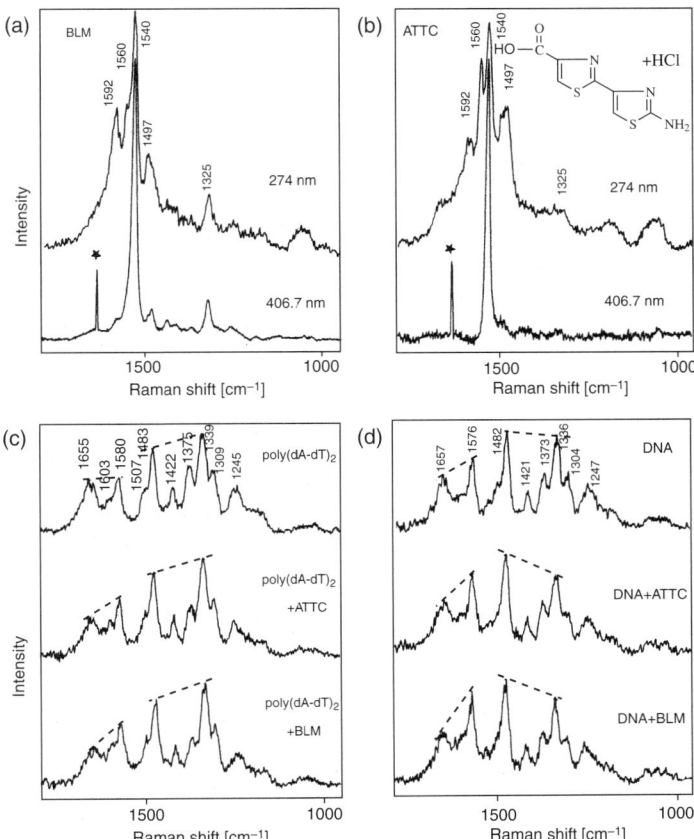

FIGURE 5.3. Resonance Raman spectra of bleomycin (BLM), its model compound ATTC, and their mixtures with poly(dA-dT)$_2$ and CT-DNA: (a) resonance Raman spectra of BLM with 274 and 406.7 nm excitation; (b) resonance Raman spectra of ATTC with 274 and 406.7 nm excitation [in (a) and (b), the asterisks indicate 1641 cm^{-1} band of mercury lamp light]; (c) UVRR spectra (274 nm excitation) of poly(dA-dT)$_2$ (3.79 × 10^{-3} M) and its mixtures with ATTC and BLM [the molar ratio (R) of base pairs of the oligomer to the ligand was 4]; (d) UVRR spectra (274 nm excitation) of CT-DNA (4 × 10^{-3} and its mixtures with ATTC and BLM (R = 4) [11].

5.1.2. Iron Complexes of Bleomycin

Several groups of workers carried out ESR studies to determine the structure of activated BLM formed in the presence of the Fe(II) ion and O$_2$ and to elucidate the pathway leading to DNA strand cleavage. Figure 5.4A shows the ESR spectra of Fe(II)-BLM-A$_2$ on oxygen bubbling (pH = 6.9 at 77 K) obtained by Sugiura [13]. At 0 s (trace **a** in Fig. 5.4A), no signals are observed since BLM-Fe(II)(d^6) is ESR-silent. After 3 s (trace **b** in Fig. 5.4A), three ESR signals with a small rhombic splitting of g values appear at g_z = 2.254, g_y = 2.171, and g_x = 1.937, indicating the formation of the low-spin Fe(III) (d^5) complex. In addition, a signal due to a free-radical species is seen at g = 2.005. After 6, 10, 30, and 90 s (Fig. 5.4A, traces **c–f**), these signals are gradually

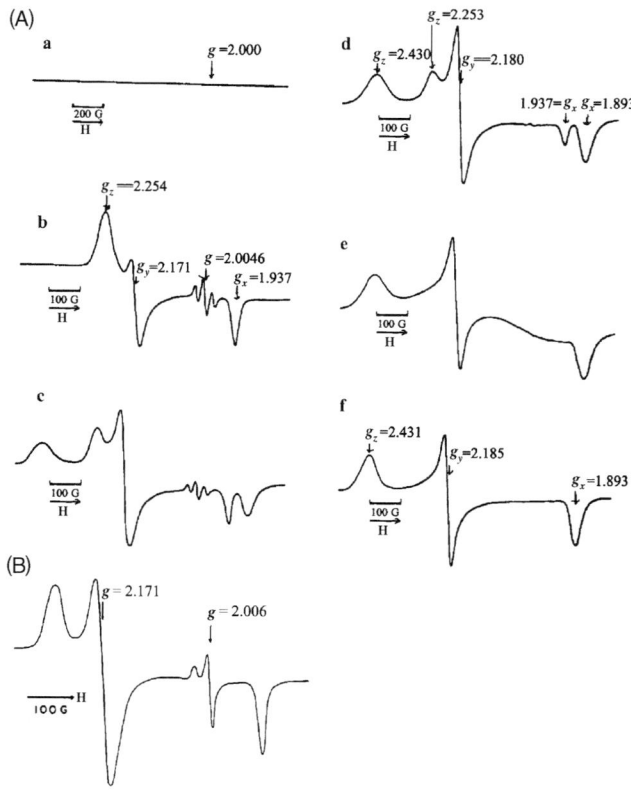

FIGURE 5.4. (A) ESR spectral changes of the Fe(II)-bleomycin complex (1.0 mM) on oxygen bubbling at pH = 6.9 as a function of time. Traces: (**a**) 0 s, (**b**) 3 s, (**c**) 6 s, (**d**) 10 s, (**e**) 30 s, and (**f**) 90 s; (B) initial ESR spectrum of the Fe(II)–bleomycin complex system obtained by oxygen bubbling in the presence of DNA. The spectrum was measured after 5 s of the reaction, and the concentrations of the complex and DNA were 1.0 and 0.02 mM, respectively [13].

replaced by a new set of signals with a large rhombic splitting of g values at $g_z = 2.431$, $g_y = 2.185$, and $g_x = 1893$ (Fig. 5.4A, trace **f**). Since these ESR parameters are similar to that of methemoglobin hydroxide ($g_z = 2.4$, $g_y = 2.2$, and $g_x = 1.9$), it was assigned to the low-spin ferric complex BLM-Fe(III)-OH$^-$. This species is stable and exhibits an absorption maximum near 370 nm (orange-yellow). On the other hand, the initial unstable intermediate was assigned to BLM-Fe(III)-OOH$^-$ (activated BLM) because its ESR parameters are close to those of hemoglobin-Fe(III)-OOH$^-$ ($g_1 = 2.30$, $g_2 = 2.16$, and $g_3 = 1.94$) and peroxidase-Fe(III)-OOH$^-$ ($g_1 = 2.31$, g_2 2.16, and $g_3 = 1.95$) [14]. This peroxide formulation was also supported by similar studies [15,16].

When CT-DNA is added to the BLM-Fe(II)-O$_2$ system (Fig. 5.4B), three signals ($g_z = 2.254$, $g_y = 2.171$, and $g_x = 1.937$) appear at the initial stage. These values are the same as those of the activated BLM alone. In addition, ESR signals at $g_\perp = 2.006$ and $g_\parallel = 2.030$ are close to those of the typical superoxide (2.006 and 2.080, respectively) which may be responsible for DNA strand cleavage.

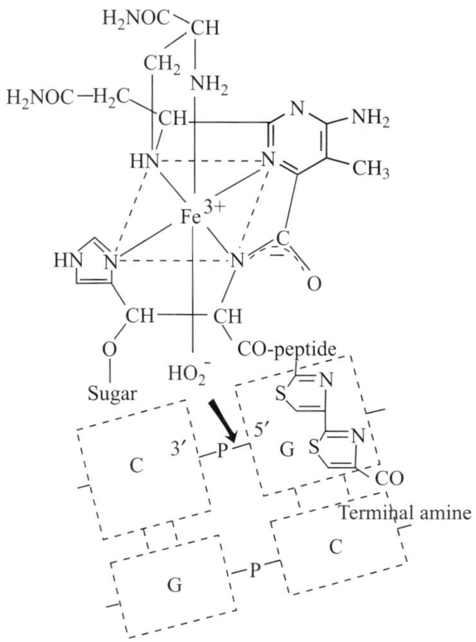

FIGURE 5.5. Structure of activated bleomycin bound to the GpC sequence [14].

Figure 5.5 illustrates the structure of the activated BLM proposed by Sugiura [14]. The coordination sphere around the Fe(III) center is a distorted square–planar pyramid formed by the five nitrogen atoms marked in Fig. 5.1, and the sixth axial position is occupied by the OOH^- ligand. This structure is supported not only by the similarity of ESR parameters between BLM-Fe(III)-OOH^- and the corresponding Fe(III) porphyrins mentioned above but also by the X-ray crystal structure of a biosynthetic intermediate of BLM complexed to Cu(II) [the 1 : 1 P-3A-Cu(II) complex] [17]. This also accounts for the preferential cleavage site at 5′-GpC-3′ by showing that the BTZ ring is intercalated between the G-C base pair to bring the OOH^- ligand close to the C4′ hydrogen of the deoxyribose linked to cytosine.

Figure 5.6a shows the electronic absorption (ABS), CD, and MCD (magnetic circular dichroism) spectra of BLM-Fe(III)-OH^- (ground state), whereas Fig. 5.6b shows the ABS and MCD spectra of BLM-Fe(III)-OOH^- (activated BLM) obtained by Neese et al. [18]. Curvefitting analysis revealed that these structureless spectra consist of many components numbered in the figure. On the basis of theoretical calculations, most of these component bands were assigned to the d–d transitions of the low-spin Fe(III) ion. However, relatively strong bands of the ground-state complex, calculated to be at 24,510 and 26,380 cm^{-1}, were assigned to the pyrimidine/amide → d and amide → d charge transfer (LMCT) bands, respectively. The corresponding bands of the activated BLM were calculated to be at 23,830 and 27,250 cm^{-1}, respectively, and its peroxide (π^*) → d LMCT transition was predicted to be at 19,510 cm^{-1}. The ligand field parameters calculated for these two

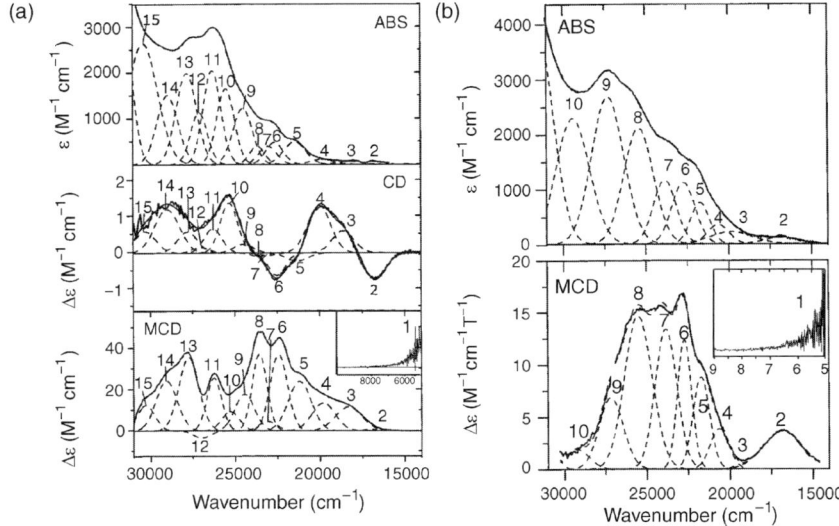

FIGURE 5.6. Absorption (ABS), CD, and MCD spectra of (a) BLM-Fe(III)-OH$^-$ (ground state) and (b) BLM-Fe(III)-OOH$^-$ (activated BLM) at low temperature. BLM denotes BLM-A$_2$ [18].

states reflect the difference in the axial ligand (OH$^-$ vs. OOH$^-$), indicating that the latter is a somewhat better π donor than the former.

Figure 5.7 compares the RR spectra (407 nm excitation) of Fe(III)-BLM[trace (a)], Fe(II)-BLM[trace (b)], CO-Fe(II)-BLM[trace (c)], and metal-free BLM[trace (d)] in

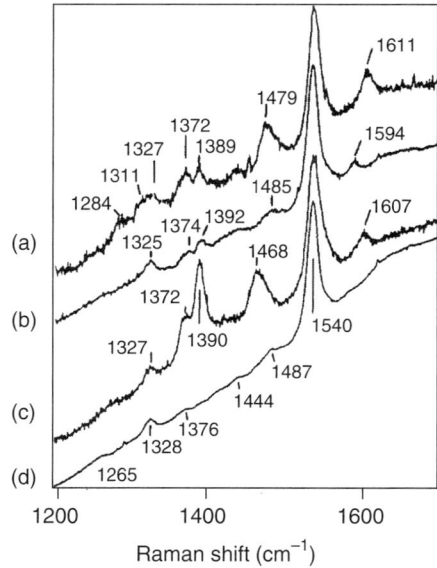

FIGURE 5.7. Raman spectra (407 nm excitation) of (a) Fe(III)-BLM, (b) Fe(II)-BLM, (c) CO-Fe(II)-BLM, and (d) metal-free BLM. BLM denotes BLM-A$_2$ [19].

H_2O obtained by Takahashi et al. [19]. As mentioned in the preceding section, the strongest band at 1540 cm^{-1} originates in the BTZ ring. The bands at 1611 and 1479 cm^{-1} of Fe(III)-BLM [trace (a)] were assigned to the amide I (mainly C=O stretching) and amide II (mainly C−N stretching) of the β-hydroxyhistidinyl amide, respectively, and a pair of bands at 1389 and 1372 cm^{-1} were assigned to the pyrimidine ring vibrations. These assignments were based on evidence that included their redox and ligation dependence, D_2O shifts, bandwidths, and resonance Raman intensities. The Fe-OH$^-$ stretching vibration of BLM-Fe(III)-OH$^-$ [trace (a), ground state] is at 561 cm^{-1}; it shifts to 554 cm^{-1} in D_2O and to 540 cm^{-1} in $H_2^{18}O$. Neese et al. [18] also measured the RR spectra of BLM-Fe(III)-OH$^-$ with excitation lines in the range of 379 and 514 nm, and noted that two bands at 1627 and 1473 cm^{-1} are strongly resonance-enhanced when the excitation was made with the 386 nm (25,906 cm^{-1}) line, which is close to the amide $\to d$ charge transfer LMCT transition. More information about activated bleomycin is available in a review article by Burger [20].

Since activated BLM is the last detectable intermediate, its pathway leading to DNA strand cleavage has been a subject of controversy. It is generally assumed that the •OH radical released from BLM-Fe(III)-OOH$^-$ is responsible for the cleavage reaction. Neese et al. [18] considered the following three mechanisms:

1. Heterolytic cleavage of the O−O bond yields BLM-Fe(V)=O and H_2O, and the former high-valence intermediate attacks the DNA, namely, BLM-Fe(III)-O−O−H + H$^+$ → BLMFe(V)=O + H_2O.
2. Homolytic cleavage of the O−O bond yields BLM-Fe(IV)=O and an •OH radical, and both species attack the DNA, namely, BLM-Fe(III)-O−O−H → BLMFe(IV)=O + •OH.
3. Direct attack of activated BLM on the DNA produces BLM-Fe(IV)=O, H_2O, and a DNA radical.

According to their theoretical calculations, mechanism 1 is not energetically favorable because it involves production of a high-valence oxo species. Mechanism 2 is also not favorable because the homolytic cleavage reaction has a very high activation barrier, and the •OH radical attack is indiscriminate despite the known sequence specificity of the reaction. Thus, they proposed mechanism 3 as a new working hypothesis, and discussed its feasibility.

5.1.3. Cobalt and Zinc Complexes of Bleomycin

Under anaerobic conditions, Sugiura [21] obtained the ESR spectrum of the 1 : 1 BLM–A$_2$-Co(II) complex at 77 K shown in Fig. 5.8a. It exhibits a nearly axial symmetry about the Co(II) ion and an eight-line parallel hyperfine structure due to the interaction with the ^{59}Co ($I = \frac{7}{2}$) nucleus. Such a spectrum is characteristic of the low-spin ($d^7, S = \frac{1}{2}$), five-coordinate square–pyramidal Co(II) complex with the unpaired electron in the d_{z^2} orbital. This is confirmed by the presence of the three-line superhyperfine splittings due to interaction with the axial nitrogen (^{14}N, $I = 1$), the

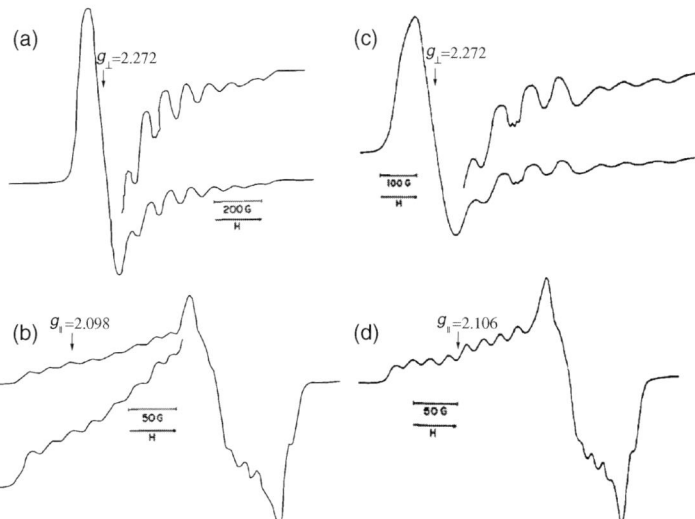

FIGURE 5.8. ESR spectra of (a) BLM-Co(II) and (b) BLM-Co(II)-O_2 at 77 K. The spectra of these complexes in the presence of DNA are shown in (c) and (d), respectively. BLM denotes BLM-A_2 [21].

observed relationship of $g_\perp > g_\parallel \sim 2.0$, and the absence of superhyperfine splittings from the in-plane ^{14}N donor atoms. Furthermore, its ESR parameters ($g_\perp = 2.272$, $g_\parallel = 2.025$, $A_\parallel^{Co} = 92.5$ G, and $A_\parallel^N = 13$ G) are close to those of five-coordinate Co (II) complexes of Schiff bases and porphyrins, which are known to assume similar structures.

On oxygenation, the ESR spectrum (at 77 K) changes drastically as seen in Fig. 5.8b. The signal at $g_\perp = 2.272$ is replaced by a new signal at $g \sim 2.0$, and the ^{59}Co hyperfine splitting becomes much smaller. The ESR parameters ($g_\perp = 2,007$, $g_\parallel = 2.098$, $A_\perp^{Co} = 12.4$ G, $A_\parallel^{Co} = 20.2$ G) are close to those of the superoxide anion radical ($\cdot O_2^-$) ($g_\perp = 2.006$ and $g_\parallel = 2.080$–2.103). This result suggests that the unpaired electron resides mostly (90%) on the dioxygen moiety. Thus, the adduct should be formulated as BLM-Co(III)-O_2^- instead of BLM-Co(II)-O_2. Similar formulation is feasible for the dioxygen adducts of the Co(II) complexes of Schiff bases and porphyrins since their ESR parameters are similar to those of BLM-Co(III)-O_2^-.

Figure 5.8 traces (c) and (d) show the ESR spectra of BLM-Co(II) and BLM-Co (II)-O_2, respectively, in the presence of DNA at 77 K. The effective g values and A tensors obtained indicate that the mode of coordination and the electron spin delocalization are essentially similar to those obtained in the absence of DNA. Nevertheless, the presence of DNA has distinct effects on the ESR spectrum of BLM-Co(II)-O_2; the $g_\parallel$ value increases from 2.098 to 2.106, and the $g_\perp$ value decreases from 2.007 to 2.004. In addition, the hyperfine structure is better resolved, and the splitting parameters, $A_\perp^{Co}$ and $A_\parallel^{Co}$ decrease by 0.9 and 1.3 G, respectively. These differences suggest that the interaction of DNA with the BTZ ring and its

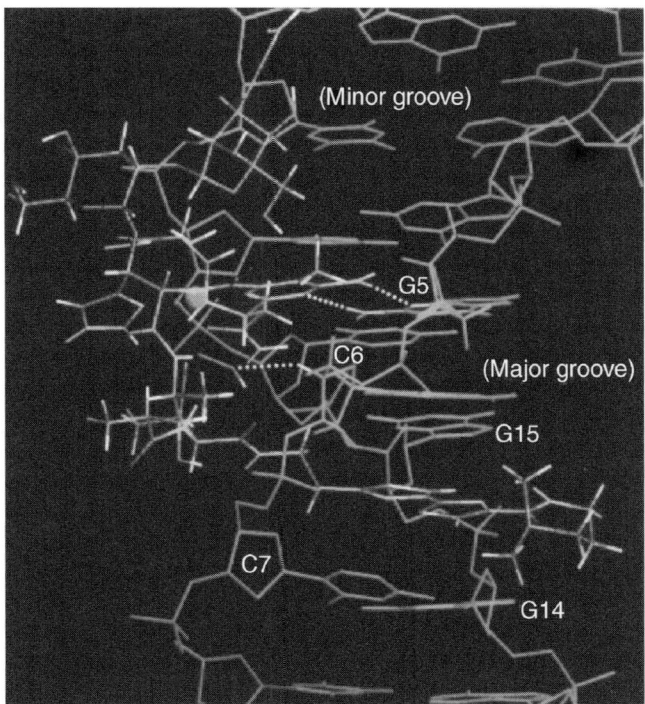

FIGURE 5.9. Structure of BLM A$_2$-Co(III)-OOH (shown in green) complexed to the 10-mer (in purple). The damaged strand is in the foreground running 5′ → 3′ from upper right to lower left corner. The dotted lines indicate hydrogen-bonding interactions between the pyrimidine moiety of BLM and G$_5$. The proximity of the distal oxygen of the OOH group to the H4′ of C$_6$ (2.5 Å) is also indicated by the dotted line [22]. (See color insert.)

terminal group (R) has an appreciable influence on the orientation of the bound O$_2$ molecule relative to the Co(II) plane.

The dioxygen adduct of Co(II)-BLM in aqueous solution is slowly converted into a mixture of BLM-Co(III)-OH and BLM-Co(III)-OOH (shown in green in Fig. 5.9). The latter can be isolated from the mixture, and serves as an ideal model of the activated BLM [BLM-Fe(III)-OOH$^-$] discussed in the preceding section. On photochemical activation, it can cleave DNA with the same sequence specificity as that of the activated BLM. Furthermore, it is diamagnetic and, at neutral pH, stable enough to measure the NMR spectra. Thus, Wu et al. [22] carried out NMR studies to determine the solution structure of green BLM-A$_2$-Co(III)-OOH complexed to the 10-mer:

	1	2	3	4	5	6	7	8	9	10	
5′-d(	C	C	A	G	G	C	C	T	G	G	)-3′
3′-d(	G	G	T	C	C	G	G	A	C	C	)-5′
	20	19	18	17	16	15	14	13	12	11	

On the basis of observed inter- and intramolecular NOEs, these workers were able to build a molecular model of the complex shown in Fig. 5.9. It shows that the terminal thiazole ring of BTZ is completely intercalated between the bases of G_{14} and G_{15} whereas its penultimate ring is only partially intercalated between those of C_6 and C_7. This structure is stabilized by hydrogen bonding between the N/NH$_2$ pyrimidine ring of BLM and the N3/NH$_2$ of G_5 (dotted lines) which is responsible for the G preference of over A. Intermolecular NOEs were also observed between the proton signals of the axial OOH ligand and those of C_6 (cleavage site) and C_7 in the minor groove. It was estimated that the distant oxygen of the hydroperoxide ligand is only 2.5 Å away from the H4' atom of C_6 (dotted line), the abstraction of which triggers the DNA cleavage reaction [23].

Their NMR studies [24] were extended to BLM-A$_2$-Co(III)-OOH complexed to another 10-mer:

$$\begin{array}{cccccccccc}
1 & 2 & 3 & 4 & 5 & 6 & 7 & 8 & 9 & 10 \\
5'\text{-d(C} & C & A & G & T & A & C & T & G & G)\text{-}3' \\
3'\text{-d(G} & G & T & C & A & T & G & A & C & C)\text{-}5' \\
20 & 19 & 18 & 17 & 16 & 15 & 14 & 13 & 12 & 11
\end{array}$$

in which the central GpC sequence of the previous 10-mer is replaced by the TpA sequence. Although the overall structures of these two complexes are similar, several differences are noted in the location and the mode of intercalation of the BTZ moiety between base pairs of DNA. In this case, the BTZ rings are between the T_5-A_{16} and A_6-T_{15} base pairs, but neither ring of the BTZ moiety is fully intercalated. Furthermore, the atomic positions in this domain were less clearly defined relative to the former complex [22] because the interaction is either not as stable as, and/or is more dynamic than, that of the former. The terminal oxygen of the axial hydroperoxide ligand of BLM-A$_2$-Co(III)-OOH is at 2.4 ± 0.2 Å from the H4' atom of T_5 (cleavage site).

Two-dimensional NMR studies on the interaction of Pep-Co(III)-OOH (Pep = pepleomycin, Fig. 5.1) with d(CGTACG)$_2$ show that the BTZ ring is intercalated between the central T-A base pairs and the OOH group is close to the H4' atom of T, where the DNA cleavage occurs [25].

Although BLMs can cleave both single- and double-stranded DNA, the main origin of its cytotoxicity is thought to be due to the cleavage of double-stranded DNA by a single BLM molecule. Vanderwall et al. [24] carried out molecular modeling studies to account for the mechanism of the double-strand cleavage by BLM-Co(III)-OOH. Starting from the structure of the complex in which the terminal oxygen of the OOH group is close to the H4' of T_5 (first cleavage site), they were able to derive a similar structure that brings the OOH group close to the H4' of T_{15} of the complementary strand (second cleavage site) without completely dissociating the ligand from the DNA duplex.

Hoehn et al. [26] proposed a molecular model to explain the mechanism of the second cleavage based on 2D NMR studies of BLM-Co(III)-OOH bound to d(CCAAAGXACTGGG)·d(CCCAGTACTTTGG) where X indicates the first

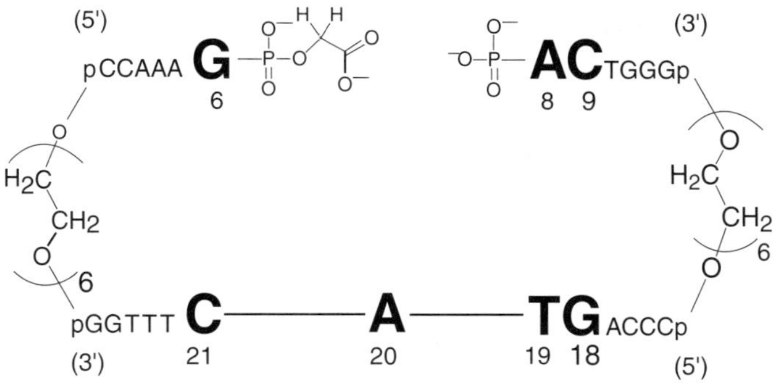

FIGURE 5.10. Schematic drawing of the duplex DNA with the 3′-PG lesion and the 5′-P. The duplex oligonucleotide was constructed as a double hairpin. Hexaethylene glycol spacers connect the two strands of the duplex. These linkers were added to prevent fraying at the ends of the short oligonucleotide pieces. The numbering of the bases of GTAC box is shown [26].

cleavage site, a 3′-phosphoglycolate lesion next to a 5′-phosphate (Fig. 5.10). A molecular model was built using the observed 729 inter- as well as intra-NOEs together with 96 dihedral angle constraints. The results indicate that the BTZ tail is partially intercalated between T_{19} and A_{20} (second cleavage site) and that the metal-binding domain is close enough to abstract the H4′ atom of T_{19} in the minor groove. Furthermore, the predominant conformation of the two thiazole rings in the BTZ moiety is *trans*, although two *cis* conformations that differ with respect to the ring-flipping direction are also observed, and ROESY spectra indicate interconversion among these conformations.

The Raman spectra (514.5 nm excitation) of DNA, BLM-DNA, and BLM-Co(II)-O_2-DNA (where DNA denotes CT-DNA) were measured, and the individual peaks obtained by curvefitting analysis were assigned by Rajani et al. [27]. The results indicate that, although DNA predominantly retains B form, local structural changes from B to Z and B to A forms occur in the O_2 adduct. For example, it exhibits a band at 623 cm^{-1} of guanosine that indicates the C3′-*endo–syn* conformation of the Z form. These structural changes have been attributed to the perturbation of structural water in the minor groove resulting from the binding of BLM-Co(II)-O_2 to DNA. Rajani et al. [28] also measured the RR spectra (406.7 nm excitation) of BLM-A_2-Co(III)-OOH in H_2O buffer, and located the ν(O—OH) at 828 cm^{-1} and ν(Co—OOH) at 545 cm^{-1}, which are highly important in characterizing the structure of the activated BLM. These vibrations are shifted to 784 and 518 cm^{-1}, respectively, by $^{16}O/^{18}O$ substitution but not shifted by H_2O/D_2O substitution.

The mode of interaction of Zn(II)-BLM-A_5 with d(CGCTAGCG)$_2$ was proposed by Manderville et al. [29] on the basis of 2D NMR studies and molecular dynamics calculations. BLM-A_5 is analogous to BLM-A_2 except that the R group in Fig. 5.1 is Sp [spermidine, NH-(CH$_2$)$_3$-NH-(CH$_2$)$_4$-NH$_2$]. Figure 5.11 illustrates intermolecular NOEs observed between Zn(II)-BLM-A_5 and the octamer duplex (arrows 1–8). Arrows 8 and 7 connect the H2 (A_5) with the CH$_2$ (at 3 of Sp) and C5H (**B** ring of BTZ), respectively. The latter suggests **B** ring stacking between A_5 and G_6 bases in the minor groove. Arrows 6 and 5 connect the C5′H (**A** ring of BTZ) to the CH6 of T_4 and

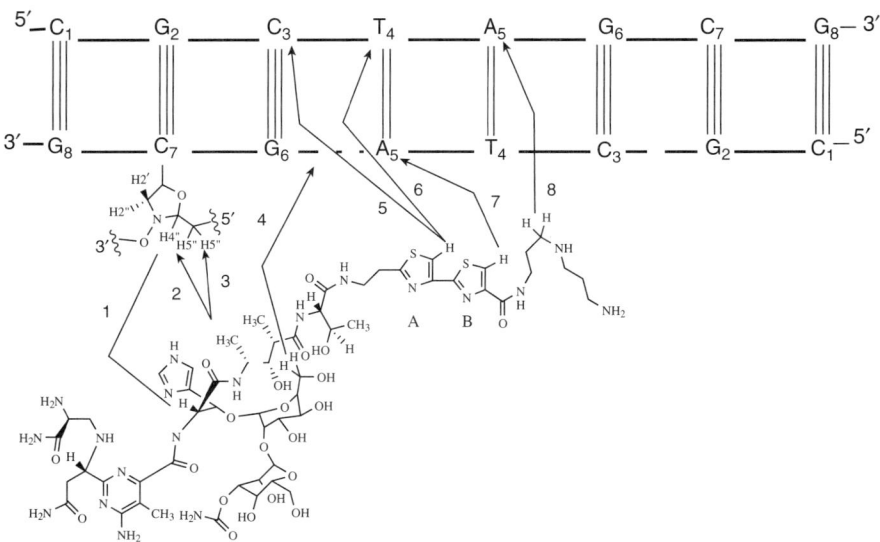

FIGURE 5.11. Schematic diagram to show eight intermolecular NOEs between Zn(II)-BLM-A_5 and d(CGCTAGCG)$_2$ [29].

CH6 of C_3 in the major groove, respectively, indicating **A** ring stacking between C_3 and T_4. However, "classical intercalation" can be ruled out since the sequential NOE assignments of the octamer are not disrupted by the interaction with Zn(II)-BLM-A_5. The NOEs shown by arrows 1–4 provide information about the orientations of the remaining domains relative to T_4 and C_7. Using these NOE distance restraints together with dihedral angle restraints, these workers carried out molecular dynamics calculations to build a molecular model. The two thiazole rings of BTZ were found to be in a *cis* conformation, and the BLM was found to have a folded conformation that favored hydrogen bonding between the NH_2 group of G_6 and the OH group of the methyl valerate residue in the linker region (Fig. 5.1).

Urata et al. [30] studied the enantiospecific recognition of DNA by BLM using ^{1}H NMR spectroscopy. The D-enantiomer of d(CGCGCG)$_2$ (a model of natural B-DNA) is cleaved by Fe(II)-BLM-B_2 (Fig. 5.1), whereas the corresponding L-enantiomer is not cleaved under the same conditions. Changes in ^{1}H NMR spectra of Zn(II)-BLM-B_2 solution titrated by both enantiomers were compared. No appreciable differences were noted in the signals from the DNA-binding domain, but different behaviors were noted in the metal-binding domain and the linker, depending on the chirality of the hexamer duplex. Thus, the former domain is nonenantiospecific, whereas the latter domain and the linker recognize the chirality of the DNA duplex.

5.1.4. Model Compounds of Bleomycins

As described in the preceding sections, bleomycins consist of the DNA-binding and the metal-binding domains that are connected by the linker. The role of the former is to

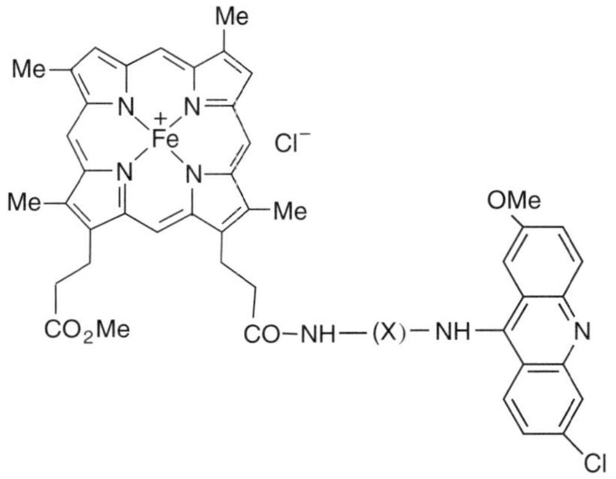

FIGURE 5.12. Structure of haemin–acridines: X = $(CH_2)_n$ (n = 2,3,4), $(CH_2)_2NH(CH_2)_2$, and $(CH_2)_3NH(CH_2)_3$ [31].

bring the latter close to the DNA duplex via (partial) intercalation, whereas that of the latter is to produce a radical species that cleaves the DNA strand in the presence of the Fe(II) ion and O_2. Several types of model compounds of bleomycins have been synthesized on the basis of this concept. For example, Lown and Joshua [31] prepared a series of haemin–acridines, shown in Fig. 5.12. Here, acridine acts as the anchor (intercalator; Chapter 2) and haemin [Fe(III) porphyrin] produces a radical species in the presence of oxygen and a reducing agent such as 3-mercaptoethanol. In the case of X = $(CH_2)_3NH(CH_2)_3$ (2.5×10^{-5} M), for example, the percentage of scission was 88% for PM2 covalently closed circular DNA (pH = 7.1, 37°C, 20 min), which was comparable to bleomycins. This work was extended to a number of analogs [32], including haemin–bis-acridines [33].

Hertzberg and Dervan [34] synthesized (methidiumpropyl-EDTA)-Fe(III) (Fig. 5.13a), which causes single- and some double-strand cleavages of DNA in the

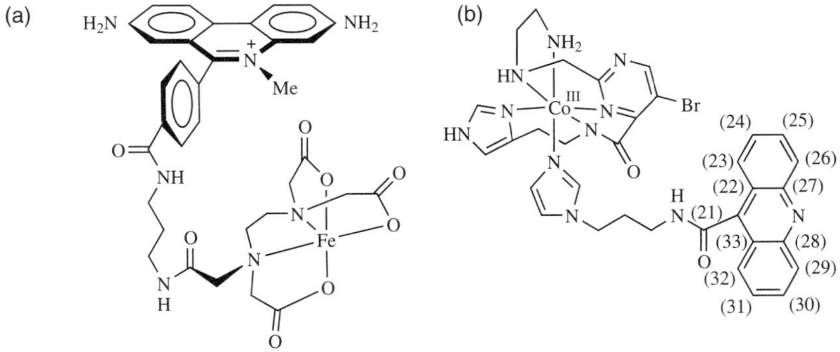

FIGURE 5.13. Structures of (a) (methidiumpropyl-EDTA)iron(II) [34] and (b) a hybrid molecule of acridine-9-carboxamido-N′-(3-propyl)imidazole and $[Co(PMA)]^{2+}$ moiety [35].

presence of dithiothreitol and molecular oxygen. At 10^{-8} M concentration, it cleaves pBR-322 plasmid DNA comparable to the efficiencies found with bleomycin. In this case, methidium is the intercalator and the O_2 adduct of Fe(II)(EDTA) is the source of a radical species responsible for cleavage reactions.

As mentioned earlier, BLM-Co(III)-OOH can cleave the DNA strand on photochemical activation. Tan et al. [35] prepared a model compound containing acridine as the intercalator and the $[Co(PMA)]^{2+}$ moiety as the DNA cleaving agent (Fig. 5.13b).

These workers studied the interaction of this model compound with the self-complementary decamer

```
           1  2  3  4  5  6  7  8  9  10
5'-d(G     A  T  C  C  G  G  A  T  C)-3'
3'-d(C     T  A  G  G  C  C  T  A  G)-5'
```

using NMR spectroscopy. It was known previously that the $[Co(PMA)]^{2+}$ moiety releases a ligand-based radical that rapidly collapses into the •OH radical responsible for DNA strand cleavage. They observed both broadening and upfield shifts of the acridine proton signals (H23–H32) on reaction with the decamer, thus confirming the intercalation of the acridine ring into the duplex. However, these changes were more conspicuous for the outer half of the protons (H25, H26, H29, and H30) than for the inner half of the protons of the acridine ring, indicating that only the outer half of the acridine ring is intercalated. The modest shifts were noted for the base signals from G_7, G_6, C_5, and C_4 of the decamer. In all 2D NMR spectra of the 1 : 1 adduct, the acridine protons showed strong NOEs to sugar H1′, H2′-H2″, and aromatic base protons in the G_6pG_7, and C_4pC_5 sequences. These and other observations provided strong evidence for the intercalation at the G_6pG_7 step, which might explain the origin of preferential photocleavage of DNA at the 5′-GpG-3′ site. Furthermore, no NOEs were observed between the protons of the $[Co(PMA)]^{2+}$ ion and the decamer, and the signals of the former were strongly affected by the salt concentration. Thus, the [Co(PMA)]$^{2+}$ moiety may bind probably via electrostatic interaction to the phosphate backbone of the decamer at points one base pair away from the GG site.

5.2. ENEDIYNE ANTIBIOTICS

Enediyne antibiotics such as calicheamicins, esperamycins, and neocarzinostatins can cleave double-stranded DNA. These drug molecules contain enediyne rings that produce biradical species when activated by a reducing agent such as thiol or by several other means. A variety of interactions between the drug and DNA bring these biradical moieties close to two bases from which their sugar protons are abstracted simultaneously. Since the cleavage of double-stranded DNA is the main origin of strong cytotoxicities, these drugs are highly cytotoxic, and their antitumor activities are reported to be more than thousands times stronger than that of Adriamycin (doxorubicin), for example.

The reaction pathway and mechanism of enediyne antibiotics have been reviewed by Smith and Nicolaou [36].

5.2.1. Calicheamicins

Calicheamicins are a family of antitumor drugs isolated from *Micromonospora echinospora*. Figure 5.14a shows the structure of calicheamicin γ_1^I. It consists of a 3-ene-1,5-diyne moiety (R ring) to which an oligosaccharide chain containing a hydroxylamino sugar (A ring), a thiosugar (B ring), a thiobenzoate (C ring), and a rhamnose sugar (D ring) is attached. In addition, an ethylamino sugar (E ring) is attached to A ring via a glycosidic linkage. Calicheamicin γ_1^I cleaves DNA at pyrimidine-rich sequences such as TCCT, CTCT, TTTT, and ACCT. Figure 5.14b shows the DNA-cleaving mechanism by calicheamicins [36]; a nucleophile (e.g., glutathione) attacks the central sulfur atom of the allylic trisulfide ($SSSCH_3$) connected to the R ring, and forms a thiol that is added to the adjacent α,β-unsaturated ketone ring. It converts a trigonal bridgehead position to a tetragonal center, and results in a drastic change in the structural geometry with a severe strain in the 10-membered

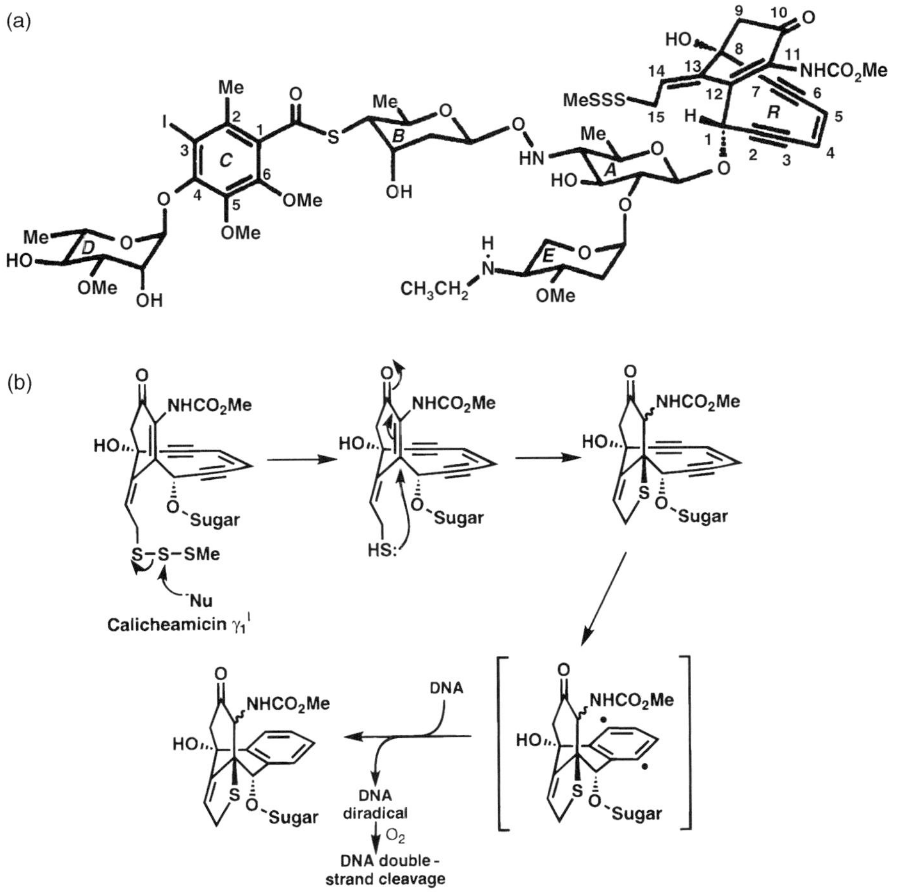

FIGURE 5.14. (a) Structure of calicheamicin γ_1^I and (b) DNA-cleaving mechanism by calicheamicin [36].

enediyne ring. This strain is relieved by forming a highly reactive 1,4-benzenoid biradical species via Bergman cyclization, and the biradical thus produced cleaves both strands of DNA simultaneously by abstracting hydrogen atoms from sugar–phosphate backbones.

Walker et al. [37] carried out 2D NMR studies to elucidate the structures of calicheamicin γ_1^I complexed to octamers containing recognition sites such as ACCT and TTTT in the absence of thiol. On formation of the 1 : 1 (drug/DNA) complex with an octamer

```
              1   2   3   4   5   6   7   8
      5'-d(G  T   G   A   C   C   T   G)-3'
      3'-d(C  A   C   T   G   G   A   C)-5'
             16  15  14  13  12  11  10   9
```

they observed ^{1}H chemical shifts of the drug and the octamer signals together with 11 intramolecular NOEs of the drug and 14 intermolecular NOEs between the drug and the octamer. The results show that the tetrasaccharide-aryl tail (A-B-C-D ring chain) of the drug is centered over the CpC step of the 5'-ACCT-3' site in the minor groove, and that the aglycone (R ring) is located in a position to abstract hydrogen atoms from the C_5 and A_{15} deoxyriboses. Conformational changes resulting from complexation were significant for the octamer, and appeared to be larger in the ACCT strand than in the complementary TGGA strand. However, conformational changes of the drug were not significant. Similar experiments with an octamer containing a TTTT site show that the drug assumes the same orientation as that observed for the ACCT site. Thus, the binding site selectivity of the drug is determined largely by the ability of the pyrimidine–purine sequence to adapt to a particular shape of the drug.

Paloma et al. [38] also studied the interaction of calicheamicin γ_1^I and its analog, θ_1^I (with the $SSSCH_3$ group in γ_1^I replaced by the $SCOCH_3$ group) with the decamer

```
              1   2   3   4   5   6   7   8   9  10
      5'-d(G  C   A   T   C   C   T   A   G   C)-3'
      3'-d(C  G   T   A   G   G   A   T   C   G)-5'
             20  19  18  17  16  15  14  13  12  11
```

containing the TCCT recognition sequence by 1D and 2D ^{1}H NMR spectroscopy. The results were virtually identical for both drugs. On drug binding, significant changes were observed in chemical shifts of the decamer in the C_5 to T_7 and G_{15} to T_{18} regions. Since similar results were obtained with calicheamicin oligosaccharide (calicheamicin without the aglycone residue), these results indicate that the calicheamycin sugar interacts with the decamer in the specific sequence of DNA regardless of the presence or absence of the aglycone residue. All the signals from the drug were observed except for those from sugar E ring, and the intermolecular NOEs between the drug and the decamer provided definitive evidence of complexation.

Molecular models were constructed using NMR distance restraints obtained from intramolecular as well as intermolecular NOEs combined with molecular dynamics calculations. Figure 5.15 shows a molecular model of the adduct of calicheamicin θ_1^I

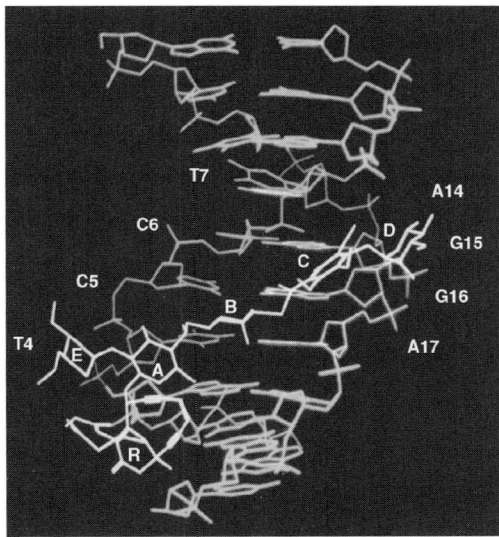

FIGURE 5.15. Molecular model of calicheamicin θ_1^1 bound to the decamer in the central region. The drug (shown in yellow) and the central region of the duplex (in red and pink) are shown by different colors [38]. (See color insert.)

bound to the central region of the abovementioned decamer. The decamer duplex assumes B form with the deoxyribose rings in a C2′-*endo* conformation except for those of T_{18} and C_5, which are in a C3′-*endo* conformation. The width of the minor groove was expanded in the range from 7.1 to 8.5 Å to accommodate the drug molecule. Strong intramolecular NOEs of the drug were observed between the CH_3 (D ring) and C5-OCH_3 (C ring) and between the CH_3 (B ring) and C6-OCH_3 (C ring), indicating that the two C-ring methoxy groups must be on the same side with the two B- and D-rings methyl groups. The orientation of the aglycone residue relative to A, B, and E rings was determined by strong NOEs between their signals. The drug molecule takes a gentle curvature in the sugar chain to fit into the minor groove, and its aglycone residue (R ring) has NOE contacts with the decamer in the regions of A_3-T_{18} and T_4-A_{17} base pairs. Detailed analysis reveals that the drug–DNA interaction involves several types of interactions, such as stacking interaction, hydrogen bonding, and possibly saltbridge formation. Temperature dependence studies show that calicheamicin θ_1^I binds to the octamer in two different modes, probably due to a movement of the aglycone residue in the minor groove.

Kumar et al. [39] determined the solution structure of calicheamicin γ_1^I complexed to the hairpin duplex

$$\begin{array}{c} T_{10} \\ C_1 - A_2 - C_3 - T_4 - C_5 - C_6 - T_7 - G_8 - G_9 \diagdown T_{11} \\ \diagdown T_{12} \\ G_{23} - T_{22} - G_{21} - A_{20} - G_{19} - G_{18} - A_{17} - C_{16} - C_{15} T_{13} \\ \diagdown T_{14} \end{array}$$

containing a TCCT sequence by 2D NMR spectroscopy coupled with molecular dynamics calculations. They were able to build a highly accurate molecular model based on a large number of intramolecular NOEs (353 among DNA protons and 52 among drug protons) and 70 intermolecular NOEs (drug–DNA). The drug molecule binds to the DNA minor groove with the aryltetrasaccharide chain in an extended form spanning over the TCCT sequence, and the B and C rings are sandwiched between the walls of the minor groove. Intermolecular hydrophobic and hydrogen-bonding interactions account for the observed sequence specificity of the drug. These include intermolecular contacts between the iodine and thiobenzoate sulfur atoms of the C ring and the exocyclic NH_2 protons of G_{18} and G_{19} of the AGGA sequence. The aglycone residue is located deep in the minor groove so that the proradical R ring C3 and C6 atoms are close to H5′ of C_5 (3.07 Å) and H4′ of T_{22} (3.43 Å), respectively. The DNA duplex has a B-type conformation at the aryltetrasaccharide binding site, and widening of the minor groove width is noted at the aglycone binding site.

5.2.2. Esperamicins

Esperamicin A_1 (Fig. 5.16) is an antibiotic and anticancer drug isolated from *Actinomadura verrucosopora*. It cleaves single-stranded as well as double-stranded DNA when activated by a reducing agent (thiol) or UV irradiation or heating. Although esperamycin A_1 does not exhibit specific sequence selectivity, the preferred cleavage sites are known to be in the order of $T > C > A > G$. Similar to calicheamicins discussed in the preceding section, the origin of its cytotoxic activity is the 3-ene-1,5-diyne moiety (R ring), which produces a biradical species. As shown in Fig. 5.16, the R ring is attached to a trisaccharide (A-B-C rings) on one end, and to a fucose (D ring)—anthranilate (E-domain) moiety on the opposite end.

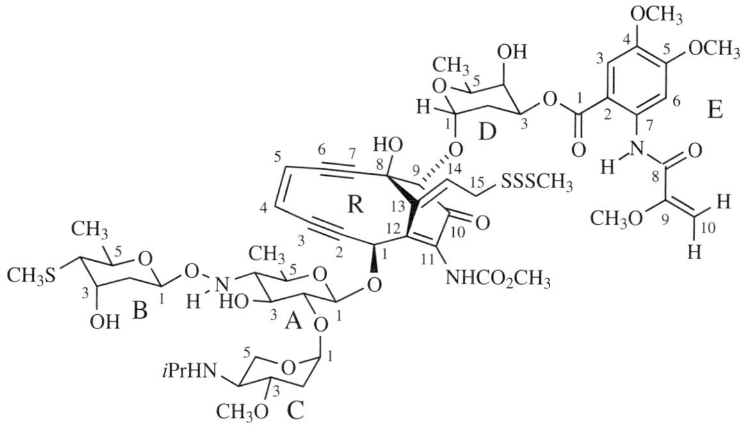

FIGURE 5.16. Structure of esperamicin A_1 [40].

To determine the mode of binding of esperamicin A_1 with DNA, Kumar et al. [40] studied the solution structure of the 1 : 1 complex of the drug with the octamer duplex

$$\begin{array}{ccccccccc} & 1 & 2 & 3 & 4 & 5 & 6 & 7 & 8 \\ 5'\text{-d(C} & G & G & A & T & C & C & G)\text{-}3' \\ 3'\text{-d(G} & C & C & T & A & G & G & C)\text{-}5' \\ & 16 & 15 & 14 & 13 & 12 & 11 & 10 & 9 \end{array}$$

using 2D NMR spectroscopy together with molecular dynamics calculations. Here, the base-pair sequence in the complementary strand is numbered differently from the original to simplify the notation. These workers observed 88 intermolecular NOEs, the majority of which were between the E domain of the drug and the G_2-C_{15} and G_3-C_{14} base pairs of the duplex. Figure 5.17 shows a stereo view of four superpositioned intensity refined structures of the complex; the methoxyacrylyl–anthranilate moiety (E domain) in the minor groove is intercalated between the G_2-C_{15} and G_3-C_{14} base pairs and that the enediyne (R) ring, together with the A-B-C sugar rings, is positioned at the central A_4-T_5-C_6/T_{13}-A_{12}-G_{11} segment. The proradical C3 and C6 atoms of the enediyne ring are aligned to face the abstractable H5′ of C_6 and H4′ of C_{14}. The thiomethyl sugar residue (B ring) is deep in the minor groove with its faces sandwiched between the groove walls, and its polarizable sulfur atom is in a position that enables it to form a hydrogen bond with the amine proton of G_{11}.

The conformation of the duplex is close to B form, although a pronounced helical unwinding near the intercalation site and widening of the minor groove in the segment covering the enediyne moiety and the intercalation site are noted. These structural

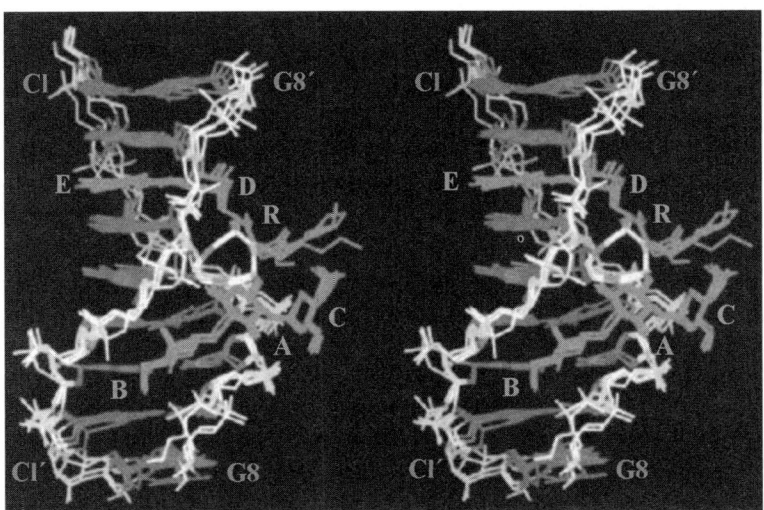

FIGURE 5.17. Stereo view of four superpositioned intensity refined structures of esperamicin A_1–d(CGGATCCG)$_2$ complex. The DNA backbone and base pairs are shown in yellow and magenta, respectively. The drug is shown in blue except for the enediyne ring, which is shown in white with its C3 and C6 carbon atoms in green [40]. (See color insert.)

features result from several factor including (1) the complementarity of the fit between the drug and the floor of the minor groove, (2) intercalation of the anthranilate ring between flanking purine bases, and (3) intermolecular hydrogen bonding of the thiomethyl sugar residue mentioned above.

5.2.3. Neocarzinostatin

Neocarzinostatin (NCS) is a natural antitumor antibiotic composed of a highly labile chromophore (NCS chrom) and a protein that stabilizes the chromophore. As shown in Fig. 5.18, NCS chrom (structure **a**) contains a highly strained enediyne ring to which a cyclocarbonate (CC), 2-*N*-methyl-fucosamine (NMF), and a naphthorate (NPH) are

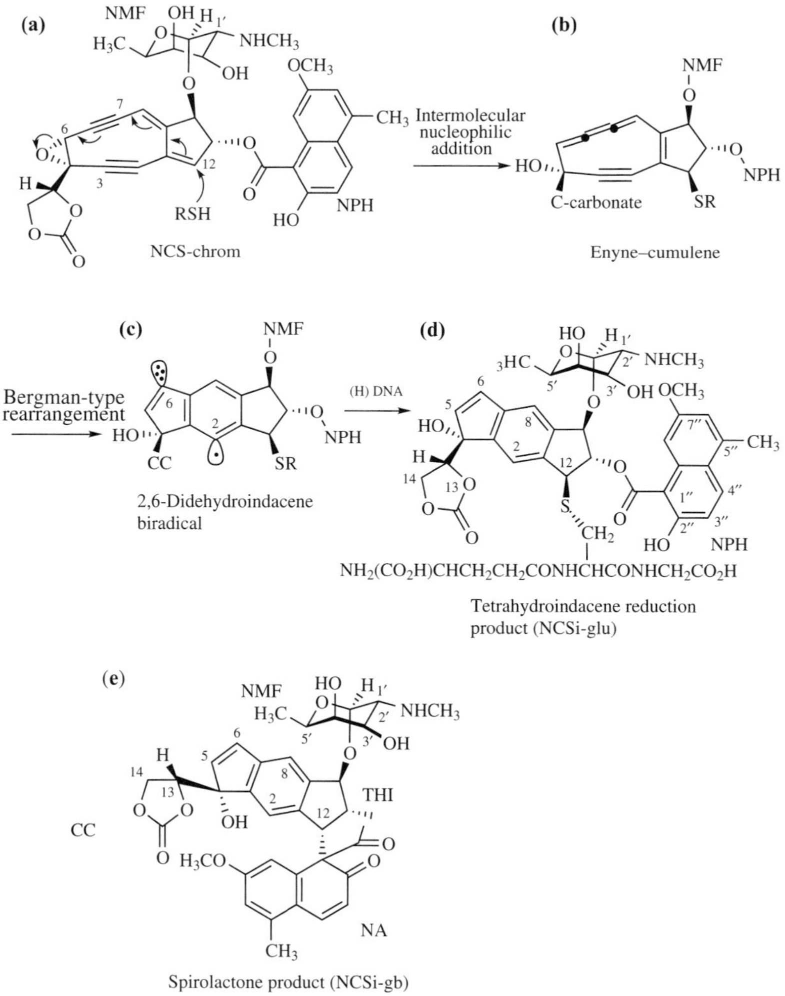

FIGURE 5.18. Formation of NCSi-glu in the presence of a thiol(RSH) (structures **a–d**) and the structure of NCSi-gb (structure **e**) [44].

attached at C4, C10, and C11 positions, respectively. When the enediyne ring is activated by nucleophilic addition of a thiol (SR) such as glutathione (glu) at the C12 atom (structure **b**), the 2,6-biradical species (structure **c**) is produced. This biradical species cleaves single-stranded DNA with base preference (T>A>C>G) and double-stranded DNA with sequence specificity such as AG**C**G**C**T and AG**T**A**C**T, where the boldfaced bases indicate cleavage sites.

Gao et al. [41] determined the 3D structure of the glutathione postactivated NCS chrom, namely NCSi-glu (structure **d** in Fig. 5.18) bonded to the heptamer duplex

$$
\begin{array}{cccccccc}
 & 1 & 2 & 3 & 4 & 5 & 6 & 7 \\
5'\text{-d(G} & \text{G} & \text{A} & \text{G} & \text{C} & \text{G} & \text{C)-3'} \\
3'\text{-d(C} & \text{C} & \text{T} & \text{C} & \text{G} & \text{C} & \text{G)-5'} \\
 & 14 & 13 & 12 & 11 & 10 & 9 & 8
\end{array}
$$

by 2D NMR spectroscopy coupled with molecular dynamics calculations. The structure of NCSi-glu is the same as the native NCS chrom except that the enediyne ring of the latter is replaced by the tetrahydroindacene ring (THI). Figure 5.19 shows a stereo view of an overlay of five calculated structures and the energy-minimized average structure in the central region of the duplex; NCSi-glu takes an extended form and interacts with the $G_{10}C_{11}T_{12}C_{13}$ sequence on one strand and with the $A_3G_4C_5$ sequence on the other strand. Geometry parameters related to the rotatable bonds of NCSi-glu in the complex show that the NPH is directed away from the enediyne ring and intercalated between the A_3-T_{12} and G_4-C_{11} base pairs. One of the phenyl rings containing the $C1''$ atom of the NPH stacks on the purine ring of G_4, and the NPH ring is aligned in the minor groove in an orientation so that the C6 atom of NCSi-glu is close to the $C5'$ atom of T_{12} and the C2 atom is close to the $C1'$ atom of C_5 (shown by arrows in Fig. 5.19). This orientation of the NPH ring is responsible for the double-strand cleavage of DNA duplex by the 2,6-biradical. Such intercalation at $5'$-ApG and $5'$-CpT sequences is unusual because intercalation generally occurs at an alternating purine–pyrimidine or pyrimidine–purine base step. The complex is also stabilized by the drug contacts with the DNA backbone and minor groove via van der Waals interactions.

The work above described was extended to the complex of NCSi-glu with the decamer d(GCCAGAGAGC) [42]. In this case, drug binding triggered a conformational change from a loose duplex to a single-stranded, tightly folded hairpin structure containing a bulged adenosine embedded within a 3-bp stem closing the hairpin loop:

$$
\begin{array}{l}
5'\text{- d}(G_1\text{ - }C_2\text{ – }C_3\text{ – }A_4 \\
\hspace{5em} \searrow G_5 \\
3'\text{- d}(C_{10}\text{ - }G_9\text{ }G_7\text{ – }A_6 \\
\hspace{5em} \searrow\nearrow \\
\hspace{5.5em} A_8
\end{array}
$$

The structure of this complex was generated by using 276 distance contraints coupled with 40 dihedral angle constraints. The NPH moiety of NCSi-glu is intercalated between the C_3-G_7 and C_2-G_9 steps flanked by the bulge site with H2 of A_8 and O6 of

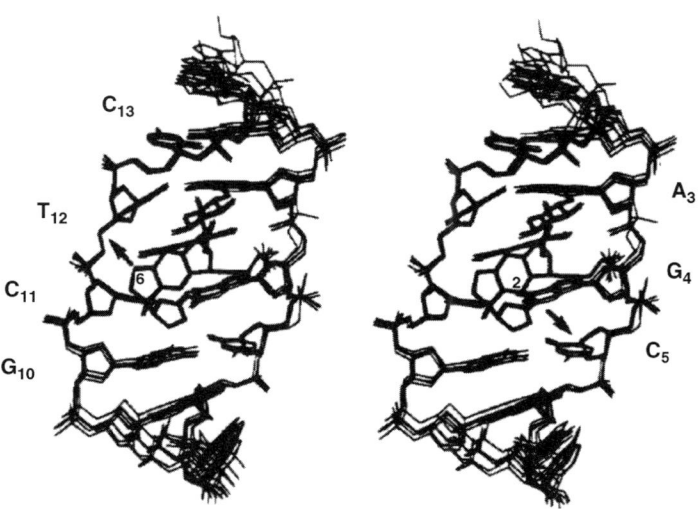

FIGURE 5.19. Overlay stereo view of five calculated structures and the energy-minimized average structure of the NCSi-glu-d(GGAGCGC)·d(GCGCTCC) complex. The view into the minor groove displays the extended, V-shaped binding site. The $C_5(C1')$ and $T_{12}(C5')$ cleavage sites by the 2,6-biradical (Fig. 5.18(**c**)) are indicated by the arrows [41].

G_9 separated by 2.69 Å. The minor groove binding portion of NCSi-glu (NMF and THI groups) assumes an extended conformation along the $C_2C_3A_4$ strand, and the SR moiety aligns in parallel with THI and NMF on one side and mainly along the $G_7A_8G_9$ strand on the other side. The complex is stabilized largely by the stacking interaction at the bulged intercalation site and van der Waals interaction (also possibly by hydrogen bonding) between NCSi-glu and the minor groove of the decamer. NCSi-glu is positioned so that only single-strand cleavage at C_3 is possible because the bases opposite to C_3 are bulged.

Another stable postactivated mimic of NCS chrom is NCSi-gb, which is formed by the general-base(gb)-catalyzed activation of NCS chrom in the absence of DNA under thiol-free conditions. As shown in Fig. 5.18(structure **e**), NCSi-gb consists of NMF, THI, and NA (dihyronaphthone), which are connected to THI via the spirolactone ring. The mode of binding of NCSi-gb with a nucleotide containing two bulged bases is shown in Fig. 5.20. Stassinopoulos et al. [43,44] determined the detailed structure of this complex using 2D NMR spectroscopy coupled with molecular dynamics calculations. These workers derived a molecular model based on 629 NMR restraints including 85 intermolecular NOEs. The rigidly held NA and THI rings form a molecular wedge penetrating the binding pocket and immobilizing the otherwise flexible bulged A_{12} and T_{13} residues. Partial stacking interactions are seen for the NA ring with the A_{11}-T_6 base pair and for the THI ring with the T_{14}-A_5 base pair. The NMF moiety attached to the THI ring is located in the center of the major groove, and determines how deeply NCSi-gb can penetrate in the major groove. Binding of NCSi-gb locks the conformation of the two bulged residues (A_{12} and T_{13}) in

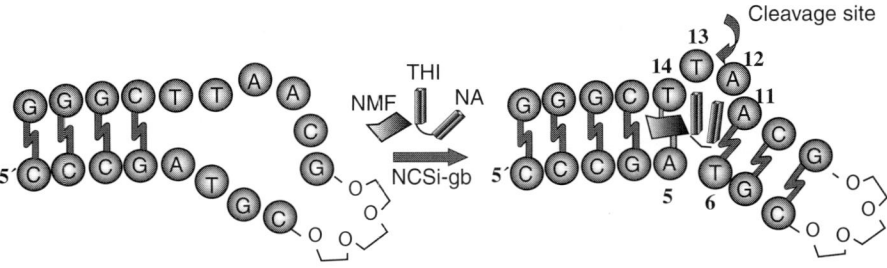

FIGURE 5.20. Schematic illustration of conformational change from free DNA to NCSi-gb bulged DNA on complex formation. The two strands are connected by a triethylene glycol [44].

the major groove with *anti* glycoside conformation, and the complex is bent by ~45° around the bulged site.

Despite similarity between the structures of NCSi-gb and NCSi-glu, there are marked differences in the structures of their DNA complexes and in their DNA cleavage specificities. It is not possible, however, to establish simple structure–activity relationships in this case because the factors affecting the binding are complicated [44]. The ability of NCS chrom derivatives to recognize bulged base sites is important because the biomedical significance of bulged DNA is well known [43]. Goldberg and co workers [45] extended the above mentioned work to the interaction of an oligonucleotide containing a two-base bulge site with a pair of synthetic spirocyclic enantiomers, one of which mimics the spirocyclic geometry of natural NCSi-gb. Differences between the enantiomers were noted in the orientation direction of the drug molecule in the minor groove and in the orientation of the two bases at the bulged site.

REFERENCES

1. Umezawa, H., Maeda, K., Takeuchi, T., and Okami, Y., "New antibiotics, bleomycin A and B," *J. Antibiot.* **19A**:200–209 (1966).
2. Petering, D. H., Byrnes, R. W., and Antholine, W. E., "The role of redox-active metals in the mechanism of action of bleomycin," *Chem.–Biol. Interact.* **73**:133–182 (1990).
3. Hecht, S. M., "RNA degradation by bleomycin, a naturally occuring bioconjugate," *Bioconj. Chem.* **5**:513–526 (1994).
4. Stubbe, J., Kozarich, J. W., Wu, W., and Vanderwall, D. E., "Bleomycins: A structural model for specificity, binding, and double strand cleavage," *Acc. Chem. Res.* **29**:322–330 (1996).
5. Calafat, A. M. and Marzilli, L. G., "Chiralities of complexes of bleomycin-type ligands, a neglected feature in structural studies relevant to anticancer drug action," *Comments Inorg. Chem.* **20**:121–141 (1998).
6. Burger, R. M., "Cleavage of nucleic acids by bleomycin," *Chem. Rev.* **98**:1153–1169 (1998).

7. Gamcsik, M. P., Glickson, J. D., and Zon, G., "NMR studies of the interaction of bleomycin with (dC-dG)$_3$," *J. Biomol. Struct. Dyn.* **7**:1117–1133 (1990).
8. Chen, D. M., Sakai, T. T., Glickson, J. D., and Patel, D. J., "Bleomycin-A$_2$ complexes with poly(dA-dT). A proton nuclear magnetic resonance study of the nonexchangeable hydrogens," *Biochem. Biophys. Res. Commun.* **92**:197–205 (1980).
9. Rajani, C., Kincaid, J. R., and Petering, D. H., "A systematic approach toward the analysis of drug-DNA interactions using Raman spectroscopy: The binding of metal-free bleomycins A$_2$ and B$_2$ to calf thymus DNA," *Biopolymers* **52**:110–128 (1999).
10. Rajani, C., Kincaid, J. R., and Petering, D. H., "The presence of two modes of binding to calf thymus DNA by metal-free bleomycin: A low frequency Raman study," *Biopolymers* **52**:129–146 (1999).
11. Weselucha-Birczynska, A., Strahan, G. D., Tsuboi, M., and Nakamoto, K., "Interaction of bleomycin A$_2$ with DNA studied by resonance Raman spectroscopy: Intercalation or groove binding?" *J. Raman Spectrosc.* **31**:1073–1077 (2000).
12. Chien, M., Grollman, A. P., and Horwitz, S. B., " Bleomycin-DNA interactions: Fluorescence and proton magnetic resonance studies," *Biochemistry* **16**:3641–3647 (1977).
13. Sugiura, Y., " Bleomycin-iron complexes. Electron spin resonance study, ligand effect, and implication for action mechanism," *J. Am. Chem. Soc.* **102**:5208–5215 (1980).
14. Sugiura, Y., Takita, T., and Umezawa, H., "Bleomycin antibiotics: Metal complexes and their biological action," in Sigel, H. and Sigel, A., eds., *Metal Ions in Biological Systems*, Marcel Dekker, New York, 1985, Vol. 19, Chap. 4, pp. 81–108.
15. Burger, R. M., Peisach, J., and Horwitz, S. B., "Activated bleomycin," *J. Biol. Chem.* **256**:11636–11644 (1981).
16. Kuramochi, H., Takahashi, K., Takita, T., and Umezawa, H., "An active intermediate formed in the reaction of bleomycin-Fe(II) complex with oxygen," *J. Antibiot.* **34**:576–582 (1981).
17. Iitaka, Y., Nakamura, H., Nakatani, T., Muraoka, Y., Fujii, A., Takita, T., and Umezawa, H., "Chemistry of bleomycin. XX. The X-ray structure determination of P-3A Cu(II)-complex, a biosynthetic intermediate of bleomycin," *J. Antibiot.* **31**:1070–1072 (1978).
18. Neese, F., Zaleski, J. M., Zaleski, K. L., and Solomon, E. I., "Electronic structure of activated bleomycin: Oxygen intermediates in heme versus non-heme iron," *J. Am. Chem. Soc.* **122**:11703–11724 (2000).
19. Takahashi, S., Sam, J. W., Peisach, J., and Rousseau, D. L., "Structural characterization of iron-bleomycin by resonance Raman spectroscopy," *J. Am. Chem. Soc.* **116**:4408–4413 (1994).
20. Burger, R. M., "Nature of activated bleomycin", in Meunier, B., ed., *Structure and Bonding: Metal-Oxo and Metal-Peroxo Species in Catalytic Oxidations*, Springer-Verlag Berlin, 2000, Vol. 97, 287–323.
21. Sugiura, Y., "Monomeric cobalt(II)-oxygen adducts of bleomycin antibiotics in aqueous solution. A new ligand type for oxygen binding and effect of axial Lewis base," *J. Am. Chem. Soc.* **102**:5216–5221 (1980).
22. Wu, W., Vanderwall, D. E., Turner, C. J., Kozarich, J. W., and Stubbe, J., "Solution structure of Co-bleomycin A2 green complexed with d(CCAGGCCTGG)," *J. Am. Chem. Soc.* **118**:1281–1294 (1996).

23. Stubbe, J. and Kozarich, J. W., "Mechanisms of bleomycin-induced DNA degradation," *Chem. Rev.* **87**:1107–1136 (1987).
24. Vanderwall, D. E., Lui, S. M., Wu, W., Turner, C. J., Kozarich, J. W., and Stubbe, J., "A model of the structure of HOO-Co·bleomycin bound to d(CCAGTACTGG): Recognition at the d(GpT) site and implications for double-stranded DNA cleavage," *Chem. Biol.* **4**:373–387 (1997).
25. Caceres-Cortes, J., Sugiyama, H., Ikudome, K., Saito, I., and Wang, A. H.-J., "Interactions of deglycosylated cobalt(III)-peplomycin (green form) with DNA based on NMR structural studies," *Biochemistry* **36**:9995–10005 (1997).
26. Hoehn, S. T., Junker, H.-D., Bunt, R. C., Turner, C. J., and Stubbe, J., "Solution structure of Co(III)-bleomycin-OOH bound to a phosphoglycolate lesion containing oligonucleotide: Implications for bleomycin-induced double-strand DNA cleavage," *Biochemistry* **40**:5894–5905 (2001).
27. Rajani, C., Kincaid, J. R., and Petering, D. H., "Raman spectroscopy of an O_2-Co(II) bleomycin-calf thymus DNA adduct: Alternate polymer conformations," *Biophys. Chem.* **94**:219–236 (2001).
28. Rajani, C., Kincaid, J. R., and Petering, D. H., "Resonance Raman studies of HOO-Co(III) bleomycin and Co(III)bleomycin: Identification of two important vibrational modes, ν(Co-OOH) and ν(O-OH)," *J. Am. Chem. Soc.* **126**:3829–3836 (2004).
29. Manderville, R. A., Ellena, J. F., and Hecht, S. M., "Interaction of Zn(II)-bleomycin with d(CGCTAGCG)$_2$. A binding model based on NMR experiments and restrained molecular dynamics calculations," *J. Am. Chem. Soc.* **117**:7891–7903 (1995).
30. Urata, H., Ueda, Y., Usami, Y., and Akagi, M., "Enantiospecific recognition of DNA by bleomycin," *J. Am. Chem. Soc.* **115**:7135–7138 (1993).
31. Lown, J. W. and Joshua, A. V., "Bleomycin models. Haemin-acridines which bind to DNA and cause oxygen-dependent scission," *J. Chem. Soc., Chem. Commun.* 1298–1300 (1982).
32. Lown, J. W., Sondhi, S. M., and Ong, C. W., "Deoxyribonucleic acid cleavage specificity of a series of acridine- and acodazole-porphyrins as functional bleomycin models," *Biochemistry* **25**:5111–5117 (1986).
33. Lown, J. W., Plenkiewicz, J., Ong, C.-W., Joshua, A. V., McGovren, J. P., and Hanka, L. J., "Models for bleomycin antitumour antibiotics: Haemin-acridines which bind to DNA, cause oxygen-dependent strand scission, and exhibit anticancer properties," *Proc. 9th Int. Union of Pharmacology Congress*, Macmillan, London, 1984, pp. 265–269.
34. Hertzberg, R. P. and Dervan, P. B., "Cleavage of double helical DNA by (methidiumpropyl-EDTA)iron(II)," *J. Am. Chem. Soc.* **104**:313–315 (1982).
35. Tan, J. D., Farinas, E. T., David, S. S., and Mascharak, P. K., "NMR evidence of sequence specific DNA binding by a cobalt(III)-bleomycin analogue with tethered acridine," *Inorg. Chem.* **33**:4295–4308 (1994).
36. Smith, A. L. and Nicolaou, K. C., "The enediyne antibiotics," *J. Med. Chem.* **39**:2103–2117 (1996).
37. Walker, S. L., Andreotti, A. H., and Kahne, D. E., "NMR characterization of calicheamicin γ_1^1 bound to DNA," *Tetrahedron* **50**:1351–1360 (1994).
38. Paloma, L. G., Smith, J. A., Chazin, W. J., and Nicolaou, K. C., "Interaction of calicheamicin with duplex DNA: Role of the oligosaccharide domain and identification of multiple binding modes," *J. Am. Chem. Soc.* **116**:3697–3708 (1994).

39. Kumar, R. A., Ikemoto, N., and Patel, D. J., "Solution structure of the calicheamicin γ_1^1-DNA complex," *J. Mol. Biol.* **265**:187–201 (1997).
40. Kumar, R. A., Ikemoto, N., and Patel, D. J., " Solution structure of the esperamicin A_1-DNA complex," *J. Mol. Biol.* **265**:173–186 (1997).
41. Gao, X., Stassinopoulos, A., Rice, J. S., and Goldberg, I. H., "Structural basis for the sequence-specific DNA strand cleavage by the enediyne neocarzinostatin chromophore. Structure of the post-activated chromophore-DNA complex," *Biochemistry* **34**:40–49 (1995).
42. Kwon, Y., Xi, Z., Kappen, L. S., Goldberg, I. H., and Gao, X., "New complex of post-activated neocarzinostatin chromophore with DNA: Bulge DNA binding from the minor groove," *Biochemistry* **42**:1186–1198 (2003).
43. Stassinopoulos, A., Ji, J., Gao, X., and Goldberg, I. H., "Solution structure of a two-base DNA bulge complexed with an enediyne cleaving analog," *Science* **272**:1943–1946 (1996).
44. Gao, X., Stassinopoulos, A., Ji, J., Kwon, Y., Bare, S., and Goldberg, I. H., "Induced formation of a DNA bulge structure by a molecular wedge ligand-postactivated neocarzinostatin chromophore," *Biochemistry* **41**:5131–5143 (2002).
45. Hwang, G.-S., Jones, G. B., and Goldberg, I. H., "Stereochemical control of small molecule binding to bulged DNA: Comparison of structures of spirocyclic enanthiomer-bulged DNA complexes," *Biochemistry* **43**:641–650 (2004).

CHAPTER 6

Metal-Containing Drugs

6.1. CISPLATIN

In 1965, Rosenberg and coworkers [1] observed the biological activity of platinum complexes during their investigation of the effect of electric field on the growth process of *Escherichia coli* cells. In the following work [2], they discovered that some platinum complexes exhibit strong antitumor activities. Among them, a square–planar platinum complex, *cis*-diamminedichloroplatinum(II), *cis*-Pt(NH$_3$)$_2$Cl$_2$ (abbreviated as *cisplatin*), was found to be highly effective in treating some cancers such as testicular and ovarian cancers. Since then, extensive studies have been conducted to search for more potent drugs with a wider range of activity and less toxic side effects than cisplatin. Thus far, more than 3000 cisplatin analogs have been prepared and their clinical activities tested. However, only a few drugs, including cisplatin, carboplatin, and oxaliplatin, (shown in Fig. 6.1) are currently in clinical use. In the following, structural and spectroscopic studies of interactions of these drugs with DNA are discussed using selected examples. Structural studies up to 1987 have been reviewed by Sherman and Lippard [3].

When cisplatin is reacted with DNA, the four types of platinum complexes shown in Fig. 6.2b are formed [4]. In all cases, the two Cl ligands of cisplatin are replaced by the N7 atoms of guanine or adenine to form covalent Pt—N bonds (Fig. 6.2a), while the two NH$_3$ ligands remain intact. Structures **a** and **b** in Fig. 6.2b illustrate 1,2-intrastrand crosslinking between two successive guanine–guanine (5′-GpG-3′) and adenine–guanine (5′-ApG-3′) sequences, respectively. Computer-generated model studies show that cisplatin cannot bind to the guanine–adenine (5′-GpA-3′) sequence because of an unfavorable contact between the NH$_3$ group of cisplatin and the exocyclic NH$_2$ group of adenine [4]. Structures **c** and **d** in Fig. 6.2b illustrate *1,3-intrastrand crosslinking* between two guanine bases separated by one base and *interstrand crosslinking* between two guanines of complementary strands, respectively. The major products of cisplatin–DNA reaction are 60% structure **a** and 25% structure **b**, while each of the minor products (structures **c** and **d**) is formed less than 10% [5].

Drug–DNA Interactions: Structures and Spectra by Kazuo Nakamoto, Masamichi Tsuboi, and Gary D. Strahan
Copyright © 2008 John Wiley & Sons, Inc.

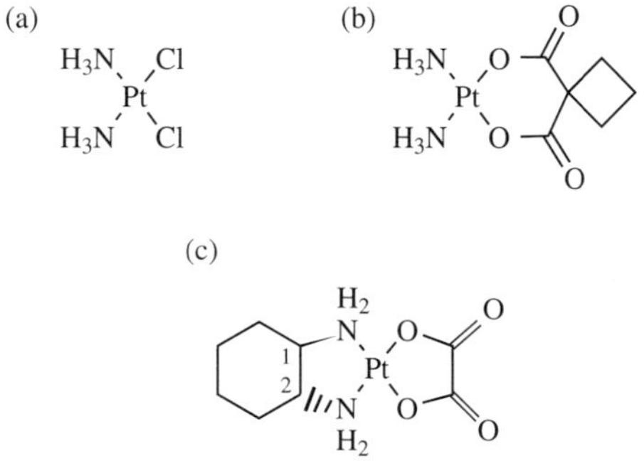

304 METAL-CONTAINING DRUGS

FIGURE 6.1. Structures of cisplatin (a), carboplatin (b), and oxaliplatin (c).

Thus, the majority of investigations described in this section are those on 1,2-intrastrand crosslinking complexes of structure **a** type.

6.1.1. Intrastrand Crosslinking Complexes

6.1.1.1. X-Ray Crystallography   In 1985 and 1988, Lippard and coworkers [6,7] reported the X-ray crystal structure of *cis*-[Pt(NH$_3$)$_2${d(pGpG)}]. The unit cell of type I crystal contains four crystallographically independent conformers in a tightly packed aggregate. The structure of one of these conformers is shown in Fig. 6.3a. In all these conformers, the geometry around the Pt atom is square–planar and the average Pt-NH$_3$ and Pt-N7 (G) distances are 2.04 and 2.00 Å, respectively. As a result, the guanine bases destack each other with the G/G dihedral angle in the 76.2 to 86.8° range by changing the torsion angles of the dinucleotide. The guanosine nucleosides are in

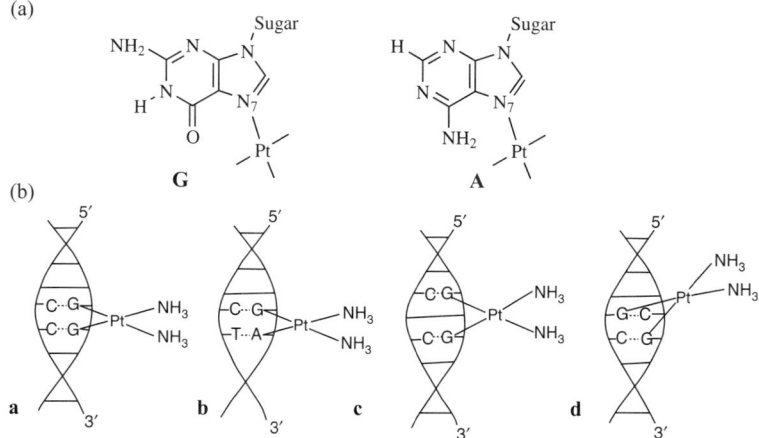

FIGURE 6.2. (a) Modes of binding of cisplatin to guanine (**G**) and adenine (**A**); (b) 1,2-intrastrand GpG (structure **a**), 1,2-intrastrand ApG (structure **b**), 1,3-intrastrand GpNpG (structure **c**) (N=any base), and 1,2- interstrand GpG (structure **d**) [4].

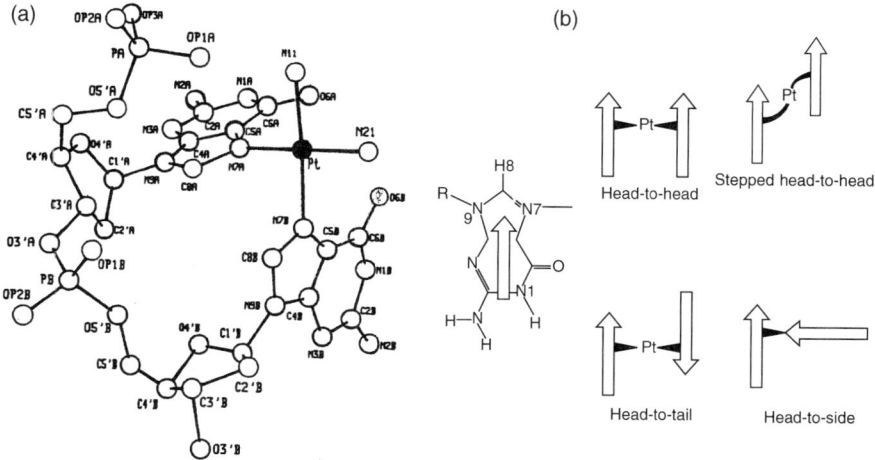

FIGURE 6.3. (a) Molecular structure of one of the four conformers of cis-[Pt(NH$_3$)$_2$ {d(pGpG)}] [6]; (b) head-to-head (HH), head-to-tail (HT), head-to-side (HS), and stepped head-to-head conformations.

the *anti* conformation, and the two guanines are orientated in the same direction (head-to-head orientation, HH) shown in Fig. 6.3b. Hydrogen bonding occurs between the NH$_3$ proton and the O atom of the terminal 5'-phosphate in most conformers. In the case of cis-[Pt(NH$_3$)$_2$(purine)$_2$], molecular modeling studies [8] suggest that the HT (head-to-tail) isomer (Fig. 6.3b) is generally more stable than the HH isomer. However, the X-ray crystal structure of cis-[Pt(NH$_3$)$_2$(9-methyladenine)(9-ethylguanine)](NO$_3$) · 2H$_2$O shows that the Pt atom is bound to the N7 atoms of the two purine bases, which assume an HH orientation [9]. X-ray analysis also shows that cisplatin complexed to the trinucleotide, [Pt(NH$_3$)$_2$(CpGpG)], forms *crosslinking* between the N7 atoms of the two guanine bases that are in a HH orientation [10].

The 1,2-intrastrand crosslinking of cisplatin to a double-stranded oligonucleotide was first confirmed by Lippard and coworkers [11,12], who carried out X-ray analysis of cisplatin complexed to the dodecamer

```
         1   2   3   4   5   6   7   8   9   10  11  12
5'-d(C   C   T   C   T   G*  G*  T   C   T   C   C)-3'
3'-d(G   G   A   G   A   C   C   A   G   A   G   G)-5'
         24  23  22  21  20  19  18  17  16  15  14  13
```

in which the N7 atoms of the two successive G bases marked by the G* coordinate to the cis-Pt(NH$_3$)$_2$ moiety. The structure was determined by isomorphous replacement methods using three brominated derivatives. Figure 6.4 shows a stereo view of a model of the platinated 12-mer in which T$_3$ is replaced by 5-bromouridine. It is seen that the 1,2-intrastrand crosslinking of cisplatin in the major groove siginificantly bends the duplex (by 39° or 55° depending on the molecule in the unit cell) at the platination site without disrupting Watson–Crick hydrogen bonding. However, this geometry

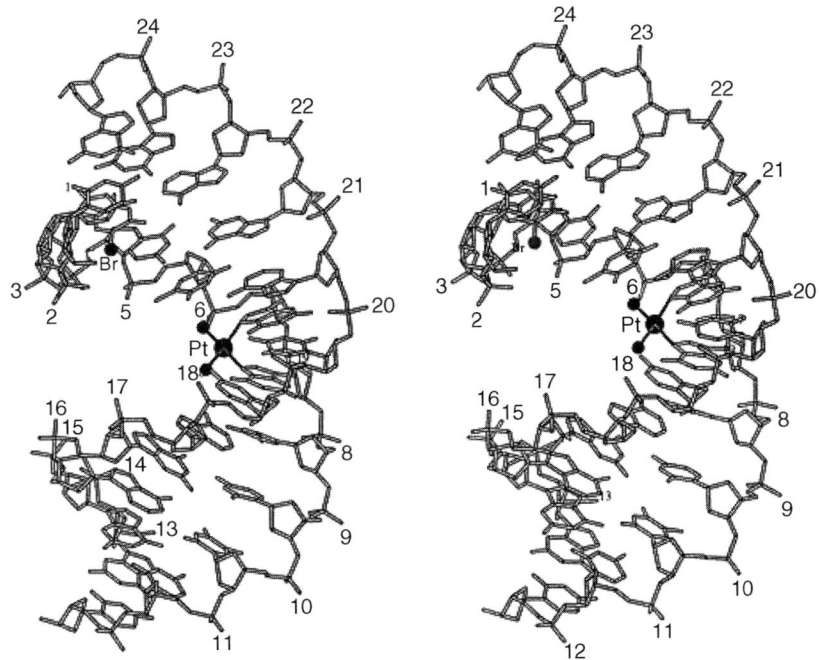

FIGURE 6.4. Stereo view of the 12-mer modified by cis-$[Pt(NH_3)_2]^{2+}$ binding [12].

compacts the major groove, and widens and flattens the minor groove. As a result, the platinum atom is displaced from the square–planar plane by $\sim$1 Å. It was also noted that the conformation of one end of the duplex (from C_1-G_{24} to T_5-A_{20} base pairs) is A-DNA-like while that of the other end is B-DNA-like. Such a severely bent structure with a widened minor groove resembles the DNA structures in complexes with proteins containing the high-mobility group (HMG). This similarity suggests that HMG-domain proteins may recognize a cisplatin–DNA complex, thus explaining the origin of the antitumor activity of cisplatin.

In 1999, Ohndorf et al. [13] added credence to the concept described above by determining the crystal structure of the cisplatinated 16-mer

$$5'\text{-d(C C T C T C T G* G* A C C T T C C)-}3'$$
$$5'\text{-d(G G A G A G A C C T G G A A G G)-}3'$$

complexed to domain A of the structure-specific HMG1 protein. As seen in Fig. 6.5, this 16-mer is severly kinked at the central (G*G*) notch, and bound to the protein via the 3' side of the platinated strand. Furthermore, the aromatic sidechain of phenylalanine residue (F37) at the amino terminous of helix II is intercalated to the hydrophobic G*G* notch. Site-directed mutagenesis experiments indicate a similar sidechain intercalation of a methionine or isoleucine residue at position 16 of helix I of domain A [14].

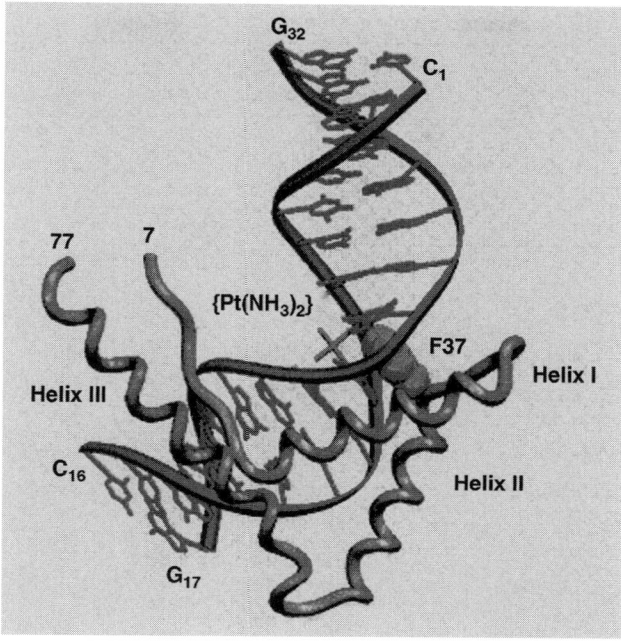

FIGURE 6.5. Overall structure of cisplatinated 16-mer complexed to domain A of the HMG1 protein. The protein backbone is shown in brown. The intercalatinphenylalanine (F37) residue is shown as van der Waals spheres, the 16-mer is shown in red and blue, and the cis-[Pt(NH$_3$)$_2$]$^{2+}$ moiety is shown in green [13,14]. (See color insert.)

Structures of cisplatin–DNA adducts and their interaction with cellular proteins have been reviewed by Jamieson and Lippard [15].

6.1.1.2. NMR Spectroscopy

Table 6.1 lists NMR studies on oligonucleotides complexed to cisplatin via 1,2- and 1,3-intrastrand crosslinking [16–23]. In all cases, 2D experiments (COSY and NOESY) were carried out to assign nonexchangeable ^{1}H signals of bases such as G(H8), A(H8,H2), C(H5,H6), and T(H6,CH$_3$), and of deoxyribose such as H1′, H2′, H2″, H3′, H4′, H5′ and H5″. In some cases, exchangeable protons of the NH and NH$_2$ groups of these bases were also assigned, and ^{31}P and ^{13}C spectra were analyzed.

Formation of 1,2- and 1,3-intrastrand crosslinking can be confirmed by large downfield shifts of G(H8) and A(H8) signals relative to other protons. As seen in Table 6.1, chemical shift values (ppm) of these protons are much larger than their normal values in B-DNA duplex (7.6–8.0 ppm for G and 8.0–8.4 ppm for A). This is anticipated because these protons are located nearest to the platination site and most strongly influenced by its inductive effect [24]. On platination, a short duplex such as d(AGGCCT)$_2$ adopts a single-stranded, base-destacked, coil structure because of severe distortion at the G*G* site [16]. Figure 6.6 shows the 2D-NOESY spectra (nonexchangeable protons) of the platinated dodecamer shown in Table 6.1 (abbreviated as **Dode1**) obtained by van Boom et al. [22]. The strong cross-peak

TABLE 6.1. NMR Studies of Cisplatin–Oligonucleotide Adducts

Nucleotide[a]	Chemical Shifts (ppm) of H8 (G/A)		Kink Angle (degree)	Ref.
d(AG*G*CCT)	G_2	8.17	—	16
d(T C C GGA)	G_3	8.84		
d(CCTG*G*TCC)	G_4	8.76	58	17
d(GGAC C AGG)	G_5	8.19		
d(CT CCG*G*CCT)	G_5	8.66	—	18
d(GAGGC C GGA)	G_6	8.39		
d(CTC A*G*CCTC)	A_4	9.01	40±5	19
d(GAGG C TGAG)	G_5	8.81		
d(CC TCTG*G*TCTCC)	G_6	8.74	78	20
d(GGAGAC C AGAGG)	G_7	8.16		
d(ATACATG* G*TACATA)	G_7	8.64	52±9	21
d(TATGTAC C ATGTAT)	G_8	8.08		
d(GAC C ATATG*G* TC)	G_9	8.66	40[b]	22[c]
d(CTG*G*TATAC C AG)	G_{10}	8.07		
d(CTCTA G*TG*CTCAC)	G_6	8.69	20–24	23
d(GAGAT C A C GAGTG)	G_8	7.77		

[a] G* and A* indicate the N(7)-platinated bases.
[b] Double-kinked duplex.
[c] Abbreviated as **Dode1**.

between G_9(H8) and G_{10}(H8) suggests that these two G bases are 1,2-crosslinked by cisplatin and take a *head-to-head* (HH) configuration. It is a double-kinked duplex containing two G^*G^* sites.

The 1,3-intrastrand crosslinking in the platinated 13-mer listed in Table 6.1 was confirmed by van Garderen and van Houte [23]. In addition to ^{1}H NMR, they compared the ^{31}P NMR of the 13-mer and its platinated complex as shown in Fig. 6.7. It is seen that on platination, two ^{31}P signals are downshifted to −3.10 and −3.72 ppm from other ^{31}P signals in the region from −3.8 to −4.5 ppm. Here, the chemical shift values were determined by using trimethylphosphate as the external standard. These two downshifted signals were assigned to the P atoms of G_6^*-p-T_7 and T_7-p-G_8^* sequences.

The solution structure of the cisplatin-12-mer complex obtained by NMR spectroscopy [20] was compared with that previously determined by X-ray crystallography [11,12]. Comparison of these two results indicates that the DNA duplex is more bent and the dihedral angle between the planes of the crosslinked guanine bases is much larger in solution than in the solid state.

Once the NMR signals were assigned, three-dimensional models of intrastrand crosslinking complexes were generated on the basis of NOE data followed by energy minimization methods. Figure 6.8 shows a stereo view of a molecular model of the platinated **Dode1** obtained by van Boom et al. [22]. As stated earlier, this complex is unique in that platination occurs at two G*G* sites, resulting in two kinks (∼40°) in one duplex.

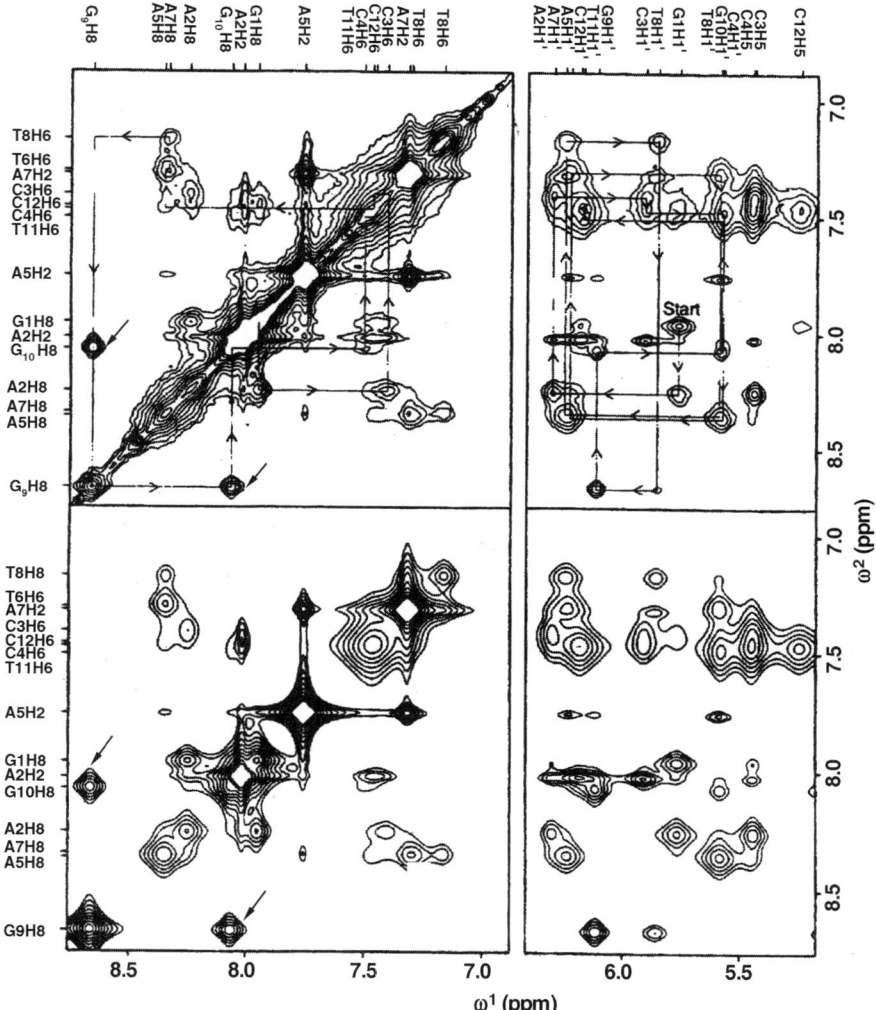

FIGURE 6.6. Two-dimensional NOESY spectra of **Dode1** in the expanded region of the nonexchangeable protons; aromatic–aromatic region (left) and aromatic-H1′/CH5 region (right). The sequential assignment passageways are shown in the upper diagrams, and the simulation spectra based on the refined structure are shown in the bottom diagrams. The arrow indicates the strong cross-peak between G$_9$(H8) and G$_{10}$(H8) [22].

6.1.1.3. UV and CD Spectra

The effect of 1,2-intrastrand crosslinking by cisplatin on DNA double helix has also been studied by UV absorption and CD spectroscopy. As shown in Fig. 6.9, the π–π* transition of the base residues of CT-DNA near 260 nm becomes bathochromic and hyperchromic on platination [25]. The bathochromic shift (259 → 264 nm) is caused by perturbation resulting from covalent bonding between the Pt atom and the N7 atoms of purine bases. Since hypochromism is caused by parallel stacking of base pairs in a double helix (Section 1.2.1), the observed hyperchromism (an increase in absorption intensity) indicates a decrease in the degree

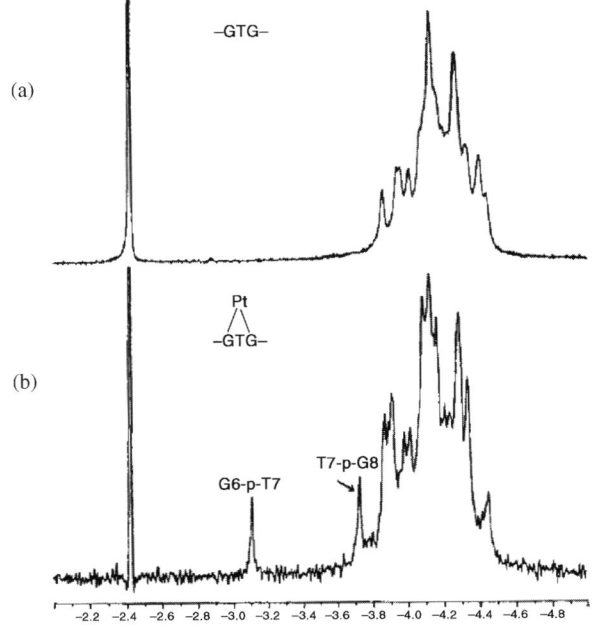

FIGURE 6.7. ^{31}P NMR spectra of (a) the 13-mer and (b) its cisplatin adduct [23].

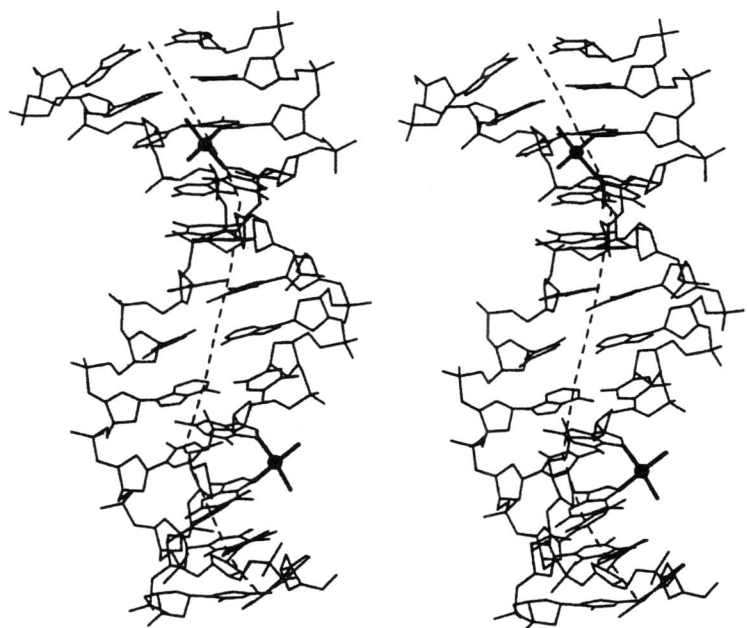

FIGURE 6.8. Stereo view of the doubly kinked structure of platinated **Dode1** [22].

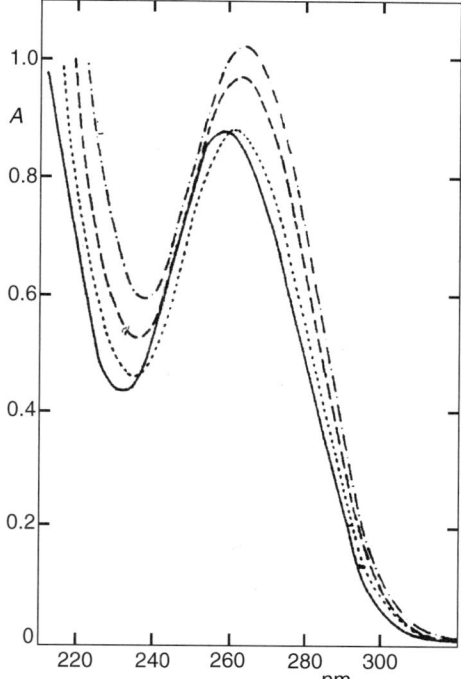

FIGURE 6.9. UV spectra of the mixture of CT-DNA and cisplatin in 5 mM NaCl after incubation at 37°C for 7 h and kept overnight at 0°. The spectra were measured at room temperature. The Pt/PO$_4$ molar ratio is 0 (—); 0.2 (···); 0.5 (— —); 1 (–•–•) [25].

of parallel stacking in a DNA double helix resulting from 1,2-intrastrand crosslinking. Similar bathochromic shifts have been reported for *cis*-[Pt(NH$_3$)$_2$(GpG)] (6 nm) and its GpTpG analog (8 nm) [26].

Breslauer and coworkers [27,28] used CD spectroscopy to study the effect of 1,2-intrastrand crosslinking on the conformation of oligonucleotides. Figure 6.10 shows the structures and CD spectra of three 15-mers and their platinated complexes. On platination, the positive band at 272 nm is redshifted with increasing intensity while the intensity of the negative band at 241 nm decreases. The band near 212 nm is weakly positive or negative, and platination tends to increase its negative intensity. In general, these changes indicate the B → A conformational change, consistent with the results of X-ray [11,12] and NMR [20] studies discussed in the preceding sections. The CD difference (platinated minus unplatinated) spectra shown in the insets suggest that the extent of the B → A conformational change increases in the order of TGGT < CGGC < AGGA.

Circular dichroism observed in the vibrational (IR and Raman) region is called *vibrational circular dichroism* (VCD). The IR and VCD spectra of the adduct of cisplatin with the octamer as previously used for NMR studies (see Table 6.1, results for Ref. 17) were measured in D$_2$O solution [29]. Comparison of these spectra between the unmodified and platinated octamer indicates that the structure of the octamer duplex is considerably distorted by platination. As expected, the major

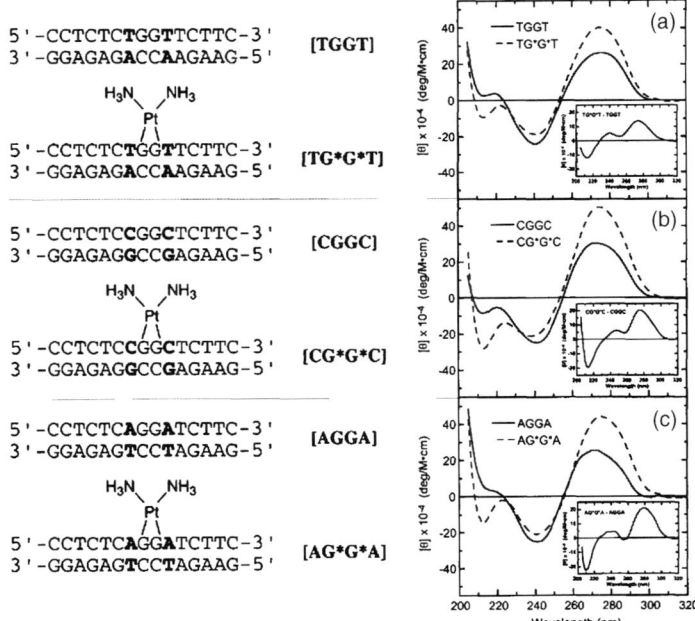

FIGURE 6.10. Structures and CD spectra (25°C) of three 15-mers and their platinated complexes. The inset shows the difference spectrum obtained by subtraction of the CD spectrum of the unmodified duplex from that of the platinated duplex [28].

changes in VCD spectra were detected in the $G_4^*pG_5^*$ step. Slow isomerization of this 1,2-intrastrand crosslinking adduct to the interstrand crosslinking adduct (between G_4 and G_9) previously observed by NMR spectroscopy [17] was also detected by VCD spectroscopy. In general, the corresponding changes in IR spectra were less clear or not observed.

6.1.1.4. IR and Raman Spectra As stated in the preceding section, the major reaction products of clinically active cisplatin with DNA are 1,2-intrastrand crosslinking complexes at 5′-GpG-3′ and 5′-ApG-3′ sequences. On the other hand, transplatin (Section 6.3) is clinically inactive, and its mode of interaction is known to be interstrand crosslinking. In all cases, however, covalent Pt−N bonds are formed by replacing the two Cl atoms of these platinum complexes with the N7 atoms of guanine (G) or adenine (A). Figure 6.11 shows the FTIR spectra of CT-DNA and its mixture with cisplatin in D_2O at low r values (molar ratio of drug/phosphate in the range of 0.03–0.2) measured and assigned by Theophanides [30]. On platination, the $\nu(C=O)$ band of the DNA bases at 1680 cm^{-1} is downshifted to 1660, and the $\nu(C-O)$ band of the ribose–phosphate linkage at 1054 cm^{-1} is reduced in intensity and upshifted to 1060 cm^{-1}. Furthermore, the decrease in intensity of the $\nu_a(P-O)$ band at 971 cm^{-1} suggests that, even at low r values, the attack of guanine bases by cisplatin locally changes the structure of DNA duplex involving the sugar-phosphate backbone.

Fourier transform IR spectroscopy was also utilized to study conformational changes of sugar rings in d(GpC) and d(CpG) on platination [31]. When d(GpC) of

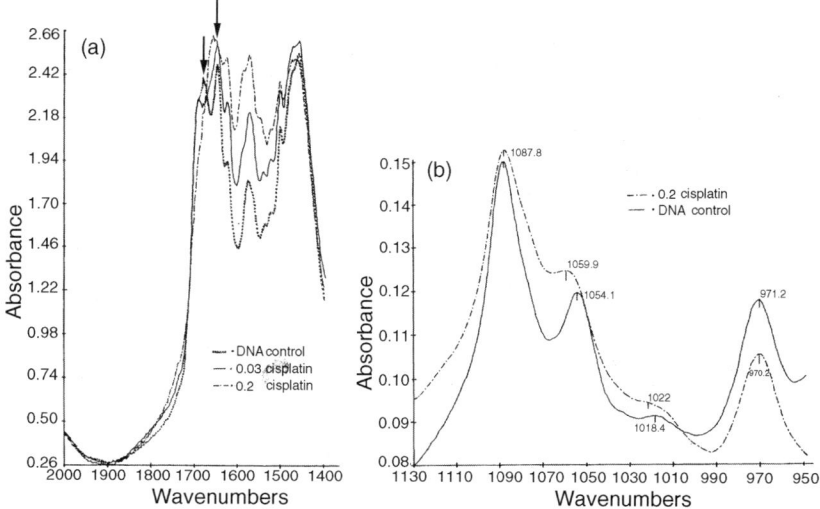

FIGURE 6.11. FTIR spectra of CT-NDA and its mixtures with cisplatin in D_2O: (a) the carbonyl region, where the bands at 1680 and 1660 cm^{-1} are indicated by arrows; (b) sugar–phosphate region, where the Pt/PO$_4$ molar ratios are shown for each figure [30].

high concentration (10^{-2} M) is reacted with cisplatin, the N7 atoms of two guanine bases are coordinated to the cis-Pt(NH$_3$)$_2$ moiety to form cis-[Pt(NH$_3$)$_2$(GpC)$_2$]$^{2+}$, and it exhibits the N (C3$'$-endo–anti)-type sugar conformation of guanosine at 797 cm^{-1}. At low concentrations (10^{-4} M), however, it forms a chelated complex, cis-[Pt(NH$_3$)$_2$(GpC)]$^+$, via coordination to the N7 atom of guanine and the N3 atom of cytosine. This complex exhibits two sugar pucker bands at 818 and 793 cm^{-1}; the former corresponds to the S (south-oriented) (C2$'$-endo)-type guanosine and the latter to the N (north-oriented) (C3$'$-endo)-type cytidine. Thus, the sugar conformation of guanosine changes from the N to S type as a result of chelation.

Complexes of nucleosides with cisplatin and its clinically inactive trans-isomer were compared using Raman spectroscopy [32,33]. No clear-cut differences were observed between cis- and trans-[Pt(NH$_3$)$_2$(Gua)$_2$]$^{2+}$ ions (Gua=guanosine), but the ν(Pt-N(Gua)) bands of both complexes were assigned in the 370–340 cm^{-1} region. This work was extended to the complexes of salmon sperm DNA with cis- and transplatin (488.8 nm excitation) [34]. In this case, the band at 1628 cm^{-1} (assigned to the ν(C=O) of G and C bases) was shifted to 1596 cm^{-1} for cisplatin, whereas no spectral changes were noted for transplatin. The former shift was attributed to 1,2-intrastrand crosslinking of cisplatin.

Although cytosine in CT-DNA is not platinated because of base pairing with guanine, the N3 atom of cytosine may bind to cis- or transplatin when the double-stranded DNA is separated into single strands during the DNA replication process. Strommen and Peticolas [35] studied the interactions of single-stranded poly(dC) with cis- and transplatin by Raman spectroscopy (488 nm excitation). They observed that the intensity of the 1236 cm^{-1} band of poly(dC) is much stronger with trans-isomer than with an equal amount of cis-isomer. The intensity of this pyrimidine ring vibration is a measure of the degree of disordered structure in poly(dC). Thus, their results indicate that the *trans*-isomer can break down the secondary (3D) structure of poly (dC) more than *cis*-isomer can, because the latter may bind to two adjacent cytosine residues at the N3 position via intrastrand crosslinking, resulting in less disruption of the secondary structure.

Surface-enhanced Raman spectra (632.8 nm excitation) of guanine (2×10^{-6} M) incubated with cisplatin (10^{-6} M) and related compounds were measured in silver sol. Density functional theory (DFT) calculations gave the best agreement with the experimentsl results when the adsorbed species was assigned to the 1:1 complex, *cis*-[Pt(NH$_3$)$_2$Cl(G)] [36].

Most of the Raman spectra mentioned above were obtained with 488 nm excitation. Since heterocyclic rings of purines and pyrimidines exhibit two $\pi-\pi^*$ transitions near 260 and 220 nm in the UV region, UV resonance Raman (UVRR) spectroscopy is expected to give more enhancement and better selectivity of their normal modes. Ziegler et al. [37] reported the UVRR spectra (266 and 213 nm excitation) of nucleoside monophosphates such as 5′-GMP and 5′-AMP and a mixture of 5′-GMP with cisplatin. In the last case, they observed that the band at 1362 cm^{-1} of 5′-GMP is shifted to 1344 cm^{-1} on platination (213 nm excitation). Thus, this vibration likely involves a motion of the N7 atom that coordinates to the Pt atom. Perno et al. [38] measured the UVRR spectra (240 and 218 nm excitation) of 5′-GMP complexed to cisplatin in H$_2$O and D$_2$O, and noted that the equilibrium mixture exhibits the sugar-conformation-sensitive band at 666 cm^{-1} (C3′-*endo*) on the shoulder of the 682 cm^{-1} (C2′-*endo*) band of free 5′-GMP.

Although most of UVRR studies are carried out with a pulse laser of high peak power and low repetition rate, its high peak power tends to cause photochemical decomposition. To avoid this problem, Benson et al. [39] generated a quasicontinuous UV excitation at wavelengths shorter than 240 nm using the frequency-doubled output of a synchronously pumped dye laser operating in the 420 nm region. Figure 6.12a, b compares the UVRR spectra(209 nm excitation) of 5′-GMP and 5′-GMP mixed with cisplatin (1:1 molar ratio). It is seen that the bands at 1686 cm^{-1} (C=O stretching coupled with out-of-phase C5=C6 stretching), 1586 cm^{-1} (ring deformation), and 1368 cm^{-1} (ring deformation) of 5′-GMP are weakened considerably as a result of the Pt-N7(G) bonding. In particular, the 1368 cm^{-1} band disappears almost completely by platination. In the 1300–1100 cm^{-1} region, the 1180 cm^{-1} of guanine (antisymmetric stretching of the five-membered ring) is shifted to higher frequencies (1245 and 1218 cm^{-1}) with increased intensity. Similar results were obtained for 5′-GMP reacted with carboplatin (Fig. 6.12c). In general, however, the changes in spectra are less pronounced compared with cisplatin. Figure 6.13 compares the UVRR spectra of d(GGCCGGCC)$_2$ with and without cisplatin. The results are similar to that obtained for cisplatin reacted with 5′-GMP, thus confirming the N7(G) coordination of cisplatin to the octamer. It is interesting to note that the 792 cm^{-1} band of cytosine (pyrimidine ring breathing) loses its intensity considerably although cytosine is not directly involved in platination. The spectra shown in Fig. 6.13 were obtained using only 1 mL of aqueous solution containing 16 μM of the octamer.

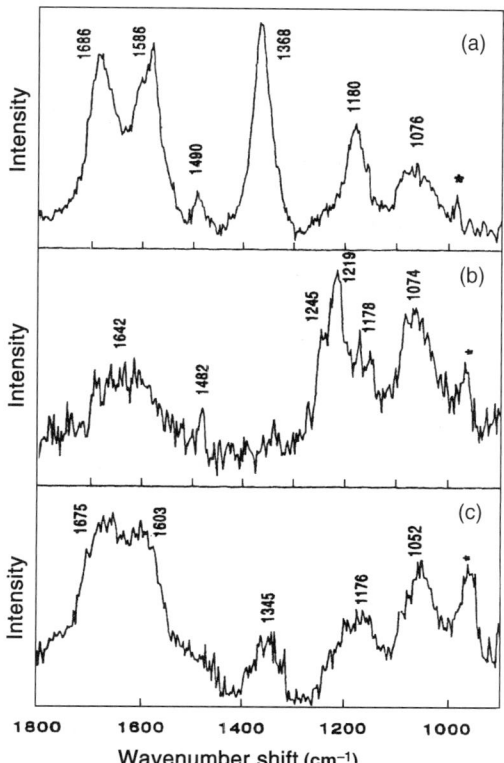

FIGURE 6.12. UVRR spectra (209 nm excitation) of (a) 5′-GMP (5 mM in H_2O), (b) 5′-GMP with cisplatin, and (c) 5′-GMP with carboplatin. The drug/nucleotide molar ratio is 1.0. The asterisk indicates the SO_4^{2-} band at 981 cm^{-1} (internal standard) [39].

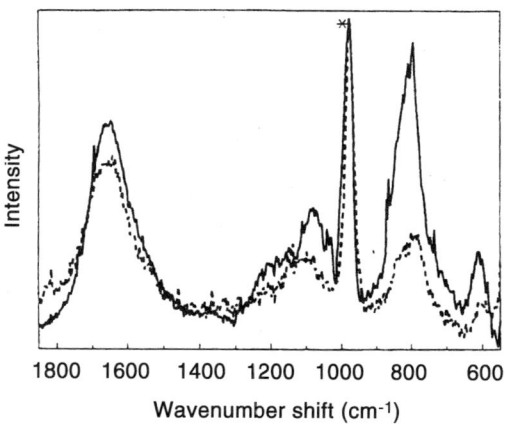

FIGURE 6.13. UVRR spectra (209 nm excitation) of d(GGCCGGCC)$_2$ (16 μM in H_2O) (solid line) and its mixture with cisplatin (dashed line). The drug/nucleotide molar ratio is 1.0 [39].

6.1.2. Interstrand Crosslinking Complexes

As mentioned in the Introduction, one of the minor products of cisplatin-DNA reaction is an *interstrand crosslinking* complex (Fig. 6.2b, structure **d**). Huang et al. [40] and Paquet et al. [41] determined the structures of the interstrand crosslinking octamer duplexes by 2D NMR spectroscopy coupled with molecular modeling techniques:

$5'$-d(CATA G* C TATG)-$3'$ [40] $5'$-d(CCT C G* C TCTC)-$3'$ [41]
$3'$-d(GTAT C G* ATAC)-$5'$ $3'$-d(GGAG C G*AGAG)-$5'$

Here, G* denotes the guanine that forms interstrand crosslinking to the $Pt(NH_3)_2$ moiety via the N7 atom. The structures obtained by these two groups are generally in good agreement, and markedly different from that of 1,2-intrastrand crosslinking complexes discussed previously. Major findings are that (1) platination occurs inside the minor groove of the duplex and the formation of interstrand crosslinking induces a kinking of 40° and an unwinding of 76° in the latter decamer duplex, (2) the two G* bases take the head-to-tail (HT) orientation (Fig. 6.3) to bring the two N7 atoms closer, and (3) base pairing between G* and C is lost. Coste et al. [42] determined the crystal structure of cisplatin complexed to the same decamer as used previously for NMR studies [41], and found that the kinking and unwinding angles of the duplex are similar to those found in solution.

6.2. CISPLATIN DERIVATIVES

As stated previously, a large number of cisplatin derivatives have been prepared and their clinical activities tested to search for more potent drugs with fewer side effects [43]. Thus far, these attempts have not been successful because the structure–activity relationship has not been fully understood. When cisplatin binds to DNA, two NH_3 ligands are retained while two Cl ligands (leaving groups) are hydrated and eventually replaced by the N7 atoms of guanine or adenine. Thus, interest has focused mainly on platinum complexes in which the NH_3 ligands are replaced by other nitrogen-coordinating ligands, particularly chelating ligands containing chiral centers. In most cases, major reaction products with DNA are 1,2-intrastrand crosslinking complexes similar to that of cisplatin.

6.2.1. Oxaliplatin and Carboplatin

In the case of oxaliplatin (Fig. 6.1c), 1,2-intrastrand crosslinking is formed by replacing the two O atoms of the oxalato (leaving) group with the N7 atoms of guanines. Since the 2-diaminocyclohexane (DACH) moiety has two chiral centers, its platinum complex can assume three stereoisomeric forms, and the order of their therapeutic activities is known to be in the order of $R,R > S,S > R,S$.

Spingler et al. [44] determined the crystal structure(2.4 Å resolution) of oxaliplatin [(R,R)-DACH] complexed to the 12-mer

```
          1   2   3   4   5   6   7   8   9   10  11  12
5'-d(C    C   T   C   T   G*  G*  T   C   T   C   C)-3'
3'-d(G    G   A   G   A   C   C   A   G   A   G   G)-5'
          32  31  30  29  28  27  26  25  24  23  22  21
```

This oligomer is the same as that used previously [11,12] to determine the crystal structure of its complex with cisplatin (the numbering scheme of the base-pair sequence of the unplatinated strand is different from those in other studies). The overall geometry and crystal packing were similar to those of the corresponding cisplatin complex. Thus, the duplex is bent by ∼30° toward the major groove at the site of platination. However, the unique feature of this complex is the formation of hydrogen bonding between the NH hydrogen atom of (R,R)-DACH and the O6 atom of G_7 (Fig. 6.14). The mechanism of action and antineoplastic activity of oxaliplatin have been reviewed by Raymond et al. [45], and differences in efficacy, mutagenicity, and tumor range between cisplatin and oxaliplatin have been discussed by Chaney et al. [46].

As stated in Section 6.1, carboplatin (Fig. 6.1b) is another anticancer drug in clinical use. Since it contains the cyclobutyldicarboxylate ion as the leaving group, the major reaction product with DNA is expected to be similar to that of cisplatin. However, the reaction of carboplatin with DNA produces intrastrand crosslinking adducts of 36% 1,3-d(GNG) (where N is any nucleotide), 30% 1,2-d(GG), and 16% 1,2-d(AG) together with 3–4% interstrand crosslinking adduct. Teuben et al. [47] isolated the 1,3-d(G*TG*) intrastrand crosslinking adduct of carboplatin with the 11-mer

```
          1   2   3   4   5   6   7   8   9   10  11
5'-d(C    T   C   T   G*  T   G*  T   C   T   C)-3'
3'-d(G    A   G   A   C   A   C   A   G   A   G)-5'
          22  21  20  19  18  17  16  15  14  13  12
```

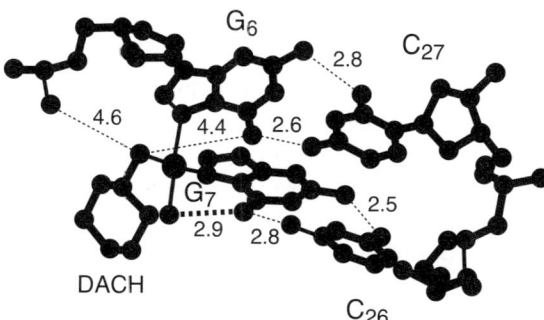

FIGURE 6.14. Close-up view of oxaliplatin bonded to the G_6pG_7 sequence of the 12-mer duplex. A hydrogen bond is formed between the pseudoequatorial NH hydrogen atom of the (R,R)-DACH and the O6 atom of G_7 (thick dotted line, 2.9 Å) [44].

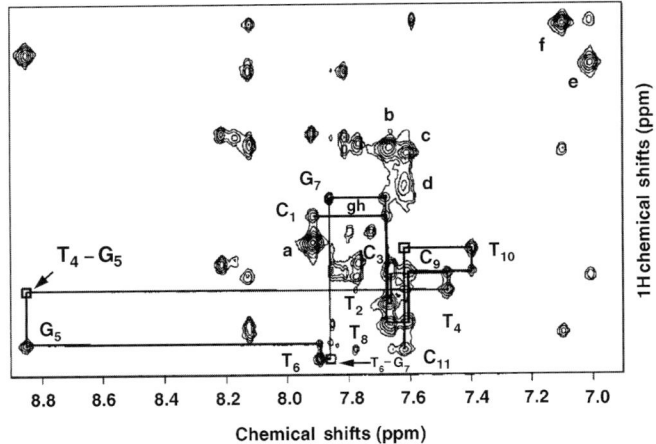

FIGURE 6.15. Experimental 2D NOESY plot in the H6/H8–H1′/H5 region. The sequential assignment passageway for d($C_1T_2C_3T_4G_5*T_6G_7*T_8C_9T_{10}C_{11}$) is indcated. The missing connectivities are indicated by empty boxes (see arrows). Cytidine H5/H6 cross-peaks are labeled (a) C_1, (b) C_3, (c) C_9, (d) C_{11}, (e) C_{16}, and (f) C_{18} Cross-peaks labeled (g) and (h) arise from the presence of a slight excess of unbound 11-mer, and do not appear in the simulated spectrum [47].

and determined its structure by ^{1}H NMR specroscopy coupled with molecular modelling techniques. As seen in the 2D NOESY plot of the platinated duplex shown in Fig. 6.15, the sequential contacts of H8/H6 signals between T_4(7.48)-G_5*(8.85) and T_6(7.90)-G_7*(7.86) (all in ppm) are missing, and the G_5*-T_6 contact is very weak. These results suggest that the stacking interactions among G_5*, T_6, and G_7* are absent, and that the platinated duplex is severely distorted at the G*TG* site. Although not shown, the NOE contact between H6(G_5*) and CH_3(T_6) is also absent. The sugar conformation deduced from the intensities of H8/H6 to H3 cross-peaks indicates that G_5* and C_{11} sugars are of N(north)-type (C3′-endo), although all other sugars take S (south)-type (C2′-endo), which is normally found in B-DNA.

Figure 6.16 shows a computer-generarted stereo views of the platinated duplex. The 1,3-intrastrand crosslinking bends the duplex toward the major groove by 27–33°, and the extruding T_6 and local unwinding cause the minor groove to become shallow and wide in the G*TG* region. In general, these distortions are similar to those found for the 1,2-intrastrand crosslinking complexes of cisplatin [8,9]. However, the distortion spreads over the entire lesion and is more severe on the 5′ side. Furthermore, the central T_6 is bulged out from the helix, which is unwound and kinked by ~30°. These differences may be responsible for the lack of recognition of the 1,3-intrastrand crosslinking adduct by the HMG1 proteins [13,14].

6.2.2. Platinum Complexes of Monofunctional Ligands

Table 6.2 lists NMR studies [48–52] of other cisplatin derivatives containing two monofunctional ligands as non–leaving groups. Since most of these compounds are of

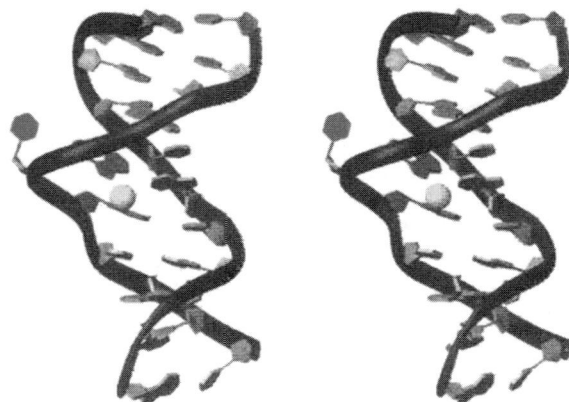

FIGURE 6.16. Stereo view into the major groove of one final structure created by molecular modeling. Structural distortions from B-DNA are illustrated by (1) the bend toward the major groove, (2) the unwinding of the helix at the site of the lesion, (3) extrusion of the central thymine, T_6 (in red) in the G*TG* lesion, (4) inversion of sugar geometry of the 5′-platinated guanine, and (5) the Pt atom (in yellow) coordination, causing the roll of the two coordinated guanines (in green) [47]. (See color insert.)

cis-[PtCl$_2$(L)(L′)] type, their adducts with nucleotides produce geometric isomers, some of which are more stable than others. For example, Bloemink et al. [48] separated two geometric isomers of cis-[PtCl$_2$(NH$_3$)(NH$_2$CH$_3$)] complexed with d(CpGpG), and determined their structures by NMR spectroscopy. In the major product (Fig. 6.17a), the NH$_2$CH$_3$ group is cis to the 3′-G residue, and the 5′-phosphate group is hydrogen-bonded to the NH$_3$ proton (dotted line). In the minor product (Fig. 6.17b), however, it is cis to the 5′-G residue. The ^{31}P chemical shift value of the GpG signal of d(CpGpG) is −3.85 ppm at pH = 6. This signal shows a large downfield shift

TABLE 6.2. NMR Studies on Cisplatin Derivatives (Monofunctional Ligands) Complexed to DNA

Cisplatin Derivatives	Nucleoside/Nucleotide	Ref.
cis-[PtCl$_2$(NH$_3$)L] L = NH$_2$CH$_3$, NH$_2$CH$_2$CH$_3$, NH(CH$_3$)$_2$	d(GpG), d(pGpG), d(CpGpG)	48
cis-[PtCl$_2$(NH$_3$)(2-picoline)]	5′-d(ATACATG*G*TACATA) - 3′ 3′-d(TATGTA C C ATGTAT) - 5′	49
cis-[PtCl I (NH$_3$)(4-AT)]^a	d(GpG)	50
cis-[PtCl$_2$(NH$_3$)(4-AT)]^a	5′- d(CTCTC G*G*TCTC) - 3′ 3′ - d(GAGAGC C AGAG) - 5′	51
cis-{PtCl$_2$L$_2$]^b cis-[PtCl$_2$(NH$_3$)$_2$L]^b	5′-GMP	52

^{a} 4-AT or 4-amino TEMPO = 4-amino-2,2,6,6-tetramethylpiperidinyloxy radical.
^{b} L = NH$_2$C(CH$_2$CH$_2$COOH)$_3$.

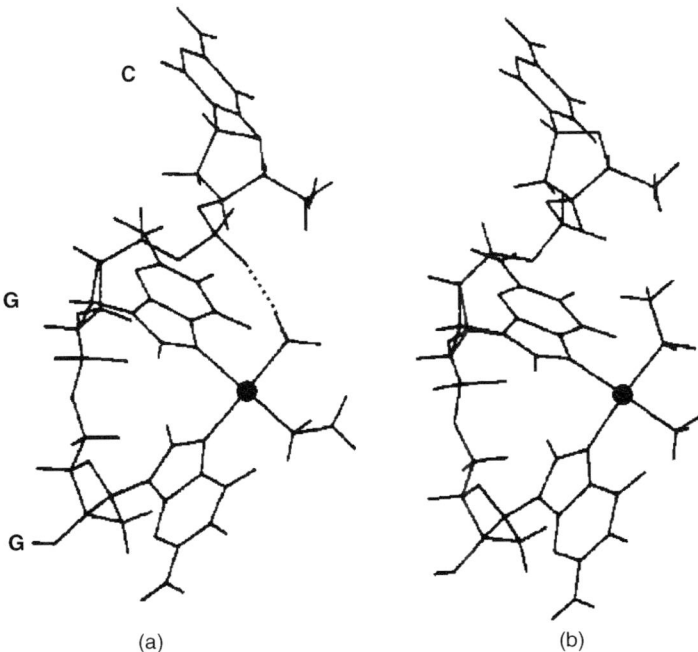

FIGURE 6.17. Schematic drawings of two geometric isomers of cis-[PtCl$_2$(NH$_3$)(NH$_2$CH$_3$)] complexed to d(CpGpG). (a) major product with the alkyl substituent toward 3'-G, where the hydrogen bond between the 5'-phosphate group and the NH$_3$ proton is shown by the dotted line; (b) minor product with the alkyl substituent toward 5'-G [48].

(to -2.62 ppm) in the major product and a moderate downfield shift (to -3.06 ppm) in the minor product. The large downshift of the former is attributed to the effect of hydrogen bonding mentioned above.

Chen et al. [49] found by 2D ^{1}H NOESY experiments that, among the four stereoisomers produced by the reaction of cis-[PtCl$_2$(NH$_3$)(2-picoline)] with the 14-mer (Table 6.2), the isomer **b** shown in Fig. 6.18 is formed predominantly. Model building studies show that (1) isomers **c** and **d** are not favorable because of steric repulsion between the 2-picoline ring and the nearby nucleotide bases, and (2) isomer **b** is favored by hydrogen-bonding interaction between the NH$_3$ ligand and the O6

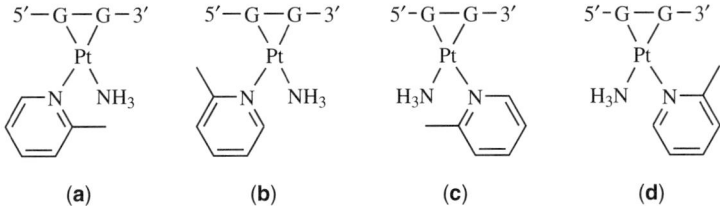

FIGURE 6.18. Structures of four isomers produced by the reaction of cis-[PtCl$_2$(NH$_3$)(picoline)] with the 14-mer. Only the G*G* sequence region is shown [49].

atom of G_8 and van der Waals interactions between the CH_3 group of 2-picoline and the CH_3 group of T_6. The latter interaction is absent in isomer **a**. Thus, isomer **b** is most stable. Both thermodynamic and kinetic factors also appear to support their conclusion.

Since distance constraints obtained from NOEs are limited to protons separated by less than 5 Å, it is not possible to elucidate the secondary structure in the region far from the site of platination. To obtain long-range distance constraints, Dunham et al. [50,51] prepared a spin-labeled platinum complex, *cis*-[PtCl I(NH$_3$)(4-AT)], where 4-AT is a nitroxyl radical, listed in Table 6.2. The reaction of this complex with d(GpG) produces two isomers that differ with respect to position of the 4-AT group toward either the 3′ or 5′ side of the phosphodiester linkage [50]. These two isomers can be differentiated by simple inspection of 1D NMR spectra since the degree of signal broadening by a paramagnetic electron depends on the electron–^{1}H (or –^{31}P) distance. The 5′- side isomer was separated, and its structure was determined by NMR spectroscopy using diamagnetic as well as paramagnetic distance constraints. Inclusion of the latter constraints was indispensable for determination of the torsional angles of the phosphodiester linkage. The same spin-labeled ligand was also utilized to determine the structure of its complex with a 11-mer (shown in Table 6.2). This study [51] included 99 long-range (10–20 Å) electron–proton restraints. The resulting structure is similar to that obtained without paramagnetic restraints except for the duplex ends, and markedly resembles to the structure of the cisplatin-modified decamer duplex previously determined by X-ray crystallography [12].

If the NH$_3$ ligand of cisplatin is replaced by a large, bulky ligand (L) such as NH$_2$C (CH$_2$CH$_2$COOH)$_3$, *cis*-[PtCl$_2$L$_2$] cannot bind to 5′-GMP, and *cis*-[PtCl$_2$(NH$_3$)L] can bind to it only monofunctionally. This was confirmed by Jansen et al. [52] by ^{1}H and ^{195}Pt NMR spectroscopy. The same conclusion was obtained from measurements of the unwinding angles of plasmid DNA. Clearly, the steric effect of the bulky ligand is responsible for these results.

6.2.3. Platinum Complexes of Chelating Ligands

Table 6.3 lists NMR studies on cisplatin derivatives containing chelating ligands as non–leaving groups. Marzilli and coworkers [53,54] found that the 12-mer and 14-mer listed in Table 6.3 remain single-stranded, and form a hairpin-like structure when reacted with PtCl$_2$(en) (where "en" is ethylenediamine). Figure 6.19 shows the structure of the 12-mer complexed to the Pt(en) moiety in the hairpin-like region [54]. In this structure, the platinum atom is bonded to the G_4 and G_5 bases with G_4 perpendicular to G_5 ("head to side"; Fig. 6.3) and A_7 is tucked on top of G_4. The ability to bind to and stabilize such hairpin-like structures may be another source of anticancer activity of this and related drugs.

NMR studies by Bloemink et al. [55] show that the reaction of PtCl$_2$(bmic) (Fig. 6.20, structure **a**) with d(GpG) yields two stereoisomers and that the isomer with the OH group of "bmic" and the O6 atom of the G bases on the same side of the Pt-N$_4$ plane is predominant. This is attributed to the hydrogen-bonding interaction between the OH group and the O6 atom. On the other hand, PtCl$_2$(bmi) (Fig. 6.20,

TABLE 6.3. NMR Studies on Cisplatin Derivatives (Chelating Ligands) Complexed to DNA

Cisplatin Derivatives	Nucleoside/Nucleotide	Ref.
PtCl$_2$(en)	5' - d(T$_1$A$_2$T$_3$G$_4$G$_5$G$_6$T$_7$A$_8$ C$_9$C$_{10}$C$_{11}$A$_{12}$T$_{13}$A$_{14}$) - 3'	53[a]
(en = ethylenediamine)	5' - d(A$_1$T$_2$G$_3$G$_4$G$_5$ T$_6$A$_7$C$_8$C$_9$ C$_{10}$A$_{11}$T$_{12}$) - 3'	54[b]
PtCl$_2$(bmic)		
bmic = bis-(N-methylimidazol-2-ylicarbinol (Fig. 6.20, structure **a**)	5'-GMP	55
PtCl$_2$(bmi)		
bmi = 1.1'-dimethyl-2,2'-biimidazol (Fig. 6.20, structure **b**)	d(GpG)	55
PtCl$_2$(Me$_2$ppz)	5'-GMP, 3'-GMP, etc.	58
Me$_2$ppz = N,N'-dimethylpiperazine	1-Methyl-5'-GMP	59
(Fig. 6.20, structure **c**)	d(GpG)	60
PtCl$_2$(pipen) (pipen = 2-aminomethylpiperidine)	5'-GMP	61
PtCl$_2$(bipip)	5'-GMP	62
(bipip = 2,2'-bipiperidine[c])	5'-GMP, 3'-GMP, etc.	63
	d(GpG)	64,65

[a] Hairpin-like structure. A$_8$ and G$_5$ or G$_6$ are proposed binding sites.
[b] Hairpin-like structure. G$_4$ and G$_5$ are proposed binding sites.
[c] The term "bipip" is used to avoid confusion with "bip" (bipyridine).

structure **b**) forms only one isomer in which the platinum atom is chelated to the N7 atoms of d(GpG). The anticancer activity of the "bmic" complex is significant, whereas the "bmi" complex is inactive. Moradell et al. [56] synthesized chelating diamine complexes of the PtCl$_2$(LL) type where LL is a chelating diamine ligand such as "dap" (Fig. 6.20, structure **c**), its six-membered ring analog, "dab" (Fig. 6.20, structure **d**), including their ethyl ethers, and studied the effects of the chelating ring size and esterification on cytotoxicity and interactions with 5'-GMP and CT-DNA by CD and NMR spectroscopy. Their work was extended to chelating diamine ligands containing an amino acid as a substituent [57]. In total, six such complexes (L-alanine, its methylester, and L-phenylalanine attached to the five- and six-membered chelate rings) were prepared. The structure of one of these Pt(II) complexes, PtCl$_2$(dap-ala), is shown in Fig. 6.20 (structure **e**). It was found that, in all cases, the interaction occurs at the N7 atoms of the purine bases, although the effects induced by complexation on the secondary structure of CT-DNA differ. The complexes containing L-alanine and L-phenylalanine exhibit cytotoxic activity in HeLa (human cervix carcinoma) and HL60 (human promyelocitic leukemia) cell

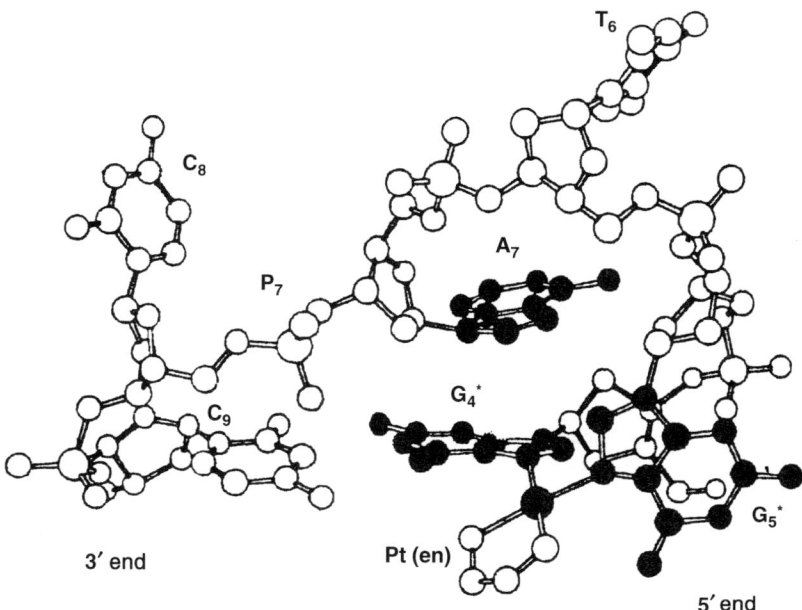

FIGURE 6.19. Structure of the loop region (from G_4^* to C_9) of the Pt(en)-12-mer complex [54].

lines, in a dose- and time-dependent manner. These complexes were less active than cisplatin, but more active than carboplatin after 24 h of incubation. However, no cytotoxic activity was detected for the methylester derivatives because of their low solubility in aqueous solution.

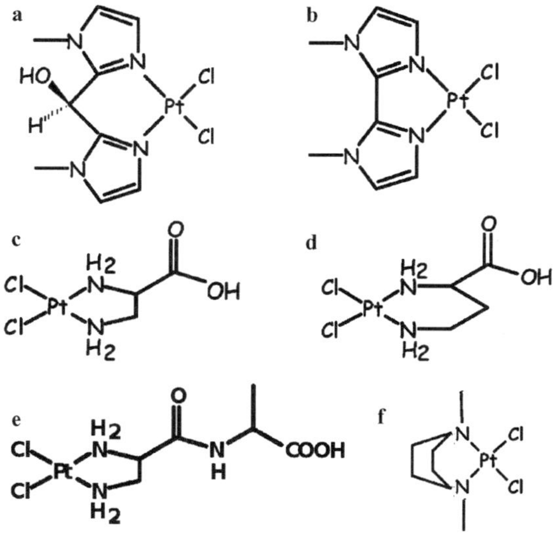

FIGURE 6.20. Structures of Pt(L-L)Cl$_2$-type complexes: (**a**) Pt(bmic)Cl$_2$; (**b**) Pt(bmi)Cl$_2$; (**c**) Pt(dap)Cl$_2$ (dap=2,3-diaminopropionic acid); (**d**) Pt(dab)Cl$_2$ (dab=2,4-diaminobutylic acid); (**e**) Pt(dap-ala)Cl$_2$; (**f**) Pt(Me$_2$ppz)Cl$_2$. Terminal methyl groups are not shown.

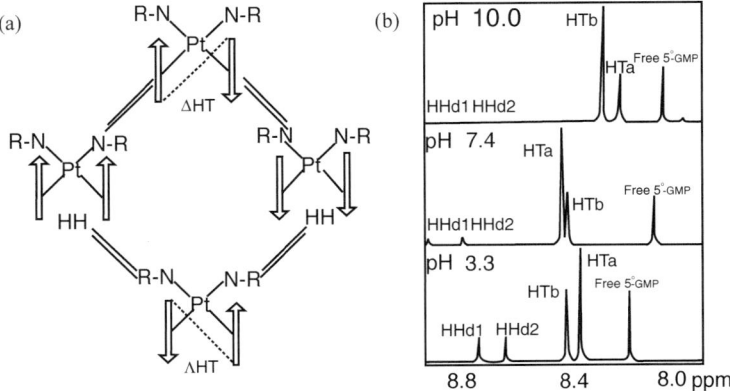

FIGURE 6.21. (a) Schematic representation of the interconversion between HT and HH atropisomers of Pt(Me$_2$ppz)(5'-GMP)$_2$ viewed with the G coordination sites forward and piperazine ligands to the rear. Arrows represent the G bases with H8 at the head (see Fig. 6.3). The two possible HT rotamers are differentiated by an imaginary line (dotted line) drawn between identical points (generally the O6 atom) of the G ligand. The resulting line (viewed from the G side of the coordination plane) has either a positive slope (ΔHT) or a negative slope (ΛHT). R represents the CH$_3$ groups of Me$_2$ppz. (b) pH dependence of NMR spectra of Pt(Me$_2$ppz)(5'-GMP)$_2$ at room temperature. The single HH conformer exhibits two signals (d_1 and d_2) of equal intensity, due to the asymmetry imposed on the complex by the sugar residue [58].

Figure 6.20 (structure **f**) shows the structure of PtCl$_2$(Me$_2$ppz), where Me$_2$ppz is N,N'-dimethylpiperazine. As shown in Fig. 6.21a, Pt(Me$_2$ppz)(5'-GMP)$_2$ exists as a slowly interconverting mixture of two head-to-tail (HT) and two head-to-head (HH) conformers [58]. Figure 6.21b shows the H8(G) signals of these conformers as a function of pH. From the integrated intensities of their signals, the percentages of HT$_a$, HT$_b$, and HH (d_1 and d_2) isomers at pH = 7.4 were estimated to be 60, 28, and 12, respectively. Although HT$_a$ predominates over HT$_b$ at pH = 7.4, an approximately equal distribution of these atropisomers was observed at pH = 7.5 in the case of 3'-GMP. Also, the percentages of the HH isomer at pH = 3.3 and 7.4 were greater for 5'-GMP than for 3'-GMP. Thus, the position of the phosphate group influences the distribution of these conformers. From CD studies, the absolute configuration of HT$_a$ was found to be Λ. Marzilli and coworkers extended this work to the reactions with 1-methyl-5'-GMP [59] and d(GpG) [60]. In general, a particular conformation is favored when hydrogen bonding occurs between the O6 atom of G base (and/or the phosphate oxygen) and the NH group of a chelating ligand. However, no such effect is present in this case because the Me$_2$ppz ligand has no NH groups. Similar NMR studies [61–65] were also carried out on several other systems containing chelating ligands, listed in Table 6.3.

Figure 6.22 shows the structures of two chiral isomers of [Pt(ampyr)(cbdca)] [where "ampyr" is aminomethylpyrrolidine and "cbdca" (cyclobutyldicarboxylate) is the same leaving group as that of carboplatin (Fig. 6.1b)]. The R-isomer is less toxic than S isomer. Bloemink et al. [66] studied the interaction of these isomers with short oligonucleotides such as d(GpG), d(pGpG), and d(CpGpG) using NMR spectroscopy and other techniques. NMR spectra and molecular modeling studies show that both

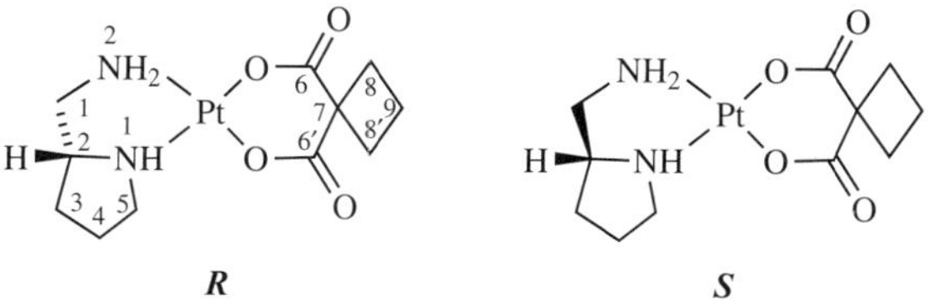

CISPLATIN DERIVATIVES 325

FIGURE 6.22. Structures of [Pt(R-ampyr)(cbdca)] and its S-isomer [66].

isomers form 1,2-intrastrand crosslinking adducts similar to cisplatin, and that, relative to the S-isomer, the adduct of the R-isomer has more steric hindrance between the bulky pyrrolidine ring and the DNA fragments. This difference may be responsible for the reduced toxicity of the R-isomer.

Bowler et al. [67] synthesized a compound in which PtCl$_2$(en) is connected to acridine orange by a hexamethylene linker (Fig. 6.23, structure **a**). Presumably, the PtCl$_2$(en) moiety forms 1,2-intrastrand crosslinking as discussed earlier, and acridine orange is intercalated between bases that are one or two base pairs apart from the platination site. Like acridine orange, this compound is also able to cleave DNA by photoactivation. Thus, it may be used to map platinum bonding sites in DNA. Two series of similar compounds using 9-anilinoacridine as the intercalator were synthesized by Palmer et al. [68]. One series is based on 1,2-ethylenediamine(en), and the other series is based on 1,3-propanediamine (Fig. 6.23, structure **b**). Coordination of these ligands to the platinum atom was verified by ^{195}Pt NMR spectroscopy, and their

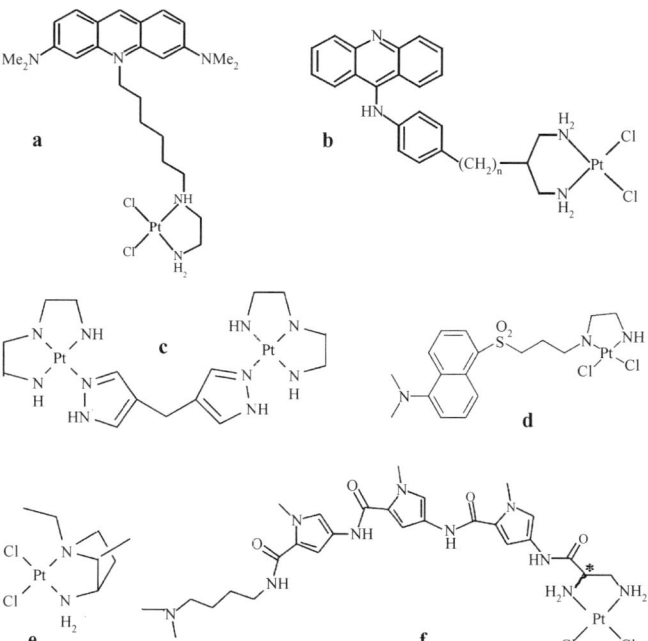

FIGURE 6.23. Structures of cisplatin analogs. Terminal methyl groups are not shown.

intercalative binding to DNA was confirmed by the observation of bathochromic and hypochromic effects on UV spectra. These acridine-linked cisplatin derivatives cause unwinding of DNA that is about twice that of $PtCl_2$(en), thus providing strong evidence for simultaneous platination and intercalation. However, the preferred intercalation site of 9-aminoacridine was not clear. The majority of these complexes were inactive *in vivo* against the wild-type P388 leukemia cell line. A series of analogous compounds in which acridine-2- and 4-carboxamides are linked to $PtCl_2$(en) were prepared, and their biological activities were tested by Lee et al. [69]. *In vitro*, 4-carboxamide complexes showed higher cytotoxicity against wild-type P388 leukemia cells than did 2-carboxamide complexes, and one of the former complexes was significantly active *in vivo* against a cisplatin-resistant cell line.

Wheate et al. [70] prepared $[\{Pt(dien)\}_2 (\mu-dpzm)]^{4+}$ (where dien = diethylenetriamine, dpzm = 4,4′-dipyrazolylmethane; Fig. 6.23, structure **c**) and *trans*-$[Pt(NH_3)_2(dpzm)_2]^{2+}$, and elucidated the structures of their adducts with the 12-mer, $d(CGCGAATTCGCG)_2$, by ^1H NMR spectroscopy. The NOE spectra indicate that these platinum complexes are in close proximity to the bases of the central AATT region in the minor groove, but do not induce any major conformational changes of the duplex. Groove binding is expected because they have no leaving groups and cannot bind covalently to the Pt atom. *In vitro* cytotoxicity studies show that the former complex has no activity while the latter complex has some activity against murine leukemia cell lines.

Figure 6.23, structure **d** shows the structure of $[PtCl_2(dansen)]$ containing a tethered fluorescent dansylated enthylenediamine (dansen) that was synthesized by Hartwig and coworkers [71]. Similar to cisplatin, this complex forms 1,2-intrastrand crosslinking with DNA without intercalation. By measuring fluoresence spectra, it was possible to monitor the cellular distribution of this complex at a level much lower than that of cisplatin.

Moreno et al. [72] studied the interaction of *cis*-$[PtCl_2(EMPyrr)]$ (EMPyrr = 1-ethyl,2-methyl,3-aminopyrrolidine) shown in Fig. 6.23, structure **e** with dinucleotides by CD spectroscopy. Figure 6.24a compares the CD spectra of free d(ApG) and its mixture with the Pt-EMPyrr complex. The former exhibits a negative band at 270 nm and two positive bands at ∼250 and 210 nm. On addition of the Pt complex, the 270 nm band was shifted to 288 nm and its sign inverted. The band near 250 nm of d(ApG) also changed sign and shifted to 268 nm. These results clearly indicate the formation of N7(A)-N7(G) crosslinking of the Pt(EMpyrr) moiety to d(ApG). On the other hand, the CD spectrum of d(ApA) showed almost no changes on addition of the Pt complex (Fig. 6.24b). In the latter case, the Pt-Empyrr complex binds to d(ApA) monofunctionally and does not distort the conformation of the dinucleotide. CD spectroscopy was also used to investigate the conformational equilibria of *cis*-PtA_2G_2-type complexes [$A_2 = (NH_3)_2$, en and tn (triethyldiamine), and G = 3′- and 5′-GMP] [73].

Figure 6.23 (structure **f**) shows $PtCl_2$(en) tethered to the AT-specific minor groove binder, distamycin (Pt-DIST). It has two stereoisomers (*R* and *S*), which have the chiral center at the C atom marked by a dot. Using various spectroscopic and biological techniques, Brabec and coworkers [74,75] compared

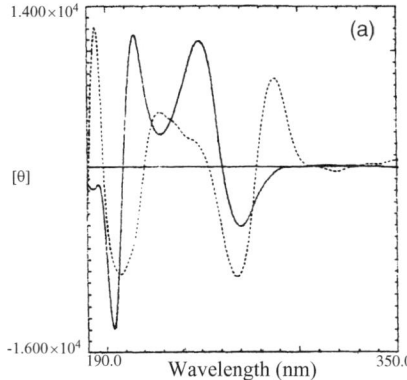 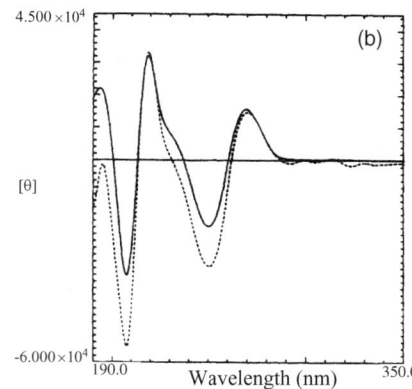

FIGURE 6.24. CD spectra of (a) d(ApG) (solid line) and its mixture with Pt-EMPyrr (dotted line) and (b) d(ApA)(solid line) and its mixture with Pt-EMPyrr (dotted line) [72].

theDNA-binding modes of Pt-DIST with those of cisplatin. Both isomers form intrastrand crosslinks preferentially at 5′-d(GpG)-3′ and 5′-d(ApG)-3′ sites. Thus, the attachment of distamycin to the Pt(en) moiety has no major effect on the capability to form intrastrand crosslinks between neighboring purine bases. However, Pt-DIST is considerably more efficient in forming interstrand crosslinks than is cisplatin. Furthermore, cisplatin forms interstrand crosslinking between two G bases (Fig. 6.2b, structure **d**), whereas Pt-DIST binds preferentially between G and C bases of the complementary strands. The CD spectra of restriction fragments of DNA (367 base pairs) modified by Pt-DIST show that the distamycin moiety also interacts with DNA to facilitate their interaction. These and other binding characteristics do not differ with respect to whether Pt-DIST takes the *R* or *S* conformation.

The reaction of $PtCl_2(en)$ with guanosine (Gua) yields at least seven complexes in aqueous solution. Götze et al. [76] separated these complexes by liquid chromatography (μg scale), and elucidated their modes of binding on the basis of IR spectra together with normal coordinate analysis. The monofunctional and bifunctional adducts obtained are abbreviated as: N1(1:1), N3(1:1), N7(1:1), N7-N1(1:2), N7-N1(2:1), N7-N3(2:1), and N7-N7(1:2). Here N1, N3, and N7 indicate the positions of guanosine nitrogens bonded to the Pt atom, and the molar ratios of Pt(en)/Gua are shown in parentheses. Figure 6.25 shows the IR spectra of N1, N7, and N7-N7 complexes in the ν(Pt−N) region. The ν(Pt−N) frequencies of two 1:1 complexes, N1 and N7, are slightly different (583 and 588 cm^{-1}, respectively), as indicated by the arrows.

6.3. TRANSPLATIN AND DERIVATIVES

Since transplatin is known to be clinically inactive, there have been only a few investigations on the interaction of transplatin and its derivatives wth DNA.

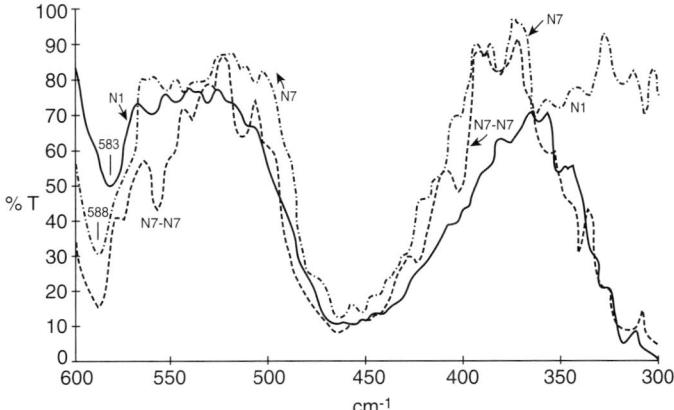

FIGURE 6.25. IR spectra of the N1 (solid line), N7(solid–dotted line), and N7–N7 (dotted line) of Pt(en)-Gua complexes in the 600–300 cm^{-1} region [76].

Paquet et al. [77] determined the structure of transplatin complexed to the 12-mer

$$\begin{array}{ccccccccccccc}
 & 1 & 2 & 3 & 4 & 5 & 6 & 7 & 8 & 9 & 10 & 11 & 12 \\
5'\text{-d(} & C & T & C & T & G^* & A & G & T & C & T & C & \text{)-}3' \\
3'\text{-d(} & G & A & G & A & C^* & T & C & A & G & A & G & \text{)-}5' \\
 & 24 & 23 & 22 & 21 & 20 & 19 & 18 & 17 & 16 & 15 & 14 & 13
\end{array}$$

by 2D NMR spectroscopy coupled with computer-aided molecular modeling techniques. Figure 6.26 shows the ^{1}H NOESY spectrum plotted the aromatic to H1' chemical shift region on the abscissa and the aromatic to H5 (cytosine) chemical shift region on the ordinate. It is seen that the cross-peak between H8 (7.83 ppm) and H1' (5.85 ppm) of G_6^* is much stronger than all other cross-peaks. Since the H8-H1' distance depends only on the glycoside torsion angle, this result suggests the *syn* conformation of G_6^*. In addition, a large distance separation between G_6^* and C_{19}^* bases is indicated by the absence of NOEs for C_5-G_6^* and T_{18}-C_{19}^*, and the weakness of NOEs for G_6^*-A_7 and C_{19}^*-G_{20}. These and other observations suggest interstrand crosslinking between G_6^* and C_{19}^*. Also the *syn* conformation of G_6^* results in a Hoogsteen-type base pairing between G_6^* and C_{19}^* that is stabilized by hydrogen bonding from the N4H(C_{19}^*) to the O6(G_6^*) as seen in Fig. 6.27a. Figure 6.27b is a stereo view of the structure of the transplatin-12-mer adduct. Although the double helix is not unwounded, it is slightly bent (by 20°) toward the minor groove, and the distortion due to platination is limited to the C_5-G_6^* and C_{19}^*-G_{20} sequences.

Lepre et al. [78] followed the reaction of transplatin with the single-stranded 12-mer:

$$d(C_1C_2T_3C_4G_5^*A_6G_7^*T_8T_9T_{10}C_{11}C_{12})$$

by NMR spectroscopy, and found two monofunctional intermediates in which platination occurred at the N7 atom of either G_5^* or G_7^*. These intermediates

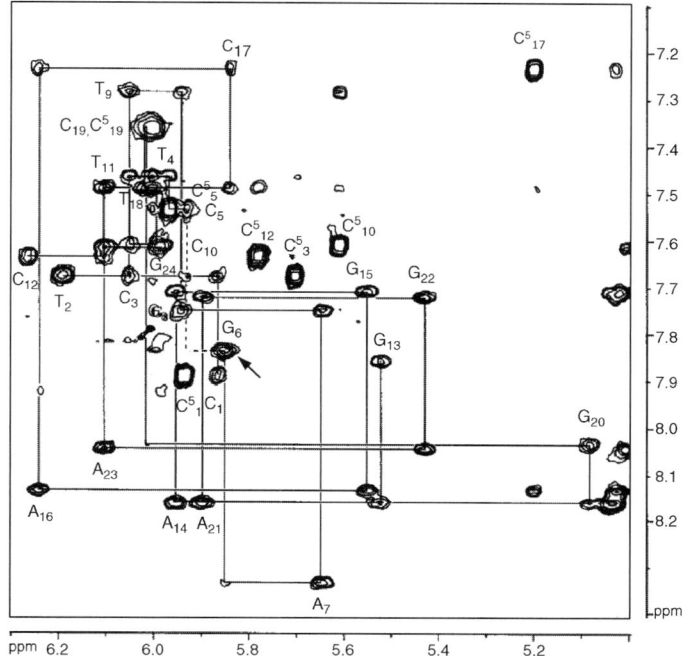

FIGURE 6.26. ^{1}H NOESY spectrum of the interstrand crosslinking transplatin-12-mer duplex, plotted with the aromatic to H1' chemical shift region as the abscissa and the aromatic to H5 (cytosine) chemical shift region as the ordinate (22°C in D_2O). The solid lines show the sequential assignment of the duplex, and the dotted line shows the lack of connectivity between C_5 and G_6* residues. The strong cross-peak between H8 and H1' of G_6* is indicated by the arrow. The very weak connectivities C_{19}*- G_{20} and G_6*-A_7, and the strong upfield shift of H1'(G_{20}) are noted. The connectvity between T_{18} and C_{19}* could not be determined [77].

eventually form bifunctional 1,3-intrastrand crosslinking at the G*AG* site. The structure of the double-stranded adduct obtained by the reaction of the platinated 12-mer with its complementary strand was studied by NMR and UV spectroscopy. The results show that platination does not cause significant changes in the duplex structure. This was confirmed by molecular modeling studies which indicated minimal distortion of the O—P—O angles with only local disruption of base pairing and destacking of the platinated bases.

DNA-binding properties of *cis*- and *trans*-[PtCl$_2$(pyridine)$_2$] with those of the corresponding NH$_3$ complexes were compared by Zou et al. [79]. As shown previously (Fig. 6.9), the UV band of DNA near 260 nm is redshifted and becomes hyperchromic on platination by cisplatin. Similar spectral changes were observed for the *cis*- and *trans*-pyridine complexes except that the band was slightly blueshifted. Figure 6.28 compares the CD spectra of CT-DNA with those modified by *cis*- and *trans*-[PtCl$_2$(pyridine)$_2$]. Platination by the *cis*-pyridine complex causes a small increase of the 280 nm band, but an appreciable decrease of the 221 nm band with a slight redshift that is concentration-dependent. This trend is similar to that observed for DNA

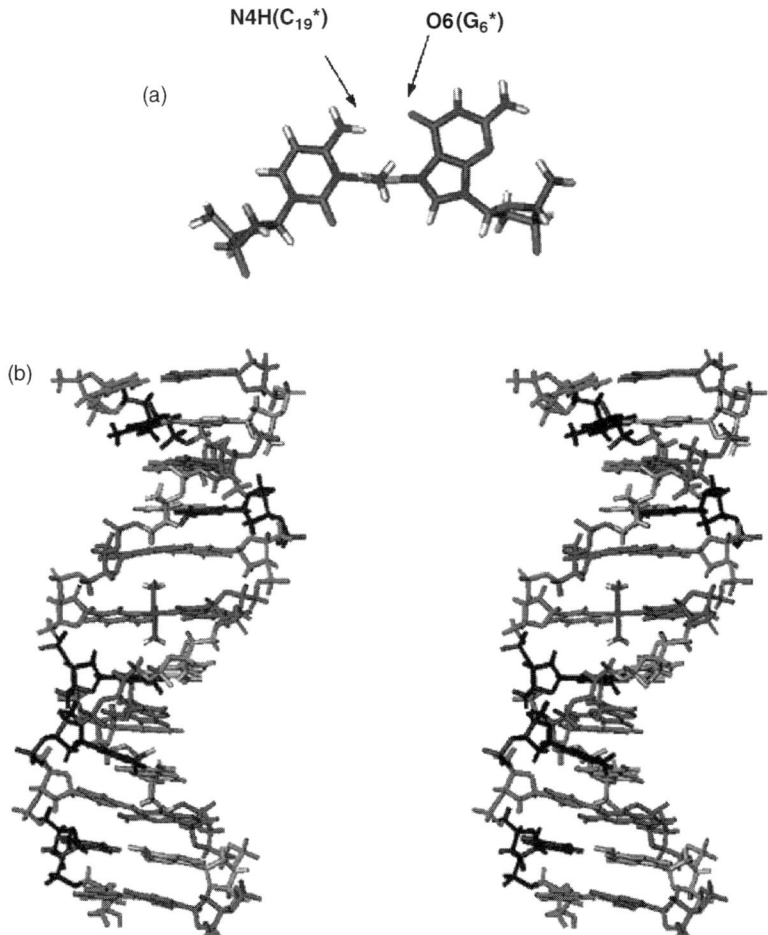

FIGURE 6.27. (a) View (from the top) of the platinated G_6^*-C_{19}^* base pair (b) Stereo view of the crosslinked complex. Both NH_3 groups of transplatin are located perpendicular to the plane formed by the crosslinked G_6^*-C_{19}^* bases. The *syn* conformation of G_6^* resulted from a Hoogsteen base pairing of G_6^*-C_{19}^* stabilized by a hydrogen bond between $O5(G_6^*)$ and N4H (C_{19}^*). Atom color codes are as follows: N (blue), C (gray), O (red), Pt (pale green) and H (light gray). (b) Stereo view of the crosslinked duplex. Color cords are as follows: Pt (blue), cytosine (green), guanine (red), thymine (black), and adenine (orange) [77]. (See color insert.)

platinated by cisplatin, suggesting intrastrand crosslinking by the *cis*-pyridine complex. In the case of *trans*-pyridine complex, the positive bands at 280 and 221 nm show large decreases in ellipticity, and little changes are observed by increasing the r_b (molar ratio of Pt complex/base pairs) value. Biochemical studies show that the efficiency of the *trans*-pyridine complex to form interstrand crosslinking and to unwind a supercoiled DNA is much higher than that of its *cis*-isomer. In fact, the unwinding angle is in the order *trans*-pyridine > cisplatin > transplatin > *cis*-pyridine. These trends may be

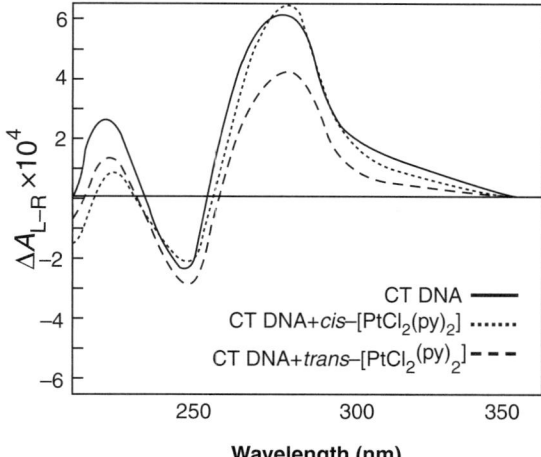

FIGURE 6.28. CD spectra of CT-DNA modified by *cis*- and *trans*-[PtCl$_2$(pyridine)$_2$] The r_b values are 0.03 [79].

responsible for the strong cytotoxicity of the *trans*-pyridine complex relative to the *cis*-pyridine complex. Thus, the *cis/trans* structure–activity relationship observed for the [PtCl$_2$(NH$_3$)$_2$] complex does not hold in this case.

Žaludová et al. [80] prepared *cis*- and *trans*-[PtCl$_2$(iminoether)$_2$] (iminoether = 1-imino-1-methoxyethane, HN=C(OMe)Me, *trans-EE*) shown in Fig. 6.29a, and compared their effects on the salt-induced B → Z transition of poly(dG-dC)$_2$ and its synthetic analogs using CD and Raman spectroscopy together with immunochemical assay. It was found that *cis-EE* facilitates the salt-induced B → Z transition in a way similar to cisplatin; whereas *trans-EE* affects it only slightly and transplatin hinders it. As mentioned earlier, transplatin is clinically inactive whereas *trans-EE* is active. Transplatin is known to bind to DNA via interstrand crosslinking. Then, the mode of interaction of *trans-EE* to DNA may be different from that of transplatin. Later, Andersen et al. [81] determined the structure of *trans-EE* bound to the decamer

```
          1   2   3   4   5    6   7   8   9   10
5'-d(C    C   T   C   G*   C   T   C   T   C)-3'
3'-d(G    G   A   G   C    G   A   G   A   G)-5'
          20  19  18  17   16  15  14  13  12  11
```

by NMR spectroscopy and found that only the H8 signal of G$_5$* is downshifted by 0.17 ppm relative to the unplatinated duplex. In this case, *trans-EE* forms a monofunctional adduct with one of the leaving groups (Cl atoms) maintained intact because the bulky ligand prevents the formation of a bifunctional adduct. Figure 6.29b is a close-up view of the platination site. The bending angle of the duplex (∼45°) is comparable to those of bifunctional adducts of cisplatin. Furthermore, the bending is toward the minor groove and not toward the major groove as

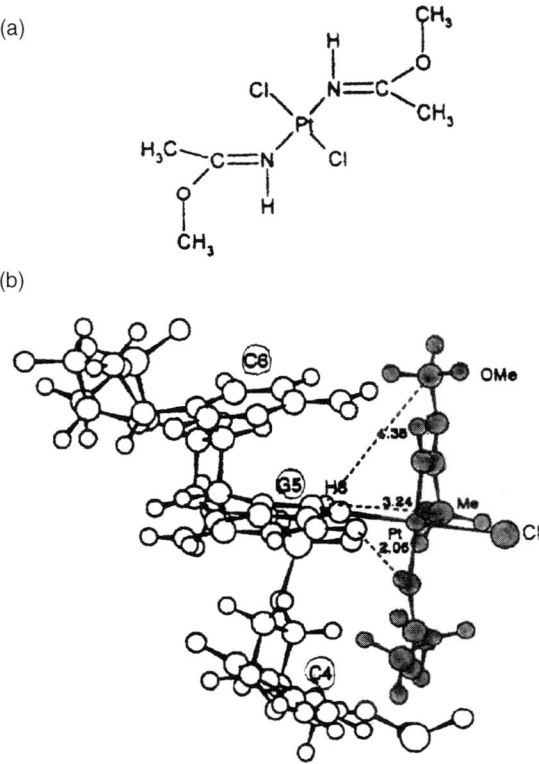

FIGURE 6.29. (a) Structure of *trans-EE* complex; (b) close-up view of *trans-EE*-decamer complex at the platination site [81].

found in intrastrand crosslinking of cisplatin. Yet this *trans-EE* complex exhibits an activity comparable to that of cisplatin in the P338 leukemia system and exerts antitumoral effects on Lewis lung carcinoma. Thus, monofunctional adducts that cause severe bending of the duplex can also be antitumor-active. The following section describes other monofunctional platinum complexes containing only one leaving group.

The complex *trans*-[PtCl$_2$(dimethylamine)(isopropylamine)] is also known to show some biological activity [82]. It forms more interstrand crosslinking adducts with pBR322 DNA than does cisplatin (~6% vs. 1.2%), shows the affinity toward alternating purine- pyrimidine sequences, and blocks the B → Z transition. Other *trans* complexes containing two different aliphatic amines may exhibit biological activity.

6.4. MONOFUNCTIONAL PLATINUM COMPLEXES

As the first step, bifunctional platunum complexes such as cisplatin form monofunctional adducts with DNA and still retain one leaving (Cl) group. These intermediate adducts may be modeled by monofunctional *cis-* and *trans-* [Pt(NH$_3$)$_2$(amine/

TABLE 6.4. Structural and Spectral Studies on Monofunctional Adducts

Monofunctional Complex	Nucleotide	Method	Ref.
[Pt(dien)Cl]Cl dien = diethylenetriamine[a]	5'-d(ApG*pA)-3'	X-ray, NMR	83
[Pt(NH$_3$)$_3$Cl] Cl	5'-d(TCTCG*TCTC)-3' and its complementary duplex	NMR	84
[Pt(NH$_3$)$_3$Cl]Cl [Pt(dien)Cl]Cl	5'-GTP and d(TpG)	NMR	85
cis-,trans-[Pt(NH$_3$)$_2$(4-Me-Py)Cl]Cl 4-Me-Py = 4-methyl-pyridine	5'-d(CCTG*TCC)-3' 3'-d(GGAC AGG)-5' duplex	NMR	86
cis-, trans-[Pt(NH$_3$)(PyAc-N,O)Cl][b] PyAc-N,O = N,O-chelating pyridine-2-yl acetate	5'-GMP 5'-d(TCG*T)-3' 5'-d(TG*CT)-3'	NMR	87
cis-, trans-[Pt (NH$_3$)$_2$Am Cl]$^+$ Am = N1-4-methyl-pyridine(Py),[d] N7-guanosine(Gua)[d], and 9-ethyl-guanine(9eG)[d]	22-mer duplex[c]	UV/CD	88

[a] NH$_2$(CH$_2$)$_2$NH(CH$_2$)$_2$NH$_2$.
[b] The structures are shown in Fig. 6.31a.
[c] Self-complementary duplex containing a single G site on one strand;

  5'-d(ATATATG*TATATATACATATAT)-3'
  3'-d(TATATAC ATATATGTATATA)-5'

[d] Abbreviation used in the text. The structures are shown in Fig. 6.31b.

pyridine)Cl]$^+$ ions bonded to DNA. Table 6.4 lists some structural and spectral studies on these monofunctional adducts [83–88].

Admiraal et al. [83] studied the structure of Pt(dien)[d(ApGpA)] obtained by the reaction of [Pt(dien)Cl]Cl with d(ApGpA). According to their X-ray analysis, the Pt(dien) moiety is coordinated to the N7 atom of the central G base, and the unit cell contains two independent molecules (**1** and **2** in Fig. 6.30) of similar conformation that are held together by stacking interactions and unusual hydrogen bonding between the G-G and A-A base pairs shown in Fig. 6.30. These interactions are stronger than the intramolecular interactions between an amine ligand and a phosphate oxygen and between another an amine ligand and the O6 atom of guanine. The solution structure obtained by NMR

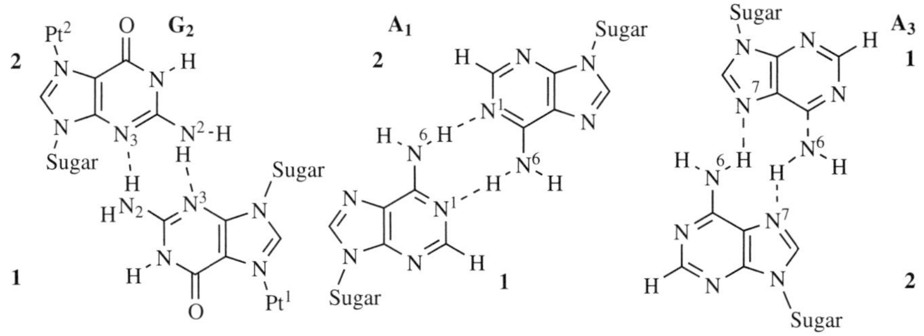

FIGURE 6.30. Schematic drawings of the G-G and A-A base pairing of two crystallographically independent molecules (**1** and **2**) in the unit cell of Pt(dien)[d(A$_1$-p-G$_2$-p-A$_3$)]. Lower and upper bases belong to **1** and **2**, respectively, for G$_2$ and A$_1$, and to **2** and **1**, respectively, for A$_3$ [83].

spectroscopy indicates that, instead of normal 80–100% S conformation of sugars, all three sugar conformations in d(ApG*pA) are in a ~50% N/S equilibrium as a result of platination at the N7 atom of the central guanine.

Reaction of [Pt(NH$_3$)$_3$Cl]Cl with single-stranded d(TCTCGTCTC) yields a monofunctional adduct in which the Pt(NH$_3$)$_3$ moiety coordinates at the N7 atom of the central G base. According to NMR studies by van Garderen et al. [84], the H8 signal of this G base shows a large downfield shift and its sugar conformation is changed toward the N type as a result of platination. When the complementary strand is added, the melting temperature of the resulting duplex is lowered by 15–20° because of monofunctional platination. However, this destabilization does not cause kinking of the double helix seen in 1,2-intrastrand crosslinking complexes. NMR spectroscopy was also utilized to study conformational changes of the sugar rings in 5′-GTP and d(TpG) induced by the reactions with [Pt(NH$_3$)$_3$Cl]Cl and [Pt(dien)Cl]Cl [85].

Both *cis*- and *trans*-[Pt(NH$_3$)$_2$(4-Me-py)Cl]$^+$ ions form monofunctional adducts with the heptamer duplex listed in Table 6.4. NMR studies by Bauer et al. [86] show that the H8(G$_4$) signal of the unplatinated heptamer duplex (Table 6.4) at 8.01 ppm is downshifted to 8.15 ppm in the *cis*-platinum adduct and to 8.64 ppm in the *trans*-platinum adduct. However, only *cis*-isomer has anticancer activity. This may be attributed to the larger structural distortion and faster reaction kinetics of the *cis*- than the *trans*-platinum adduct.

NMR spectra of water-soluble *cis*- and *trans*-[Pt(NH$_3$)(PyAc-N,O)Cl] (Fig. 6.31a) bound to single-stranded d(TCGT) and d(TGCT) show typical downfield shifts of the H8 signals of the guanine bases (0.5–0.6 ppm) indicating monofunctional platination at the N7 site of G in both tetramers [87]. In the adduct of *trans*-isomer bound to d(TCGT), the N-type character of the sugar conformations of T$_4$ and G$_3$ increased, and pronounced upfield shifts and broadening of H5 (C) and H6(C,T) signals were noted. This is probably due to the ring current effect of the pyridine ligand. The trans-complex exhibits cytotoxicity in L1210 leukemia cells comparable to that of cisplatin, whereas the *cis* complex is inactive. This result may suggest that the *trans* complex binds to DNA "pseudo bifunctionally" via covalent/intercalative bonding that mimics 1,2-intrastrand crosslinking. The *cis* complex does not show cytotoxicity probably

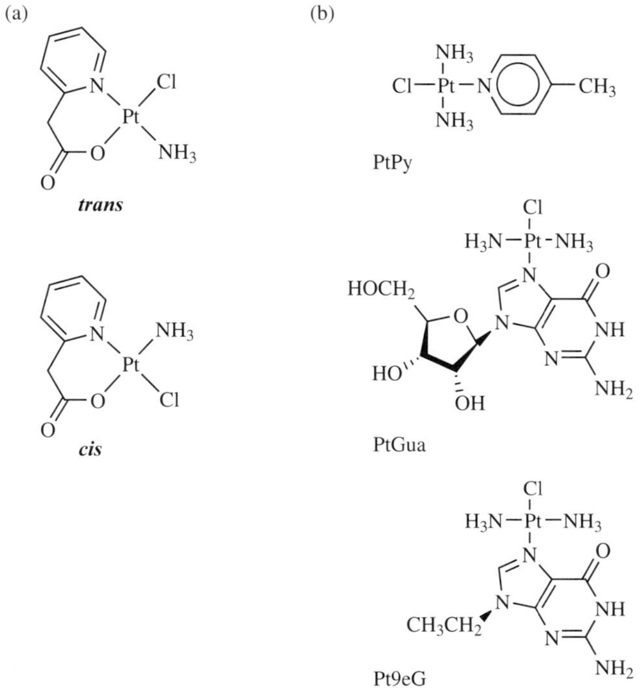

FIGURE 6.31. Structures of (a) *trans*- and *cis*-[Pt(NH$_3$)(PyAc-N,O)Cl] and (b) [Pt(NH$_3$)$_2$AmCl]$^+$. Am is in the order of Py, Gua, and 9eG from the top For the abbreviations used, see Table 6.4.

because the pyridine moiety in the monofunctional adduct is oriented *trans* to the guanine and cannot interact with the bases.

Peleg-Shulman et al. [88] studied the effects of platination by *cis*- and *trans* isomers of PtPy, PtGua, and Pt9eG ligands (Fig. 6.31b) on the conformation and thermal stability of the 22-mer duplex (Table 6.4) containing one G base (platination site). In all cases, the *cis*-isomers showed more pronounced effects than did the *trans* isomers. As shown in Fig. 6.32, the CD spectrum of the unplatinated 22-mer exhibits a positive peak at 286 nm and a negative peak at 243 nm, and platination by [Pt(NH$_3$)$_3$Cl]$^+$ does not cause appreciable changes because the NH$_3$ ligand is not bulky. However, platination by PtPy causes blueshifts of the positive peak and makes it more positive, and these changes are more conspicuous with the *cis*-isomer than with the *trans*-isomer. Essentially the same results were obtained for Pt9eG complexes. In contrast, the *cis*-isomer of PtGua makes the positive peak less positive, and the *trans*-isomer gives no appreciable changes (not shown). Also, the effect of platination by PtGua is different from that of PtPy and Pt9eG; the former raises the melting point of the 22-mer, whereas the latter lowers it. These results indicate the unique role of the sugar moiety in *cis*-PtGua. Model-building studies suggest the possibility of hydrogen bonding between the sugar and the double helix.

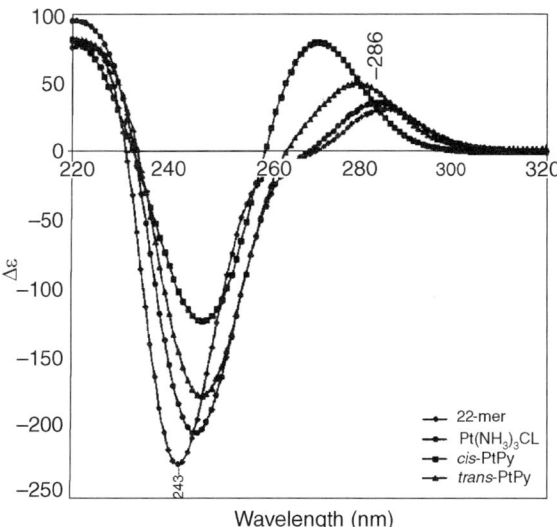

FIGURE 6.32. CD spectra of the 22-mer (Table 6.4) and its platinated adducts by [Pt(NH$_3$)$_3$Cl]$^+$ ion, *cis*-Pt-py and *trans*-Pt-py [88].

As discussed above, the CD spectrum of the 22-mer duplex is changed more markedly by *cis*-isomers than by *trans*-isomers. Thus, conformational constraints on the double-helix structure are more severe in *cis* geometry than in *trans* geometry because of a more pronounced unwinding caused by the former. In various murine tumor screening, the *trans*-isomers of all three complexes show no anticancer acitivity whereas *cis*-PtPy is cytotoxic but *cis*-PtGua is inactive. Thus, no simple correlation between *cis* geometry and biological activity exists in this case.

6.5. POLYNUCLEAR AND HIGH-VALENT PLATINUM COMPLEXES

The modes of binding of mononuclear platinum complexes with DNA have been discussed in the preceding sections. To search for new types of binding and better drugs, Farrell and coworkers [89] synthesized a series of dinuclear platinum complexes shown in Fig. 6.33. These complexes consist of two monofunctional Pt(NH$_3$)$_2$Cl or two bifunctional Pt(NH$_3$)Cl$_2$ or one monofunctional and one bifunctional moieties connected by the NH$_2$-(CH$_2$)$_n$-NH$_2$ linker ($n = 2$–6) in the *cis* or *trans* conformation. Convenient abbreviations of these structures shown in the same figure are used throughout this section. Figure 6.34 shows conceptual drawings of 1,4-intra- and interstrand crosslinking structures formed by these dinuclear complexes.

Table 6.5 [90–96] lists NMR studies on their adducts with nucleotides. In all cases, the N7 atoms of guanine bases are bonded to the platinum via intrastrand or interstrand cross linking, as indicated by the downfield shifts of the H8 signals of the guanine bases. Qu et al. [90] derived a molecular model of the dinuclear platinum complex, $\{(1,1/t,t), n = 6\}$, bonded to d(GpG) on the basis of NMR data. As seen in Fig. 6.35, it

2,2/*c*,*c*

1,1/*t*,*t*

1,1/*c*,*c*

1,2/*t*,*c*

1,2/*c*,*c*

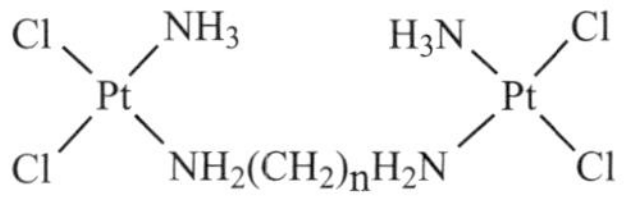

FIGURE 6.33. Structures of dinuclear platinum complexes and their abbreviations. The numbers refer to the number of the Cl^- ligand on each Pt atom. In the case of one Cl^- ligand on each Pt atom, *cis* and *trans* geometries are indicated by *c* and *t*, respectively. This convention is also used when one Pt atom is coordinated by two Cl^- ligands and the other Pt atom is coordinated by one Cl^- ligand, Thus, (1,1/*t*,*t*) denotes two monofunctional ("1") moieties of trans (*t*) geometry [e.g., *trans*-$[Pt(NH_3)_2Cl]$ connected by the NH_2-$(CH_2)_n$-NH_2 linker] [89].

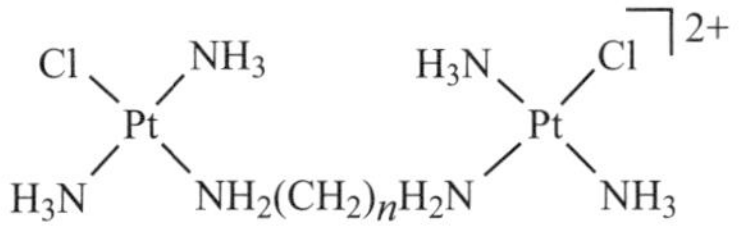

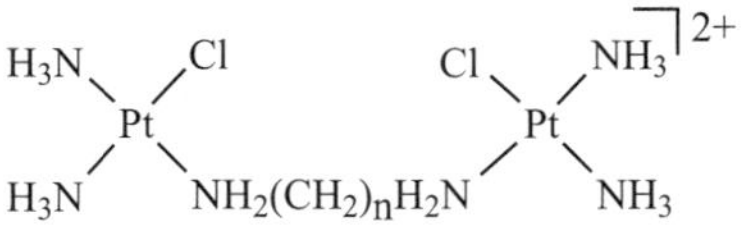

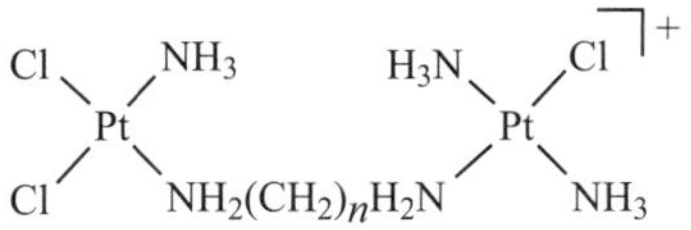

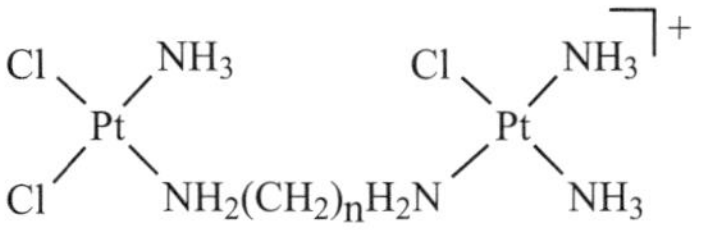

forms a large chelate ring between the two guanine bases, and their relative orientation is called the "stepped head-to-head" rather than the "head-to-head" type (Fig. 6.3) because the position of one guanine plane is shifted in the parallel direction relative to the other. Marked differences were noted between the d(GpG) adducts of this dinuclear complex and cisplatin; (1) the glycosidic bonds in the former adduct have *anti* ($5'$-G_1) and *syn* ($3'$-G_2) orientations, whereas those in the cisplatin adduct are all in *anti* orientation; and (2) the sugar conformations in the adduct of dinuclear complex ($n = 6$) are 55% *S* for $5'$-G_1 and 30% *S* for $3'$-G_2, whereas those in the adduct of cisplatin are 100% *N*-type. These differences may result from the flexible bending of

(Pt,Pt) intrastrand crosslink **(Pt,Pt) interstrand crosslink**

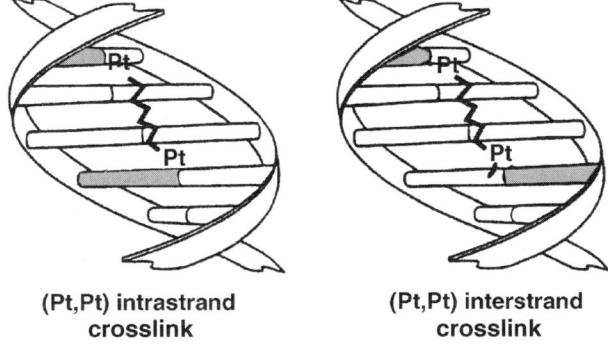

FIGURE 6.34. Conceptual drawings of 1,4-intrastrand and 1,4-interstrand crosslinkings. Bases on the same strand are darkened [89].

TABLE 6.5. NMR Studies on Dinuclear Platinum Complexes

Dinuclear Complex[a]	Nucleotide[b]/Nucleoside	Binding	Ref.
$(1,1/t, t)$, $n = 2\sim 6$	d(GpG)	1,2-Intra	90
$(1, 1/t, t)$, $n = 2,4,6$	Double-stranded 15–22-mers containing central d(TGGT) sequence	1,2-Intra	91
$(1,1/t, t)$, $n = 4$	d(CATG*CATG)$_2$	Hairpin[c]	92
$(1,1/c, c)$, $n = 4, 6$	d(GpG), r(GpG) single-stranded d(TGGT)	1,2-Intra	93
$(1,1/t, t)$, $n = 4$ $(2,2/c, c)$, $n = 4$	5′-GMP	—	94
$(1,1/t, t)$, $n = 6$	d(ATATG*TAC ATAT)· d(TATAC ATG*TATA)	1,4-Inter	95
[{PtCl(en)}$_2$(L-L)]$^{2+}$ [L-L=bifunctional thiourea derivative	5′-GMP r(GpG)	1,2-Intra[d]	96

[a] Structures and abbreviations are shown in Fig. 6.33.
[b] G* indicates platination site.
[c] See Fig. 6.36.
[d] Two Cl ligands of the dinuclear complex shown in Fig. 6.37 (structure **3**) are replaced by the N7 atoms of two Gs of r(GpG) to form an 1,2-intrastrand N7(G)-p-N7(G) bridge.

the DNA duplex induced by the dinuclear platinum complex in contrast to the rigid bending directed into the major groove caused by cisplatin.

Yang et al. [92] found that the $((1,1/t, t), n = 4)$ complex bonded to self-complementary octamer, d(CATGCATG), takes an unusual hairpin structure in which the

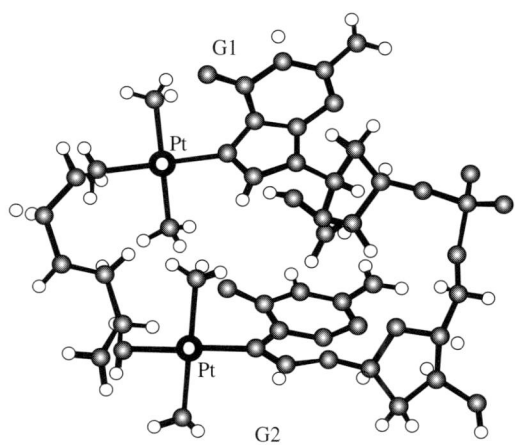

FIGURE 6.35. Molecular model of the adduct of dinuclear platinum complex [(1,1/*t*,*t*), (*n* = 6)] with d(GpG) [90].

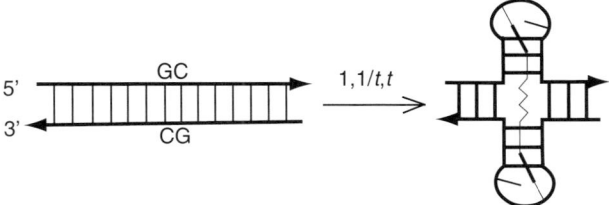

FIGURE 6.36. Schematic diagram showing the formation of a "four-way DNA junction" on platination of a GpC site by the [(1,1/t,t), (n = 6)] complex [92].

central G base is platinated and adopts a *syn* orientation in the minor groove. Two such hairpin structures are linked by the butanediamine tether to form a dumbbell structure, shown in Fig. 6.36. The dinuclear complexes ((1,1/c, c), n = 4 and 6) form 1,2-intrastrand adducts with single-stranded d(TGGT) but not with double-stranded DNA, indicating that the formation of 1,2-intrastrand links by these dinuclear platinum complexes is sterically more demanding for the double-stranded than for the single-stranded oligonucleotides [93].

The stepwise substitution of the Cl ligands in ((1,1/t, t), n = 4) and ((2,2/c, c), n = 4) by 5'-GMP was studied by NMR spectroscopy [94]. In the case of the former complex, the intermediate species was easily observed. In contrast, no intermediate species were detected for the latter complex, although all the Cl ligands were replaced by GMP in the final products. It was possible, however, to detect the species in which one GMP is bound to each platinum atom when the $NH_2(CH_2)_4NH_2$ linker of the latter complex was replaced by the $NH_2C(CH_3)_2(CH_2)_2C(CH_3)_2NH_2$ linker. This may indicate that, in the second step of the reaction, the complex of the methylated linker is sterically more demanding than the nonmethylated analog.

1,4-Interstrand crosslinking of the ((1,1/t,t), n = 6) complex bonded to the self-complementary 12-mer

$$5'\text{-d(A T A T G* T A C A T A T)-}3'$$
$$3'\text{-d(T A T A C A T G* T A T A)-}5'$$

was confirmed by Cox et al. [95] from NMR and molecular modeling studies. Unlike 1,2-intrastrand crosslinking with d(GpG) [90], 1,4-interstrand crosslinking requires minimal helical distortion localized near the binding sites.

Bierbach et al. [96] synthesized the mononuclear and dinuclear platinum complexes containing sulfur ligands shown in Fig. 6.37, and studied their reactions with 5'-GMP and r(GpG) by NMR spectroscopy. The mononuclear complexes (structures **1** and **2**) form only monofunctional adducts with 5'-GMP, and, compared with cisplatin and its analogs, the rates of hydrolysis of the Pt—Cl bond and nucleotide binding are slow because of the bulkiness of the tmtu (tetramethylthiourea) ligand. The dinuclear complexes (**3** and **4** in Fig. 6.37) disproportionate unusually rapidly in solution to yield [Pt(en)Cl$_2$] and [Pt(en)(L-L)]$^{2+}$. Here, L-L denotes the *S,S'*-chelating tmtu-(CH$_2$)$_n$-tmtu ligand. As a result, the reactions of **3** with 5'-GMP and r(GpG) yield

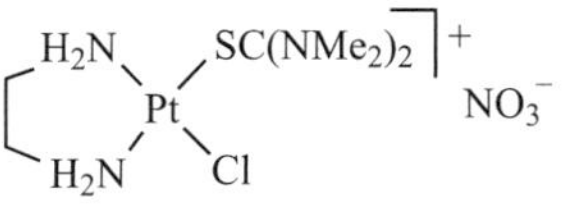

FIGURE 6.37. Structures of mononuclear and dinuclear platinum complexes in which alkyl thiourea coordinates to the platinum via the sulfur atom. The dots in structure **2** indicate the chiral centers [96].

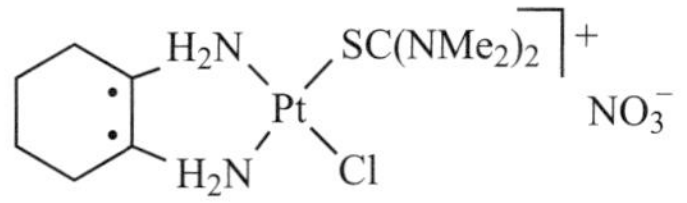

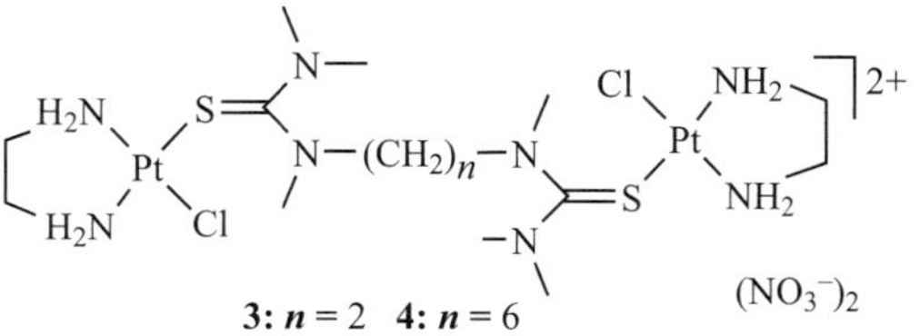

[Pt(en)(5'-GMP)$_2$] and [Pt(en){r(GpG)}], respectively. The reaction of **4** with r(GpG) proceeds much more slowly, and yields a mixture of the mononuclear complex [Pt(en){r(GpG)}] and dinuclear complex in which two Pt atoms are doubly bridged by the bifunctional L-L ligand as well as by the two G bases of r(GpG). Cytotoxicities of these complexes against cisplatin-sensitive (L1210/0) and cisplatin-resistant (L1210/DDP) leukemia cells were compared. In both cells, complexes **1** and **2** showed no response at drug concentrations of less than 50 (ID$_{50}$ value in μM) while **3** and **4** were fairly active, although their ID$_{50}$ values were still larger than those of cisplatin.

According to CD studies by Wu et al. [97], both (1,0/t), $n = 4$) and ((1,1/t, t), $n = 4$) complexes induce B → Z conformational change of poly(dG-dC)$_2$. Here, the former denotes a +3-charged dinuclear complex in which the *trans*-Pt(NH$_3$)$_2$Cl and Pt(NH$_3$)$_3$ moieties are linked by a butanediamine bridge. Since this monofunctional complex can induce B → Z conformational change, bifunctional binding is not a prerequisite for Z-form stabilization. Farrell and coworkers [98] studied the effects of *cis/trans*-isomerism and ligand substitution [NH$_3$ by py(pyridine)] on binding properties and conformational changes in DNA. Kinetics studies of binding show that the (1,1/c,c/NH$_3$) and (1,1/t,t/py) complexes bind less rapidly than the (1,1/t,t/NH$_3$) complex due to steric effect caused by the geometry around the leaving group (Cl) in the former two complexes. However, the (1,1/t,t/py) complex has a distinct binding preference for B-form poly(dG-dC)$_2$ compared with both NH$_3$ isomers. Figure 6.38 shows the CD spectra of poly(dG-dC)$_2$ and its adducts with these dinuclear complexes. It is seen that the B → Z transition is induced by both isomers of the NH$_3$ ligand, but not by the *trans*-isomer of the py ligand.

The Z form of poly(dG-dC)$_2$ induced by [{Pt(NH$_3$)$_3$}$_2$(NH$_2$-(CH$_2$)$_n$-NH$_2$)]$^{4+}$ ion (no leaving group) can be converted to the B form by adding ethydium

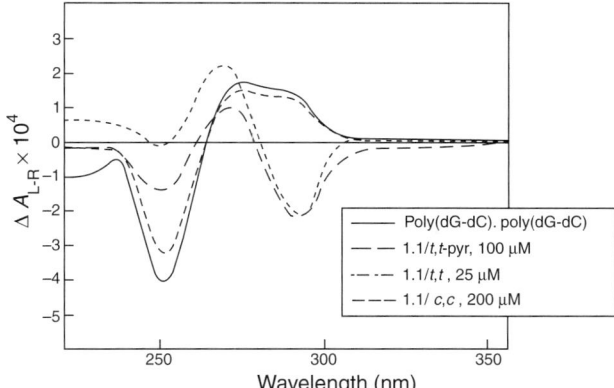

FIGURE 6.38. CD spectra of poly(dG-dC)$_2$ and its adducts with dinuclear platinum complexes The r_b values are 0.15, 0.12, and 0.135, respectively, for the three dinuclear complexes listed horizontally in the figure [98].

bromide (an intercalator, Chapter 2). However, such conversion does not occur for the Z form induced by the (1,1/t,t) complex. These results may indicate that the interaction between poly(dG-dC)$_2$ and the former complex is only electrostatic, whereas the interaction with the (1,1/t,t) complex is interstrand crosslinking (covalent bonding) [99].

Farrell and coworkers [98] noted that DNA binding parameters such as unwinding angle, percentage of interstrand crosslinking, and occurrence of B → Z transition are not directly related to cytotoxicity, as shown in Table 6.6. These dinuclear platinum complexes are highly cytotoxic, and their resistance factors are much smaller than that of cisplatin. Although they are poor recognizers of HMG protein, they are as cytotoxic

TABLE 6.6. Comparison of Mono- and Dinuclear Platinum Complexes in DNA Binding and Biological Activity

Complex	Unwinding Angle (degrees)	Interstrand Cross-linking (%)	B → Z Transition Induction	HMG Protein Recognition[a]	ID$_{50}$[b](μM)
Cisplatin	11	<5	No	1.0	0.42(28.6)
Transplatin	9	<5	No	0	—
tr[PtCl$_2$(py)$_2$]	17	25	No	ND	1.2(0.92)
(1,1/t,t)/NH$_3$	10	52	Yes	0.3	3.4(0.5)
(1,1/c,c)/NH$_3$	12	80	Yes	0.15	0.25(1.44)
(1,1/t,t)/py	42	41	No	ND	>10

[a] See Ref. 98 (ND is not determined).
[b] Data were obtained in L1210 (mouse leukemia) cells, which are sensitive and resistant to cisplatin. ID$_{50}$ value indicates the dose at 50% inhibition. Resistant factor (in parentheses) is defined as ID$_{50}$ (resistant)/ID$_{50}$ (sensitive). Thus, (1,1/t,t)/NH$_3$ is most effective in overcoming resistance in L1210 cells.
Source: Ref. 98.

as cisplatin. Thus dinuclear platinum complexes may bind to cellular DNA in a manner fundamentally different from that of cisplatin.

McGregor et al. [100] studied the conformational changes in DNA caused by three polynuclear platinum complexes: (**a**) $((1,1/t,t), n=4)$, (**b**) $((1,1/c,c), n=4)$, and (**c**) $((1,0,1/t,t,t), n=6,6)$ by CD spectroscopy. Here, the last abbreviation denotes the trinuclear cationic (+4-charged) complex in which two *trans*-Pt(NH$_3$)$_2$Cl units are coordinated to the central *trans*-Pt(NH$_3$)$_2$ moiety via the NH$_2$-(CH$_2$)$_6$-NH$_2$ linker. All these complexes induce B → Z transition in poly(dG-dC)$_2$ as well as B → A transition in poly(dG)·poly(dC). Although the B → A transition of CT-DNA is induced by ethanol, it is inhibited by these polynuclear complexes, and the order of inhibition efficiency is **b** > **a** > **c**. This order is the same as that of the relative amounts of interstrand crosslinks produced in CT-DNA by these complexes. Rauter et al. [101] prepared $(1.1/t,t)$ complexes having long linkers such as NH$_2$-(CH$_2$)$_4$-NH$_2$-(CH$_2$)$_3$-NH$_2$ ("spermidine") and NH$_2$-(CH$_2$)$_3$-NH$_2$-(CH$_2$)$_4$-NH$_2$-(CH$_2$)$_3$-NH$_2$ ("spermine"), which show *in vitro* cytotoxicity against murine L1210 leukemia cells comparable to that of cisplatin. Furthermore, the former complex, [{*trans*-(PtCl(NH$_3$)$_2$}$_2$(μ-spermidine)]Cl$_3$, exhibits cytotoxicity in the cisplatin-resistant cell line (L1210/DDP).

The interaction of Pt(II)-berenil complex., [Pt$_2$Cl$_4$(μ-berenil)$_2$]Cl$_4$·4H$_2$O (berenil = diaminazene aceturate, shown in Fig. 6.39a) with DNA was studied by CD spectroscopy and other techniques [102]. The LC$_{70}$ value [drug concentration(μM) to kill 70% of cells] against human leukemic cell lines (HL60) was about a half that of cisplatin.

The binding of cationic [(en)Pt(μ-dpzm)$_2$Pt(en)]$^{4+}$ ion (dpzm: 4.4′-dipyrazolylmethane (Fig. 6.39b) to two 12-mers, d(CGCGAATTCGCG)$_2$ and d(CAATCCGGATTG)$_2$, was studied by 2D NMR spectroscopy [103]. This platinum complex binds noncovalently to the A/T-rich regions in the minor grooves of both 12-mers. If oligonucleotides contain only G and C bases such as d(GCGCGCGCGC)$_2$ [104], the complex described above is associated noncovalently with the central C$_4$-G$_7$ region in the minor groove. Thus, this noncovalent groove binding may be regarded as the first step to form covalent bonds at GC sequences of DNA. NMR studies suggest that *trans*-[{PtCl(NH$_3$)$_2$}$_2$(μ-dpzm)]$^{2+}$ binds covalently to the N7 and N3 positions of adenine[105].

The trinuclear platinum complex $((1,0,1/t,t,t), n=6.6)$ mentioned earlier was named BBR 3464 [100], and found to be consistently more effective than cisplatin by preclinical evaluation. Qu et al. [106] determined the structure of its adduct with self- complementary octamer duplex; d(ATG*TACAT)·d(TACATG*TA), by NMR and other spectroscopic methods. It was found that the two noncentral Pt atoms are bonded to the N7 atoms of two guanine bases (marked by G*) in the major groove via 1,4-interstrand crosslinking and the central part of the complex is located in or close to the minor groove. Although A$_1$ and A$_7$ bases are not directly involved in crosslinking, the strong H8/H1′ intraresidue cross-peaks suggest their *syn* conformation, which is unusual for bases not directly involved in crosslinking. Thus, the 1,4-interstrand crosslinking induces conformational changes beyond the bonding site. However, severe distortions of the DNA duplex such as that seen in the case of cisplatin were not observed. Such a unique conformational perturbation caused by a long-range

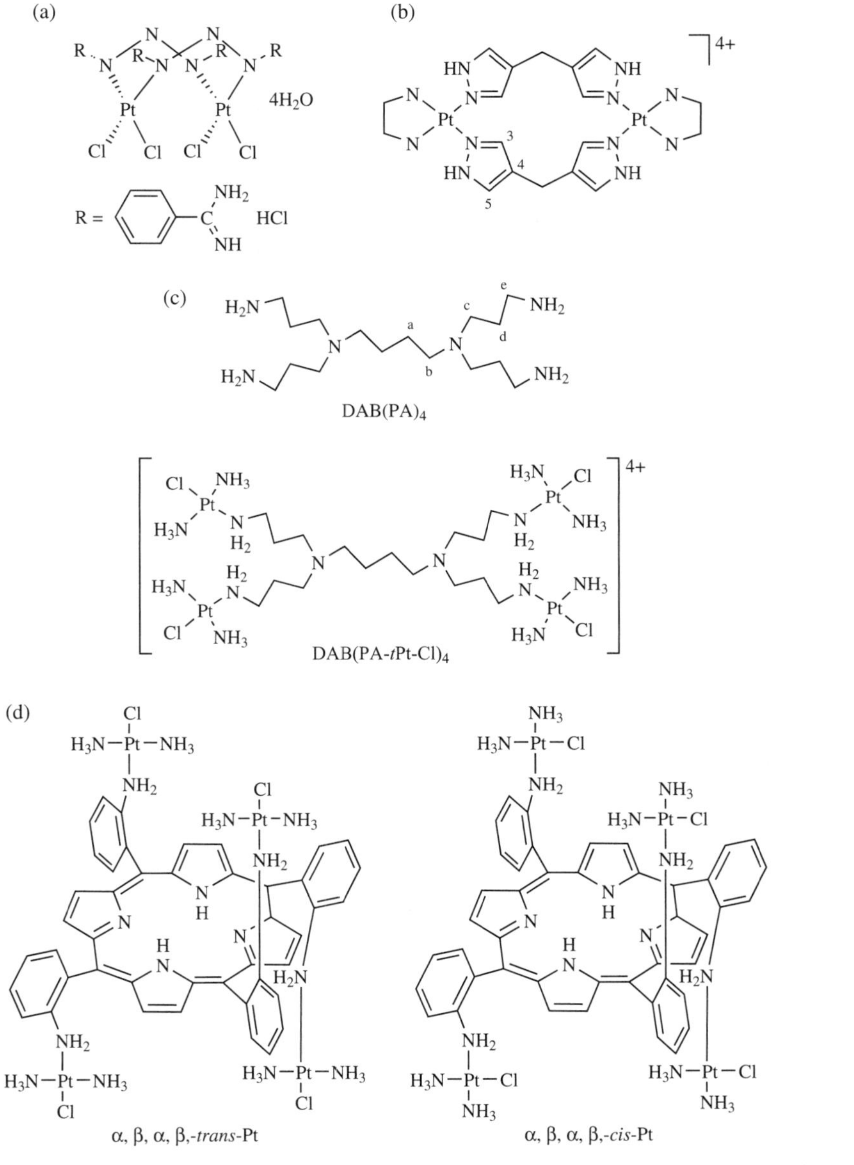

FIGURE 6.39. Structures of two dinuclear (a, b) and two tetranuclear (c, d) platinum complexes.

crosslinking may be related to its increased cytotoxicity relative to cisplatin [105]. The structure of the adduct of the trinuclear complex was compared with that of the dinuclear complex $((1,1/t,t), n = 6$, BBR 3005) [95] bonded to the same octamer. Mass spectroscopy (MS), UV, and NMR studies show that the structure of the latter adduct closely resembles that of the former adduct [107].

The first tetranuclear platinum complex shown in Fig. 6.39c was synthesized by Jansen et al. [108]. This complex, DAB(PA-tPt-Cl)$_4$, consists of four trans-[Pt(NH$_3$)$_2$Cl] moieties connected by a DAB(PA)$_4$ linker, and is capable of binding to a maximum of four nucleotide bases. Cytotoxicity tests were performed against seven human tumor cell lines, and, in all cases, the IC$_{50}$ values were larger than those of cisplatin. The low cytotoxicity of this platinum complex may be attributed to its high positive charge (+4) and branched structure, which hamper the crossing of the cell membranes.

Munakata and coworkers [109] prepared the water-soluble porphyrins appended by four [Pt(NH$_3$)$_3$Cl]$^+$ ions of cis and trans geometries shown in Fig. 6.39d. The trans complex forms a 1 : 1 adduct, while the cis complex forms a 1 : 3 adduct with poly(dA)·poly(dT). Here, 1 : 1 and 1 : 3 indicate the molar ratios of the Pt complex/base pair unit of the polynucleotide. The presence of these tetranuclear platinum complexes did not cause significant changes in the melting temperature of poly(dA)·poly(dT). This result suggests Coulombic interaction between them, and the bonding ratios may be determined by the difference in their abilities to recognize poly(dA)·poly(dT). Their interaction with poly(dG)·poly(dC) was suggested to be Coulombic as well as coordinative.

Cervantes et al. [110] synthesized [Pt$_2$Cl$_2$(Spy$^-$)$_4$] (Spy$^-$ deprotonated 2-mercaptopyridine), shown in Fig. 6.40, and confirmed its symmetric structure containing two Pt(III) centers by ^{195}Pt NMR spectroscopy. This complex exhibits cytotoxicities against U-937 and HeLa tumor cell lines comparable to or stronger than those of cisplatin. There are many other multinuclear platinum complexes, and their DNA binding, cytotoxicity, and structure–activity relationships have been reviewed by Wheate and Collins [111].

Six-coordinate platinum(IV) complexes have attracted attention because they can be orally administered and may have fewer side effects than cisplatin. Hartwig and Lippard [112] found that the major metabolite of cis,trans,cis-[Pt(IV)Cl$_2$(NH$_3$)(C$_6$H$_{11}$NH$_2$)(C$_3$H$_7$COO)$_2$] formed after ingestion is a Pt(II) complex, namely,

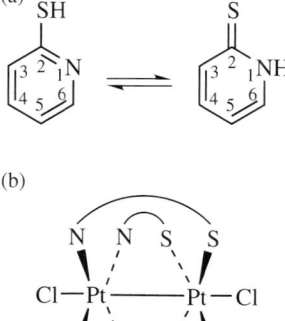

FIGURE 6.40. Structures of (a) 2-mercaptopyridine and (b) [Pt$_2$(Spy$^-$)$_4$Cl$_2$] (Spy$^-$=deprotonated 2-mercaptopyridine) [110].

TABLE 6.7. Structures and Cytotoxicities of Pd(II) Complexes[a]

Structure	Cytotoxicity[b]	Ref.
[Pd L (S$_2$CNEt$_2$)] NO$_3$ (L = 2,2'-bipyridine, etc.)	ID$_{50}$ value lower than that of cisplatin (P388 lymphocytic leukemia cell lines)	116
[Pd$_2$(bpy)$_2$(DADA)] (NO$_3$)$_2$ (bpy = 2,2'-bipyridine; DADA = α,ω-diaminoalkane-N,N'-diacetate dianion)	ID$_{50}$ value lower than that of cisplatin (P388 lymphocytic leukemia cell lines)	117
[Pd(C$_9$H$_{11}$N$_3$S)Cl$_2$] (ligand = phenylacetaldehyde thiosemicarbazone)	IC$_{50}$ value similar to that of cisplatin (human and murine cell lines)	118
[Pd(p-isoTSCN)]$_4$ (p-isoTSCN = p-isopropylbenzaldehyde thiosemicarbazone)	IC$_{50}$ value $\frac{1}{33}$ rd that of cisplatin (cisplatin-resistant Pam-ras cell lines)	119
[Pd(p-isoTSCN)(μ-Cl)]$_2$	IC$_{50}$ value $\frac{1}{3}$ rd that of cisplatin (cisplatin resistant Pam-ras cell lines)	120
[Pd(bip)(cbdca)] (cbdca = cyclobutyldicarboxylate ion)	ID$_{50}$ value less than that of cisplatin (P388 lymphocytic leukemia cell lines)	121

[a] Both Pt(II) and Pd(II) complexes were studied in all the investigations cited above.
[b] ID$_{50}$ and IC$_{50}$ indicate the dose (**D**) and the concentration (**C**), respectively, at 50% inhibition.

cis-[PtCl$_2$(NH$_3$)(C$_6$H$_{11}$NH$_2$)]. Here, C$_6$H$_{11}$NH$_2$ is cyclohexylamine. The latter Pt(II) complex forms an 1,2-intrastrand crosslinking adduct at the d(GpG) sequence of DNA similar to cisplatin but much less d(ApG) adduct compared with cisplatin. NMR studies show that the d(GpG) adduct is a mixture of two isomers in which the cyclohexyl group is orientated toward the 3' or 5' direction of the phosphate linkage.

On irradiation of 410 nm light, trans,cis-[Pt(IV)(en)Cl$_2$I$_2$] is decomposed to cytotoxic [Pt(en)Cl$_2$] and I$_2$ via reductive elimination [113]. It is also known that other Pt(IV) complexes can directly interact with DNA. For example, NMR studies [114] show that [Pt(IV)(en)(CH$_3$COO)$_4$] reacts with 5'-GMP to form [Pt(IV)(en)(CH$_3$COO)$_3$(5'-GMP)].

6.6. COMPLEXES CONTAINING OTHER METALS

In the preceding sections, the structures of a variety of platinum complexes and their cytotoxicities have been discussed. This section deals with metal complexes containing Pd(II), Au(III), and other metals. Some of these metal complexes exhibit cytotoxicity comparable to or stronger than cisplatin. Although their strong cytotoxicities suggest the formation of intrastrand- or interstrand crosslinking, detailed structural information on their DNA adducts is unavailable in most cases.

TABLE 6.8. Structures and Cytotoxicities of Metal Complexes Containing Other Metals

Structure[a]	Cytotoxicity	Ref.
[Ru(azpy)$_2$Cl$_2$] (α-form) (azpy = 2-phenylazopyridine)	ID$_{50}$ value comparable to that of cisplatin (MCF7 human tumor cell lines)	125,126
mer—[Ru(terpy)Cl$_3$] (terpy = 2,2′:6′,2″-terpyridine)	ID$_{50}$ value close to cisplatin (L1210 leukemia cell lines)	127
[Ru(cyclam)(φ)](ClO$_4$)$_2$ (cyclam = 1,4,8,11-tetraazacyclotetradecane; φ = 9,10-phenanthroquinonediimine)	IC$_{50}$ values for human carcinomas KB-3-1 cells and its multi-drug-resistant subclone KB-V1 cells	128
mer,cis-[RhCl$_3$(Me$_2$SO)$_2$(NH$_3$)] Me$_2$SO = dimethylsulfoxide	IC$_{50}$ value $\frac{1}{2}$ that of cisplatin (cisplatin-resistant A2780 human tumor cell lines)	129
[Rh(tpy)$_2$(Him)]Cl$_3$ 3H$_2$O[b] (tpy = 2,2′,2″-terpyridine Him = imidazole)	ID$_{50}$ value $\frac{1}{3}$rd that of cisplatin (*HCV29T* tumor cell line)	130
Cu(NBAB)$_2$(NO$_3$)$_2$ [NBAB = nicotinamido-4-bis-(2-chloroethyl)aminobenzaldimine]	ID$_{50}$ value less than $\frac{1}{2}$ that of nitrogen mustard	131
Cd (HL^1)Cl$_2$ (HL^1 = Me 2-pyridyl ketonethiosemicarbazone)	IC$_{50}$ value $\frac{1}{43}$rd that of cisplatin (cisplatin-resistant Pam-ras cell lines)	132
Zn(HL^2)Cl$_2$ (HL^2 = *p*-iso-Pr benzaldehydethiosemicarbazone)	IC$_{50}$ value $\frac{1}{3}$rd that of ciaplatin (cisplatin-resistant Pam-ras cell lines)	132
Cp$_2$MX$_2$ (Cp = cyclopentadienyl radical; M = Ti, Mo; X = a halogen)	Exhibit antitumor activities against leukemia P388, L1210 cell lines, etc.	133

[a] Cl$^-$ and NO$_3^-$ are the leaving groups.
[b] Three N atoms of one tpy, two N atoms of second tpy and one N atom of Him coordinate to the Rh atom.

6.6.1. Palladium(II) and Gold(III)

Since Pd(II) and Au(III) are isoelectronic with Pt(II) (d^8 configuration), a number of square–planar complexes of Pd(II) and Au(III) have been prepared to compare their cytotoxicities with that of cisplatin.

Both *cis*-[Pd(NH$_3$)$_2$Cl$_2$] and its derivative, such as [Pd(en)Cl$_2$], are known to be biologically inactive [115]. However, more recent investigations listed in Table 6.7 [116–121] show that some Pd(II) complexes exhibit cytotoxicity comparable to or stronger than that of cisplatin. Gold(III) complexes, [Au(esal)Cl$_2$] (esal = *N*-ethylsalycilaldiminate) and its *N*-methyl derivative, exhibit strong cytotoxicity against human tumor cell lines such as A2780/*S* (which is sensitive to cisplatin), and their cytotoxicity against A2780/*R* (which is resistant to cisplatin) is stronger than that of cisplatin [122]. This work has been extended to Au(III) complexes containing polyamine ligands [123] and bipyridine ligands [124].

6.6.2. Other Metals

Table 6.8 lists structures and cytotoxicities of metal complexes containing other metals [125–133]. The first six compounds are six-coordinate octahedral complexes, whereas the last three compounds are four-coordinate tetrahedral complexes. All these complexes exhibit strong cytotoxicity, suggesting the formation of intra- or interstrand crosslinking with DNA bases. Other types of interaction such as hydrogen bonding and Coulombic interaction may also be involved depending on the structure and orientation of the metal complex. Interactions of metallocene dihalides, Cp_2MX_2 (Cp = cyclopentadienyl anion; M = Ti, Mo; X = a halogen) with nucleic acid constitutents have been studied by NMR spectroscopy. In aqueous solution, Cp_2MoX_2 produces the Cp_2Mo^{2+} ion, which coordinates to the nitrogen atoms of nucleotide bases and/or the oxygen atom of the phosphodiester group [134]. Chemical properties and biomedical activities of Ru(II,III) complexes have been reviewed by Clarke [135].

REFERENCES

1. Rosenberg, B., Van Camp, L., and Krigas, T., "Inhibition of cell division in *Escherichia coli* by electrolysis products from a platinum electrode," *Nature* **205**:698–699 (1965).
2. Rosenberg, B., Van Camp, L., Trosko, J. E., and Mansour, V. H., "Platinum compounds: A new class of potent antitumor agents," *Nature* **222**:385–386 (1969).
3. Sherman, S. E. and Lippard, S. J., "Structural aspects of platinum anticancer drug interactions with DNA," *Chem. Rev.* **87**:1153–1181 (1987).
4. Hambley, T. W., "What can be learnt from computer-generated models of interactions between DNA and Pt(II) based anti-cancer drugs?" *Comments Inorg. Chem.* **14**:1–26 (1992).
5. Eastman, A., "Characterization of the adducts produced in DNA by *cis*-diamminedichloroplatinum(II) and *cis*-dichloro(ethylenediamine)platinum(II)," *Biochemistry* **22**:3927–3933 (1983).
6. Sherman, S. E., Gibson, D., Wang, A. H.-J., and Lippard, S. J., "X-ray structurte of the major adduct of the anticancer drug cisplatin with DNA: cis-[Pt(NH$_3$)$_2${d(pGpG)}]," *Science* **230**:412–417 (1985).
7. Sherman, S. E., Gibson, D., Wang, A. J.-H., and Lippard, S. J., "Crystal and molecular structure of *cis*-[Pt(NH$_3$)$_2${d(pGpG)}], the principal adduct formed by *cis*-diamminedichloroplatinum(II) with DNA," *J. Am. Chem. Soc.* **110**:7368–7381 (1988).
8. Hambley, T. W., "Molecular mechanics analysis of the influence of interligand interactions on isomer stabilities and barriers to isomer interconversion in diammine- and bis(amine)bis(purine)platinum(II) complexes," *Inorg. Chem.* **27**:1073–1077 (1988).
9. Schröder, G., Kozelka, J., Sabat, M., Fouchet, M.-H., Beyerle-Pfnür, R., and Lippert, B., "Model of the second most abundant cisplatin-DNA cross-link: X-ray crystal structure and conformational analysis of *cis*-[(NH$_3$)$_2$Pt(9-MeA-*N*7)(9-EtGH-*N*7)](NO$_3$)·2H$_2$O(9-MeA = 9-Methyladenine; 9-EtGH = 9-ethylguanine)," *Inorg. Chem.* **35**:1647–1652 (1996).
10. Admiraal, G., van der Veer, J. L., de Graaff, R. A. G., den Hartog, J. H., and Reedijk, J., "Intrastrand bis(guanine) chelation of d(CpGpG) to cis-platinum: An X-ray single-crystal structure analysis," *J. Am. Chem. Soc.* **109**:592–594 (1987).

11. Takahara, P. M., Rosenzweig, A. C., Frederick, C. A., and Lippard, S. J., "Crystal structure of double-stranded DNA containing the major adduct of the anticancer drug cisplatin," *Nature* **377**:649–652 (1995).
12. Takahara, P. M., Frederick, C. A., and Lippard, S. J., "Crystal structure of the anticancer drug cisplatin bound to duplex DNA," *J. Am. Chem. Soc.* **118**:12309–12321 (1996).
13. Ohndorf, U.-M., Rould, M. A., He, Q., Pabo, C. O., and Lippard, S. J., "Basis for recognition of cisplatin-modified DNA by high-mobility-group proteins," *Nature* **399**:708–712 (1999).
14. He, Q., Ohndorf, U.-M., and Lippard, S. J., "Intercalating residues determine the mode of HMG1 domains A and B binding to cisplatin-modified DNA," *Biochemistry* **39**:14426–14435 (2000).
15. Jamieson, E. R. and Lippard, S. J., "Structure, recognition, and processing of cisplatin-DNA adducts," *Chem. Rev.* **99**:2467–2498 (1999).
16. Caradonna, J. P. and Lippard, S. J., "Synthesis and characterization of [d (ApGpGpCpCpT)]$_2$ and its adduct with the anticancer drug cis-diamminedichloroplatinum(II)," *Inorg. Chem.* **27**:1454–1466 (1988).
17. Yang, D., van Boom, S. S. G. E., Reedijk, J., van Boom, J. H., and Wang, A. H.-J., "Structure and isomerization of an intrastrand cisplatin-cross-linked octamer DNA duplex by NMR analysis," *Biochemistry* **34**:12912–12920 (1995).
18. Marzilli, L. G., Saad, J. S., Kuklenyik, Z., Keating, K. A., and Xu, Y., "Relationship of solution and protein-bound structures of DNA duplexes with the major intrastrand cross-link lesions formed on cisplatin binding to DNA," *J. Am. Chem. Soc.* **123**:2764–2770 (2001).
19. Fouchet, M.-H., Guittel, E., Cognet, J. A. H., Kozelka, J., Gauthier, C., Le Bret, M., Zimmermann, K., and Chottard, J.-C., "Structure of a nonanucleotide duplex cross-linked by cisplatin at an ApG sequence," *J. Biol. Inorg. Chem.* **2**:83–92 (1997).
20. Gelasco, A. and Lippard, S. J., "NMR solution structure of a DNA dodecamer duplex contatining a *cis*-diammineplatinum(II) d(GpG) intrastrand cross-link, the major adduct of the anticancer drug cisplatin," *Biochemistry* **37**:9230–9239 (1998).
21. Parkinson, J. A., Chen, Y., del Socorro Murdoch, P., Guo, Z., Berners-Price, S. J., Brown, T., and Sadler, P. J., "Sequence-dependent bending of DNA induced by cisplatin: NMR structures of an A·T-rich 14-mer duplex," *Chem. -Eur. J.*, **6**:3636–3644 (2000).
22. van Boom, S. S. G. E., Yang, D., Reedijk, J., van der Marel, G. A., and Wang, A. H.-J., "Structural effect of intra-strand cisplatin-crosslink on palindromic DNA sequences," *J. Biomol. Struct. Dyn.* **13**:989–998 (1996).
23. van Garderen, C. J. and van Houte, L. P. A., "The solution structure of a DNA duplex containing the *cis*-Pt(NH$_3$)$_2$[d(-GtG-)-*N*7(G), *N*7(G)] adduct, as determined with high-field NMR and molecular mechanics/dynamics," *Eur. J. Biochem.* **225**:1169–1179 (1994).
24. Lemaire, D., Fouchet, M.-H., and Kozelka, J., "Effect of platinum N7-binding to deoxyguanosine and deoxyadeninosine on the H8 and H2 chemical shifts. A quantitative analysis," *J. Inorg. Biochem.* **53**:261–271 (1994).
25. Horáček, P. and Drobnik, J., "Interaction of *cis*-dichlorodiammineplatinum(II) with DNA," *Biochim. Biophys. Acta* **254**:341–347 (1971).
26. Zenker, A., Galanski, M., Bereuter, T. L., Keppler, B. K., and Lindner, W., "Capillary electrophoretic study of cisplatin interaction with nucleoside monophosphates, di- and trinucleotides," *J. Chromatogr. A* **852**:337–346 (1999).

27. Poklar, N., Pilch, D. S., Lippard, S. J., Redding, E. A., Dunham, S. U., and Breslauer, K. J., "Influence of cisplatin intrastrand crosslinking on the conformation, thermal stability, and energetics of a 20-mer DNA duplex," *Proc. Natl. Acad. Sci. USA* **93**:7606–7611 (1996).
28. Pilch, D. S., Dunham, S. U., Jamieson, E. R., Lippard, S. J., and Breslauer, K. J., "DNA sequence context modulates the impact of a cisplatin 1, 2-d(GpG) intrastrand cross-link on the conformational and thermodynamic properties of duplex DNA," *J. Mol. Biol.* **296**:803–812 (2000).
29. Tsankov, D., Kalisch, B., van de Sande, H., and Wieser, H., "Cisplatin-DNA adducts by vibrational circular dichroism spectroscopy: structure and isomerization of d(CCTG*G*TCC)·d(GGACCAGG) intrastrand cross-linked by cisplatin," *J. Phys. Chem. B* **107**:6479–6485 (2003).
30. Theophanides, T., "Fourier transform infrared spectra of calf thymus DNA and its reactions with the anticancer drug cisplatin," *Appl. Spectrosc.* **35**:461–465 (1981).
31. Okamoto, K., Benham, V., and Theophanides, T., "FT-IR spectroscopic evidence of sugar ring conformational changes in GpC and CpG on platination and intercalation," *Inorg. Chim. Acta* **135**:207–210 (1987).
32. Theophanides, T., Hadjiliadis, N., Berjot, M., Manfait, M., and Bernard, L., "Raman studies of platinum-nucleoside complexes," *J. Raman Spectrosc.* **5**:315–323 (1976).
33. Mansy, S., Chu, G. Y. H., Duncan, R. E., and Tobias, R. S., "Heavy metal nucleotide interactions. 12. Competitive reactions in systems of four nucleotides with *cis*- and *trans*-diammineplatinum(II). Raman difference spectrophotometric determination of the relative nucleophilicity of guanosine, cytidine, adenosine, and uridine monophosphates as well as the analogous bases in DNA," *J. Am. Chem. Soc.* **100**:607–616 (1978).
34. Alix, A. J. P., Bernard, L., Manfait, M., Ganguli, P. K., and Theophanides, T., "Binding of *cis*- and *trans*-dichlorodiammineplatinum(II) to nucleic acids studied by Raman spectroscopy. Part I. Salmon sperm DNA," *Inorg. Chim. Acta* **55**:147–152 (1981).
35. Strommen, D. P. and Peticolas, W. L., "Interaction of poly C with *cis*- and *trans*-diamminedichloroplatinum(II): Secondary structure effects," *Biopolymers* **21**:969–978 (1982).
36. Giese, B. and McNaughton, D., "Interaction of anticancer drug cisplatin with guanine: Density functional theory and surface-enhanced Raman spectroscopy study," *Biopolymers* **72**:472–489 (2003).
37. Ziegler, L. D., Hudson, B., Strommen, D. P., and Peticolas, W. L., "Resonance Raman spectra of mononucleotides obtained with 266 and 213 nm ultraviolet radiation," *Biopolymers* **23**:2067–2081 (1984).
38. Perno, J. R., Cwikel, D., and Spiro, T. G., "Ultraviolet resonance Raman study of deoxyguanosine monophosphate and its complexes with *cis*-$(NH_3)Pt^{2+}$, Ni^{2+} and H^+," *Inorg. Chem.* **26**:400–405 (1987).
39. Benson, R. L., Iwata, K., Weaver, W. L., and Gustafson, T. L., "Improvements in the generation of quasi-continuous, tunable ultraviolet excitation for Raman spectroscopy: Applications to drug/nucleotide interactions," *Appl. Spectrosc.* **46**:240–245 (1992).
40. Huang, H., Zhu, L., Reid, B. R., Drobny, G. P., and Hopkins, P. B., "Solution structure of a cisplatin-induced DNA interstrand cross-link," *Science* **270**:1842–1845 (1995).
41. Paquet, F., Pérez, C., Leng, M., Lancelot, G., and Malinge, J.-M., "NMR solution structure of a DNA decamer containing an interstrand cross-link of the antitumor drug *cis*-diamminedichloroplatinum(II)," *J. Biomol. Struct. Dyn.* **14**:67–77 (1996).

42. Coste, F., Malinge, J.-M., Serre, L., Shepard, W., Roth, M., Leng, M., and Zelwer, C., "Crystal structure of a double-stranded DNA containing a cisplatin interstrand cross-link at 1.63 Å resolution: Hydration at the platinated site," *Nucl. Acids Res.* **27**:1837–1846 (1999).
43. Pasini, A. and Zunino, F., "New cisplatin analogues—on the way to better antitumor agents," *Angew. Chem. Int. Ed.* **26**:615–624 (1987).
44. Spingler, B., Whittington, D. A., and Lippard, S. J., "2.4 Å crystal structure of an oxaliplatin 1, 2-d(GpG) intrastrand cross-link in a DNA dodecamer duplex," *Inorg. Chem.* **40**:5596–5602 (2001).
45. Raymond, E., Faivre, S., Woynarowski, J. M., and Chaney, S. G., "Oxaliplatin: Mechanism of action and antineoplastic activity," *Semin. Oncol.* **25**:4–12 (1998).
46. Chaney, S. G., Campbell, S. L., Temple, B., Bassett, E., Wu, Y., and Faldu, M., "Protein interactions with platinum-DNA adducts: From structure to function," *J. Inorg. Biochem.* **98**:1551–1559 (2004).
47. Teuben, J.-M., Bauer, C., Wang, A. J.-H., and Reedijk, J., "Solution structure of a DNA duplex containing a *cis*-diammineplatinum(II) 1,3-d(GTG) intrastrand cross-link, a major adduct in cells treated with the anticancer drug carboplatin," *Biochemistry* **38**:12305–12312 (1999).
48. Bloemink, M. J., Heetebrij, R. J., Inagaki, K., Kidani, Y., and Reedijk, J., "Reactions of unsymmetrically substituted derivatives of cisplatin with short oligodeoxynucleotides containing a-GpG-sequence: H-bonding interactions in pGpG moieties cross-linked by an asymmetric platinum complex enhancing the formation of one geometrical isomer," *Inorg. Chem.* **31**:4656–4661 (1992).
49. Chen, Y., Parkinson, J. A., Guo, Z., Brown, T., and Sadler, P. J., "A new platinum anticancer drug forms a highly stereoselective adduct with duplex DNA," *Angew. Chem. Int. Ed.* **38**:2060–2063 (1999).
50. Dunham, S. U. and Lippard, S. J., "Long-range distance constraints in platinated nucleotides: Structure determination of the $5'$ orientational isomer of *cis*-[Pt(NH$_3$)(4-aminoTEMPO){d(GpG)}]$^+$ from combined paramagnetic and diamagnetic NMR constraints with molecular modeling," *J. Am. Chem. Soc.* **117**:10702–10712 (1995).
51. Dunham, Sh. U., Dunham, St. U., Turner, C. J., and Lippard, S. J., "Solution structure of a DNA duplex containing a nitroxide spin-labeled platinum d(GpG) intrastrand cross-link refined with NMR-derived long-range electron-proton distance restraints," *J. Am. Chem. Soc.* **120**:5395–5406 (1998).
52. Jansen, B. A. J., Pérez, J. M., Pizarro, A., Alonso, C., Reedijk, J., and Navarro-Ranninger, C., "Sterically hindered cisplatin derivatives with multiple carboxylate auxiliary arms: Synthesis and reactions with guanosine-$5'$-monophosphate and plasmid DNA," *J. Inorg. Biochem.* **85**:229–235 (2001).
53. Kline, T. P., Marzilli, L. G., Live, D., and Zon, G., "NMR studies of an oligonucleotide with an unusual structure induced by platinum anticancer drugs," *Biochem. Pharm.* **40**:97–113 (1990).
54. Iwamoto, M., Mukundan, S., Jr., and Marzilli, L. G., "DNA adduct formation by platinum anticancer drugs. Insight into an unusual GpG intrastrand cross-link in a hairpin-like DNA oligonucleotide using NMR and distance geometry methods," *J. Am. Chem. Soc.* **116**:6238–6244 (1994).
55. Bloemink, M. J., Engelking, H., Karentzopoulos, S., Krebs, B., and Reedijk, J., "Synthesis, crystal structure, antitumor activity, and DNA–binding properties of the

new active platinum compound (bis(N-methylimidazole-2-yl)carbonyl)- dichloroplatinum(II), lacking a NH moiety, and of the inactive analog dichloro(N^1 and $N^{1'}$-dimethyl-2,2'-biimidazole)platinum(II)," *Inorg. Chem.* **35**:619–627 (1996).

56. Moradell, S., Lorenzo, J., Rovira, A., Robillard, M. S., Avilés, F. X., Moreno, V., de Llorens, R., Martinez, M. A., Reedijk, J., and Llobet, A., "Platinum complexes of diaminocarboxylic acids and their ethyl ester derivatives: The effect of the chelate ring size on antitumor activity and interactions with GMP and DNA," *J. Inorg. Biochem.* **96**:493–502 (2003).

57. Moradell, S., Lorenzo, J., Rovira, A., van Zutphen, S., Avilés, F. X., Moreno, V., de Llorens, R., Martinez, M. A., Reedijk, J., and Llobet, A., "Water-soluble platinum(II) complexes of diamine chelating ligands bearing amino-acid type substituents: The effect of the linked amino acid and the diamine chelate ring size on antitumor activity, and interactions with 5'-GMP and DNA," *J. Inorg. Biochem.* **98**:1933–1946 (2004).

58. Sullivan, S. T., Ciccarese, A., Fanizzi, F. P., and Marzilli, L. G., "Cisplatin-DNA cross-link models with an unusual type of chirality-neutral chelate amine carrier ligand, N,N'-dimethylpiperazine (Me$_2$ppz): Me$_2$ppzPt(guanosine monophosphate)$_2$ adducts that exhibit novel properties," *Inorg. Chem.* **39**:836–842 (2000).

59. Sullivan, S. T., Ciccarese, A., Fanizzi, F. P., and Marzilli, L. G., "Cisplatin-DNA cross-link retro models with a chirality-neutral carrier ligand: Evidence for the importance of 'scond-sphere communication'" *Inorg. Chem.* **40**:455–462 (2001).

60. Sullivan, S. T., Ciccarese, A., Fanizzi, F. P., and Marzilli, L. G., "A rare example of three abundant conformers in one retro model of the cisplatin-DNA d(GpG) intrastrand cross link. Unambiguous evidence that guanine O6 to carrier amine ligand hydrogen bonding is not important. Possible effect of the Lippard base pair step adjacent to the lesion on carrier ligand hydrogen bonding in DNA adducts," *J. Am. Chem. Soc.* **123**:9345–9355 (2001).

61. Wong, H. C., Coogan, R., Intini, F. P., Natile, G., and Marzilli, L. G., "New steric hindrance approach employing the hybrid ligand 2-aminomethylpiperidine for diminishing dynamic motion problems of platinum anticancer drug adducts containing guanine derivatives," *Inorg. Chem.* **38**:777–787 (1999).

62. Ano, S. O., Intini, F. P., Natile, G., and Marzilli, L. G., "Viewing early stages of guanine nucleotide attack on Pt(II) complexes designed with in-plane bulk to trap initial adducts. Relevance to *cis*-type Pt(II) anticancer drugs," *J. Am. Chem. Soc.* **119**:8570–8571 (1997).

63. Ano, S. O., Intini, F. P., Natile, G., and Marzilli, L. G., "Retro models of Pt anticancer drug DNA adducts: Chirality-controlling chelate ligand restriction of guanine dynamic motion in (2,2'-Bipiperidine)PtG$_2$ complexes (G = guanine derivative)," *Inorg. Chem.* **38**:2989–2999 (1999).

64. Ano, S. O., Intini, F. P., Natile, G., and Marzilli, L. G., "A novel head-to-head conformer of d(GpG) cross-linked by Pt: New light on the conformation of such cross-links formed by Pt anticancer drugs," *J. Am. Chem. Soc.* **120**:12017–12022 (1998).

65. Marzilli, L. G., Ano, S. O., Intini, F. P., and Natile, G., "New concepts relevant to cisplatin anticancer activity from unique spectral features providing evidence that adjacent guanines in d(GpG), intrastrand-cross-linked at N7 by a *cis*-platinum(II) moiety, can adopt a head-to-tail arrangement," *J. Am. Chem. Soc.* **121**:9133–9142 (1999).

66. Bloemink, M. J., Pérez, J. M. J., Heetebrij, R. J., and Reedijk, J., "The new anticancer drug [(2*R*)-aminomethylpyrrolidine](1,1-cyclobutanedicarboxylato)platinum(II) and its toxic *S* enantiomer do interact differently with nucleic acids," *J. Biol. Inorg. Chem.* **4**:554–567 (1999).

67. Bowler, B. E., Hollis, S., and Lippard, S. J., "Synthesis and DNA binding and photonicking properties of acridine orange linked by a polymethylene tether to (1,2-diaminoethane)dichloroplatinum(II)," *J. Am. Chem. Soc.* **106**:6102–6104 (1984).
68. Palmer, B. D., Lee, H. H., Johnson, P., Baguley, B. C., Wickham, G., Wakelin, L. P. G., McFadyen, W. D., and Denny, W. A., "DNA-directed alkylating agents. 2. Synthesis and biological activity of platinum complexes linked to 9-anilinoacridine," *J. Med. Chem.* **33**:3008–3014 (1990).
69. Lee, H. H., Palmer, B. D., Baguley, B. C., Chin, M., McFadyen, W. D., Wickham, G., Thorsbourne-Palmer, D., Wakelin, L. P. G., and Denny, W. A., "DNA-directed alkylating agents. 5. Acridinecarboxamide derivatives of (1,2-diaminoethane)dichloroplatinum (II)," *J. Med. Chem.* **35**:2983–2987 (1992).
70. Wheate, N. J., Webster, L. K., Brodie, C. R., and Collins, J. G., "Synthesis, DNA binding and cytotoxicity of isohelical DNA groove binding platinum complexes," *Anti-Cancer Drug Design* **15**:313–322 (2000).
71. Hartwig, J. F., Pil, P. M., and Lippard, S. J., "Synthesis and DNA-binding properties of a cisplatin analogue containing a tethered dansyl group," *J. Am. Chem. Soc.* **114**:8292–8293 (1992).
72. Moreno, V., Cervantes, G., Onoa, G. B., Sampedro, F., Santaló, P., Solans, X., and Font-Bardia, M., "Platinum(II) substituted pyrrolidine complexes. Crystal structure of dichloride 1,2 diethyl 3-aminopyrrolidine Pt^{II}. Reactions with DNA and dinucleotides," *Polyhedron* **16**:4297–4303 (1997).
73. Wong, H. C., Shinozuka, K., Natile, G., and Marzilli, L. G., "Second-sphere 'communication' between two *cis*-bound guanine nucleotides. Factors influencing conformations of dynamic adducts of *cis*-type platinum anticancer drugs with guanine nucleotides as deduced by circular dichroism spectroscopy," *Inorg. Chim. Acta* **297**:36–46 (2000).
74. Loskotová, H. and Brabec, V., "DNA interactions of cisplatin tethered to the DNA minor groove binder distamycin," *Eur. J. Biochem.* **266**:392–402 (1999).
75. Kostrhunova, H. and Brabec, V., "Conformational analysis of site-specific DNA crosslinks of cisplatin-distamycin conjugates," *Biochemistry* **39**:12639–12649 (2000).
76. Götze, H.-J., Jursch, M., and Kahmann, A., "Liquid chromatographic and infrared studies on ethylenediamineplatinum(II) guanosine complexes," *Anal. Chim. Acta.* **324**:155–163 (1996).
77. Paquet, F., Boudvillain, M., Lancelot, G., and Leng, M., "NMR solution structure of a DNA dodecamer containing a transplatin interstrand GN7–CN3 cross-link," *Nucl. Acids Res.* **27**:4261–4268 (1999).
78. Lepre, C. A., Chassot, L., Costello, C. E., and Lippard, S. J., "Synthesis and characterization of *trans*-[Pt(NH$_3$)$_2$Cl$_2$] adducts of d(CCTCGAGTCTCC)·d(GGAGACTCG-AGG)," *Biochemistry* **29**:811–822 (1990).
79. Zou, Y., Van Houten, B., and Farrell, N., "Ligand effects in platinum binding to DNA. A comparison of DNA binding properties for *cis*- and *trans*-[PtCl$_2$(amine)$_2$] (amine = NH$_3$. pyridine)," *Biochemistry* **32**:9632–9638 (1993).
80. Žaludová, R., Natile, G., and Brabec, V., "The effect of antitumor *trans*-[PtCl$_2$(E-iminoether)$_2$] on B→Z transition in DNA," *Anti-Cancer Drug Design* **12**:295–309 (1997).
81. Andersen, B., Margiotta, N., Coluccia, M., Natile, G., and Sletten, E., "Antitumor transplatinum DNA adducts: NMR and HPLC study of the interaction between

a *trans*-Pt iminoether complex and the deoxy decamer d(CCTCGCTCTC)·d(GAGAGCGAGG)," *Metal-Based Drugs* **7**:23–32 (2000).

82. Pérez, J. M., Montero, E. I., González, A. M., Solans, X., Font-Bardia, M., Fuertes, M. A., Alonso, C., and Navarro-Ranninger, C., "X-ray structure of cytotoxic trans-[PtCl$_2$(dimethylamine)(isopropylamine)]: Interstrand cross-link efficiency, DNA sequence specificity, and inhibition of the B→Z transition," *J. Med. Chem.* **43**:2411–2418 (2000).

83. Admiraal, G., Alink, M., Altona, C., Dijt, F. J., van Garderen, C. J., de Graaff, R. A. G., and Reedijk, J., "Conformation of Pt(dien)[d(ApGpA)-N7(2)] in the solid state and in aqueous solution, as determined with single-crystal X-ray diffraction and high-resolution NMR spectroscopy in solution," *J. Am. Chem. Soc.* **114**:930–938 (1992).

84. van Garderen, C. J., Altona, C., and Reedijk, J., "Conformational changes in a single- and double-stranded nonanucleotide upon complexation of a monofunctional platinum compound as studied by ^{1}H NMR, ^{31}P NMR, and CD methods," *Inorg. Chem.* **29**:1481–1487 (1990).

85. Talman, E. G., Myers, D. P., and Reedijk, J., "Conformational changes in d(TpG) and 5′-GTP induced by monofunctional binding of [PtCl(NH$_3$)$_3$]Cl and [PtCl(dien)]Cl are small," *Inorg. Chim. Acta* **240**:25–28 (1995).

86. Bauer, C., Peleg-Shulman, T., Gibson, D., and Wang, A. J.-H., "Monofunctional platinum amine complexes destabilize DNA significantly," *Eur. J. Biochem.* **256**:253–260 (1998).

87. Bierbach, U., Sabat, M., and Farrell, N., "Inversion of the *cis* geometry requirement for cytotoxicity in structurally novel platinum(II) complexes containing the bidentate N,O-donor pyridin-2-yl-acetate," *Inorg. Chem.* **39**:1882–1890 (2000).

88. Peleg-Shulman, T., Katzhendler, J., and Gibson, D., "Effects of monofunctional platinum binding on the thermal stability and conformation of a self-complementally 22-mer," *J. Inorg. Biochem.* **81**:313–323 (2000).

89. Farrell, N., "DNA binding of dinuclear platinum complexes", in Hurley, L. H. and Chaires, J. B. eds., *Advances in DNA Sequence Specific Agents*, JAI Press, Greenwich, CT, 1996, Vol. 2, pp. 187–216.

90. Qu, Y., Bloemink, M. J., Reedijk, J., Hambley, T. W., and Farrell, N., "Dinuclear platinum complexes form a novel intrastrand adduct with d(GpG), *anti−syn* conformation of the macrochelate as observed by NMR and molecular modeling," *J. Am. Chem. Soc.* **118**:9307–9313 (1996).

91. Kašpárková, J., Mellish, K. J., Qu, Y., Brabec, V., and Farrell, N., "Site-specific d(GpG) intrastrand cross-links formed by dinuclear platinum complexes. Bending and NMR studies," *Biochemistry* **35**:16705–16713 (1996).

92. Yang, D., van Boom, S. S. G. E., Reedijk, J., van Boom, J. H., Farrell, N., and Wang, A. H.-J., "A novel DNA structure induced by the anticancer bisplatinum compound cross-linked to a GpC site in DNA," *Nature Struct. Biol.* **2**:577–586 (1995).

93. Mellish, K. J., Qu, Y., Scarsdale, N., and Farrell, N., "Effect of geometric isomerism in dinuclear platinum antitumour complexes on the rate of formation and structure of intrastrand adducts with oligonucleotides," *Nucl. Acids Res.* **25**:1265–1271 (1997).

94. Qu, Y. and Farrell, N., "Interaction of bis(platinum) complexes with the mononucleotide 5′-guanosine monophosphate. Effect of diamine linker and the nature of the bis(platinum) complexes on product formation," *J. Am. Chem. Soc.* **113**:4851–4857 (1991).

95. Cox, J. W., Berners-Price, S. J., Davies, M. S., Qu, Y., and Farrell, N., "Kinetic analysis of the stepwise formation of a long-range DNA interstrand cross-link by a dinuclear

platinum antitumor complex: Evidence for aquated intermediates and formation of both kinetically and thermodynamically controlled conformers," *J. Am. Chem. Soc.* **123**:1316–1326 (2001).

96. Bierbach, U., Roberts, J. D., and Farrell, N., "Modification of platinum(II) antitumor complexes with sulfur ligands. 2. Reactivity and nucleotide binding properties of cationic complexes of the types [PtCl(diamine)(L)]NO$_3$ and [{PtCl(diamine)}$_2$(L-L)](NO$_3$)$_2$ (L = monofunctional thiourea derivative; L-L = bifunctional thiourea derivative) in relation to their cytotoxicity," *Inorg. Chem.* **37**:717–723 (1998).

97. Wu, P. K., Qu, Y., Van Houten, B., and Farrell, N., "Chemical reactivity and DNA sequence specificity of formally monofunctional and bifunctional bis(platinum) complexes," *J. Inorg. Biochem.* **54**:207–220 (1994).

98. Farrell, N., Appleton, T. G., Qu, Y., Roberts, J. D., Soares Fontes, A. P., Skov, K. A., Wu, P., and Zou, Y., "Effects of geometric isomerism and ligand substitution in bifunctional dinuclear platinum complexes on binding properties and conformational changes in DNA," *Biochemistry* **34**:15480–15486 (1995).

99. Wu, P. K., Kharatishvili, M., Qu, Y., and Farrell, N., "A circular dichroism study of ethidium bromide binding to Z-DNA induced by dinuclear platinum complexes," *J. Inorg. Biochem.* **63**:9–18 (1996).

100. McGregor, T. D., Balcarová, Z., Qu, Y., Tran, M.-C., Zaludová, R., Brabec, V., and Farrell, N., "Sequence-dependent conformational changes in DNA induced by polynuclear platinum complexes," *J. Inorg. Biochem.* **77**:43–46 (1999).

101. Rauter, H., Di Domenico, R., Menta, E., Oliva, A., Qu, Y., and Farrell, N., "Selective platination of biologically relevant polyamines. Linear coordinating spermidine and spermine as amplifying linkers in dinuclear platinum complexes," *Inorg. Chem.* **36**:3919–3927 (1997).

102. González, V. M., Amo-Ochoa, P., Pérez, J. M., Fuertes, M. A., Masaguer, J. R., Navarro-Ranninger, C., and Alonso, C., "Synthesis, characterization, and DNA modification induced by a novel Pt-Berenil compound with cytotoxic activity," *J. Inorg. Biochem.* **63**:57–68 (1996).

103. Wheate, N. J. and Collins, J. G., "A ^{1}H NMR study of the oligonucleotide binding of [(en)Pt(μ-dpzm)$_2$Pt(en)]Cl$_4$," *J. Inorg. Biochem.* **78**:313–320 (2000).

104. Wheate, N. J., Cutts, S. M., Phillips, D. R., Aldrich-Wright, J. R., and Collins, J. G., "The binding of [(en)Pt(μ-dpzm)$_2$Pt(en)]$^{4+}$ to G/C-rich regions of DNA," *J. Inorg. Biochem.* **84**:119–127 (2001).

105. Collins, J. G. and Wheate, N. J., "Potential adenine and minor groove binding platinum complexes," *J. Inorg. Biochem.* **98**:1578–1584 (2004).

106. Qu, Y., Scarsdale, N. J., Tran, M.-C., and Farrell, N. P., "Cooperative effects in long-range 1,4 DNA-DNA interstrand cross-lonks formed by polynuclear platinum complexes: An unexpected *syn* orientation of adenine bases outside the binding sites," *J. Biol. Inorg. Chem.* **8**:19–28 (2003).

107. Qu, Y., Scarsdale, N. J., Tran, M.-C., and Farrell, N. P., "Comparison of structural effects in 1. 4 DNA-DNA interstrand cross-links formed by dinuclear and trinuclear platinum complexes," *J. Inorg. Biochem.* **98**:1585–1590 (2004).

108. Jansen, B. A. J., van der Zwan, J., Reedijk, J., den Dulk, H., and Brouwer, J., "A tetranuclear platinum compound designed to overcome cisplatin resistance," *Eur. J. Inorg. Chem.* **9**:1429–1433 (1999).

109. Munakata, H., Imai, H., Nakagawa, S., Osada, A., and Uemori, Y., "Water-soluble porphyrins appending platinum(II) complexes as binders for synthetic nucleic acid polymers," *Chem. Pharm. Bull.* **51**:614–615 (2003).
110. Cervantes, G., Marchal, S., Prieto, M. J., Pérez, J. M., González, V. M., Alonso, C., and Moreno, V., "DNA interaction and antitumor activity of a Pt(III) derivative of 2-mercaptopyridine," *J. Inorg. Biochem.* **77**:197–203 (1999).
111. Wheate, N. J. and Collins, J. G., "Multi-nuclear platinum complexes as anti-cancer drugs," *Coord. Chem. Rev.* **241**:133–145 (2003).
112. Hartwig, J. F. and Lippard, S. J., "DNA binding properties of *cis*-[Pt(NH$_3$)(C$_6$H$_{11}$NH$_2$)Cl$_2$], a metabolite of an orally active platinum anticancer drug," *J. Am. Chem. Soc.* **114**:5646–5654 (1992).
113. Kratochwil, N. A., Bednarski, P. J., Mrozek, H., Vogler, A., and Nagle, J. K., "Photolysis of an iodoplatinum(IV) diamine complex to cytotoxic species by visible light," *Anti-Cancer Drug Design* **11**:155–171 (1996).
114. Galanski, M. and Keppler, B. K., "Is reduction required for antitumor activity of platinum (IV) compounds?" *Inorg. Chim. Acta* **300–302**, 783–789, (2000).
115. Cleare, M. J. and Hoeschele, J. D., "Studies on the antitumor activity of group VIII transition metal complexes. Part I. Platinum(II) complexes," *Bioinorg. Chem.* **2**:187–210 (1973).
116. Mital, R., Jain, N., and Srivastava, T. S., "Synthesis, characterization and cytotoxic studies of diamine and diimine palladium(II) complexes of diethyldithiocarbamate and binding of these and analogous platinum(II) complexes with DNA," *Inorg. Chim. Acta* **166**:135–140 (1989).
117. Jain, N., Mittal, R., Srivastava, T. S., Satyamoorthy, K., and Chitnis, M. P., "Synthesis, characterization, DNA binding, and cytotoxic studies of dinuclear complexes of palladium(II) and platinum(II) with 2,2'-bipyridine and α,ω-diaminoalkane-N,N'-diacetic acid," *J. Inorg. Biochem.* **53**:79–94 (1994).
118. Quiroga, A. G., Pérez, J. M., Montero, E. I., Masaguer, J. R., Alonso, C., and Navarro-Ranninger, C., "Palladated and platinated complexes derived from phenylacetaldehyde thiosemicarbazone with cytotoxic activity in *cis*-DDP resistant tumor cells. Formation of DNA interstrand cross-links by these complexes," *J. Inorg. Biochem.* **70**:117–123 (1998).
119. Quiroga, A. G., Pérez, J. M., López-Solera, I., Masaguer, J. R., Luque, A., Román, P., Edwards, A., Alonso, C., and Navarro-Ranninger, C., "Novel tetranuclear orthometalated complexes of Pd(II) and Pt(II) derived from *p*-isopropylbenzaldehyde thiosemicarbazone with cytotoxic activity in *cis*-DDP resistant tumor cell lines. Interaction of these complexes with DNA," *J. Med. Chem.* **41**:1399–1408 (1998).
120. Quiroga, A. G., Pérez, J. M., López-Solera, I., Montero, E. I., Masaguer, J. R., Alonso, C., and Navarro-Ranninger, C., "Binuclear chloro-bridged palladated and platinated complexes derived from *p*-isopropylbenaldehyde thiosemicarbazone with cytotoxicity against cisplatin resistant tumor cell lines," *J. Inorg. Biochem.* **69**:275–281 (1998).
121. Mansuri-Torshizi, H., Ghadimy, S., and Akbarzadeh, N., "Synthesis, characterization, DNA binding and cytotoxic studies of platinum(II) and palladium(II) complexes of the 2,2'-bipyridine and an anion of 1,1-cyclobutadienedicarboxylic acid," *Chem. Pharm. Bull.* **49**:1517–1520 (2001).
122. Calamai, P., Carotti, S., Guerri, A., Mazzei, L., Messori, T., Mini, E., Orioli, P., and Speroni, G. P., "Cytotoxic effects of gold(III) complexes on established human tumor cell lines sensitive and resistant to cisplatin," *Anti-Cancer Drug Design* **13**:67–80 (1998).

123. Carotti, S., Guerri, A., Mazzei, T., Messori, L., Mini, E., and Orioli, P., "Gold(III) compounds as potential antitumor agents; cytotoxicity and DNA binding properties of some selected polyamine-gold(III) complexes," *Inorg. Chim. Acta* **281**:90–94 (1998).
124. Marcon, G., Carotti, S., Coronnello, M., Messori, L., Mini, E., Orioli, P., Mazzei, T., Cinellu, M. A., and Minghetti, G., "Gold(III) complexes with bipyridyl ligands: Solution chemistry, cytotoxicity, and DNA binding properties," *J. Med. Chem.* **45**:1672–1677 (2002).
125. Velders, A. H., Kooijman, H., Spek, A. L., Haasnoot, J. G., de Vos, D., and Reedijk, J., "Strong differences in the *in vitro* cytotoxicity of three isomeric dichlorobis(2-phenylazopyridine)ruthenium(II) complexes," *Inorg. Chem.* **39**:2966–2967 (2000).
126. Hotze, A. C. G., van der Geer, E. P. L., Caspers, S. E., Kooijman, H., Spek, A. L., Haasnoot, J. G., and Reedijk, J., "Coordination of 9-ethylguanine to the mixed-ligand compound α-[Ru(azpy)(bpy)Cl$_2$] (azpy = 2-phenylazopyridine and bpy = 2,2'-bipyridine). An unprecedented ligand positional shift, correlated to the cytotoxicity of this type of [RuL$_2$Cl$_2$] (with L = azpy or bpy) complex," *Inorg. Chem.* **43**:4935–4943 (2004).
127. van Vliet, P. M., Tockimin, S. M. S., Haasnoot, J. G., Reedijk, J., Nováková, O., Vrána, O., and Brabec, V., "*mer*-[Ru(terpy)Cl$_3$] (terpy = 2,2':6',2''-terpyridine) shows biological activity, forms interstrand cross-links in DNA and binds two guanine derivatives in a *trans* configuration," *Inorg. Chim. Acta* **231**:57–64 (1995).
128. Chan, H.-L., Liu, H.-D., Tzeng, B.-C., You, Y.-S., Peng, S.-M., Yang, M., and Che, C.-M., "Syntheses of ruthenium(II) quinonediimine complexes of cyclam and characterization of their DNA-binding activities and cytotoxicity," *Inorg. Chem.* **41**:3161–3171 (2002).
129. Mestroni, G., Alessio, E., Sessanta o Santi, A., Geremia, S., Bergamo, A., Sava, G., Boccarelli, A., Schettino, A., and Coluccia, M., "Rhodium(III) analogues of antitumour-active ruthenium(III) compounds: The crystal structure of [ImH][*trans*-RhCl$_4$(Im)$_2$] (Im = imidazole)," *Inorg. Chim. Acta*, **273**:62–71 (1998).).
130. Pruchnik, P. F., Jakimowicz, P., Ciunik, Z., Zakrzewska-Czerwińska, J., Opolski, A., Wietrzyk, J., and Wojdat, E., "Rhodium(III) complexes with polypyridyls and pyrazole and their antitumor activity," *Inorg. Chim. Acta* **334**:59–66 (2002).
131. Li, C., Wu, J., Liufang, W., Min, R., Naiyong, J., and Jie, G., "Synthesis, characterization and antitumor activity of copper(II) complex with nicotinamido-4-bis(2-chloroethyl) aminobenzaldimine," *J. Inorg. Biochem.* **73**:195–202 (1999).
132. Pérez, J. M., Matesanz, A. I., Martín-Ambite, A., Navarro, P., Alonso, C., and Souza, P., "Synthesis and characterization of complexes of *p*-isopropyl benzaldehyde and methyl 2-pyridyl ketone thiosemicarbazones with Zn(II) and Cd(II) metallic centers. Cytotoxic activity and induction of apoptosis in Pam-*ras* cells," *J. Inorg. Biochem.* **75**:255–261 (1999).
133. Murray, J. H. and Harding, M. M., "Organometallic anticancer agents: The effects of the central metal and halide ligands on the interaction of metallocene dihalides Cp$_2$MX$_2$ with nucleic acid constituents," *J. Med. Chem.* **37**:1936–1941 (1994).
134. Kuo, L. Y., Kanatzidis, M. G., Sabat, M., Tipton, A. L., and Marks, T. J., "Metallocene antitumor agents. Solution and solid-state molybdenocene coordination chemistry of DNA constituents," *J. Am. Chem. Soc.* **113**:9027–9045 (1991).
135. Clarke, M. J., "Ruthenium metallopharmaceuticals," *Coord. Chem. Rev.* **236**:209–233 (2003).

APPENDIXES

APPENDIX A.1. RAMAN TENSOR–DEPOLARIZATION RATIO

The tensor associated with a Raman band plays an important role in determining the intensity and structural significance of the band. A Raman tensor describes the relationship between two electric vectors, that of the incident, or exciting radiation (i.e., laser photon) and that of the Raman scattered radiation (i.e., the inelastically scattered photon that results from the exchange of a vibrational quantum between the exciting photon and the molecule). The Raman tensor is obtained formally as the first derivative of the molecular polarizability tensor with respect to the vibrational normal coordinate. In other words, the Raman tensor associated with a vibrational Raman band is an indicator of how the polarizability of the molecule oscillates with the normal mode of vibration.

In general, a Raman tensor has six components, α_{xx}, α_{yy}, α_{zz}, α_{xy}, α_{xz}, and α_{yz}, where x, y, and z refer to an imaginary Cartesian coordinate axes that is fixed to the molecule, but otherwise chosen arbitrarily. If the principal axes of the Raman tensor, which, are unique for a given Raman band, are selected as the xyz coordinate system, then the six nonzero tensor components are reduced to the three diagonal components, α_{xx}, α_{yy}, and α_{zz}. The orientation of a principal axis of a Raman tensor is given by three parameters (Euler angles), and the relative magnitudes of its components are given by another set of two parameters, for example, by

$$r_1 = \frac{\alpha_{xx}}{\alpha_{zz}} \quad \text{and} \quad r_2 = \frac{\alpha_{yy}}{\alpha_{zz}} \tag{A.1}$$

Thus, five experimental parameters are necessary for the unique determination of the Raman tensor for a given Raman band. The experimental procedure for determining a Raman tensor generally requires two types of measurements: (1) the polarized Raman

---

Drug–DNA Interactions: Structures and Spectra by Kazuo Nakamoto, Masamichi Tsuboi and Gary D. Strahan
Copyright © 2008 John Wiley & Sons, Inc.

scattering intensities (I_{aa}, I_{bb}, I_{cc}, etc.) must be obtained for a single crystal of known structure with a defined orientation, and (2) the Raman depolarization ratio

$$\rho = \frac{I_\perp}{I_\parallel} \tag{A.2}$$

must be measured for a completely random ensemble of the same molecular species; this is generally done by measuring the solution spectra where the molecules are isotropic in their orientation. In this case, $I_\parallel$ is the intensity of the scattered light polarized parallel to that of the exciting light and $I_\perp$ is the intensity of the scattered light polarized perpendicular to the exciting light (e.g., see Fig. 1.39a, b). ρ is related to r_1 and r_2 as

$$\rho = \frac{3[(r_1-r_2)^2+(r_2-1)^2+(1-r_1)^2]}{10(r_1+r_2+1)^2+4[(r_1-r_2)^2+(r_2-1)^2+(1-r_1)^2]}. \tag{A.3}$$

This relation is shown graphically in Fig. A.1 [1].

The depolarization ratios ρ of all the Raman bands of 5′-GMP, 5′-CMP, 5′-AMP, 5′-dAMP, 5′-dTMP, and 5′-UMP were measured in H_2O and in D_2O and are listed by Ueda et al. [2]. Raman tensors of some Raman bands of nucleotides are given by Tsuboi and Thomas [1].

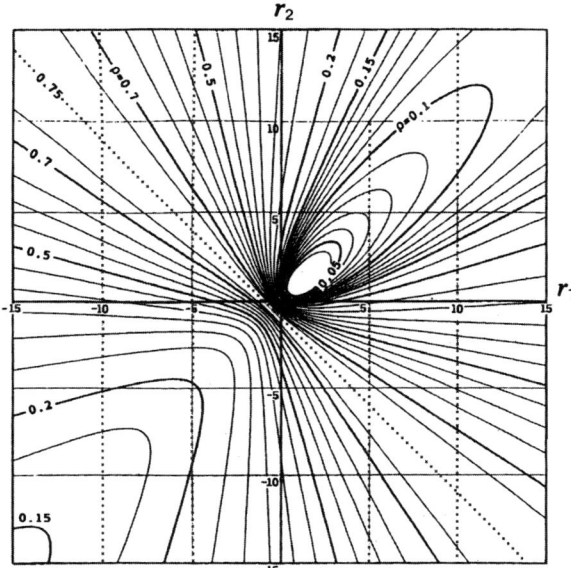

FIGURE A.1. Contour diagram relating the Raman depolarization ratio ρ of a randomly oriented molecular subgroup to the parameters r_1 (ordinate) and r_2 (abscissa) according to Eq. A.4 [1].

APPENDIX A.2. NMR SPECTROSCOPY

A.2.1. Origins of the Phenomenon

As noted in Section 1.4, NMR derives its name from its reliance on the quantum mechanical properties of atomic nuclei. Atoms are composed of the subatomic particles, electrons, neutrons, and protons, each of which have an intrinsic quantum mechanical property **I**, called the *nuclear spin*, which is expressed by the nuclear spin quantum number I. The nuclear spin produces an angular momentum **L**, which has a magnitude of

$$L = \frac{h}{2\pi[I(I+1)]^{1/2}} \tag{A.4}$$

where h is Planck's constant. Notice that L is limited to discrete (quantized) values that are determined by the spin quantum number I, which can have values of $0, \frac{1}{2}, 1, 1\frac{1}{2}, 2$, and so on. The actual value of I for a given nuclide depends on its atomic mass and atomic number (see Section 1.4 and Table 1.17).

If a nucleus has a nonzero spin, it will have a small magnetic moment (**μ**), which arises from the quantum mechanical angular momentum **L**. Thus, protons can be regarded as very small bar magnets. This is given by the equation

$$\boldsymbol{\mu} = \gamma \mathbf{L} \tag{A.5}$$

in which γ is a proportionality constant called the *gyromagnetic ratio*. As discussed in Section 1.4, those nuclides with large values of γ are easier to detect, and are said to be "sensitive," while the converse is true of nuclides with small γ values (see Table 1.18b). Notice that **μ** and **L** are in boldface, which indicates that they are vectors, having both magnitude and direction. Spinning objects that carry a magnetic field will also generate an electrical charge. For protons, this charge is $+1e$. This is also true of electrons, with the exception that their charge is $-1e$.

For a given value of the spin quantum number I, a nucleus will have a magnetic spin quantum number m, which has $2I+1$ possible values, such that $m = (-I, -I+1, \ldots, I-1, +I)$. Thus, for an $I = \frac{1}{2}$ nucleus (such as ^{1}H or ^{13}C), there are two possible magnetic spin states: $m = +\frac{1}{2}$ and $-\frac{1}{2}$. In the absence of a magnetic field, the ensemble of nuclei will equally populate all the magnetic spin states available to them, as they are of equal energy. When nuclei are placed in a very strong, external magnetic field $\mathbf{B_0}$, the energy levels of these different spin states become different, with the $m = +\frac{1}{2}$ lower in energy than $m = -\frac{1}{2}$. The "+" spin states are lower in energy because they correspond to being aligned parallel to the magnetic field, whereas the "−" spin states are higher in energy because these states are aligned opposed, or antiparallel, to the magnetic field (see Fig. 1.48 of Section 1.4.) Notice that the direction of the external magnetic field is used to define the z axis; hence $\mathbf{B_0} = \mathbf{B}_z$. When a nucleus (or an electron) is in this external magnetic field, its spin states become oriented and have energies given by

$$E_m = \omega_0 I_z = -\gamma \mathbf{I_z} \mathbf{B}_0 = -\gamma m \frac{h}{2\pi} \mathbf{B}_0 \qquad (A.6)$$

where

$$I_z = m \frac{h}{2\pi}, \qquad (A.7)$$

and

$$m = (-I, -I+1, \ldots, I-1, +I). \qquad (A.8)$$

Here, ω_0 is the Larmor frequency, which is equal to $-\gamma \mathbf{B_0}$, and is given in angular frequency (radians per second). The Larmor frequency (or resonant frequency) also can be expressed in Hertz (cycles/s) as ν_0, where $\omega_0 = 2\pi\nu_0$.

The energy difference between these states is very small. In a bulk sample, the number of nuclei that populate each state is nearly equal, dictated by a Boltzmann distribution. In the presence of an external magnetic field, the population in the lower energy state is slightly greater, causing a net magnetization in the $\mathbf{B_0}$ direction. This net magnetization of the sample, $\mathbf{M}$, is the bulk magnetic momentum of the collection of identical nuclei. This is simply the sum of all the $\boldsymbol{\mu}$ vectors for each nucleus (Fig. 1.48). The Larmor frequency ω_0 describes the rate of the precession of that bulk magnetic momentum around the magnetic field $\mathbf{B_0}$ (Fig. 1.48). Hence, the vector $\mathbf{M}$ is a function of the relative populations of the different spin states. At equilibrium with $\mathbf{B_0}$, it is aligned along the z axis.

In NMR experiments, the nuclei are subjected to very short pulses of a circularly polarized radiofrequency (RF) field ($\mathbf{B_1}$), the vector of which rotates about the z axis at a rate equal to the Larmor frequency ω_0, of the nuclei [3]. (In actual practice, the frequency of the RF pulse is slightly larger than ω_0, and slightly tilted with respect to the z axis.) When this RF pulse is turned on, $\mathbf{M}$ will precess around the new $\mathbf{B_1}$ field. Thus, these pulses perturb the initial alignment with $\mathbf{B_0}$ by producing a torque on $\mathbf{M}$, and rotating it away from the z axis. When the RF pulse is turned off, the net magnetization will remain where it is. The duration of a pulse determines how far $\mathbf{M}$ is rotated. When $\mathbf{M}$ is rotated into the x, y plane it is called a "90° pulse" (Fig. 1.49), which causes the populations of the spin states to become equal. The populations become inverted when $\mathbf{M}$ is rotated by 180° into the -z direction (via a "180° pulse"). If a pulse is on long enough, $\mathbf{M}$ will rotate completely around to where it began (a "360° pulse").

After the net magnetization has been perturbed by an RF pulse, it cannot instantaneously return to its initial alignment with $\mathbf{B_0}$, but will relax back to that equilibrium state via a number of different mechanisms (as described in Section A.2.3). NMR experiments involve a combination of RF pulses to manipulate $\mathbf{M}$, and delay times, called *evolution periods*, during which these different relaxation processes take place. By using an appropriate sequence of RF pulses (with judiciously chosen pulse durations), specific relaxation properties can be selected, revealing different properties of the molecule. Often, the last pulse in an NMR experiment ensures that the net magnetization (or a portion of it) is in the *x*, *y* plane. If **M** has any magnitude in that plane, it can be detected by the receiver coils located along the *x* and *y* axes. This signal is a time-dependent decay function, called a *free induction decay* (FID). By using the mathematical Fourier transform algorithm, the signal is converted from a time-domain function into a frequency-domain function, which is the NMR spectrum.

It is important to remember that both during and after the RF pulses, the net magnetization of the sample continues to precess around the magnetic field at the Larmor frequency ω_0. As this is difficult to visualize, these events are usually depicted in a Cartesian coordinate frame which also rotates at ω_0. When viewed from this rotating frame, the RF field appears to be static and the spins do not appear to precess around $\mathbf{B_0}$, which allows us to ignore it for a while. (See Figs. 1.48 and 1.49.)

A.2.2. Chemical Shift

The nuclei of a given type of isotope do not all resonate at exactly the same frequency. Instead, within each molecule, each atom has its own unique electronic environment, arising from the effects of solvent, neighboring atoms within the molecule, and its own electron orbitals. These cause each nucleus to be shielded from the external field by slightly varying amounts. This can be expressed mathematically as

$$\mathbf{B}'_0 = (1+\sigma)\mathbf{B}_0 \tag{A.9}$$

where σ is the shielding constant and $\mathbf{B}'_0$ is the *effective* local magnetic field experienced by a given proton. This shielding constant usually has a very small negative value (-10^{-6}). In practice, however, an NMR spectrum is plotted in terms of *chemical shift* δ, rather than in σ, where

$$\delta = \frac{\mathbf{B_S} - \mathbf{B_R}}{\mathbf{B_R}} \times 10^6 \tag{A.10}$$

Here, $\mathbf{B_S}$ is the field experienced by a nucleus in the sample and $\mathbf{B_R}$ is the field experienced by a nucleus in a reference molecule, which is usually added to the experimental sample. The chemical shift is, therefore, the difference between the reference and the sample, given in parts per million (ppm), or 10^{-6}. Since δ is a dimensionless parameter, chemical shift measurements expressed in this way remain unchanged regardless of the strength of the external magnetic field. For ^{1}H or ^{13}C, the standard reference compounds are either tetramethylsilane (TMS) for organic solvents, or 3-(trimethylsilyl)propionate (TSP) for biological samples in aqueous solvent.

A.2.3. Magnetization Relaxation Processes

In the presence of a strong, applied magnetic field **B**$_0$, the net magnetization vector of the nuclear spins **M** is aligned in the same direction as that field; this is called the *equilibrium*, or *longitudinal magnetization,* **M**$_0$. Since the applied field defines the z axis, the z component M$_z$ must be equal to M$_0$. At equilibrium, there is no magnetization in the x or y directions; that is, its transverse magnetization, $M_x = M_y = 0$. In NMR, an electromagnetic RF pulse is used to perturb this equilibrium, and change the direction and magnitude of the net magnetization vector. After the pulse is turned off, it returns to its equilibrium condition through a variety of relaxation processes (Fig. 1.49) [3,4,5]. These fall into two general categories, spin–lattice (T_1) and spin–spin relaxation (T_2). Together, they both contribute to the natural linewidths of the peaks in the spectra. Their unique properties are particularly important as they constitute the basis of all NMR experiments. *Spin–spin relaxation* refers to the direct interaction of spins with each other. *Spin–lattice relaxation* refers to the interaction of a nuclear spin with its environment through its physical structures (the *lattice*) and consists of several relaxation processes, including dipole-dipole, spin–rotation, quadrupolar, paramagnetic, scalar, and chemical shift anisotropy.

A.2.3.1. Longitudinal or Spin–Lattice Relaxation Time (T1)

If a system is irradiated with enough electromagnetic energy, it is possible to saturate the spin states such that the populations of the $+\frac{1}{2}$ and $-\frac{1}{2}$ spins are reversed, rendering $M_z = -M_0$. This can be regarded classically as rotating the macroscopic magnetization vector by 180°, and putting all the magnetization into the $-z$ direction. The spins then relax back to the equilibrium condition, $M_z = M_0$. The time required for this relaxation to occur is described by a time constant called the *spin–lattice relaxation time*, or T_1. Typically, T_1 is expressed in units of seconds for carbon atoms, but it can be as long as several minutes, and even longer for other nuclides. The equation governing this relaxation behavior as a function of time t is

$$M_z = M_0(1 - e^{-t/T_1}). \quad (A.11)$$

T_1 is called the *spin–lattice relaxation time* because the magnetization is transferred to the surroundings, or lattice, often in the form of heat. One of the primary measures of a molecule's interactions with its surroundings is its rotational tumbling in solution. This tumbling rate of a molecule in solution is called its *rotational correlation time* (τ_c), which is a measure of the time it takes for the molecule to rotate 1 radian.

In general, $\tau_c \propto 1/T_1$. However, many other mechanisms contribute to T_1 relaxation. For example, intramolecular flexibility can cause the effective correlation time for a given atom, or group of atoms, to differ from that of the molecule as a whole. When a flexible portion of a molecule undergoes localized motion (such as in a sidechain of a protein, or the rotation of the methyl group in thymine), then the T_1 values for those atoms will exceed those of the rest of the molecule. This is because the motion can slow down dipolar relaxation by shortening the effective correlation time τ_c. In addition, each individual atom has a slightly different surrounding due to its molecular connectivity. For these reasons, not all atoms in a molecule will have the same T_1 relaxation times.

Dipole–Dipole, or Dipolar Coupling. Whenever two magnetic dipoles are in close proximity, there is an opportunity for them to interact and enhance each other's relaxation rate. This is the primary mechanism of T_1 relaxation, and it can be observed through NOE analysis. This effect can be observed in the interaction of the magnetic dipoles between carbon atoms and directly attached hydrogen atoms, which can greatly shorten ^{13}C relaxation rates; the carbons in CH, CH$_2$ and CH$_3$ groups have T_1 values between 0.1 ∼ 10 s, while quaternary carbons, which lack attached hydrogens, have much longer T_1 values. This effect can also be observed whenever two atoms are within 5 Å of each other, even if they are not directly bonded, and is the basis of the 2D NOESY experiment.

Chemical Shift Anisotropy. The chemical shift of a nucleus is dependent on its local field, which is determined by the orientation of the bond relative to the static field. Chemical shift anisotropy (CSA) results from an atom which has an inherently unsymmetric, or anisotropic, distribution of electron density in its chemical bonds. Thus, in solution, rapid tumbling of the molecule produces a fluctuating local field. This tumbling averages the CSA, producing a single (averaged) chemical shift. However, if the fluctuating field is sufficiently strong, it can induce significantly increased relaxation rates. Nuclei with large chemical shift ranges generally have the greatest CSA, especially ^{31}P, ^{19}F, and many metals.

Quadrupolar Relaxation. Quadrupolar relaxation is the primary relaxation mechanism for nuclei with a quantum spin number, I, greater than $\frac{1}{2}$. Besides their magnetic dipole moment, quadrupolar nuclei possess an electric quadrupole moment that gives rise to an electric field gradient at the nucleus. This relaxation mechanism is extremely efficient and can result in very large natural linewidths (100–1000 Hz).

When possible, quadrupolar nuclei are avoided in NMR experiments, as their broad linewidths make them difficult to study. For example, biomolecular NMR samples are generally modified by replacing ^{14}N with ^{15}N, which has a spin of ½ and hence does not suffer from quadrupolar relaxation. This is relatively easily done synthetically for nucleic acids, and by growing bacteria on ^{15}N enriched media, in the case of proteins. Nevertheless, it is not uncommon for biomolecular NMR experiments to be performed with quadrupolar nuclei, especially, ^{2}H, ^{14}N, and sometimes ^{17}O and ^{23}Na.

A.2.3.2. Transverse or Spin–Spin Relaxation Time (T2)

When magnetization is placed in the x,y plane, via a 90° pulse (Fig. 1.49), it rotates around the z axis at its Larmor frequency, which is proportional to the energy difference between the two energy levels of the spin. In addition, the net magnetization **M** begins to dephase as each of the protons in the ensemble experiences a slightly different magnetic environment. This is an entropy-driven process, and, given enough time, the vectors spread out evenly, leaving no net magnetization in the x,y plane. This is true even without T_1 relaxation back to the z axis. The equation describing this process is

$$\mathbf{M}_{xy,(t)} = \mathbf{M}_{xy,(0)}(e^{-t/T_2}). \tag{A.12}$$

where $\mathbf{M}_{xy}$ is the projection of magnetization vector of a nucleus onto the x,y plane. T_2 is always shorter than T_1, with the difference arising from this loss of phase coherence. Hence, T_2 is the primary contributor to the linewidth of the NMR peaks, especially for viscous liquids, solids, and large biomacromolecules.

As noted above, it is the T_2 relaxation process that produces the coupling patterns in NMR spectra, and as such it is also very useful in determining molecular structure. T_2 is also correlated with dynamic processes within the molecule, and it decreases as the isotropic tumbling rate of the molecule decreases (i.e., with increasing molecular size).

APPENDIX A.3. MOLECULAR MECHANICS

Many books and papers have been written on molecular modeling and molecular mechanics. What follows is a general description designed to address basic questions. The interested reader is referred to the books cited in the Section 1.6.

In most cases, molecular modeling uses a type of semiempirically parameterized classical (Newtonian) mechanics called *molecular mechanics* (MM). Hybrid approaches are also continuously under development, including some level of quantum mechanics (called "QM/MM"). Molecular mechanics is based largely on Newton's laws of motion, especially the second law, $\mathbf{F} = m\mathbf{a}$. In the standard molecular mechanics approach, atoms are treated as if they were charged balls connected together by springs. The atomic mass and electrostatic charge are located at a single point in the center of the atom, and each different type of atom has a uniquely defined radius. Although the atoms have partial electrostatic charges, the electron clouds surrounding them and comprising the chemical bonds are rarely polarizable. Such polarizable "forcefields" do exist, especially as in the QM/MM approach, but tend to be computationally more difficult. Bond lengths and vibrations are described by force constants in the form of the Hooke's law potential for the energy of a spring, or simple harmonic oscillator

$$V(x) = \tfrac{1}{2}Kx^2 \tag{A.13}$$

where V is the potential energy, x is the displacement of the atom(s) from its equilibrium position, and K is the empirical force constant. The frequency of a vibration in cycles/second ν, is related to K by the equation

$$\nu = (2\pi)^{-1}\left(\frac{K}{m}\right)^{1/2}. \tag{A.14}$$

The bending of bond angles and dihedral torsion angles are similarly treated by force constants. There are also *cross-terms* that describe the coupling of bond stretching motions with bond angles, dihedral torsions, constrained ring systems, and the like.

The general form for the total molecular energy is given as

$$E_{\text{total}} = E_{\text{covalent}} + E_{\text{noncovalent}} \tag{A.15}$$

$$E_{\text{covalent}} = E_{\text{bond}} + E_{\text{angle}} + E_{\text{dihedral}} \tag{A.16}$$

$$E_{\text{noncovalent}} = E_{\text{electrostatic}} + E_{\text{van der Waals}} \tag{A.17}$$

where the term *noncovalent* is sometimes alternated with *nonbonded*. The exact form of these equations varies among different approaches, but certain key physical properties are always similar.

When molecular models are refined using NMR distance and dihedral torsion angle information (derived from NOE and *J*-coupling data, respectively), the experimental data are expressed as additional terms in the potential energy:

$$\text{Total } E_{\text{covalent}} = E_{\text{covalent}} + E_{\text{bond distance}}^{\text{NMR}} + E_{\text{dihedral}}^{\text{NMR}} \tag{A.18}$$

The force of the electrostatic interaction of atoms is defined by Coulomb's law, which relates the atomic charges (q_n) to the inverse of the square of the distance (r_{ij}) between them: $F \propto q_i q_j / r_{ij}^2$. The non-electrostatic interactions between atoms and molecules are usually described by the Lennard-Jones potential (or L-J, or 6–12 potential). This equation has two components: (1) an interatomic attractive force arising from van der Waals interactions, which operates at short distances, and (2) a repulsive force arising from the Pauli exclusion principle, due to the overlap of electron orbitals, which operates at longer distances. The equation consists of an $(1/r_{ij})^6$ term that is used to describe the attraction component, and an $(1/r_{ij})^{12}$ term that is used to describe the repulsion component. These terms have long-range effects, and can affect the molecular structure over many angstroms. They greatly increase the amount of computational work required to model a molecular system. In addition, in real systems, these interactions die off more quickly than do those given by the L-J potential as a result of the shielding affects from other atoms. Therefore, optimization of cutoff values and solvent dielectric effects has been the subject of many studies.

Often these calculations include explicit solvent molecules as well as the molecule of interest. The total energy of an entire molecular system is therefore the sum of all of these interatomic interactions, which must be calculated separately between each pair of atoms in both the molecule and the solvent. For large molecular systems, where there can be many thousands of atoms, all of these pairwise calculations can be computationally strenuous for anything except the most powerful supercomputers. Consequently, finding the most time and computationally efficient means to realistically simulate a solvated biomacromolecular structure is an ongoing effort. The use of experimental data to guide these calculations, such as from NMR or X-ray studies, can greatly enhance their effectiveness.

REFERENCES

1. Tsuboi, M. and Thomas, Jr., G. J., "Raman Scattering Tensors in Biological Molecules and Their Assembly," *Appl. Spectrosc. Rev.* **32**:263–299 (1997).
2. Ueda, T., Ushizawa, K., and Tsuboi, M., "Depolarization of Raman Scattering from Some Nucleotides of RNA and DNA," *Biopolymers* **33**:1791–1803 (1990).
3. Wehrli, F. W. and Wirthlin, T., *Interpretation of Carbon-13 NMR Spectra*, Heyden & Son, London, 1976.
4. Claridge, T. D. W., *High Resolution NMR Techniques in Organic Chemistry*, Pergamon, Elsevier, Amsterdam, 1999.
5. Friebolin, H., *Basic One- and Two-Dimensional NMR Spectroscopy*, translated by J. K. Becconsall, Wiley-VCH, Weinheim, Germany, 2005.

INDEX

aclacinomycin   122, 165, 167, 186
acridines   120, 122, 126, 128, 134, 142
acridine derivatives   127, 325
acridine orange   120
actinomycins   186–188
adenine(A)   2, 47, 49–51
adenosine   3, 6
A-DNA   10, 17, 67
adriamycin   122, 179, 184, 193, 268–270
alkavinone   165
alkylating drugs   245
altromycin B   268
aminoacridines   123, 125, 130, 134, 136, 189
anthelvencin A   223
anthramycin   254–255
arabinosyl cytosine   268
atomic force microscopy   161–165

bases   2, 5, 6
  IR   42
  NMR   76
  Raman   49
B-DNA   9, 10, 12, 24, 63, 67
base pairing   25–26, 89, 91
berenil   92
bis-acridines   137, 194–198
bizelesin   253, 254
bleomycins   275
  activated BLM   281–283
  bithiazole(BTZ) ring   276
  cobalt and zinc complexes   283–288
  iron complex   279–283

metal-free   275–279
model compounds   288–290
Bragg diffraction   97
bulged double helix   26–27

calicheamicins   291–294
Calladine-Dickerson rule   181, 193
carboplatin   304, 316–317
(+) CC1065   245–248
  derivatives   249
chromomycin   230–232
cinerulose   165
circular dichroism(CD) spectra
  drug-DNA complexes   141, 151, 213, 230, 262, 282, 312, 327, 331, 336, 341
  magnetic (MCD)   282
  nucleotides   41–42, 220
  vibrational(VCD)   41
cisplatin   303–316
cisplatin derivatives
  chelating   321–327
  monofunctional   318–321
closed circular DNA   25, 159–164
covalent bonding drugs   245
$CPI-CDPI_2$ 250
crosslinking
  interstrand   304, 330–332
  intrastrand   304
cytidine   6
cytosine(C)   2, 43
cytotoxicity   249, 341, 345, 346

dansyl group   221, 223
daunomycin (see daunorubicin)

Drug–DNA Interactions: Structures and Spectra by Kazuo Nakamoto, Masamichi Tsuboi, and Gary D. Strahan
Copyright © 2008 John Wiley & Sons, Inc.

daunorubicin(DAU)   179, 180, 182,
    193, 268
2'-deoxyfucose   165
deoxyribose   2–3, 7, 13–16, 18
2,7-DAM   263
dihedral torsion angle
    syn and, anti   10, 20–21, 32
    C2'-endo (S) and C3'-endo(N)   14, 22,
       57, 125
distamycin   210, 212, 215
    furan derivatives   223
    imidazole derivatives   217–220
distance geometry   92, 100
DNA (see A-, B- and Z-DNA)
    structure   2, 4, 29
    IR spectra   61
    melting temperature   34, 135
    Raman spectra   63, 67
ditrisarubicin   179
DNase I footprinting   178–181
doxorubicin(DOX) (see adriamycin)
duocarmycins   250–253

ecteinascidins (Ets)   256–259
electric linear dichroism(ELD)
    236–237
electrophoresis   159, 177
electron spin resonance(ESR) spectra   95,
    280, 284
enediyne antibiotics   290–299
4'-epiadriamycin   193
esperamicins   294–296
ethidium bromide   143, 146, 148,
    154–159

fiber X-ray diffraction   145
flow linear dichroism(FLD)   40, 140, 222
fluorescence spectra   37–39, 137, 152
footprinting   178–181
furanose sugar ring   7

GC-binders   230
glycosidic bond angles (syn, anti)   22
gold complexes   346
G-quadplex structure   31–34, 131, 151
groove-binding drugs   209
groove-binding hybrids   236, 237
groove-binding intercalation   234, 235

guanine(G)   2, 47, 52
guanosine (Gua)   6, 313

haemin-acridines   289
hairpin nucleotides   28, 250, 293
hedamycin   264–267
Hoechst drugs   224–228
Hoogsteen base pairing   29, 82, 90
Hooke's law   364
hybrid drugs (combilexins)   236
hydration (spine of water)   18
hydroxypyrrole   220, 222
hypochromism   36

infrared(IR) spectra
    bases   42–49
    DNA   61–63
    drug-DNA complexes   229, 262, 313, 328
    nucleosides   54–57
    nucleotides   57–59, 214
    phosphodiester group   59–61
intercalating alkylators   264
intercalating drugs   119, 182–192
internal dihedral angle(see dihedral
    torsion angle)

Karplus equation   79
kikumycin B (dihydro)   223

Larmor precession   74
linear dichroism(LD)   38
linking number   25, 159

major and minor grooves   10, 11
McGee-von Hippel equation   149, 150, 167
melting temperature (Tm)   36
metal-containing drugs   303, 345
minor groove binders   230
mismatched base pairs   25–26
mithramycin   230–232
mitomycins   259–264
molecular dynamics   100, 364
molecular modeling   99–100
monofunctional(monochloro)
    Pt complexes   332–336

neighbor exclusion rule   121, 147
neocazinostatin(NCS)   296–299

netropsin 209–214, 217
nitroxyl radical 96, 321
NMR spectra 72–75, 134, 359
  backbone dihedral angles 83–85
  chemical shifts 76, 361
  correlation time 93, 362
  COSY 81
  coupling constants 77–80
  gyromagnetic ratio 72, 359
  HECTOR 82
  ^{1}H NMR 77, 134, 168–174
  hydrogen bonds 82, 89–92
  Larmor frequency 360
  magnetization relaxation (T1, T2) 73, 75, 362
  NOESY 85, 172, 309, 318, 329
  NOESY walk 87
  nuclear Overhauser effect (NOE) 85, 92–93
  nuclear spin 72–73, 359
  ^{31}P NMR 76, 83, 167, 168, 235, 310
  residual dipolar coupling 94
  ring current shifts 135
  ROESY 86
  spin-spin (*J*) coupling 74, 77, 80
  TOCSY 81
nogalamycin 190–192
norfloxacin 239
normal vibration 42
nucleosides 3, 54, 76
nucleotides 3, 12, 57, 76

oxaliplatin 304, 317

palladium complexes 346
pepleomycin 276, 286
peptide nucleic acid (PNA) 142
phosphodiester 5, 8, 57, 59
platinum complexes
  bifunctional 303–332
  chelating 321–327
  high-valent 344
  monofunctional 332–336
  polynuclear 336–345
polynucleotides 4
proflavine 120, 129, 132
pyrrolobenzodiazepines 256

quadruplex DNA 30–34, 131, 151
quinacridine 139

Raman depolarization ratio 357
Raman spectra
  bases 49–54
  DNA 63–71
  drug-DNA complexes 154–155, 157–158, 215, 234, 263, 278, 279, 282, 315
  drugs 175
  nucleosides 54–57
  nucleotides 57–59, 215
  phosphodiester group 59–61
  polarized 157–158
Raman tensor 357
Rayleigh scattering 50
resonance Raman spectra 59
ribose 2–3
ring puckering 7, 13
RNA 4

Scatchard plot 149, 150, 166, 233, 264
sedimentation 132–133
sequence preference 178
SN 07 270
SN 6999 228–230
SN 7167 229
SN 18071 230
stacking interactioins 6
strand-breaking drugs 275
sugar conformation (*endo* and *exo*) 13
superhelical DNA 23–25, 133

telomeric DNA 30, 131
Thia-Net-GA 237
thymidine 54
thymine(T) 2, 43, 56
tomaymycin 254–256
topotecan (Tpt) 233, 234
transplatin 327–332
tris-acridines 194
triple helix 30–34, 270

uracil(U) 2
uridine 6
unwinding angle 131–132, 161, 176–178

UV absorption spectra   34–36, 165, 265, 311
UV resonance Raman(UVRR) spectra   314–315

vibrational circular dichroism(VCD)   41, 311
vibrational spectra (see infrared and Raman spectra)
visible absorption spectra   148, 165, 282

water molecules in B-DNA   16, 18

Watson-Crick base pairing   9, 11
writhing number   25, 159

X-ray structures   96
  DNA (A, B and Z)   10, 18, 21, 24, 70
  drug-DNA complexes   121, 123, 125, 226, 232, 247, 255, 258, 261, 267, 269, 285, 295, 298, 305, 306, 307, 310, 317, 319, 323, 330, 332
  nucleotides   17, 27, 28

Z-DNA   10, 21, 24, 70